# GUNNER'S MATE MISSILE M1 & C

**NAVAL EDUCATION AND TRAINING COMMAND**

**RATE TRAINING MANUAL AND NONRESIDENT CAREER COURSE**

**NAVEDTRA 10200-D**

©2013 Periscope Film LLC
All Rights Reserved
ISBN#978-1-937684-30-3
www.PeriscopeFilm.com

# PREFACE

This rate training manual and nonresident career course (RTM/NRCC) form a self-study package for the Gunner's Mate Missile 1 & C. It will serve as an aid for enlisted personnel of the Navy and Naval Reserve who have a Gunner's Mate Missile classification and are preparing for increased responsibility in the rate of Gunner's Mate Missile 1 & C. One of a series, this course is designed to give enlisted personnel background information necessary for the proper performance of their duties and training in consonance with the Navy's goal of training for victory at sea. Designed for individual study rather than formal classroom instruction, the RTM provides subject matter that deals with the underlying principles of guided missile launching systems. This approach provides for specialized knowledge requirements of the Gunner's Mate Missile.

The NRCC provides a way of satisfying the requirements for completing the RTM. The assignments in the NRCC include learning objectives and supporting items designed to lead students through the RTM.

This training manual and nonresident career course was prepared by the Naval Education and Training Program Development Center, Pensacola, Florida, for the Chief of Naval Education and Training. Acknowledgment is made to the Naval Sea Systems Command, Washington, D.C.; Naval Ship Weapon System Engineering Station, Port Hueneme, California; and the Gunner's Mate School, Great Lakes, Illinois, whose technical review comments, ideas, and suggestions have been most helpful.

Revised 1979
Reprinted 1981

Stock Ordering No.
0502-LP-051-0022

Published by
NAVAL EDUCATION AND TRAINING
PROGRAM DEVELOPMENT CENTER

UNITED STATES
GOVERNMENT PRINTING OFFICE
WASHINGTON, D.C.: 1979

# THE UNITED STATES NAVY

## GUARDIAN OF OUR COUNTRY

The United States Navy is responsible for maintaining control of the sea and is a ready force on watch at home and overseas, capable of strong action to preserve the peace or of instant offensive action to win in war.

It is upon the maintenance of this control that our country's glorious future depends; the United States Navy exists to make it so.

## WE SERVE WITH HONOR

Tradition, valor, and victory are the Navy's heritage from the past. To these may be added dedication, discipline, and vigilance as the watchwords of the present and the future.

At home or on distant stations we serve with pride, confident in the respect of our country, our shipmates, and our families.

Our responsibilities sober us; our adversities strengthen us.

Service to God and Country is our special privilege. We serve with honor.

## THE FUTURE OF THE NAVY

The Navy will always employ new weapons, new techniques, and greater power to protect and defend the United States on the sea, under the sea, and in the air.

Now and in the future, control of the sea gives the United States her greatest advantage for the maintenance of peace and for victory in war.

Mobility, surprise, dispersal, and offensive power are the keynotes of the new Navy. The roots of the Navy lie in a strong belief in the future, in continued dedication to our tasks, and in reflection on our heritage from the past.

Never have our opportunities and our responsibilities been greater.

# CONTENTS

| CHAPTER | Page |
|---|---|
| 1. The Gunner's Mate M 1 and C | 1 |
| 2. Missile Handling | 17 |
| 3. Preparing for a Firing Exercise | 55 |
| 4. Loading, Unloading, and Dud Jettisoning | 79 |
| 5. Electricity and Electronics | 106 |
| 6. Hydraulics and Pneumatics | 164 |
| 7. Weapon Surveillance and Stowage | 230 |
| 8. Fire Control and Alignment | 250 |
| 9. Repair and Test | 279 |
| 10. Administration and Supply | 299 |
| INDEX | 323 |

Nonresident Career Course follows Index

# CHAPTER 1

# THE GUNNER'S MATE M

This rate training manual (RTM)/ nonresident career course (NRCC) has been prepared for personnel of the Regular Navy and the Naval Reserve. It is organized to give you a broader understanding of your job as a Gunner's Mate Missile (GMM), and to expand your knowledge of the equipment and operating procedures for the various launching systems.

The RTM provides information based on the occupational standards for advancement to GMM1 and GMMC. These standards are promulgated in the *Manual of Navy Enlisted Manpower and Personnel Classifications and Occupational Standards,* NAVPERS 18068D. Changes in the GMM standards occur periodically. It is important that you obtain and study a set of the most recent standards.

The NRCC (formerly called an enlisted correspondence course) in the back of this manual is a self-study enlisted training course. It includes a comprehensive series of questions based on the material in the RTM and is designed to assist you in studying for advancement. It must be satisfactorily completed before you can advance to GMM1 or GMMC, whether you are in the Regular Navy or the Naval Reserve.

## SENIOR PETTY OFFICER RESPONSIBILITIES

With each advancement, you will accept an increasing responsibility in military matters and in matters relating to the occupational requirements of the Gunner's Mate M rating. In large measure the extent of your contribution to the Navy depends upon your willingness and ability to accept these responsibilities.

Your responsibilities for military leadership are similar to those of petty officers in other ratings; your responsibilities for technical leadership are directly related to your work as a GMM. The effective operation and maintenance of a ship's guided missile launching system requires a leader who has a high degree of technical competence and a deep sense of personal responsibility. As a senior GMM, you must strive to improve your leadership ability and technical knowledge through study, observation, and practical application.

## EFFECTIVE COMMUNICATION

Every phase of your duties as a leading GMM involves the use of language, spoken or written. You are expected to translate general orders given by officers into specific orders that can be understood and executed by relatively inexperienced personnel. At the same time, you must be able to explain to officers the important needs or problems of enlisted personnel. An understanding of your own language will enable you to say exactly what you mean in the appropriate terms.

Effective communication demands the precise use of technical terms. A command of the technical language of the Gunner's Mate M rating will enable you to receive and convey information accurately and to exchange ideas with others. If you cannot understand the exact meanings of technical terms, you are handicapped when you try to read work-related official publications. You are also at a disadvantage when taking a written examination for advancement in rating.

Particularly when training lower rated personnel, it is important that you use correct terminology. The importance of learning and using technical terms should be emphasized.

Practically everything in the Navy—policies, procedures, equipment, publications, systems—is subject to change and development. Some changes will be pointed out to you; others you must find independently.

As a GMM1, and more so as a GMMC, you must keep informed of all changes and new developments which might affect your rating or your work. Try to develop a special alertness for new information and keep up to date on all available sources of technical information. Above all, keep an open mind on the subject of missile launchers and associated equipment. New types of missiles are being designed and tested constantly, and existing types of launching systems are subject to modification. A number of important changes in missile launchers has occurred and will continue to occur. Greater sophistication in missile guidance and control has required greater cooperation between the GMM and FT ratings. The addition of nuclear warheads to many of the missiles has increased the security problem and aggravated the safety problem. These few changes have been noted here merely to indicate the variety of changes that can be expected in the field of missile systems and associated equipment.

## TRAINING

As a GMM1 or GMMC you will have regular and continuing responsibilities for training others. Even if you are fortunate enough to have highly skilled and well-trained Gunner's Mates, you will find that continuous training is still necessary. As examples, you will always be responsible for training lower rated personnel for advancement in rating; some of your personnel may be transferred and inexperienced or poorly trained personnel may be assigned to replace them; or a particular job may call for skills that none of your personnel have. Such circumstances will require you to plan and conduct formal and informal training.

At times this training will affect a large number of personnel not in your division or even in your department. It becomes increasingly important, therefore, to understand the duties and responsibilities of personnel in other ratings. The more you know about related ratings, the more complete and comprehensive will be your training plans—especially those requiring an interdivisional effort to achieve their goals.

An example of a related rating is Gunner's Mate Guns (GMG). At the E-8 level the responsibilities of the GMM and GMG are merged. Other ratings that are, in part, related to the Gunner's Mate M are Fire Control Technician (Missiles), Sonar Technician, and Data Systems Technician.

### On-the-Job Training

Most readers of this RTM will be concerned with on-the-job training. A number of instructional methods, ranging from the lecture to the demonstration-doing method, to the planned or impromptu discussion can be adapted to shipboard conditions.

Because there never seems to be enough time for training, odd moments between drills, at quarters while entering and leaving port, in the mess line, and so on, can be used to good advantage. Personnel should be encouraged to conduct training "bull" sessions wherein they ask each other questions about their work, Navy and ship orientation matter, examinations taken, and other training matter. On-watch time in the missile batteries during condition III watches is a good time for the launcher captain to conduct equipment operation training of watch personnel.

### New Training Techniques

Navy training is changing in several ways; for example, it is becoming more individualized. This change has been brought about by the introduction of scores of programmed instruction courses and a few audiovisual courses. These self-paced courses permit the trainee to choose the medium of instruction and to proceed at his or her own speed. The new instructional material is job related and "system designed"; that is, it teaches the trainee to perform a task, following specific steps which include defining the need, and planning, developing, and evaluating the course. Thus, all elements required for a complete course are included in each unit.

In the future, training for men and women in many ratings will be planned from the time they enter the Navy until they retire. Since the length of many, if not all, A-schools will be reduced and others eliminated, more training will be done aboard the ship or station. To expedite on-board training, "on-board training packages," many of them multimedia, will be produced. The objective is to provide all the training needed but to eliminate the irrelevant training sometimes used in the past.

Personnel Qualification
Standards Program

The personnel qualification standards (PQS) program is a method for qualifying officer and enlisted personnel to perform specific duties. A personnel qualification standard is a written compilation, derived from task analyses, of knowledges and skills required to qualify for a specific watch station, maintain a specific equipment or system, or perform as a team member within the assigned unit.

"Watch station," as it applies to PQS, refers to a position that is normally assigned by a watch bill, is usually of 4 hours' duration, and in the majority of cases, is operator oriented. "To maintain" means to perform tasks which pertain to technical upkeep of equipment and systems. "Performance as a team member within the unit" refers to the use of the knowledges and skills appropriate for standardized qualification which are not peculiar to a specific watch station or equipment, but apply more broadly within the unit.

At present, very little PQS material has been developed for the GMM. However, current plans include the further development of this material. More detailed information on PQS will be found in *Shipboard Training,* OPNAVINST 3500.32 and *Personnel Qualification Standards (PQS) Program,* OPNAVINST 3500.34.

Once PQS material is available, the senior GMM will be responsible for the efficient functioning of the program. The GMM will conscientiously sign off those qualifications of deserving personnel, monitor individual progress, and conduct the necessary training to ensure that personnel qualify in the time allotted.

## THE GUNNER'S MATE RATING

The Gunner's Mate (GM) rating was first established in 1797. In May of 1864, General Order 37 established the pay grade of chief petty officer. It was not until 1894, by General Order 409 that the pay grades of third class through first class petty officers were established. The GM at all rate levels became the jack of all trades in the ordnance field. As new gun systems were developed, the need for special training and a system of shipboard billets became necessary. In July of 1903, General Order 137 established the rates of Turret Captain first class and chief. From WWI to WWII the GM rating stucture changed very little. During WWII two new ratings were established: the armourer and powederman. The rating structure was changed again in 1947 to three new ratings: the GMM (mount), the GMA (armourer), and the GMT (turret). In 1948 all personnel in the Gunner's Mate rating were combined into one general GM rating. Each member of the GM rating was assigned a job code number which reflected a specific type of weapon or weapon system and was used as a guide for shipboard assignments. In 1958 the pay grades of E-8 and E-9 were established for all naval ratings. It was not until 1961 that the present GM rating structure was developed.

## THE GUNNER'S MATE
## MISSILE RATINGS

As aircraft performance (speed, maneuvering and altitude capabilities) increased, the efficiency of gunfire against them decreased correspondingly. This situation led to the development of a surface-to-air missile system which became operational in the fleet in 1955. In 1944 the Navy assigned the development of a surface-to-air missile project to John Hopkins University. This project, known as Bumblebee, produced the Navy's 3Ts: Terrier, Talos, and Tartar missiles.

The Terrier medium range surface to air missile became the Navy's first operational missile system aboard the USS *Boston* (CA-69) in 1955. The first guided missile ship, the USS *Gyatt* (DDG-1), was equipped with the Mk 8 GMLS and was used as a test frame for evaluating Terrier missiles.

The long range Talos surface-to-air missile became the Navy's second operational missile system aboard the USS *Galveston* late in 1958 and gave the fleet a missile nuclear weapon capability against aircraft.

The Tartar short range surface-to-air missile was the last of the 3Ts to become operational and was designed to be used aboard DDGs and DEGs.

As weaponry changed in types and complexity, so did the Gunner's Mate rating. When guided missile systems were added to the fire power of the fleet, selected personnel of the general rating of Gunner's Mate were given the responsibility for operating and maintaining the missile launching systems. As missile systems multiplied and became more sophsiticated and the working knowledge of electricity and electronics became more extensive, the decision was made to separate the general service rating into three allied ratings. In 1961 the GM rating was split into three groups: the GMG (Guns); the GMT (Nuclear); and the GMM (Missile). Each group is now responsible for maintaining, operating, training, and repairing the equipment of specific types of weapon systems.

The missile Gunner's Mate is required to operate, inspect, and perform organizational and intermediate maintenance on guided missile launching groups and missile handling equipment; make detailed casualty analyses; inspect and repair electric, electronic, hydraulic, pneumatic, and mechanical systems and servosystems in missile launching systems; supervise personnel in handling and stowing missiles; and supervise wing and fin assemblymen in their duties. To obtain all the skills and technical background necessary for the maintenance, operation, and repair of guided missile launchers, the GMM must have an extensive knowledge of hydraulics, be able to use a wide variety of tools and test equipment, and have a working knowledge of electricity and electronics as well as all explosives associated with a surface launched missile.

The separation of the ratings holds true up to and including E-7. At the E-8 and E-9 levels, the GMG and GMM requirements are combined. This means that the E-7 Gunner's Mate M, to be advanced to E-8, must be prepared to maintain the conventional weapons. An E-7 GMM, taking the Gunner's Mate examination for E-8, will be examined on qualifications expected of both a GMG and a GMM.

The GMM rating can be further subdivided into classes; each class is assigned to a code number. The purpose of these codes is to assist in identifying personnel in a rating when a broad definition (such as GMM) is not sufficient to identify a special skill. These are called Navy Enlisted Classification (NEC) codes. The codes are changed to suit the needs of the Navy. At present time, all trainees are given NEC code numbers. Occasionally it becomes necessary to cancel certain codes; personnel in them are then recoded. A complete list of the codes is in the *Manual of Navy Enlisted Manpower and Personnel Classifications and Occupational Standards* NAVPERS 18068 (latest revision).

## CLASSIFICATION CODES

The Navy enlisted classification (NEC) structure supplements the enlisted rating structure in identifying personnel on active duty and billets in manpower authorizations. These classifications reflect special knowledges and skills that identify personnel and requirements when the rating structure is insufficient by itself for manpower management purposes. There are three types of NECs.

### Entry Series

These NECs identify aptitudes and qualifications not discernible for rates alone. They are used to code personnel who are not yet identified as strikers or who are in training for change of rating or status. Rating conversion NECs parallel rating entry NECs but are assigned only to identify petty officers or identified strikers who are assigned in-service training for change of rating or status under approved

## Chapter 1—THE GUNNER'S MATE M

programs. The rating conversion NEC for the GM rating is GM-0899 Gunner's Mate Basic.

**Rating Series**

These NECs are related to specific source ratings. They are used to identify billet requirements which are not sufficiently identified by rates, and to identify the personnel who are qualified to be detailed to fill these requirements. The following is a list of the rating series NECs for the GMM rating, showing source ratings, applicable courses, and a brief description of the jobs involved.

Priority

Number

4     GM-0986 TERRIER MISSILE AND GMLS (MK 4/10) MAINTENANCE TECHNICIAN.
Applicable Course: Terrier Guided Missile Launching System MK 10, Class C (A-121-0046)

4     GM-0987 TARTAR MISSILE AND GMLS (MK 11) MAINTENANCE TECHNICIAN.
Applicable Course: Tartar Guided Missile Launching System MK 11, Class C (A-121-0043)

4     GM-0988 TARTAR MISSILE AND GMLS (MK 13/22) MAINTENANCE TECHNICIAN.
Applicable Course: Tartar Guided Missile Launching System MK 13, (A-121-0044)

4     GM-0989 GMLS MK 26 MAINTENANCE TECHNICIAN.
Applicable Course: Not assigned

4     GM-0991 GUIDED MISSILE LAUNCHING SYSTEM MK 13 MOD 4/MK 22 MOD 1 MAINTENANCE TECHNICIAN.
Applicable Course: Not assigned

3     GM-0993 GMLS MK 29 (NATO SEASPARROW) MAINTENANCE TECHNICIAN.
Applicable Course: Guided Missile Launching System MK 29, (A-121-0041)

2     GM-0981 ORDNANCE SYSTEMS TECHNICIAN.—Administers the test, maintenance, and repair of guided missile launching systems 5"/54 rapid fire gun mounts, and the Mk 16 ASROC launching group with primary concern for overall systems maintenance: Ensures that tests and interface with other component systems properly reflect readiness of subsystem, system and integrated weapons system; and supervises organization and administration of weapons department.
Source Ratings: GM
Applicable Course: Not assigned

**Special Series**

These NECs are not related to any particular rating. They are used to identify billet requirements which are not sufficiently identified by rates, and to identify the personnel who are qualified to be distributed and detailed to fill these requirements. Special Series NECs may be assigned to personnel in any rating, provided they are otherwise eligible to receive the training involved. It is not practical to present all of the Special Series NECs in this rate training manual.

**Assignment Priorities**

To facilitate central control of NEC assignments by allowing computer programmed instructions to position NECs as primary, secondary, and lower positions, a sequence code number is assigned to each NEC. These sequence code numbers (formerly priority numbers) range from 1 to 8 and only one such sequence code number is assigned to an NEC.

Entry Series NECs are assigned sequence code number 1. All other NECs are assigned a sequence code number of 2 through 8. The lowest sequence code number has the highest priority in sequencing NECs; e.g., sequence 2 takes precedence over sequence 3. In those cases where an individual earns two or more NECs with the same sequence code number, the NEC code that the individual most recently qualified for will take precedence, except in unusual circumstances; e.g., cost of training and length of course earning the NEC.

If a Gunner's Mate is qualified for an Ordnance System Technician (GM-0981 NEC) and a Tartar Missile and GMLS Mk 13/22 Maintenance Technician (GM-0988 NEC), the Gunner's Mate will be coded GM-0981/GM-0988. The primary NEC is priority 2 and the secondary NEC is priority 4. The fact that one NEC is priority 2 and another is priority 4 does not imply that the priority 2 skill is a higher level skill than the priority 4 skill. Priority numbers are based on the need to retain NEC identification in any given instance and this need varies for each rating.

## GMM DUTY ASSIGNMENTS

All ships having surface-to-air missiles as armament, or ammunition supply ships carrying surface-to-air missiles, and shore activities where missiles are repaired, assembled, tested, and/or stored will have a GMM assigned for duty. A GMM's duties naturally will vary according to the NEC and the type of ship in which a guided missile launching system (GMLS) is part of a surface missile system (SMS). Some large ships may carry more than one GMLS and require a large number of GMMs; smaller ships require fewer personnel. Since the number of GMMs varies according to ship type, so does the rate level of each ship. The senior GMM aboard a large ship (CG) could be an E-8 or E-9; on a small ship (DDG) the senior GMM could be an E-6 or E-7. As a senior Gunner's Mate, your enlisted job code number is the primary factor in determining your assignability.

In training assignments ashore, GMMs serve as instructors in the Gunner's Mate School at Great Lakes, Ill., and as instructors in the Mk 29 GMLS at either of the Guided Missile Schools at Dam Neck, Virginia Beach, Va., or Vallejo, Calif. GMMs are also assigned to duty at the Naval Education and Training Program Development Center (NAVEDTRAPRODEVCEN), Pensacola, Fla., as subject matter experts/writers for fleetwide examinations, rate training manuals, and correspondence courses. In fulfilling duties in training billets, the knowledge gained afloat will be put to use in preparing training material. Personnel selected for training billets are carefully chosen and are expected to be experts in specific missile systems.

In addition to training billets, the GMM can be assigned duty ashore at naval weapons stations where missiles are assembled and tested prior to shipboard loading.

The billets mentioned above comprise a portion of the billets ashore for GMMs. In some instances primary duties will be military rather than technical or occupational although such assignments are seldom made for personnel in critical ratings.

### Out-of-Rating Assignments

Unfortunately, there are just not enough shore billets written for the GMM to allow shore assignments to those billets alone. Therefore, GMMs (as well as personnel in many other ratings) are assigned to billets out of their rating (Shore Patrol, Armed Forces Police, MAA, Career Counseling, Drug Abuse Specialist, Recruiting, Courier, Security Groups, etc.) to alleviate this situation. Without these out-of-rating billets, shore duty opportunity for GMMs would be considerably reduced. Keep this in mind if you receive orders ashore which are not quite what you desired—chances are you were assigned close to the area of your choice.

### Tour Lengths

The tour lengths for GMMs are determined by analysis of the distribution of billets between sea and shore duty, retention rate, and related manning level projections. As billet structures and activity allowances change, so do tour lengths. For this reason, specific tour lenths cannot be included in this manual. However, you can determine the tour lengths for your rate by contacting your personnel office or by calling your detailer.

## Chapter 1—THE GUNNER'S MATE M

### PERSONNEL ASSIGNMENT SYSTEM

The basic concept of centralized assignment is simple. It provides the capability of matching total enlisted personnel resources against total requirements. Under this method all enlisted members, with the exception of nondesignated SN/FN/AN, are under the exclusive detailing control of the Director of Naval Military Personnel Command (NMPC), (formerly CHNAVPERS), who determines the member's ultimate assignment. Nondesignated SN/FN/AN are detailed by the Enlisted Personnel Management Center (EPMAC) in New Orleans.

### PERSONNEL REQUISITION SYSTEM

The Personnel Requisition System was developed to enable NMPC to determine personnel requirements for activities Navywide. The requisition is prepared by the manning control authorities (MCAs) and lists the consolidated personnel requirements for each of their areas of cognizance.

Each requirement (by rate, rating, and NEC, if appropriate) is identified by a requisition number which must appear on orders issued to satisfy that requirement. The requirements are assigned priorities by the MCAs who consider such factors as command employment (deployment, CNO SPEC OPS, etc.), one-for-one billets, and on-board population in a given unit (15 GMMs on board; 25 GMMs allowed).

Although the requirements are projected for up to 8 months, the requisition is updated monthly by the MCA in response to each command's Enlisted Distribution Verification Report (EDVR), Form 1080. Through the 1080, and individual requirement may be generated in the requisition for any of the following reasons:

1. To fill an existing vacancy.
2. To relieve an individual on board.
3. To fill a new billet.

These new requirements are assigned appropriate priorities and are integrated into the requisition by the MCAs. Detailers are required to make assignments in strict accordance with requisition priorities.

### TYPES OF DUTY

For rotation purposes, six types of duty currently exist.

Type 1—Shore Duty: CONUS, fleet shore duty, and certain fleet activities.

Type 2—Sea Duty: Arduous sea duty in deploying ships/units homeported in CONUS, Hawaii, or Alaska.

Type 3—Overseas Shore Duty: Duty ashore outside CONUS where prescribed DOD tours are 36 months or less. Counts as sea duty for rotation purposes, but may not completely fulfill an individual's sea tour.

Type 4—Non-rotated Sea Duty: Sea duty in ships/staffs/units listed in OPNAVINST 4600.16 (for example, units homeported in Japan). Tour lengths are defined by the area tour for the locale of assignment. Counts as sea duty for rotation purposes, but may not completely fulfill an individual's total sea tour obligation.

Type 5—Neutral Duty: Duty in CONUS, Hawaii, or Alaska homeported ships/staffs/units which normally do not operate away from homeport, except for brief periods. Not credited for either sea or shore rotation. Sea duty commencement date (SDCD) or shore duty commencement date (SHDCD) is adjusted to compensate for the neutral time. You should expect another sea assignment upon completion of neutral duty.

Type 6—Preferred Overseas Shore Duty: Duty overseas where the prescribed DOD tours are from 36 to 48 months. Three possible alternatives exist within the application of this range of tour length.

1. Individual's normal shore tour is less than the area tour. In this case, the shore tour will be extended to coincide with the area tour. Other extensions normally will not be granted.

2. Individual's shore tour coincides with the area tour. No adjustments are necessary. Extensions will be considered on case basis.

3. Individual's shore tour is longer than the area tour. Initial orders will be issued for the area tour. Upon request, an extension up to normal shore tour limits or 48 months' total time in billet, whichever is less, may be granted. The total time counts as shore duty for rotation

purposes; extension requests will be viewed accordingly.

## DUTY PREFERENCES

An accurate and up-to-date Enlisted Duty Preference form, NAVPERS 1306/63, should be on file with the detailers for each individual under their detailing responsibility. If you have never submitted a preference form, do it now. Without it, your detailer's only assumption is that your preference is simply "anywhere world." You are encouraged to submit a new form at least 10 months prior to your projected rotation date (PRD). If you neglect to submit a new form prior to transfer, your detailer will assume that your old one is still correct.

Accurate information is important since it may affect your next duty assignement. However, after a few years, you probably will not remember what you put on your form. Make a copy of it for your files and break it out occasionally to check the information. As changes occur concerning service schools completed, number of dependents, desires, etc., submit a new form. Even minor changes should be reported. For example, most of us consider a change in the number of dependents to mean a new baby has been born. However, it is also a decrease in the number of dependents when your daughter gets married. With fewer dependents, you may be eligible for a wider range of duty assignments. Your assignability definitely improves with fewer dependents.

Eligibility for duties should be carefully investigated. If you are not qualified for the duties you have requested, your request is essentially invalid. If you are not sure of the qualifications required, check with your career counselor or personnel office for directives or instructions which may apply to the assignment you desire.

You should indicate a different choice in each of the preference boxes. This will maximize the possibility of your detailer to assign you the duty you desire. The detailer will consider your first choice first, and if at all possible, will assign you to that duty. Therefore, nothing is gained by repeating your preferences. If assignment to your top choice is not possible, the detailer has to know your alternate selections in order to best fulfill your desires.

The locations you indicate as duty preferences may be broad (anywhere 11th Naval District) or quite specific (Seattle, NTC, San Diego). Your chances, however, are much better with broad preferences, so it makes sense to include both. If your first choice is specific, add a broad area as the second or third alternative.

The remarks section is particularly important. Include in your remarks any information which may influence your assignments. Following are a few important items that should be explained in the remarks section:

Is your wife a teacher?
Do you speak a foreign language?
Do you have a handicapped child?
Do you have any special qualifications?
Do you have any deficiencies or physical problems that do not appear in your service record which might be a handicap in an instructor or shore patrol billet?

Do not confuse the Enlisted Duty Preferences form with the Enlisted Transfer and Special Duty Request (NAVPERS 1306/7). These forms are not interchangeable; they serve two completely different functions. Essentially, the preference form is a permanent record maintained on file for use when you become available for routine assignment. On the other hand, the Enlisted Transfer and Special Duty Request form is used to request a single, specific program. It is answered "yes" or "no" one time and is not retained for future reference. Some of the types of duty that require form NAVPERS 1306/7 or a letter request with specific information are New Construction, Recruiting, Drug Abuse Specialist, Minority Affairs, Courier, and special projects (as they are initiated). If you are in doubt as to which form to use, call your detailer who will be happy to answer your questions.

## DETAILING

The job of a detailer is a thankless one. A detailer must treat each decision as if it is the most important decision he or she will ever

## Chapter 1—THE GUNNER'S MATE M

make. Considering the instructions and the restrictions placed upon a detailer, the task seems impossible. However, a surprisingly high percentage of individuals are assigned to billets of their choice. Regretfully, every individual cannot be assigned to an area or billet of preference, and the detailer must be prepared to justify any decision made.

The basic policy governing the assignment of enlisted personnel is that every effort will be made by rating control officers and detailers to assign personnel to duty stations in accordance with their preferences. Assignments must also be commensurate with eligibility, availability of billets desired, and with the priorities established by the Chief of Naval Operations and the manning control authorities.

There are at least 20 potential variables that the detailer must consider in making a specific assignment. Taken together, they constitute a complex decisionmaking process which may frequently give the appearance of a mystic rite. The following lists are the variables most frequently considered by the detailer.

Individual Variables:

1. Rate, rating, NEC—obviously the first considerations in any assignment.
2. Previous performance and evaluation of potential performance—particularly when considering assignment to special duties such as instructor and recruiter.
3. Specialized training or qualifications held, other than NEC-identified skills.
4. Service history, active duty base date, obligated service career status, and advancement status.
5. Career rotation status—sea/shore duty history and policy concerning current duty station (neutral duty, etc.).
6. Security clearance or clearability.
7. Training required to satisfy next assignment and career training pattern.
8. Applicable guarantees or incentives—first reenlistment incentive, twilight tour, etc.
9. Any special considerations known to the detailer—dependent's status, medical problems, etc.

Billet Variables:

1. First there must be a valid billet vacancy to be filled, requiring a certain rate and NEC.
2. The assignment, when considered with all others of the rate involved, must contribute to maintaining a balance in manning levels between the two fleets and other manpower users.
3. The assignment must respond to the established priorities.
4. The manning rules of the activity involved.
5. Activity employment schedule.
6. Any special rotation plans which apply to the activity.
7. Any special billet requirements—factory training, foreign language skill, special screening, security clearance, etc.

External Variables:

1. Cost of the move. All other factors being equal, the least costly PCS move is ordered. Detailers must assess such factors as distance involved, number of dependents, and location of household goods in arriving at cost judgments.

2. Time frames of PRDs, billet vacancies, class convening dates, leave and travel time, must all coincide.

3. Constraints imposed by higher authority. Overseas area tour lengths, number of dependents allowed, sole surviving son restrictions, etc.

4. Tour lengths for the type of duty, activity, location, and rating must dovetail with obligated service and career history of the individual.

These lists do not include all possible variables, but they do indicate that much thought, time, and effort go into making actual assignments. In matching the right person for the right job, the detailer continues to strive for more efficient manpower utilization which brings billets, skills, and personal desires into harmony and provides timely response to the Navy's changing manpower requirements.

## ADVANCEMENT

The advancement system is another area in which you and the Navy have a mutually beneficial relationship. You benefit through higher pay, greater prestige, more interesting and challenging work, and the satisfaction of getting ahead in your chosen career. The Navy benefits from your advancement in two ways. First, you become more valuable as a technical specialist in your own rating. Second, you become more valuable as a person who can supervise, lead, and train others. It is important that you understand the advancement system and are aware of frequent changes. BUPERS Notices 1418 will usually keep you up to date.

The normal system of advancement may be easier to understand if it is broken into two parts:

1. Those requirements that must be met before you may be considered for advancement.
2. Those factors that actually determine whether or not you will be advanced.

## QUALIFYING FOR ADVANCEMENT

In general, to qualify (be considered) for advancement to first class or chief, you must first fulfill the following:

1. Have a certain amount of time in service.
2. Have a certain amount of time in pay grade.
3. Demonstrate knowledge of material in your mandatory rate training manuals by achieving a suitable score on your command's test, successfully completing the appropriate NRCCs or, in some cases, successfully completing an appropriate Navy school.
4. Satisfy the requirements of the Personnel Advancement Requirement (PAR), NAVPERS 1414/4, program.
5. Be recommended by your commanding officer.
6. Demonstrate knowledge of the technical aspects of your rating by passing a Navywide advancement examination based on the occupational standards applicable to your rate (from NAVPERS 18068 series, those standards listed at and below your rate level).

If you meet all the above requirements satisfactorily, you become a member of the group from which advancements will be made.

## WHO WILL BE ADVANCED?

Advancement is not automatic. Meeting all the requirements makes you eligible but does not guarantee your advancement. Some of the factors that determine which qualified persons will actually be advanced in rate are the score made on the advancement examination, the length of time in service, the performance marks earned, and the number of vacancies being filled in a given rate.

If the number of vacancies in a given rate exceeds the number of qualified personnel, all of those qualified will be advanced. More often, the number of qualified people exceeds the number of vacancies. For those instances, the Navy has devised a procedure for advancing those who are best qualified. This procedure is based on combining three personnel evaluation systems:

Merit rating system (annual evaluation and CO recommendation).
Personnel testing system (advancement examination score—with some credit for passing previous advancement exams).
Longevity (seniority) system (time in rate and time in service).

Simply, credit is given for how much the individual has achieved in the three areas of performance, knowledge, and seniority. A composite, known as the final multiple score, is generated from these three factors. All the candidates from a given advancement population who have passed the examination are then placed on one list. Based on the final multiple score, the person with the highest multiple score is ranked first, and so on, down to the person with the lowest multiple score. For candidates for E-6, advancement authorizations are then issued, beginning at the top of the list, for the number of persons needed to fill the existing vacancies. Candidates for E-7 whose final mutliple scores are high enough are designated PASS SELBD ELIG (pass selection board eligible). This means that their names will be placed before the Chief Petty Officer Selection

Board, an OPNAV board charged with considering all so-designated eligible candidates for advancement to CPO. Advancement authorizations for those being advanced to CPO are issued by this board.

Who, then, are the individuals who are advanced? Basically, they are the ones who achieved the most in preparing for advancement. They were not content to just qualify; they went the extra mile in their training, and through that training and their work experience developed greater skills, learned more, and accepted more responsibility.

While it cannot guarantee that any one person will be advanced, the advancement system does guarantee that all persons within a particular rate will compete equally for the vacancies that exist.

A change in promotion policy, starting with the August 1974 examinations, changed the passed-but-not-advanced (PNA) factor to the high quality bonus point (HQP) factor. Under this policy, a person who passed the examination but was not advanced could gain points toward promotion in subsequent attempts. Up to three multiple points can be gained in a single promotion period. A maximum of 15 points can be accumulated over six promotion periods. The addition of the HQP factor, with its 15-point maximum, raises the number of points possible on an examination multiple from 185 to 200. This gives the examinee added incentive to keep trying for promotion in spite of repeated failure to gain a stripe because of quota limitations.

## HOW TO PREPARE FOR ADVANCEMENT

To prepare for advancement, you must study specific references for detailed, authoritative, up-to-date information on all subjects related to the naval requirements and to the occupational standards of the Gunner's Mate M rating. Some of the publications are absolutely essential for anyone seeking advancement and other, although not essential, are extremely helpful. The publications described here should be available at the educational services office (ESO) of your unit.

### Sources of Information

Some of the publications discussed here are subject to change or revision from time to time; some at regular intervals, others as the need arises. When using any publication that is subject to revision, be sure you have the latest edition. When using any publication that is kept current by means of changes, be sure you have a copy in which all official changes have been entered.

Official publications and directives carry abbreviations and numbers which identify the source and subject matter of each document. This training manual, for instance is NAVEDTRA 10200-C, which means that it is a publication of the Chief of Naval Education and Training. The letter following the numerals designates the edition and indicates that significant revisions were made to the previous edition (NAVEDTRA 10200-B). Your ESO can tell you the latest edition of any publication or directive.

NAVPERS 18068.—The *Manual of Navy Enlisted Manpower and Personnel Classifications and Occupational Standards*, NAVPERS 18068 (with changes), contains in Section I, Navy Enlisted Occupational Standards, the occupational and naval standards for advancement to each pay grade in each rating. Section II is Navy Enlisted Classifications. This manual replaces the "quals manual" and the NEC manual.

Naval standards are expressed as minimum skills that apply to all ratings rather than to any particular rating. Occupational standards are defined as tasks directly related to the work of each rating.

Each Navywide occupational examination for pay grades E-6 and E-7 contains 150 questions, all related to the appropriate occupational areas. For pay grade E-8, only 50 of the 150 questions are so related; and for E-9, only 45 of the 150.

NAVPERS 1414/4.—NAVPERS 1414/4 is used in the new Personnel Advancement Requirement (PAR) program and replaces the Record of Practical Factors (NAVEDTRA 1414/1).

In NAVEDTRA 1414/1 "quals" were stated in terms of practical factors and knowledge factors. The new occupational standards are presented only as task statements in NAVPERS 1414/4. This new system allows a command to evaluate the overall abilities of an individual in a day-to-day work situation and eliminates the need to complete a mandatory, lengthy, and detailed checkoff list.

The E-8 and E-9 are exempted from the program as there are other means of selection for advancement to these pay grades. The E-3 apprenticeships are so broad as to make the development of a single PAR impractical.

Each rating PAR lists the requirements for advancement to pay grades E-4 through E-7 in one pamphlet. It contains descriptive information, instructions for administration, special rating requirements, and advancement requirements in the following sections:

Section I—Administration Requirements

Section II—Formal School and Training Requirements

Section III—Occupational and Military Ability Requirements

Section I contains the individual's length of service, time in rate, and checkoff for the individual having passed the E-4/E-5 military leadership examination.

Section II contains a checkoff entry to indicate that the individual has completed the Navy's applicable training courses for military requirements and the rating.

Section III is a checkoff list of task statements. Items in this section are to be interpreted broadly and do not demand actual demonstration of the item or completion of an alternate local examination, although demonstration is a command prerogative. Individuals are evaluated on their ability to perform the task whether the evaluation is based on observation of their ability in related areas, training received, or if desired, demonstration.

There is currently a pilot program which includes the PQS watch station qualifications and preventive maintenance actions as a separate section of the PAR form. Section III under this program lists task statements required of the rating which are not reflected in the PQS qualifications. As PQS qualifications are developed, PAR forms will be revised.

PAR forms for the Gunner's Mate M may be ordered through the Navy supply system.

A current copy of the PAR form for GMMs should be kept in each person's service record and should be forwarded with the service record to the next duty station. Each GMM should also keep a copy of the form for personal use.

NAVEDTRA 10052.—*Bibliography for Advancement Study,* NAVEDTRA 10052, is the most important single source when preparing for advancement. This pamphlet, based on the naval and occupational standards, lists the training manuals and other publications which enlisted personnel should study in preparation for advancement examinations. The same publications are used by the examination writer in preparing the advancement examination.

The first few pages of the pamphlet contain the references which cover naval standards. The remainder of the booklet contains reference listings by ratings, using the five-column format shown in figure 1-1.

Column (1) Ratings: This column contains the names and abbreviations of both service and general ratings.

Column (2) Publication Titles: Titles and applicable parts of rate training manuals and other publications are shown in this column. Notice that, when only certain parts of a book apply as references for a particular rate, just those parts (paragraphs or chapters) are listed.

Column (3) Text Identification Number: This column identifies the references by specific numbers. Make sure the publication you use is the most recent edition.

Column (4) Course Identification Number: The NAVPERS/NAVEDTRA/NAVTRA number(s) of the correspondence course(s) covering the subject matter of the prescribed text(s) described in columns 2 and 3 is/are listed. For courses which combine the RTM and NRCC, such as this one, the RTM and NRCC have the same identification number.

Column (5) Applicable Pay Grade Levels: Except for training manuals that are mandatory, only the lowest pay grade level for which a publication is applicable is listed in this column. If a publication is mandatory, all the

Chapter 1—THE GUNNER'S MATE M

| Ratings (1) | Publications Titles (2) | Text Identification Number (3) | Course Identification Number (4) | Applicable Pay Grade Levels (5) |
|---|---|---|---|---|
| \multicolumn{5}{c}{Occupational Field 9 - Ordnance Systems} ||||
| GUNNER'S MATE M (GMM) | *Gunner's Mate M 3 & 2 | NAVEDTRA 10199-C | 10199-C | E-4, E-5 |
| | *Gunner's Mate M 1 & C | NAVPERS 10200-B | 91380-C | E-6, E-7 |
| | Department of the Navy Information Security Program Regulation (Ch 5, Sec 1-2; Ch 7, Sec 1, Para 7-100, 7-101G, 7-102, and 7-103, and Sec 2, Para 7-204 thru 7-208, 7-210 thru 7-212; Sec 3; Ch 9) | OPNAVINST 5510.1E | | E-5 |
| | Ships' Maintenance and Material Management (3-M) Manual (Vol I, Ch 1, 2; Ch 5, Para 5-1 thru 5-2.8.3 Vol II, Ch 9, Para 9-1 thru 9-7) (Vol I, Ch 5, Para 5-2.9 thru 5-4.2) (Vol I, Ch 3, 4; Vol II, Ch 9, Para 9-8 thru 9-11) (Vol I, Ch 6, 7) | OPNAVINST 4790.4 | | E-4  E-5   E-6 E-7 |
| | Synchro, Servo, and Gyro Fundamentals | NAVPERS 10105 | 91333 | E-5 |
| | Instruction Book, Magazine Sprinkler System | NAVSHIPS 0348-078-1000 | | E-5 |
| | Principles of Naval Ordnance and Gunnery (Less Ch 1-4, 10) | NAVEDTRA 10783-C | 10922-D1 | E-7 |
| | Principles of Guided Missiles and Nuclear Weapons (Ch 1-11) | NAVTRA 10784-B | 10924-C2 | E-7 |

Figure 1-1.—Partial bibliography listing for the GMM rating from NAVEDTRA 10052.

pay grade levels for which it is mandatory are shown. All higher pay grade levels may be held responsible for the material contained in publications listed for lower levels in their paths of advancement.

Asterisks which appear through the listings indicate the rate training manuals and correspondence courses whose mandatory completion is specified by the *Manual of Advancement,* BUPERSINST 1430.16.

RATE TRAINING MANUALS.—Rate training manuals are written for the specific purpose of helping personnel prepare for advancement. Some manuals are general and are intended for use by more than one rating; others (such as this one) are specific to the particular rating. For a complete listing of rate training manuals and to ensure that you have the latest edition consult the most recent edition of *List of Training Manuals and Correspondence Courses,* NAVEDTRA 10061.

Each time a rate training manual is revised, it is brought into conformance with the official publications and directives on which it is based. However, during the life of any edition, discrepancies between the manual and the official sources are almost sure to arise because of changes to the latter which are issued in the interim. In the performance of your duties, always refer to the appropriate official publication or directive. If the official source is listed in NAVEDTRA 10052 and, therefore, is a source used by the NAVEDTRAPRODEVCEN in preparing the advancement examinations, the NAVEDTRAPRODEVCEN will resolve any discrepancy by using the material which is most recent.

There are four rate training manuals specially prepared to present information on the naval standards for advancement. They are as follows:

*Basic Military Requirements,* NAVEDTRA 10054.
*Military Requirements for Petty Officer 3 & 2,* NAVEDTRA 10056.
*Military Requirements for Petty Officer 1 & C,* NAVEDTRA 10057.
*Military Requirements for Senior and Master CPO,* NAVEDTRA 10115.

Each of the naval standards manuals is mandatory at the indicated pay grade levels. In addition to giving information on the naval standards, these books give a good deal of useful information on the enlisted rating structure; how to prepare for advancement; how to supervise, train, and lead other people; and how to meet your increasing responsibilities as you advance.

Rate training manuals prepared for other ratings are often useful sources of information. Reference to these training manuals will increase your knowledge of the duties and skills of other personnel in the weapons department. The manuals prepared for the Fire Control Technician M and Gunner's Mate G ratings will be of particular interest to you.

OFFICER CORRESPONDENCE COURSES.—Officier correspondence courses you may find helpful when preparing for advancement to GMM1 and GMMC include *Principles of Guided Missiles and Nuclear Weapons,* NAVEDTRA 10924; *Leadership,* NAVTRA 10903; *Personnel Management,* NAVEDTRA 10968; and *U.S. Navy Regulations, Part II,* NAVEDTRA 10427.

NONRESIDENT CAREER COURSES.— Most rate training manuals are used as the basis for correspondence courses. Completion of a mandatory training course can be accomplished by passing the nonresident career course that is based on the training manual. You will find it helpful to take other courses, as well as those that are based on mandatory training manuals. Taking a course helps you to master the information given in the training manual or text and also gives you an idea of how much you have learned from studying the book.

NAVEDTRA 10061.—*List of Training Manuals and Correspondence Courses,* NAVEDTRA 10061, is published annually.

The quarterly publication TIPS amplifies the information in NAVEDTRA 10061 and NAVEDTRA 10052. Both TIPS and NAVEDTRA 10061 should be available at the ESO of your unit.

## Chapter 1—THE GUNNER'S MATE M

MANUAL FOR NAVY INSTRUCTORS.—Another training publication which you may find useful in connection with your responsibilities for leadership, supervision, and training is the *Manual for Navy Instructors,* SP 107.

### How to Study

The following suggestions will help you to make the best use of Navy training publications when you are preparing for advancement.

1. Study the naval standards and the occupational standards for your rating before you study the training manual, and refer to the standards frequently as you study. Remember, you are studying the manual primarily to meet these standards.

2. Set up a regular study plan. It will probably be easier for you to stick to a schedule if you plan to study at the same time each day. If possible, schedule your studying for a time when you will not have too many interruptions or distractions.

3. Before you begin to study any part of the manual intensively, become familiar with the entire book. Read the preface and the table of contents. Check through the index. Thumb through the book without any particular plan, looking at the illustrations and reading bits here and there as you see things that interest you.

4. Look at the training manual in more detail to see how it is organized. Look at the table of contents again. Then, chapter by chapter, read the introductions, the headings, and the subheadings. This will give you a pretty clear picture of the scope and content of the book. As you look through the book in this way, ask yourself some questions:

    a. What do I need to learn about this subject?

    b. What do I already know about this subject?

    c. How is this information related to information given in other chapters?

    d. How is this information related to the occupational standards?

5. When you have a general idea of what is in the training manual and how it is organized, fill in the details by intensive study. In each study period, try to cover a complete unit—it may be a chapter, a section of a chapter, or a subsection. The amount of material you can cover at one time will vary. If you know the subject well, or if the material is easy, you can cover quite a lot at one time. Difficult or unfamiliar material will require more study time.

6. In studying any one unit—chapter, section, or subsection—write down questions as they occur to you. Many people find it helpful to make a written outline of the unit as they study, or at least to write down the most important ideas.

7. As you study, relate the information in the training manual to the knowledge you already have. When you read about a process, a skill, or a situation, try to see how this information ties in with your own past experiences.

8. When you have finished studying a unit, take time to review what you have learned. Without looking at the training manual, write down the main ideas that you have gotten from studying this unit. Do not just quote the book. If you cannot give these ideas in your own words, the chances are that you have not really mastered the information.

9. Use nonresident career courses whenever you can. The courses are based on rate training manuals or on other appropriate texts. You will probably find it helpful to take other courses as well as those based on mandatory manuals. Taking a course helps you to master the information given in the training manual and helps you see how much you have learned.

10. Think of your future as you study rate training manuals. Anything you learn now will help you to advance both now and later.

### ADVANCEMENT OPPORTUNITIES FOR SENIOR PETTY OFFICERS

Making chief is not the end of the advancement line. Advancement to senior chief (E-8), master chief (E-9), warrant officer, and commissioned officer are among the opportunities available to qualified petty officers. These special paths of advancement are open to personnel who have demonstrated outstanding professional ability, the highest order of leadership and military repsonsibility, and unquestionable moral integrity.

## Advancement to Senior or Master Chief

Chief petty officers may qualify for the advanced grades of senior and master chief. These advanced grades provide for substantial increases in pay, with increased responsibilities and additional prestige. The requirements for advancement to senior and master chief are subject to change but, in general, include a certain length of time in grade and in the naval service, a recommendation by the commanding officer, and a sufficiently high mark on the Navywide examination. The final selection for senior and master chief is made by a regularly convened selection board.

The satisfactory completion of the nonresident career course *Military Requirements for Senior and Master Chief Petty Officer,* NAVEDTRA 10115, is required of all personnel advancing to E-8 and E-9.

## Advancement to Warrant or Commissioned Officer

The warrant officer (WO) program provides the opportunity for E-7 and above enlisted personnel to advance to warrant rank.

The limited duty officer (LDO) program provides a path of advancement from warrant officer to commissioned officer. LDOs are limited, as are WOs, to duty in the broad technical fields associated with their former ratings. E-6s, to be eligible, must have passed an E-7 rating exam prior to selection.

If you are interested in becoming a warrant or commissioned officer, ask your educational services officer for the latest requirements that apply to your particular case. See figure 1-2 for an illustration of the WO/LDO career path.

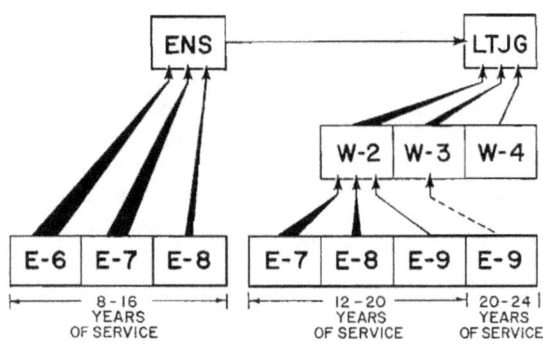

200.29
Figure 1-2.—Senior petty officer career path to LDO/WO.

NOTE: As this text was going to press, the Navy announced that BUPERS was to be replaced by two commands—the Deputy Chief of Naval Operations for Manpower, Personnel, and Training, and the Navy Military Personnel Command (NMPC). When the BUPERS or NAVPERS forms, notices, and instructions, referred to in chapter 1 are reissued, they may appear as OPNAV or NMPC forms, notices, and instructions.

# CHAPTER 2

# MISSILE HANDLING

Early in your career as a GMM you learned to handle missiles from dockside to ship to stowage area, or from ship to ship. You became familiar with standard and special handling equipment, the operation of strikedown and strikeup machinery, and the equipment or tools needed for mating and checkout of the missile. General and special rules of safety as applied to missile handling were called to your attention many times, so you would not forget them. You performed operational tests on the handling equipment; lubricated, disassembled, inspected, cleaned, and reassembled mechanical, electric, pneumatic, and hydraulic handling equipment.

As you advanced in rate, you not only had to know how to operate the handling equipment, but you trained others as individuals or teams in the use of the equipment. If your missiles required wing and fin assembly, you were expected to train the teams to do the work with the speed and accuracy required for that weapon system.

If something went wrong with the electrical or electronic parts of the handling equipment, it was your responsibility as a GMM2 to trace the trouble to its source with the use of test equipment and the aid of wiring diagrams.

What is left for the GMM1 and Chief to learn regarding the handling and stowage of missiles and missile components? Planning of the work and supervision and teaching of lower rated personnel are important parts of your duties. To teach others you must have knowledge that is broader and deeper than that of those you teach. This knowledge is also necessary for intelligent planning of handling and stowage operations. From experience and study you should know about different missiles and different ships. Your crew may have had experience with only one or two types of missiles.

Lower rated personnel perform the routine preventive maintenance and the simpler repairs. Adjustment, overhaul, and the more difficult repairs are the province of the GMM1 and C. In addition to the ability to troubleshoot and repair the equipment, you must be able to plan and carry out a maintenance and repair program for the equipment.

Rules for stowage of supplies are the proper concern of the supply department or the supply officer, but the GMM must know and apply the special rules applicable to the stowage of guided missiles and their components. Because guided missiles contain explosives, the GMM needs to know the properties of explosives and the rules for handling and stowing the explosives safely. The GMM1 and C see to it that the weapons and their components are properly stowed and insist on observance of safety regulations all through the process of handling and stowing.

## LOADING AND STOWAGE PLANS

In this chapter the loading operation discussed is that of putting the missile on the ship, whether from dockside or from another ship. Chapter 3 discusses the loading of missiles onto the launcher for firing.

## KNOWLEDGE REQUIRED FOR PLANNING WORK

Before you can plan a loading operation you need to know a great deal, not only about the missile, but about the ship, its handling equipment, and its stowage areas. How much responsibility for planning will be yours depends on the size of the ship, the personnel of the ship, and other factors. On a small ship you may be the leading petty officer in the GMM rating; prepare yourself to accept responsibility.

### About Your Ship

If you have been on your ship for some time, it is assumed that you have learned the location of strikedown hatches, missile elevators, missile stowage areas, and strikedown equipment. If, however, you have not had the opportunity to become acquainted with these details, you need to make an active effort to know your ship. When a loading operation is impending, you need to know whether the loading will be from dockside or from another ship. This is information you must have in order to plan the handling of the missiles in getting them on the ship.

Find out which of the stowage spaces are to be used for this particular load, then determine which elevator or strikedown equipment is best or most convenient to use.

### About The Weapon

The planner needs to know the number of weapons of each type to be taken aboard. If only one type of weapon is being received, the matter is greatly simplified. At the present time Terrier, Standard ER/MR, Tartar, Harpoon, Tomahawk, and Sparrow III missiles, and ASROC weapons are received aboard ship in the desired configuration. Handling requirements may vary widely for different weapons. You need to know the configuration of the weapon, the number of containers per weapon, the size and weight of each container (shape may be important, too), the places on the containers where attachments are to be made, and the special handling equipment to be used with each. How many crewmembers will you need for each type of weapon, and what will be their specific posts and duties? These are the things that you must find out before the loading operation begins.

Safety rules for handling of explosives are applicable to guided missiles, but there are some additional rules for handling of particular missiles. What are the rules for grounding of the missile components during handling and stowage? What are the temperature and moisture limits of the explosive components? Do they have to be kept in the shade while waiting to be struck down?

### Knowledge of Operation

The experience you gained in using handling equipment during your years as a striker, a GMM3, and then a GMM2 is invaluable and may be sufficient to enable you to manage the present situation with a higher degree of efficiency. A good planner cannot just hope everything will work out all right. If there is equipment which you have not used before, find out how to operate it and pass the information on to your subordinates. Know the safety precautions that apply. Find out where each missile is to be stowed. This is especially necessary if you are fairly new on the ship. Considerable confusion can result, for example, if you discover after a missile has been brought down to the magazine that it belongs in a magazine at the other end of the ship. This can be particularly bad if the missiles are of the type in which the components are sent down in a specific sequence so that they will be in the correct order for assembly. If you are in charge of the handling operations, the blame for the confusion is yours. Careful preplanning prevents such mixups.

## SCHEDULING OF WORK

On any ship cooperation among divisions is necessary for a loading or offloading operation, even though the load consists entirely of missiles. When other material besides missiles is being loaded, the time for using certain of the ship's gear has to be allotted. If missile loading is scheduled for 1000, be ready with your crew to swing into action and to do your work on schedule. Loading of explosives should be done in daylight hours if at all possible, and the ship's plan for the loading will conform to this rule. Unexpected foulups can throw the plan off schedule. Plan your part of the work so there will be no such delays. Remember, however,

## Chapter 2—MISSILE HANDLING

that missile handling must not become a speed contest.

Usually you will have information several days in advance of the actual loading date. Have your personnel check the operation of the special handling equipment to be used with the missiles. If any of the equipment does not operate as it should, locate the cause of the trouble with the use of test equipment, wiring diagrams, hydraulic, schematic, and troubleshooting techniques, beginning with the simplest method; then make the necessary repairs and adjustments. The checking of the ship's cargo handling gear is the responsibility of the Boatswain's Mates. They rig the lines and other cargo gear, but before you entrust any of your missile cargo to the gear be sure it has been checked out for handling the weights required. It is your and your crew's responsibility to cycle the equipment for striking down the weapons, or the special gear for moving them to on-deck launchers. If the equipment does not operate properly, you must repair and adjust it so it will be ready to use on the day required. As a GMM2 you have had some experience in locating the trouble spots in such equipment; now you must learn to make more difficult repairs on the equipment and adjust it to working condition.

### SECURITY

In addition to the safety provisions that must be observed during handling of any explosives to prevent fire or explosions, provision must also be made for the security of the weapons against theft, damage, destruction, or access to enemies. Knowledge of the transfer, type, number, and design of the weapons, etc., is information that must be concealed from enemies. Access to nuclear warheads must be especially guarded against. The commanding officer sets the security watch on the pier and on the ship. You learned about sentry and watch duties and security of classified documents in *Seaman*, NAVPERS 10120 series, in *Basic Military Requirements*, NAVTRA 10054 series, and in *Military Requirements for Petty Officer 3 & 2*, NAVEDTRA 10056 series.

Although the stenciled information on containers conceals any classified nature of the contents, personnel handling the containers usually need to be aware of what they are handling so they will use adequate precautions. If the individuals who handle fuzes, for example, know that is what they are handling, they will be much more careful than if they do not know.

### PLANNING SEQUENCE OF OPERATION

As soon as you know your working party assignment in the loading or unloading operation, think through the work sequence as you and your personnel are going to accomplish it. Roughly sketching in your plan of action on paper may be very helpful in filling in the details of the plan. Where are you going to spot your crew? How are you going to manage the handling of the missile boosters so they will be placed in the correct order without delay or confusion? What receipt inspections are necessary before the missiles are struck below? How much assembly, if any, is to be performed before stowing the components? Have your personnel trained for this work or will you have to schedule a practice session before the day of loading arrives? If such a session is necessary, check to be sure the crew is not already scheduled to be doing something else during the time you want them. Consult with the training officer of the division on this.

### STOWAGE AREAS

Before the loading day arrives, check the stowage areas that will be needed. All of them should be clean, and no unnecessary material should be there. The sprinkler systems must be in operating condition. Check all other firefighting systems or equipment in and adjacent to the magazines. Be sure the alarm systems are working. In addition to fire warning systems, continuous operation of a radiation detection device with an automatic alarm is required at shore stations and on submarines in spaces where nuclear missiles, weapons, or warheads are stowed. In air-conditioned spaces check to be sure the system is maintaining the space at the required temperature and humidity.

Extra checks are needed after a magazine has been painted. The areas for attaching ground wires should be clear of paint. Make sure that holes in the sprinkler head valves and sprinkler pipes are not clogged with paint. Hooks, latches, pins, straps, and similar gear may have been made inoperable by painting over them. Free all such fittings so they can be used. Check openings such as ventilation ducts and outlets to be sure they can be opened. Inspect tiedown, blocking and bracing gear, chocks, and other means for stowing and holding missiles and their components. Examine the movable parts of trolley conveyors, such as switches, portable tracks, and trolleys, and make sure they operate freely.

Much of the work of checking the magazines can be delegated to your crew, but you must be sure the spaces are in the best condition possible. The simpler repairs may be done by lower rated personnel, but you need to approve the results. Make a checklist to be sure nothing is overlooked, and that the stowage areas and stowing equipment are ready to receive missiles and components.

Most of our missiles are stowed completely assembled (except for wings, fins, etc.) in the magazine or ready service ring.

The ship's plans show the designated stowage areas for all the ammunition, missile, and missile components that are allotted to the ship. You should be familiar with these plans before attempting to stow the missiles.

If for some reason a magazine is not available when it is needed (it could be undergoing repairs, etc.), and some other place must be found to use instead, consult the chart entitled "Permissible Stowage of Ammunition and Explosives," in OP 4, volume 2, Ammunition Afloat, to determine the next best place for stowage. Study the explanation of how to use the chart. You will not find missiles listed on this chart, but components such as boosters, JATO units, and primers are listed.

## TERRIER MISSILE HANDLING AND STOWAGE

Terrier missile systems are operational on DDGs, CVs, CGs, and CGNs. The number and the location of the launchers, the location of the magazines, assembly areas, and checkout areas are different for each type ship. That is why you were reminded to check the location of the magazines, the hatches, and elevators to be used when preparing to load Terrier missiles on the ship.

## SPECIAL PROBLEMS WITH TERRIER

Although the Terrier is not the largest nor the heaviest of our missiles, its size and weight make special handling equipment necessary. The extra length of the BT-3 missile makes special handling care essential in moving it to the mating area. A crack or strain in the propellant grain can cause missile failure through uneven burning when fired. In striking down boosters and missiles to the mating area, a booster must precede every missile through the strikedown hatches to be in the correct order of assembly. A set of complementary items must follow in order to make a complete missile. The order must be maintained througout the strikedown operation.

It might seem more efficient to transfer all items that require the same handling equipment before breaking out other equipment. However, the reason for requiring transfer of all parts of a weapon should be obvious. Suppose you transferred all the missiles and boosters first because they required the same handling equipment. Should anything occur to break off the loading operation, such as a severe storm or the appearance of an enemy, your ship might have all the missiles and boosters and the other ship all the complementary items, all equally useless when not put together. The rule, therefore, is that for every booster a missile must be transferred, followed by all the complementary parts needed to complete the missile.

Safety rules for handling high explosives and propellants must be observed during handling and strikedown. Flash units are treated as pyrotechnic items; they must not be dropped and must not be brought into areas where they will be exposed to RF energy from radars or communication transmitters, or beams from operating missile radars.

## Chapter 2—MISSILE HANDLING

### INITIAL RECEIPT

Missile components may be received from dockside, from barges or lighters, or from another ship in transfer at sea. Replenishment by helicopter is also possible in some instances. The equipment used has to be adaptable to the method of transfer used.

### Handling Equipment Needed

On board the supply ship or at the supply depot, the assembled missiles and boosters are stowed in Mk 199 and Mk 200 containers. Before transfer to a combatant ship, they are removed from their containers and attached to transfer dollies (figure 2-1). After the missiles and boosters have been struck below on the receiving ship, the dollies must be returned to the supply ship or depot.

The Mk 6 Mod 1 and/or Mod 3 transfer dollies are used for the transfer of Terrier, Standard ER missiles and Terrier boosters, and the Mk 6 Mod 2 and/or Mod 4 transfer dollies are used for Tartar and Standard MR missiles.

TRANSFER AT SEA.—For underway replenishment from an ammunition ship (AE), the standard tensioned replenishment alongside method (STREAM) is the preferred system of transfer.

All STREAM rigs use a tensioned highline to support the trolley and load. There are different rigging arrangements used for lowering the load to the deck.

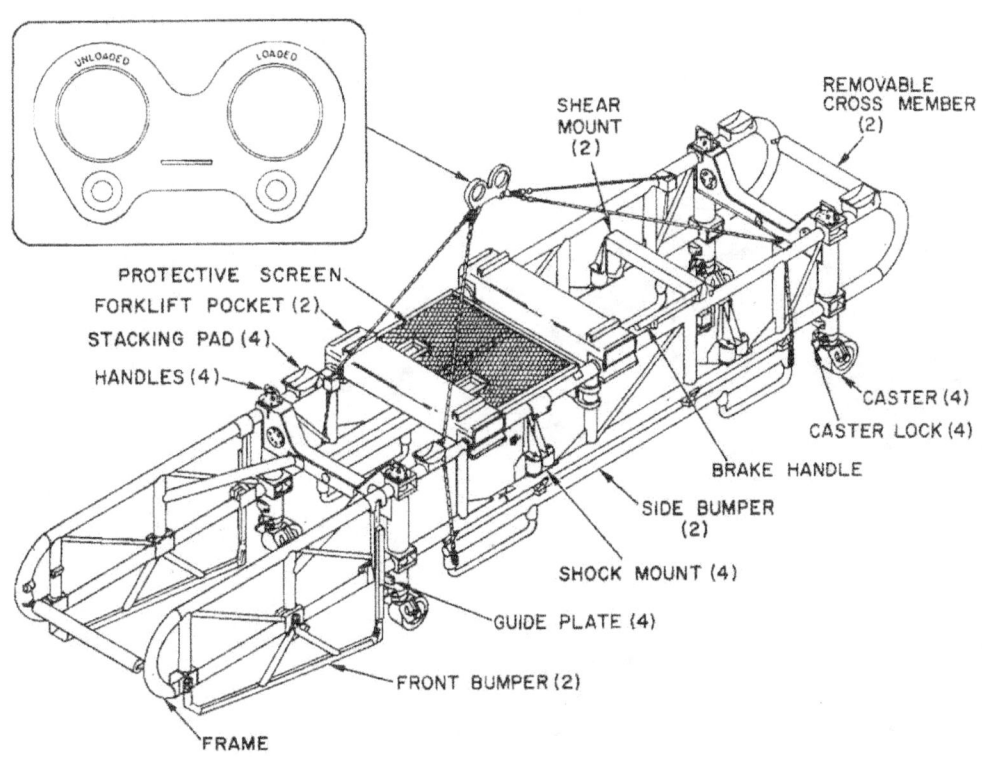

94.198

Figure 2-1.—Mk 6 Mod 3 transfer dolly (Terrier).

STREAM rigs can be rigged to a sliding pad eye (figure 2-2) or to another STREAM station. If the AE is delivering to a sliding pad eye or to another STREAM delivery station, a cargo hook or STREAM strongback will be attached to the STREAM trolley.

If the STREAM rig is rigged to a fixed pad eye, kingpost, or outrigger (figure 2-3), a cargo drop reel is attached to the trolley to lower the load to the receiving ship's deck. An alternate method that can be used for lowering or raising loads at the receiving station is tension/detension. The tensioning/detensioning is controlled by the sending ship.

When using the STREAM system for transfer, regardless of the rigging arrangements, either the Mk 6 Mod 1 or Mod 3 transfer dollies (Terrier) or Mk 6 Mod 2 or Mod 4 transfer dollies (Tartar) are utilized. A STREAM strongback adapter is being installed on all dollies as ORDALT 8481. The purpose of this strongback adapter is to provide more control of the dolly during at sea transfer operations using the STREAM system. These dollies must be kept under positive control at all times. To maintain control, a minimum of six personnel is required for the movement from the replenishment area to the strikedown area. In rough seas you may need to add personnel. An empty dolly weighs approximately 1100 pounds.

The modified housefall and double burton highline are still used to a limited degree for transfer of weapons, and will not be covered in this text. *Gunner's Mate M 3 & 2*, NAVEDTRA 10199-C, chapter 14 covers these methods.

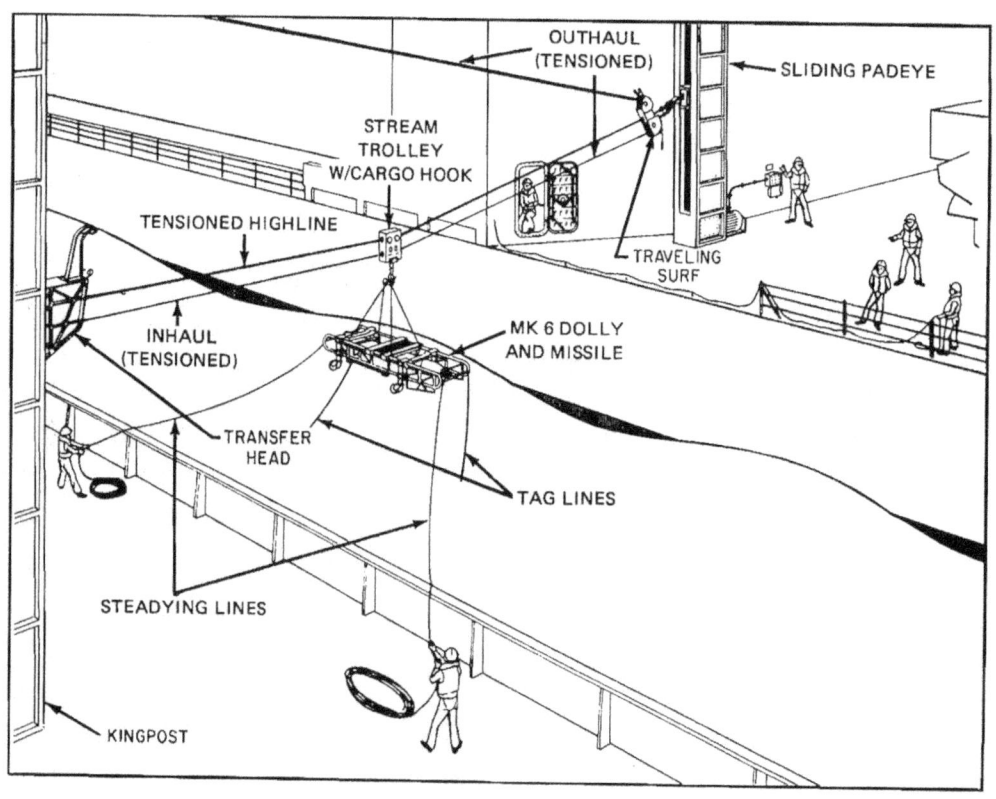

Figure 2-2.—Missile transfer in Mk 6 dolly.

## Chapter 2—MISSILE HANDLING

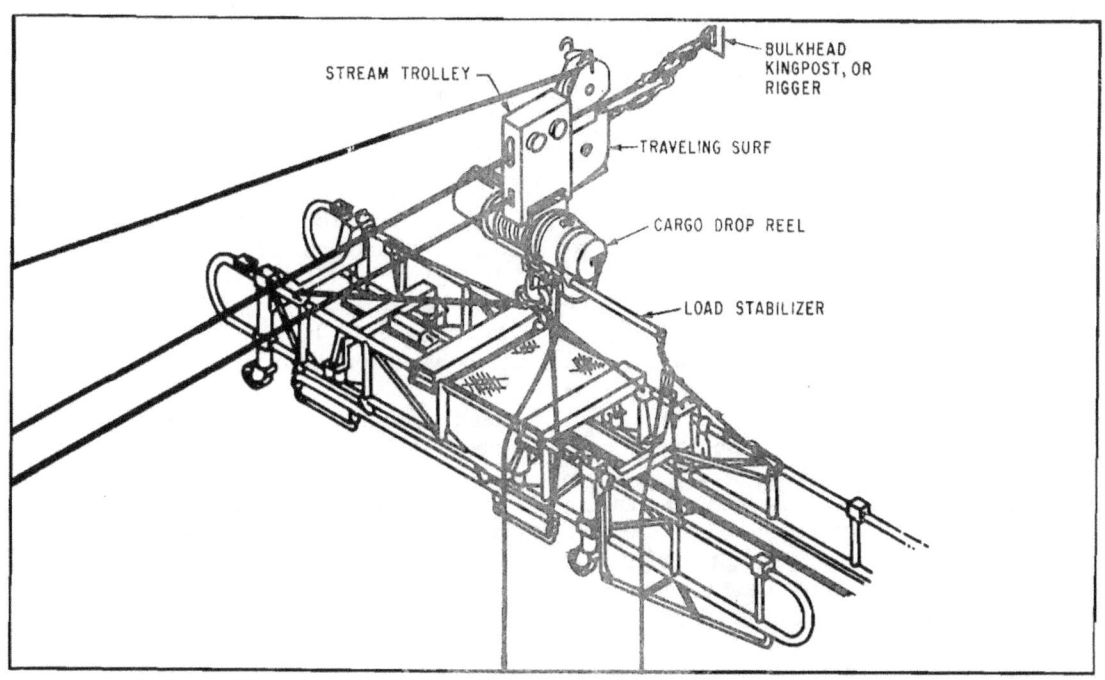

Figure 2-3.—Mk 6 transfer dolly attached to cargo drop reel.

The work of handling the lines and rigging the stations may be done by the boatswain's crew but a trained and experienced GMM needs to be at hand to ensure that the explosive components will be handled with proper care. Skip boxes and dollies must be lowered gently to the deck. No smoking is permitted while any ammunition is being handled. The petty officer in charge should have all crewmembers leave their smoking materials, including lights and matches, at a specific checkpoint before coming to the loading area. Lead the way by putting yours there first.

TRANSFER FROM LIGHTER OR BARGE.—A dockside or floating crane is used to transfer boosters and missiles in their transfer dollies. Handling beams may be used, however, if the missile/boosters are stowed in Mk 199 and Mk 200 containers, respectively. The complementary items may be carried aboard manually or landed aboard by crane. Subsequent handling is the same as in transfer at sea. While handling explosives on a pier or in a building, ship's personnel are under the authority of the commanding officer of the ordnance facility.

Safety Precautions
in Handling

A Terrier missile in a cradle dropped from a height of 2 inches could adversely affect the flight. The same is true with a Terrier missile in a dolly dropped 18 inches, or a missile dropped 1 foot in its container. Under no circumstances should the missile be used; it must be returned to a weapons station. Further safety precautions for Terrier and Standard (ER) guided missile rounds can be found in OP 3063. Do not permit loaded transfer dollies to bump bulkheads or railings or drop to the deck; use a steadying line if necessary.

Firefighting equipment should be readied on deck before beginning the handling operation.

Since the tragic fire on the USS *Oriskany*, the rules for handling explosive items, especially pyrotechnics, have been reexamined. New, stricter and more comprehensive rules have been promulgated to help prevent such catastrophies in the future. Adding new rules, however, will not prevent accidents. Only strict adherence to the rules will achieve that. You not only need to observe that your crew obeys the rules, but you need to strive constantly to get them to believe in the need for the rules.

When planning movement of missiles and components on deck, plot the movements to avoid RF radiation. The beams from radars and other electronic transmitting equipment can cause detonation of some components, and they are also harmful to people. The commanding officer will order inactivation of all possible radar and electronic equipment during ammunition handling, but usually some units have to be kept operating at all times. Flash units are very susceptible to detonation by RF radiation. The radiation hazards (RAD HAZ) and hazards of electromagnetic radiation on ordnance (HERO) programs on the effects of radiation from electronic equipment are discussed in chapter 7.

If you detect any odor of ether or nitroglycerine in the magazines and stowage spaces or ready service rings, report it immediately to the officer in charge. These fumes exude from doublebase propellants which are used in sustainers and boosters, are highly combustible, and are harmful when inhaled. Keep heat, sparks, and fire away from all explosive components. The Terrier warhead contains a considerable quantity of composition B, which is a high explosive; observe the safety precautions for high explosives. Remember that a shock or blow can cause their detonation.

## Strikedown

As the missile components are landed on the deck, your crew must strike them down below. On cruisers and destroyers, usually both port and starboard strikedown hatches are used. A transfer dolly with its missile or booster is placed on the closed hatch. The shipping band (third handling band) should have been removed from the booster prior to placing the booster into the Mk 6 dolly. In case the shipping band hasn't been removed, you can remove it topside or after the booster has been lowered on the elevator. When the dolly is in place over the hatch, the hatch is opened by the operator at the pushbutton station. A strikedown elevator rises beneath the dolly and latches onto the handling bands on the missile or booster (figure 2-4). The handling bands are then manually released from the dolly and the elevator lowers the missile or booster through the hatch in response to pushbutton operation at the control panels. As soon as the load has cleared the hatch, the hatch door closes automatically. The empty dolly is wheeled to the transfer line or crane for return to the supply ship.

A booster must precede a missile. From the strikedown elevator the booster is moved to the loader rail by means of the transfer car (figure 2-4), which is moved athwartship on rails. As soon as a missile is brought to the checkout area (by the same means), it is aligned with the booster and mated to it. The booster shoes must be engaged with the loader rail during alignment, and afterward for transfer to the magazine. Two or three personnel are required in the checkout area to manipulate the transfer car and mate the missile and booster. The carriage of the transfer

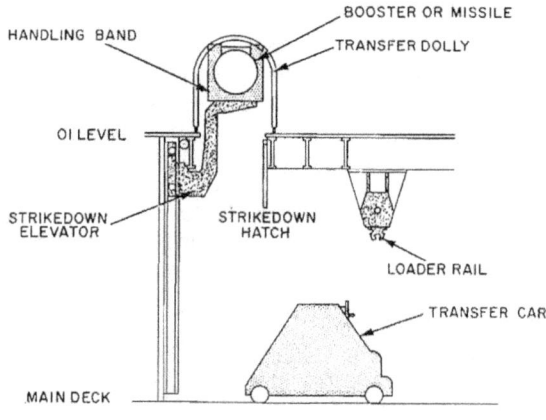

83.149
Figure 2-4.—Terrier strikedown operation; elevator raised to receive missile.

## Chapter 2—MISSILE HANDLING

car can be tilted and rotated as needed to align the missile with the booster. The mated missile and booster, called a round, is retracted along the loader rail to the assembly area. The transfer car is returned to carry the next unit. Transfer code plugs are removed and replaced by the appropriate ones for the ship's assigned guidance codes (BT missiles only). The round is then ready to be moved to the ready service area and to be inserted in the proper tray of the ready service ring (figure 2-5).

The step-by-step operation to be used in moving, aligning, and mating of the missile and booster is described in the OP for the launching system and the OP for the mark/mod missile on your ship. Study these and prepare a checksheet for you and your crew to follow when doing the work. The method of mating does not vary with the Terrier/Standard ER missiles and boosters presently in use, but the equipment of the launching system varies with the mark and mod of the system.

Before the rounds can be moved into the ready service rings, the ready service rings must be indexed so the correct round can be selected by pushbutton when it is wanted for firing or exercise. This is done by the panel operator setting the pushbuttons according to the plan. The actual arrangement of the weapons in the service ring is a tactical decision. A Terrier missile ship may carry one, two, or even three types of Terrier missiles.

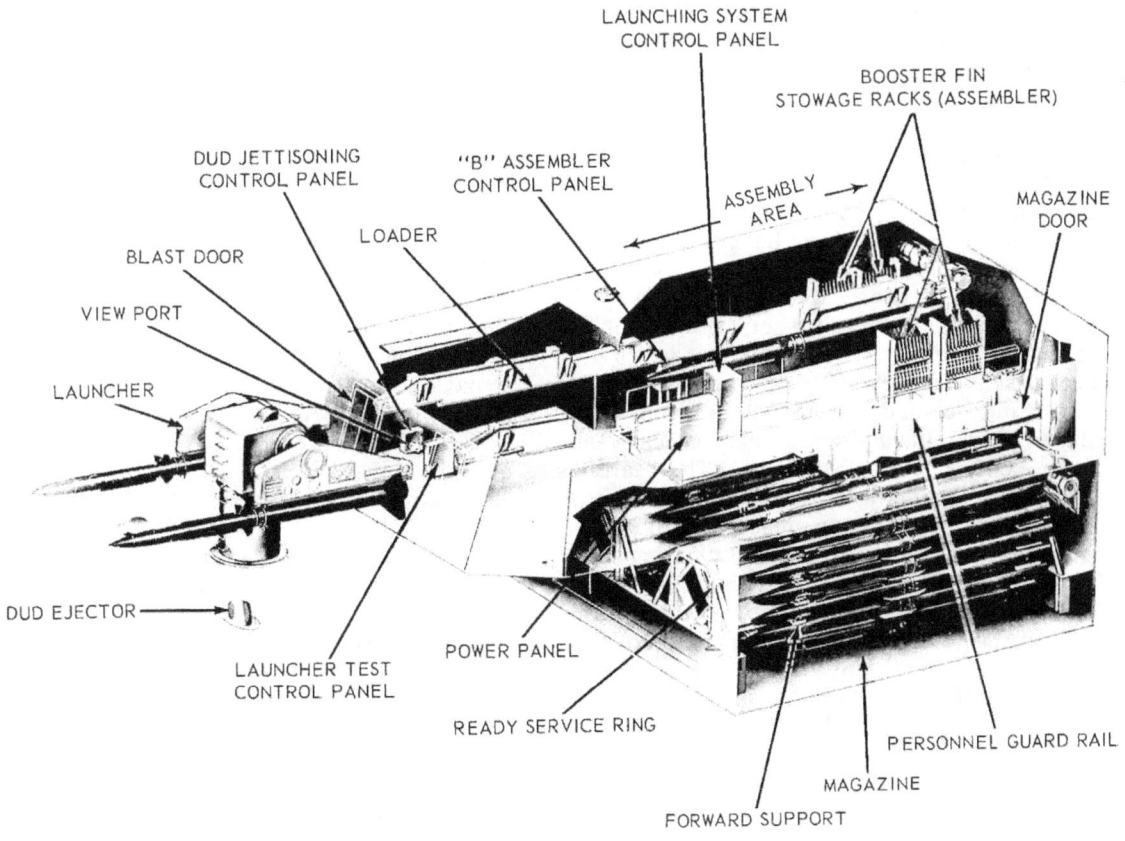

Figure 2-5.—Mk 10 Mod 0 guided missile launching system.

The assembler panel (figure 2-5) has a light for each tray, with colors and lettering to indicate what is loaded into the tray. The operator of the panel can select the weapon required by pushing the correct button for the tray wanted.

When the round is ready to put into the ready service rings, it is done by step control. Step control requires operation of pushbuttons for each step. The tray with the code designation of the round in the assembly area moves to the hoist position, the magazine doors open, and the hoist raises to the loader rail. The loader chain pawl moves the round from the loader rail onto the hoist. The hoist lowers the round into the ready service ring tray; the tray shifts the round free of the hoist; the booster shoes engage in the ready service ring structure, and the magazine doors close. As each round is unloaded to the ready service ring, the lamp (on the control panel) associated with the tray goes out. The magazine for Terrier rounds is (figure 2-5) belowdeck.

STRIKEDOWN OF ASROC WEAPON.—The Terrier Mk 10 Mods 7 and 8 guided missile launching systems (GMLSs) stow both Terrier and ASROC weapons. They have three stowage mechanisms: Mk 5 Mods 12, 13, and 14 guided missile magazines with ready service mechanisms, hoist mechanism, and magazine doors. Either of the two upper ready service mechanisms can stow 20 Terrier missiles, or 10 Terrier missiles and 10 ASROC weapons with adapters. The lower, or auxiliary mechanism, stows only Terrier missiles. The missile strikedown equipment is located in the strikedown and checkout area. The strikedown equipment was designed and installed on most Mk 10 GMLSs by NAVSHIPS before NAVSHIPS and NAVORD were merged to form NAVSEA. The description and operation of this equipment, unless recently updated, is presented in a NAVSHIP publication. *Gunner's Mate M 3 & 2*, NAVEDTRA 10199-C contains an illustration of the Mod 7 launching system showing Terriers and ASROCs placed in the ready service rings. The Mod 8 is almost identical, but it has no tilting rail in its feeder system. The operational sequence of loading and unloading is the same in the two mods but because of the increased length of the loader rail in the Mod 8, it requires a longer time to complete its load and unload cycles.

ASROC rounds cannot be located next to each other in the ready service ring. Consequently, the table of assignments of missiles to the tray must be followed carefully. The actual arrangement of the missiles is a tactical decision. The number and type of missiles for loading usually are known in advance of the strikedown procedure.

Since the ASROC weapons are not the same configuration as Terrier missiles/boosters, and ASROC handling fixture has to be placed on the transfer car carriage. This fixture is also used on the strikedown elevators when receiving or transferring ASROC weapons. The handling fixture can be transferred from the transfer car to the strikedown elevators or vice versa, with or without an ASROC weapon loaded in it.

ASROC weapons are received on board in Mk 183 containers. The Mk 42 or Mk 45 handlift trucks are used to position the container over the strikedown elevator. After removing the container cover, an ASROC decanning hoist is used to remove the ASROC from the container. Once the ASROC is lifted from the container, it is suspended above it. At this time the container is moved away from the strikedown hatch. The strikedown elevator is then raised to the 01 level. The ASROC weapon is then lowered by the decanning hoist onto the strikedown elevator and then the decanning hoist is manually released from the weapon. The elevator can be lowered into the assembly area at this time, and the strikedown hatch will automatically close as the elevator reaches its lower limits.

Figure 2-6 shows a Terrier missile and an ASROC weapon, without the ASROC adapter. The adapters often are installed in the ready service rings at the time the system is installed. Placing the ASROC adapters in the trays of the ready service rings is not considered a part of the strikedown operation.

Because there are two configurations of the ASROC weapon, an insert is used on each snubber before loading the adapter with X- or Z-type ASROC weapons (RTTs). It is necessary to remove the insert when loading a Y ASROC (RTDC) into the adapter rail.

Chapter 2—MISSILE HANDLING

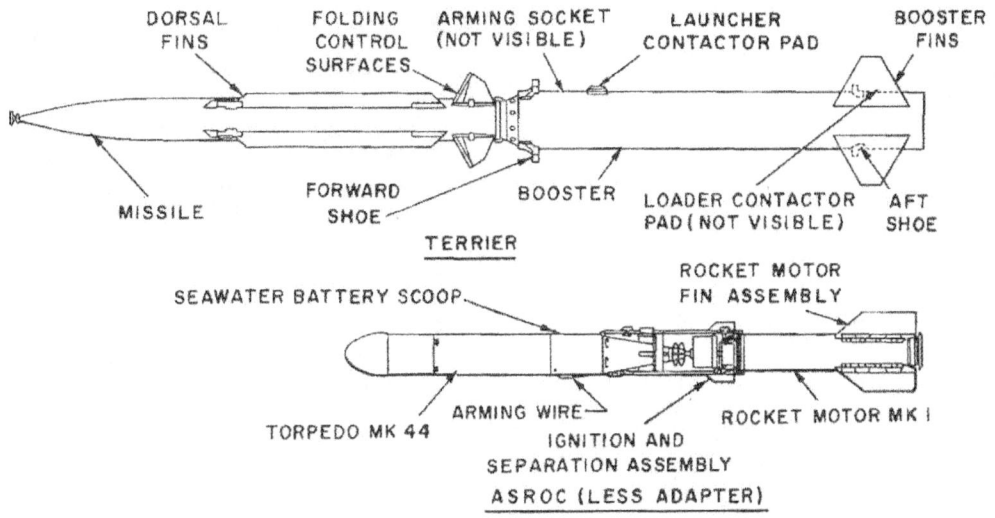

Figure 2-6.—Terrier missile/booster and ASROC weapon types.

In ASROC operations, strikedown is a step control and manual operation to load assembled weapons into adapters and then to stow the loaded adapters on the ready service rings. In the strikedown and checkout area, a NAVSHIPS transfer car operates on rails athwartship to move the weapon to positions for performing tests, checks, and adjustments. The ASROC weapons arrive on board assembled, so there is no mating process as with Terriers. An ASROC attached to an adapter rail is shown in the text *Gunner's Mate M 3 & 2*, NAVEDTRA 10199-C, page 138, figure 5-13. A special ASROC adapter loader assembly is in the strikedown area. The ASROC weapon is brought to the strikedown and checkout area on the transfer car. It is positioned into the adapter by means of the adapter loading fixture and latched into the adapter. The snubbers on the adapter are hydraulically operated, and need to be unlocked with care to avoid casualties. Everyone must be clear of the snubbers before they are unlocked. After the weapon is attached to the adapter, stow the transfer car, close the snubbers on the adapter rail, connect the Mk 10 cable, retract the loaded adapter rail to the assembly area, then unload to the magazine.

STRIKEDOWN OF COMPLEMENTARY ITEMS.—The complementary item you will be most interested in is the Terrier booster fin. The Terrier missiles HT, BT, and Standard ER have dorsal fins and folding control surfaces (tails) installed. These are serviced at a naval weapons station, normally not aboard ship, although it is likely that the ship's supply department will stock folding control surfaces (tails) and the attaching snaprings in case emergency repair is necessary. The Terrier booster fins (set of four) are shipped in the Mk 205 container. The containers are hand-carried away from the deck landing area and stacked in a convenient place until they can be unpacked. After unpacking, the containers are returned to the supply ship (or depot), and the fins are stacked in fin racks in the assembly area. If the number of personnel available is sufficient, this should all be done at the time of strikedown to avoid pileup of material. The ASROC motor fin assemblies are installed at a weapons station and remain with the weapon, but must be removed prior to striking the ASROC down on the strikedown elevator. The fin assemblies will be re-installed after loading the weapon into the ASROC adapter rail. Spare fin assemblies should be in the supply department. The rocket motor fins can be easily replaced if damaged.

**TESTS AND INSPECTIONS**

At the present time no tests of Terrier missile/booster components are made aboard ship by GMMs.

**Receipt Inspection**

Terrier missiles, boosters, and booster fins must be inspected upon receipt. This inspection should be conducted as soon as possible.

The following is a partial listing of acceptance criteria obtained from OP 4093. A more detailed inspection checklist is required to properly hold a receipt/acceptance inspection.

1. Missile sustainer SAFE-ARM indicator is on SAFE. Indicator window on Mk 7 sustainer indicates S; the window is visible through a hole in the handling band. Indicator on Mk 30 rocket motor indicates S; the indicator is located on the after end of the rocket motor nozzle.
2. Mk 7 Mod 0 safety switch actuator on the Mk 22 warhead section is positioned to SAFE.
3. Booster arming mechanism is positioned to SAFE.
4. Missile explosive sections are properly color coded.
5. Ram pressure orifice seal is in place and undamaged (BT only).
6. Exterior surface of target detection device section are clean and free of cracks, aluminum deposits, and scrapes.
7. Exterior surfaces of guidance section are clean and free of cracks, aluminum deposits, and scrapes.
8. Wheel 1 humidity indicating plug color indicates less than 20 percent relative humidity.
9. Control surfaces, latching mechanisms, and latch pinholes are clean and undamaged.
10. Control surfaces are securely attached in place, safety wired, and folded properly.
11. Dorsal fins and tips are securely attached and undamaged.
12. Missile exterior surfaces are free of corrosion and have no missing, broken, or improperly installed screws, plugs, caps or covers.
13. Aft section hydraulic sump level within acceptable range (BT/HT).
14. No hydraulic leakage from aft section (BT/HT).
15. Autopilot battery vent cap must be installed [Standard (ER) missile].
16. Correct code plug in missile and code plug cover in place (BT only).
17. Booster properly color coded with brown band.
18. Exterior surface of booster is free of damage (dents, gouges, breakage) and arming socket decal and index mark installed.
19. Booster nozzle closure is present and intact.
20. Fin support mounting snaprings are painted black or have black dot.
21. Fins are free of damage or corrosion.
22. Booster shoes are free of visible cracks.
23. Booster exterior surfaces are free of corrosion and have no missing, broken, or improperly installed screws or covers.
24. Clamp links of the clamp assembly are seated on smaller diameter of the pivot pin and not on the larger diameter.
25. Missile and booster containers show no evidence of being dropped or damaged.
26. Desiccant indicator on missile/booster container is blue.

Acceptance of a complete round is also limited by the age of the explosive components in the missile and booster. The expiration firing date for the complete round is dependent on the explosive component with the earliest expiration date in the round. Refer to the Propulsion Unit History Sheet to determine the expiration firing date of the round. Do not accept the round if the earliest expiration date is less than 9 months. The 9-month requirement is waived if the specific round is scheduled to be fired before the earliest explosive component expiration date. Do not fire the round if the expiration date of any explosive component has been exceeded. Contact NAVSEA for disposition instructions.

When, for any reason, it is not possible or practical to immediately return a rejected

## Chapter 2—MISSILE HANDLING

missile/booster to the issuing activity, do the following:

1. Attach a tag indicating rejection.
2. Place missile/booster in restricted (dud) stowage.
3. Record reason for rejection on Propulsion Unit History Sheet/Guided Missile Service Report.
4. Contact NAVSEA for disposition in accordance with NAVORDINST 8025.1 series.

Since Terrier and Standard ER missiles are under a no-test program on board ship, a great deal of dependence is placed on careful inspections, careful handling, and controlled stowage temperatures and humidity requirements.

### Radiation and Monitoring

At the present time, only naval weapons stations and tenders have IC/T2-PAB-M air monitors avaiable. Possibly later the IC/T2-PAB-Ms will be reissued to surface combatants. Until that time you will only use the IC/T2-PAB-M air monitor when assigned to a naval weapons station or a tender. The radiation monitoring equipment aboard ship will be the AN/PDR-27, 56 and 43.

If there is an accident/incident with a nuclear weapon, monitoring must be done immediately. The space in which the accident occurred must be closed off to prevent spread of contamination to other parts of the ship. Every effort should be made to move the weapon to a naturally ventilated place. All unnecessary personnel should be evacuated. If the space has access to the atmosphere, all accesses should be opened. Open the emergency ventilation exhaust, then the emergency ventilation air supply. Get out of the space as quickly as possible (the actions above should take very little time), secure the space, then notify the appropriate personnel. The trained decontamination group of which you may be a member, dressed in protective clothing and each wearing an oxygen breathing apparatus (OBA), will reenter the room and remove the source of contamination.

Afterward, the elevator used and the path followed must be decontaminated.

The order in which the above actions are accomplished will differ with the location of the contamination, the severity of the radiation, and whether the radiation is detected immediately or upon preparing to enter a closed space. If the accident occurs on an open deck, the radiation will be carried away into the atmosphere; personnel should be evacuated from the immediate area. The IC/T2-PAB-M portable air monitor can be operated by internal battery, or an external power cord can be plugged into a 115-volts a.c. 60-hertz single-phase source. The internal battery is checked monthly, and if not fully charged, can be charged by attaching the external power cord to the monitor, plugging the cord into a 115-volts a.c., 60-hertz source, then rotating the selector switch to CHARGE. Charge time is 24 hours. Technical manual for Tritium Air Monitor, Portable, Type IC/T2-PAB-M NAVSHIPS 0969-00-120-9012, describes the new model, tells how to use it, and how to maintain it. The instrument is designed to detect tritium contamination in the air, but it is also sensitive to gamma radiation and to gaseous or particulate activity in the air.

To use it, select mode of operation—either battery or external power—then turn the monitor on. Let it run for 5 minutes, then calibrate it using a checksheet. After calibration the monitor is ready for use. As air is drawn through the ion chamber of the monitor, the level of radioactivity is measured. The reading shows on the meter, and when the amount exceeds the normal setting, the alarm sounds. If you need to monitor a space without entering it, perhaps because of suspected high contamination, attach the "sniffing hose" to the instrument and insert it into the space.

The monitor is a delicate electronic instrument and should be handled with care, not dropped or abused in any way. If it becomes contaminated with radioactive particles, it must be decontaminated. Careful wiping of the outside with a cloth dampened with water and detergent will remove light contamination. Be careful not to get any water on the inside of the monitor.

Personnel who have been exposed to radiation must be sent to the medical department for evaluation. The effects of radiation are not noticeable except in extreme cases, as in a bomb explosion, so medical evaluation is necessary to detect and evaluate the exposure and provide possible treatment.

The air samplers do not measure the amount of radiation, but detect it if any is present. If there is a radiation leak, secure or safe the item you are working on, turn off the ventilation, get out of the magazine, and close the door. Hold your breath while doing this. Before reentering the contaminated space, don OBA or equivalent breathing apparatus, wear rubber gloves, and carry an operating IC/T2-PAB-M portable air monitor or call the decontamination team to take care of the situation. The methods of decontamination described in your military requirements courses are applicable here, though on a smaller scale than after a nuclear attack.

The need for extreme care to avoid inhalation or ingestion of nuclear particles should be impressed on your subordinates. Any detectable amount of tritium is potentially dangerous. Although the chances of a leak occurring in a nuclear weapon are small, the danger is ever present and you must teach your crew how to act in case it happens. The reason for turning off the ventilation aboard ship is to prevent the spread of radioactive particles through the ventilation system. Where ventilation to the open air is possible (as at shore stations), activate ventilation systems and open windows.

The danger of unauthorized personnel gaining access to a nuclear weapon during an alarm incident makes it essential to secure the area quickly. Two technically trained personnel must enter the area as soon as possible to secure it.

## SHIPBOARD ASSEMBLY AND DISASSEMBLY

The mating of the missile to the booster before stowing of the round, and the unmating for offloading are the extent of assembly and disassembly authorized at this time although fins are assembled to the booster and the control surfaces (tails) are unfolded on the missile prior to loading onto the launcher.

Any additional assembly or disassembly of the missile is done at a repair activity. Assembly or disassembly of the handling and launching equipment is done as part of the maintenance and overhaul program, and will be discussed in another chapter.

Maintenance procedures are discussed in later chapters, specifically in chapter 9.

When the missile is mated to the booster, the suitcase latch is closed, and the torque screw is tightened with 480 in-lb of torque. The clamp links should make contact with the ends of the pivot pin to ensure that the latch will not accidentally spring open under torque conditions and possibly cause injury to personnel. Do not release the latch without first removing all torque from the torque screw. Keep the missile supported on the checkout car during the entire time. Do not use the locking ring to pull the sections together.

The missile-booster release mechanism is illustrated in *Gunner's Mate M 3 & 2*, NAVEDTRA 10199-C, where the suitcase latch is shown fully closed and fully opened.

## DEPOT HANDLING AND STOWAGE

At depots, missile parts are received in sealed containers from the manufacturers. They are placed in receipt stowage according to the type of component. Sustainers, boosters, and auxiliary power supply gas generators and igniters are placed in the smokeless powder and projectile magazines. The warhead, destructor charge (if any), fuze booster, and the S&A device are placed in a high explosive magazine. Flash signals are stowed in the pyrotechnic magazine. Inert missile components are stowed in the guided missile service unit checkout building.

On shipboard, the work of GMMs is focused on care and operation of the launching systems. Note, however, that one of your occupational standards requires the E-7 to have a knowledge of methods of handling and stowing of missiles ashore. Few GMMs are assigned to ammunition depots, but naval weapons stations require many GMMs.

The volumes of OP 5 are pertinent references: volume 1, Ammunition and Explosives Ashore: Safety Regulations for

## Chapter 2—MISSILE HANDLING

Handling, Storing, Production, Renovation, and Shipping; volume 2, Ammunition and Explosives Ashore: Stowage Data, and volume 3, Ammunition and Explosives Ashore: Advanced Bases. Volume 1 contains much information on the properties of different explosives, and how they must be stowed and handled because of these properties. Numerous sketches illustrate the quantity-distance requirements for different types of ammunition. On shipboard, the quantity-distance requirements cannot be followed because there simply is not room enough to stow ammunition at the separate distances specified. At shore bases the requirements must be observed. The purpose of the requirements is to keep the quantity of ammunition per building small enough so that a fire or explosion in one building will not spread to adjacent buildings.

Guided missiles are considered a mass detonation hazard, but assembled missiles present several types of hazards. Therefore, regulations and instructions for stowing, shipping, handling, and marking of guided missiles and their major components are not covered in Op 5. The OP for the particular missile must be consulted for the specific instructions. The hazard classification and stowage requirements of some components are mentioned in OP 5. The distance requirements vary with the number stowed in the building, and the type of magazine. Fuzes, for instance, must be stowed in special magazines, either earth-covered or with equivalent protection.

The rate training manual *Gunner's Mate M 3 & 2*, NAVEDTRA 10199-C, mentions in several places that a defective component or a missile is to be returned to a depot for repair or destruction. OP 5, volume 1, contains a chapter of instructions on how to dispose of damaged or dangerous explosives of different kinds: some are burned, some a dumped in deep water. Maintenance and surveillance instructions are given for various small components such as fuzes, but missiles and boosters (except rocket boosters) are not covered. A missile is not destroyed except as a last resort if the missile cannot be made safe. A publication available from NAVSEASYCOM is Safety Regulations for Guided Missile Propellants. Request specific instructions for each missile, in the event that destruction seems to be necessary.

Conditions for shipping missiles, boosters and missile components by truck and by railroad are covered in MIL-STD-1320 and 1325. Many of the rules are applicable to transportation of any type of explosive. For example, passengers are not permitted on trucks carrying explosives. OP 5 lists references you will need if you have to pack and/or ship missiles and missile components. With on-land shipments you not only have to follow Navy regulation, but also state and interstate rules.

Security regulations, firefighting, lighting protection, static grounding, and industrial safety, health, and hygiene procedures given in OP 5 are applicable to missiles and missile components. Revisions to OP 5 added rules for quantity-distance stowage of missile propellants according to their hazard classification. The application of the rules to Navy missiles containing liquid propellant (Harpoon, Tomahawk, Bullpup) is described in the OPs for the missile.

The four routine missile operations at an ammuniution depot are initial receipt, retest, loadout, and missile return processing. Many variations are possible in each of those operations. While initial receipt is defined as receipt of the missile components from the manufacturers, the components may be delivered by different methods of transportation (railroad car, truck, etc.). The containers are designed for particular missiles and their components; some handling equipment is designed for particular missiles and their components.

All components must pass inspection when onloaded, but the extent of testing varies. On shipboard, a booster is never tested. At a weapon station, the booster is unpacked, transported to the igniter test cell where it is given an igniter squib check, then is repackaged in a Mk 200 container. If it is to be shipped with a missile, it is placed in ready issue stowage until the missile is assembled and ready for shipment. If it is to be stowed for some future time, it must be repackaged and sealed with desiccant.

Let us assume that you are going to assemble a missile to be sent to a ship for tactical use or ready issue. Assume that each package has been given receipt inspection, tests when applicable, and has been repackaged awaiting assembly. The booster remains in ready issue stowage until the last, when it is to accompany the assembled missile, but not assembled to it. The sustainer is brought to the igniter test cell and is given a continuity test, then is repacked and taken to the assembly area. As the components are inspected and checked out, they are brought to the assembly area to be assembled into the missile. Present weapon stations perform only pneumatic missile system tests (MST), which require a missile assembled without the warhead section, S&A device, and fuze booster. After the missile has successfully passed the initial MST and has been transported back to the assembly area, the nose section, the target detection device (TDD), and the tactical missile test spacer are removed to prepare the missile for tripak storage or for tactical assembly.

The tripak configuration for Terrier missiles consists of the electronic section, sustainer, and aft section assembled with dorsal fins, and placed in a Mk 199 Mod 0 container (figure 2-7). By the use of different blocking and bracing and cushioning materials, this container can be used for bipak storage (electronic section, aft section, nose section, and dorsal fins), or for a ready issue missile, sustainers, or spare parts. There are two types of handlift trucks for use with the Mk 199 and Mk 200 containers—the Mk 42 or the Mk 45 handlift trucks. Both types require installation of a Mk 26 adapter when handling these containers. Regardless of which types of handlift trucks are available, two are required and should be the same type. A forklift truck may also be used. The tripak must be grounded and the sustainer arming device must be in SAFE position during all handling operations. The loaded container is closed with 20 latches

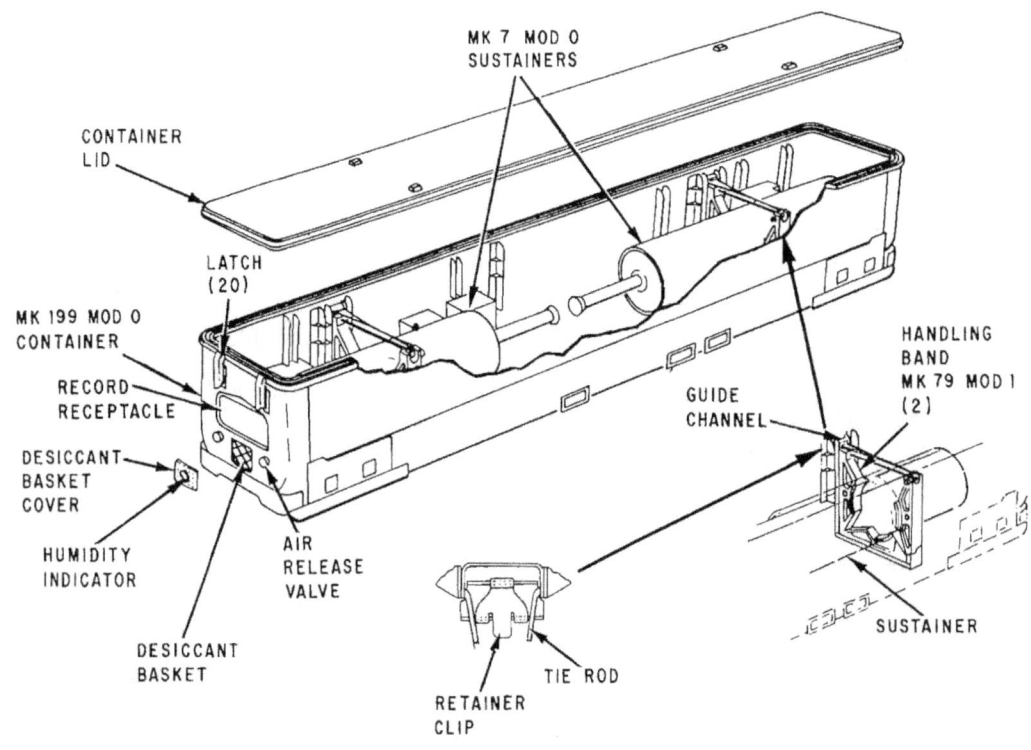

Figure 2-7.—Mk 199 Mod 0 container with two sustainers and showing placement of handling band for lifting sustainer.

## Chapter 2—MISSILE HANDLING

on the container lid. Fresh desiccant is placed in the desiccant basket; the air release valve on the container is closed; the missile log and records are placed in the records receptacle on the end of the container; and security seals are placed on it and on two of the latches. The container then is ready to be transported to stowage.

When the missile is to be assembled for ready issue or tactical assembly, the warhead section, target detection device (TDD), and nose section are added to the already assembled tripak configuration. If a Mk 7 warhead is used, a fuze booster and an S&A device are required to complete the fuse section. All these contain high explosives and must be handled as such.

The assembled missile is shipped from the depot in a Mk 199 Mod 0 or 1 container, its booster is in a Mk 200 Mod 0 or 1 container, and the booster fins are in a Mk 205 Mod 0 container. They are moved from ready issue stowage and loaded on trucks or railroad cars by means of handlift trucks or forklift trucks. The trucks or railway cars are moved to the loading dock, where the missiles are moved for ship loadout.

For loadout to an ammunition ship the missiles are transferred in Mk 20 Mod 0 and 1 cradles. If a combatant ship is loaded directly from the dock, the missiles are removed from their containers on the dock and transferred to the ship in a transfer dolly, or a Mk 15 handling beam can be used to lift the missiles or boosters directly from their containers. On the ship the missiles are struck down to the magazine, and the dollies are returned to the dock.

### NATO SEASPARROW SURFACE MISSILE SYSTEM (NSSMS)

The Mk 29 GMLS, NATO Seasparrow surface missile system (NSSMS), consists of the Mk 132 launcher, a Mk 108 launcher control cabinet, Mk 8 maintenance interconnecting cabinet, Mk 147 power supply, Mk 14 loader, and the Mk 109 deck station launcher control.

A loaded Mk 470 container includes an RIM 7H missile, four folding wings, four fins, one shear-off connector, and one vibration dampener.

### LAUNCHER

The launcher shown in figure 2-8 consists of two, four-cell clusters and a pedestal which positions the missiles in train and elevation. Sliprings allow for continuous rotation in train. The interface between the launcher cell and the missile is through a removable launch rail, which contains mechanical, electrical, and safety features. Thermal protection for the missiles, and for firing operations, is provided by cell heaters, blowers, and deicing heaters. The launcher cells are covered by frangible covers which in no way impair the missile's functions. Inadvertent firing protection is provided by a missile holdback mechanism in each launch rail and by a quick reaction brake mechanism.

#### Launcher Control Cabinet

The launcher control cabinet contains the missile control functions, including no-firing restrictions, missile select and runup controls, and missile-available or missile-away intelligence. The no-firing zone restrictions have right/left cell cluster distinction and, if necessary, provide for multiple no-firing zones.

#### Launcher Interconnection Maintenance Cabinet

The launcher interconnection maintenance cabinet provides the electrical interface between the fire control subsystem and the launcher subsystem, and contains controls for performing built-in test equipments (BITE) tests.

#### Power Supply

The peak load requirements of the guided missile launcher subsystem are provided by the power supply.

### LOADER

The loader shown in figure 2-9 is capable of rapid attachment to and disengagement from the launcher cells, and provides positive control of the missile in loading or unloading operations.

# GUNNER'S MATE M 1 & C

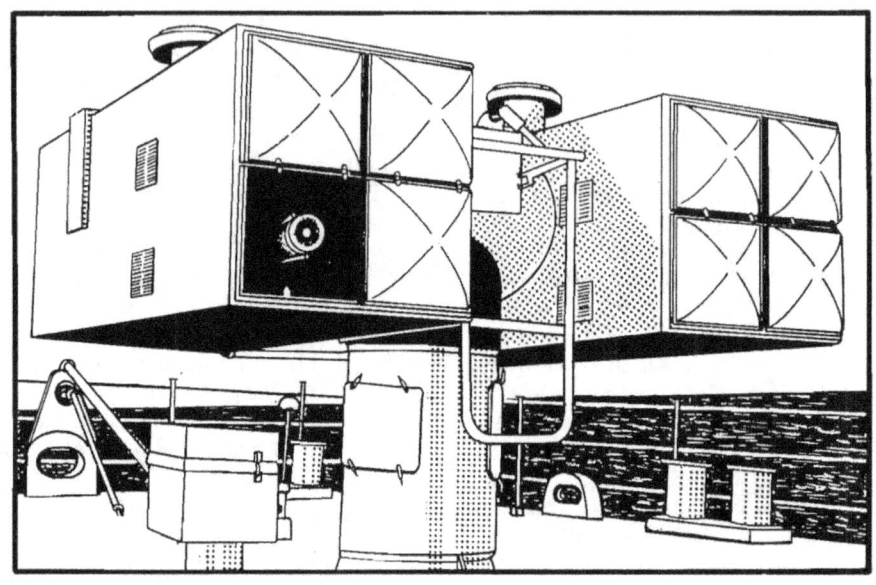

Figure 2-8.—Mk 132 launcher.

94.201

## DECK STATION LAUNCHER CONTROL

The simple handheld deck control's station, shown in figure 2-10, provides on-deck control of the launcher for loading and maintenance operations.

## ONLOADING OPERATION

Prior to missile onload, each empty cell is run through a programmed test to ensure all functions are, in fact, being supplied to the launcher rail umbilical connector. The Mk 567 missile system test set (MSTS) serves basically the same purpose as the guided missile simulator DSM-159D in the Terrier launching systems. The Mk 567 MSTS has a cable supplied with it to facilitate hookup between the test set and the rail umbilical connector.

The onload operation consists of three basic steps as follows:

1. Removing the rail assembly, shown in figure 2-11, from the launcher cell.
2. Mating the rail to the Sparrow III missile.
3. Returning the rail, with missile attached, to the launcher cell.

Before removing the rail from the launcher cell, the loader is attached to the rail assembly and the rail is manually unlatched from its cell. The rail is unlatched by rotating a rail release ring located on the after end of the rail which is reached with a rod provided for that purpose. The rail is then removed from its cell onto the loader using the loader handcrank (see figure 2-9) or is power assisted using an electric motor. With the rail still on the loader, install a shorting plug in the rail-to-cell connector, and a shear connector to the rail umbilical connector.

NOTE: The shear connectors are in the Mk 470 container with the missile. If the container is not available at this stage of the onloading operation, the shear connector will have to be installed after the rail assembly has been lowered to the deck and before mating the rail to the missile.

## Chapter 2—MISSILE HANDLING

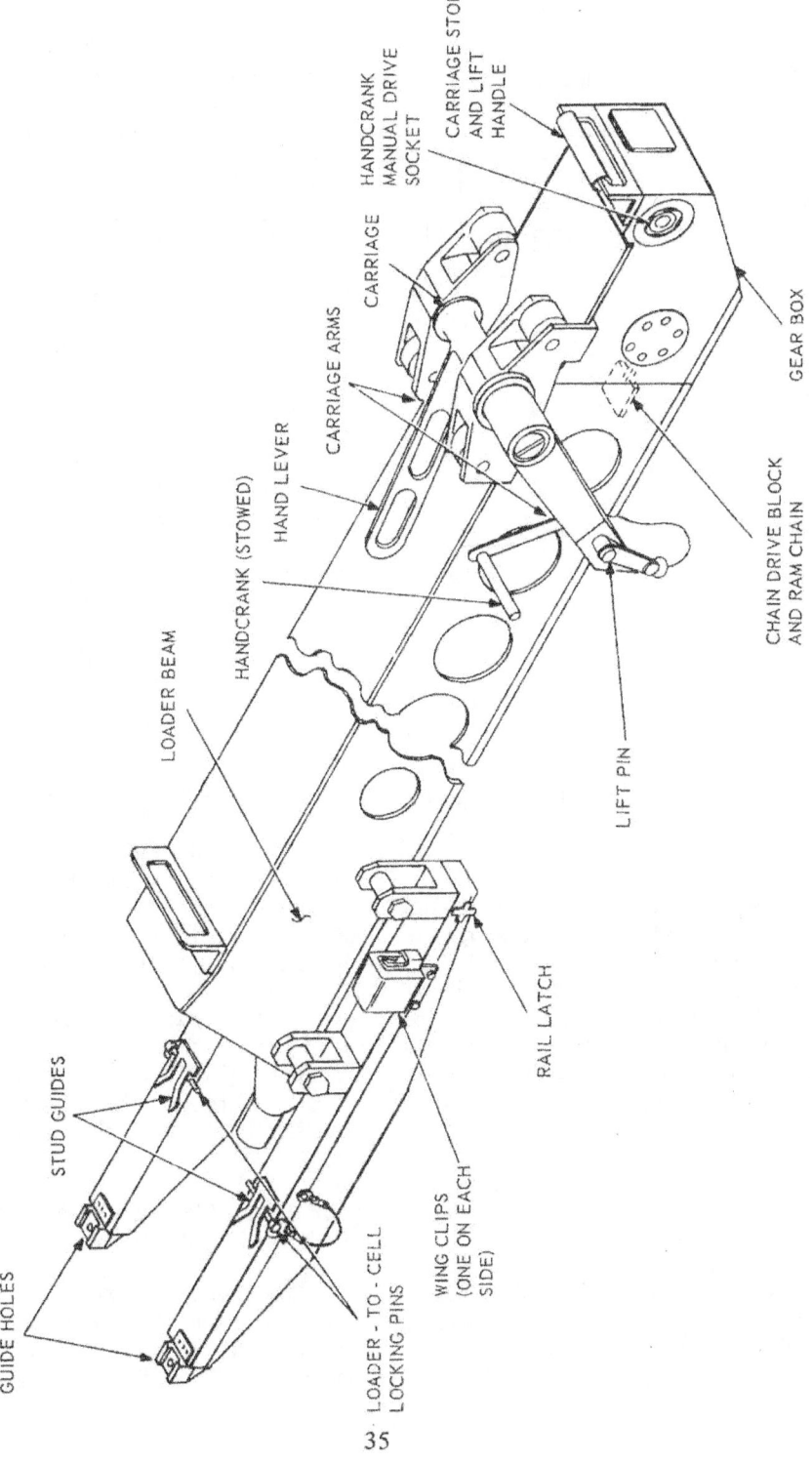

Figure 2-9.—Mk 14 guided missile loader.

# GUNNER'S MATE M 1 & C

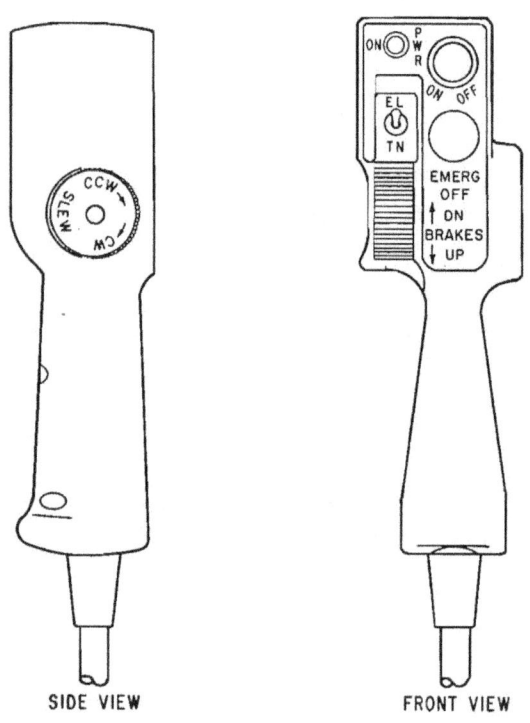

94.203
Figure 2-10.—Mk 109 deck station launcher control unit.

Using the launcher deck control unit the launcher is elevated, under power, to lower the loader with rail attached to the deck. The rail can now be disengaged from the loader by removing a pin. Assuming your loadout is for more than one missile, it is more practical to remove all applicable rail assemblies before commencing the onload operations.

With the rail assemblies on deck, the Mk 470 containers, with missiles, are positioned for onloading using the Mk 45 handlift trucks. Remove the container cover and manually position a rail assembly on the missile. Now engage the shear connector to the missile umbilical connector and insert the vibration damper in the rail. Using the launcher deck control unit, position the loader to the rail, reengage the loader to the rail, and lift the rail with missile attached out of the container.

As soon as the missile is clear of the container the loader drive chain and the rocket motor-fire connector are engaged to the rail. Align the rail and missile with their launcher cell and, by means of the handcrank or electric motor, move them a short distance into the cell. The shorting plug is then removed, and the missile/rail combination is driven further into the cell until the rail is supported by the cell hangers, at which time the missile wings are installed, folded, and held by removable clips until fully into the cell. The missile fins are then installed, the loader-to-rail pin is removed, and the missile/rail combination is driven fully into the cell. The rail is locked into the cell by rotating the rail release ring using the handheld rod. Now the loader chain can be retracted and the loader removed from the cell. The missile's rocket motor can then be armed in accordance with ship's policy.

Unloading of the Mk 132 launcher is shown in figure 2-12. The illustration will give you a general idea as to the physical alignment of the launcher rail to the loader and the loader to the launcher cell.

## HARPOON LAUNCHING GROUP

The installation of the Harpoon weapon system (HWS) provides the canister-using ships with the capability of launching eight surface attack guided missiles RGM-84A-3 (Harpoon missiles). The Harpoon missile launching group consists of weapon control console (WCC), weapon control indicator panel (WCIP), launcher switching unit (LSU), Harpoon canister launcher (HCL) (2), Harpoon causalty panel (HCP), Harpoon trainer module (HTM), and canister frame assemblies. See Figures 2-13 and 2-14.

## HARPOON WEAPON CONTROL CONSOLE (HWCC)

The HWCC provides power, data processing and interface functions for Harpoon ship command launch control set (HSCLCS) operation. The HWCC consists of a cabinet

Chapter 2—MISSILE HANDLING

Figure 2-11.—Guided missile launcher rail.

37

Figure 2-12.—Unloading the Mk 132 launcher.

containing an upper drawer, a lower drawer, and a midshelf upon which subassemblies are mounted. The upper and lower drawers slide out for access to the subassemblies. Interlock switches, located behind the upper and lower drawer front panels, remove power from the HWCC when either drawer is pulled out. Cooling is provided by circulation fans with intake vents on the sides of the cabinet and exhaust vents on the back of the cabinet above the cable entry panel.

The upper drawer contains the dimmer assembly, relay assembly, two circulation fans, and four power supplies (a +5.25-volt d.c. supply, a +15-volt d.c. supply, and two +28-volt d.c. supplies). Circuit breakers for the four power supplies are attached to a mounting bracket in this drawer.

The lower drawer contains the data conversion unit (DCU), data processor computer (DPC), three circulation fans, and elapsed time meter (HSCLCS timer) and the missile built-in test BIT switchlight.

The data conversion unit (DCU) consists of electronic tray No. 1 (ET-1) and electronic tray No. 2 (ET-2). The trays contain solid state plug-in circuit cards. ET-1 contains 133 cards and ET-2 contains 27 cards. ET-1 contains the WCIP electronics and HWCC BIT electronics. It also reformats the digital inputs from ships systems from a negative 15-volts d.c. level to a positive level for use by the DPC. ET-2 provides an interface between the digital signal format of the DPC and the synchro and discrete signals generated or used by the interfacing ship's systems.

The data processor computer (DPC) is a special-purpose, read-only, memory digital computer that computes missile orders from target data (range and bearing) and own-ship parameters (pitch, roll, speed and heading) and provides a missile built-in-test (BIT) to verify missile status. The DPC also has a BIT feature that performs the following tests.

1. Complete self-test when power is first applied to the HWCC.
2. Continuous on-line HWCC test.
3. Off-line HWCC test when selected.

The front panel of the lower drawer contains power supply and drawer interlock switch indicators.

The midshelf is located between the upper and lower drawers. Mounted on the shelf are the surge suppressor and the EMI filter.

## HARPOON WEAPON CONTROL INDICATOR PANEL (HWCIP)

The HWCIP consists of three major sections: test status, data entry, and fire control status. The test status section provides controls and indicators for power on, lamp test, BIT off-line test, fault indication, and fault numeric (code) display. The data entry section provides a keyboard for manual input, missile mode search pattern and fuze selection, true target bearing numeric dispay, target range display, and bearing only minimum range numeric display. The fire control status section provides the information and controls to select a missile for firing.

## LAUNCHER SWITCHING UNIT (LSU)

The LSU is the interface between the HWCC and the launcher. The LSU provides the necessary controls and circuitry to select the port or starboard launcher. The front panel of the LSU consists of a key operated function switch, launcher select section, circuit breakers, and indicator lights.

## HARPOON GRADE B CANISTER LAUNCHERS

Two launchers are provided for each ship. Each launcher provides the support structure and control circuitry to accommodate up to four missile/canisters. A launcher consists of one grade B support structure assembly, two lower canister frame assemblies, eight wiring harness assemblies, and the launcher relay assembly (LRA). To enhance the launcher structural support to each canister, additional canister frame assemblies are required. These additional canister frame assemblies are described in the following paragraph.

## CANISTER FRAME ASSEMBLIES

The canister frame assemblies are used to secure the forward portion of each grade B canister to the grade B launcher. This greatly enhances the missile/canister capability to withstand the excessive shock forces encountered during naval operations in high seas.

## HARPOON LIGHTWEIGHT (LTWT) CANISTER LAUNCHER

The Harpoon LTWT canister launcher is installed aboard some of the ships in the ship classes that use Harpoon missile/canisters. The LTWT launcher provides support for either two or four Harpoon missile/canisters. The LTWT canister launcher consists of one LTWT launcher support structure (LSS), the launcher relay assembly (LRA), and eight wiring harness assemblies.

## CANISTER

The canister is a cylindrical shell structure containing launch rails, restraint mechanism, and closures, and stacking frames. The launch rails support the Harpoon missile shoes. The restraint mechanism holds the Harpoon missile in place during transport and installation aboard ship, and restrains the missile in the event of inadvertent booster ignition. The end closures provide a weather seal as well as RF shielding. Each canister contains one Harpoon missile. After a Harpoon missile is fired, the canister is removed and replaced with a canister containing another Harpoon missile.

## HARPOON CASUALTY PANEL (HCP)

The HCP is used to launch a Harpoon missile when the missile cannot be launched using the WCC. The HCP interfaces with the HWCC and LSU when operating in the normal mode. The HCP operates independently of the HWCC when the causalty operation is being performed.

## HARPOON TRAINER MODULE (HTM)

The HTM allows training on the HSCLCS to be performed by simulating cell status signals that can be verified by observing the HWCIP indicators. The HTM is interfaced with the LSU and the test set simulator.

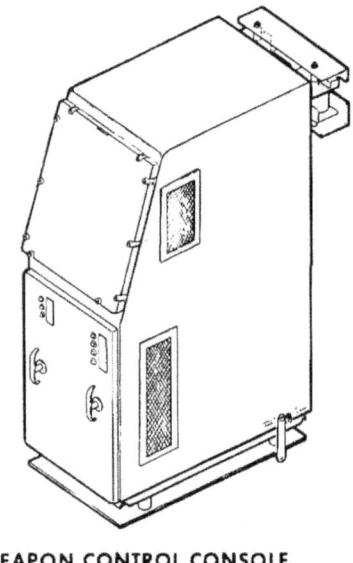

**WEAPON CONTROL CONSOLE**

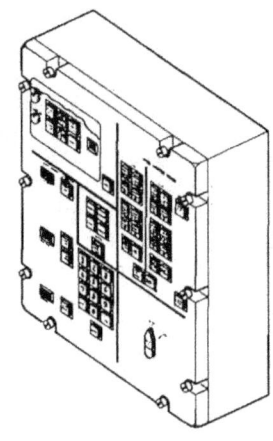

**WEAPON CONTROL INDICATOR PANEL**

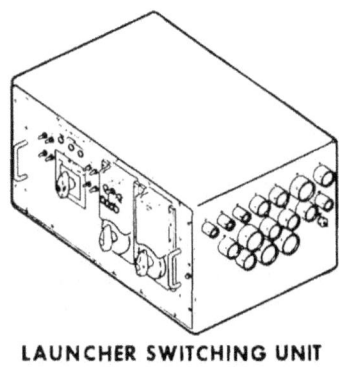

**LAUNCHER SWITCHING UNIT**

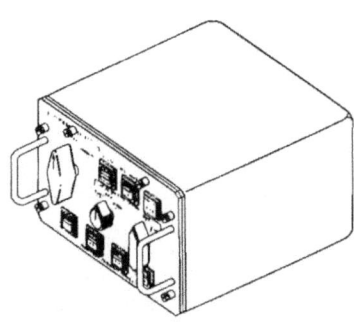

**HARPOON CASUALTY PANEL**

94.206

Figure 2-13.—Harpoon canister launching group.

## ONLOADING OPERATION

The canisterized surface attack guided missile RGM-84A-3 (Harpoon missile/canister) is now on four classes of ships: CG-16, CG-26, DD-963, and DDG-39. The responsibility for handling, stowage, and minor maintenance varies among ships, but as a prospective GMM1 or Chief chances are you will be involved with Harpoon at one time or another and most assuredly as you progress to E-8 and E-9.

The Mk 631 Mod 0 shipping container is used to transport the Harpoon missile/canister from the naval weapons station to the dockside

Chapter 2—MISSILE HANDLING

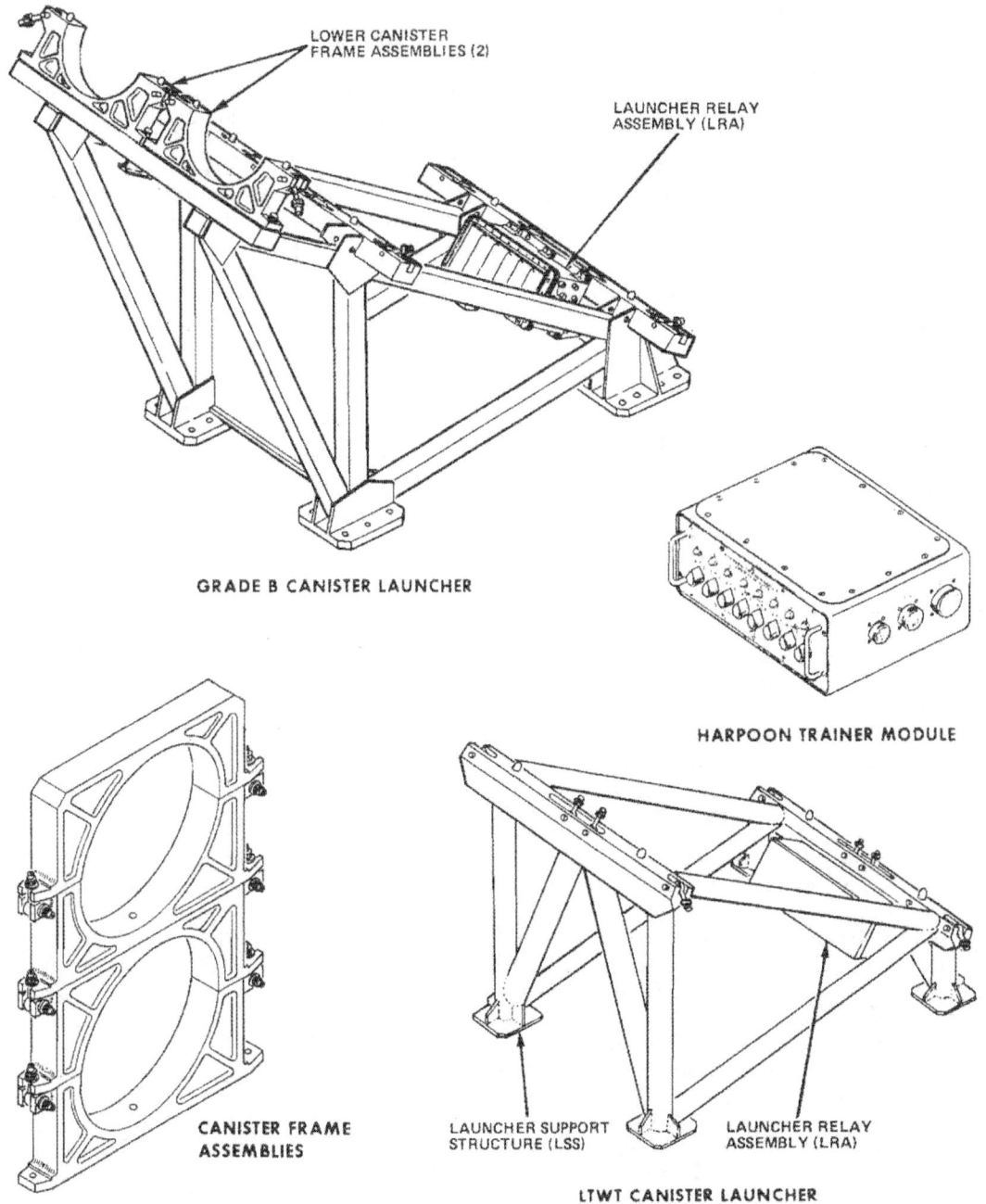

Figure 2-14.—Harpoon canister launching group—Continued.

loading area. Figure 2-15 illustrates the Mk 631 shipping container with a Mk 6 Harpoon lightweight missile canister. Figure 2-16 shows the lightweight canister being removed or installed on a lightweight (LTWT) launcher. This launcher holds four missiles/canisters. The Mk 31 hoist rotation beam is used on both the lightweight and grade B canisters. Tag lines are an essential safety requirement when handling Harpoon missile/canisters and must be manned the entire time the canister is suspended.

In figure 2-17, the Mk 631 Mod 0 shipping container with the Mk 7 grade B missile canister is shown. Figure 2-18 shows a grade B canister being installed or removed from a grade B launcher.

The Mk 31 Mod 0 hoist rotation beam is a lifting beam used at dockside to handle empty or loaded Harpoon canisters during onload/offload operations. The Harpoon launcher platform has a fixed elevation angle of 35 degrees. The hoist rotation beam can be manually adjusted to a 35-degree angle by operating a grip hoist while at dockside. After lowering the canister onto the launcher and bolting it into place, release the hoist rotation beam from the canister. The beam can then be returned to zero by the grip hoist before hooking up another canister at dockside.

After canister onload, install four control and four booster ignition cables to the launcher relay assembly (LRA). You will at this time perform a stray voltage checkout as per applicable systems manual. After completion of the stray voltage checkout, disconnect shorting plug and protective cap from connectors on canisters. Connect launcher cables to applicable canister and stow shorting plug and protective cap on dummy receptacles on the lower side of the relay assembly. The final procedure is to perform missile BIT as per your systems manual. The Harpoon missile is physically and mechanically compatible with the Mk 11 and Mk 13 Tartar launchers and their magazines with minor modifications to the fin retainers on Mk 11 and minor modifications to the anti-icing system and retractable rail on the Mk 13. On Mk 11 launching systems, Harpoon missiles will be loaded only on the outer ring of the magazine.

The Harpoon missile is directly compatible with Tartar strikedown equipment without any modifications.

The Mk 6 Mod 2/4 transfer dolly can be used with the Harpoon missile and requires no modification. No additional shipboard handling equipment is required for the Harpoon missile.

Electrically, some modifications are required for Harpoon after incorporation of U.S. Navy ORDALTs 8204 (Mk 11) or 8217 (Mk 13). The launcher control panel and missile identification box must be modified to provide Harpoon loading orders, Harpoon interlocks, and power sequencing. On the Mk 11 launching system two relay panels and the loader control panel must also be modified for Harpoon. The Harpoon relay transmitter (HRT) must be added to provide switching of the umbilical circuits from Tartar to Harpoon control. Harpoon firing interlocks are also included in the HRT in order to minimize the ship alterations.

## TOMAHAWK WEAPON SYSTEM

At the present time the type of launcher to be used for the Tomahawk weapon system has not been finalized. The Harpoon canister for the Armoured Box Launcher (ABL) is similar to the Tomahawk canister although longer to allow the spring loaded wings to extend immediately after launch. One of the proposed Tomahawk surface launchers (figure 2-19) can house four Tomahawk missile/canisters, or two Tomahawk missile/canisters and two Harpoon missile/canisters. Since the Harpoon missile/canister can be integrated into the Tomahawk surface launcher, there will be interface required to the Harpoon weapon system.

## TARTAR MISSILE HANDLING AND STOWAGE

Strikedown operation of Tartar missiles is the process of either onloading or offloading a missile. Onloading transfer a missile into the launching system; offloading transfers a missile out of the launching system. These operations involve not only launching system personnel and

# Chapter 2—MISSILE HANDLING

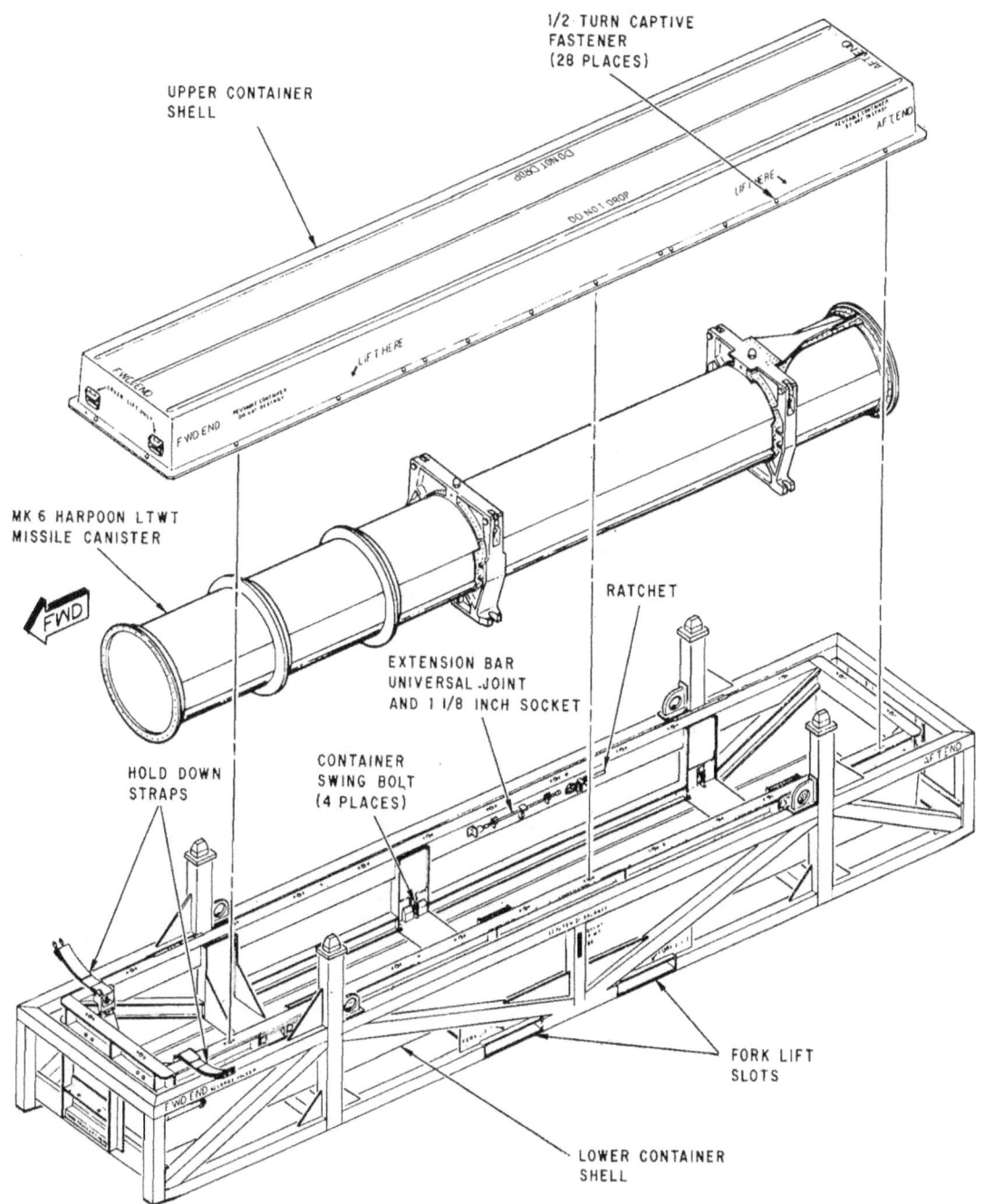

Figure 2-15.—Mk 631 Mod 0 shipping container with a Mk 6 Harpoon lightweight missile canister.

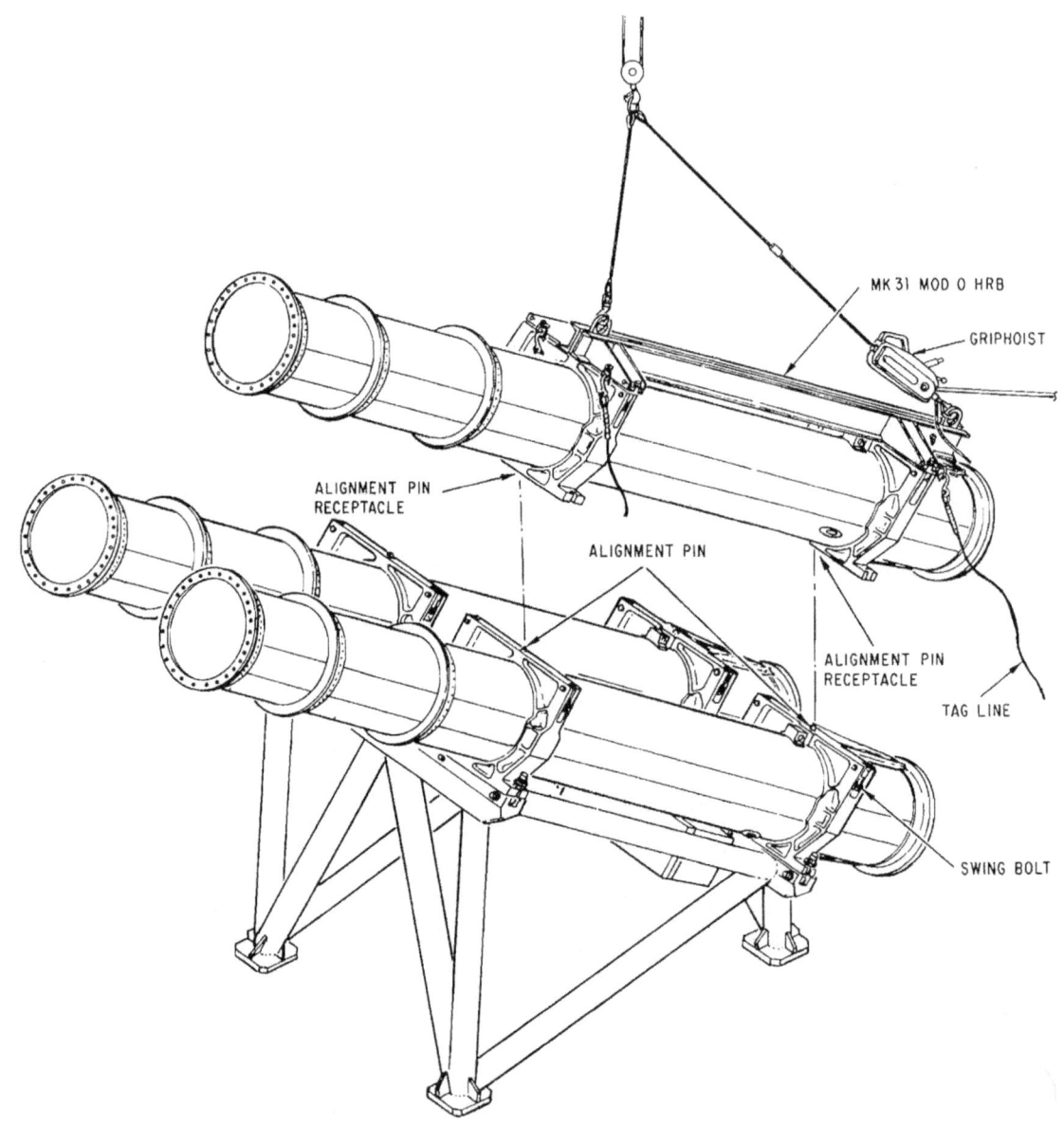

Figure 2-16.—Upper Harpoon LTWT canister installation/removal using Mk 31 Mod 0 HRB.

other crewmembers but also dockside crews, crewmembers of a second ship, or a helicopter crew.

The Tartar missile is brought aboard ship as a complete weapon. When transferred from dockside or from a barge, the missile is mounted in a Mk 6 Mod 2/4 transfer dolly. Instead of being struck below on an elevator as with Terrier, the missile is transferred from the transfer dolly to the launcher guide arm rail

Chapter 2—MISSILE HANDLING

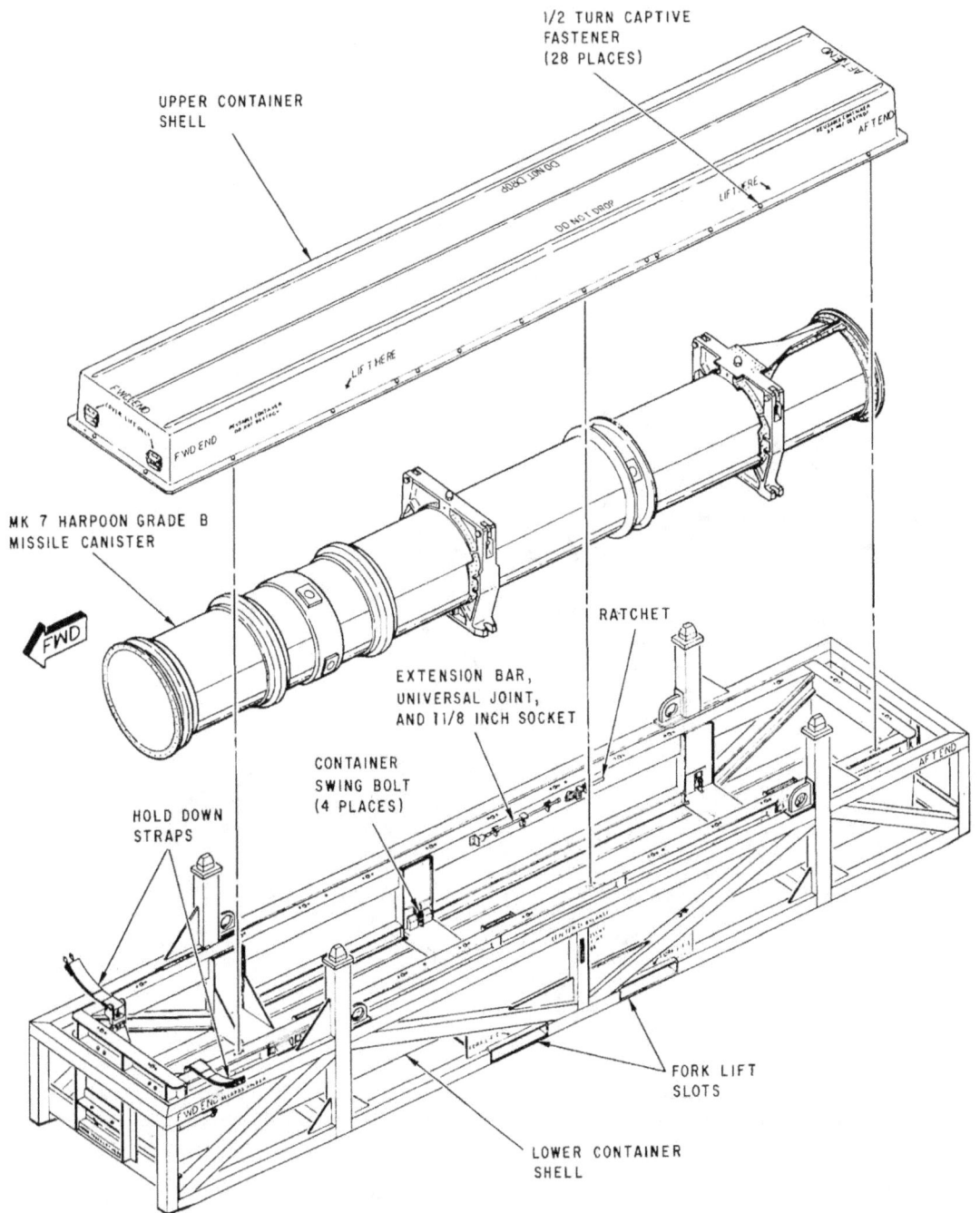

Figure 2-17.—Mk 631 Mod 0 shipping container with a Mk 7 grade B missile canister.

# GUNNER'S MATE M 1 & C

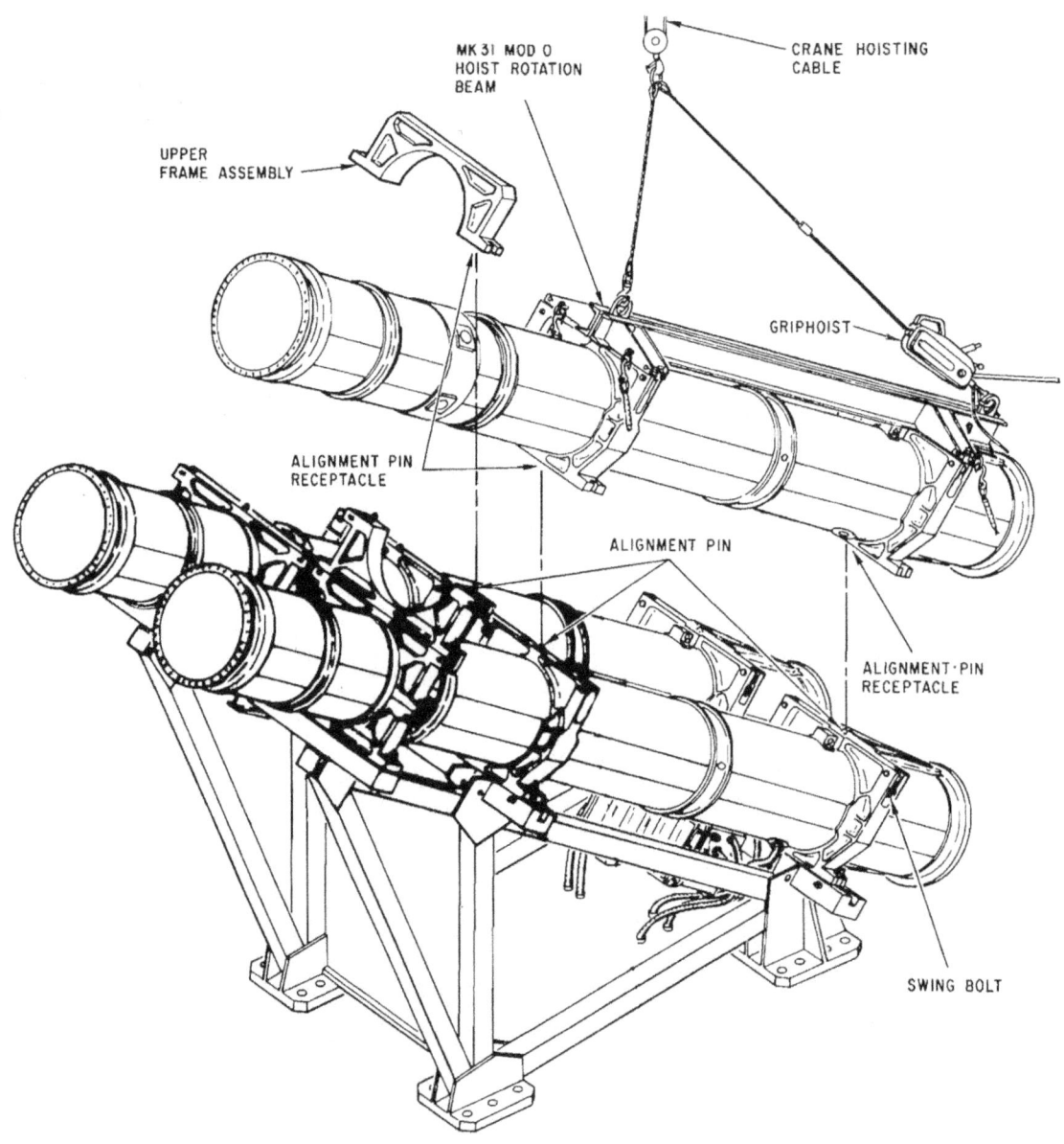

Figure 2-18.—Upper grade B canister installation/removal using Mk 31 Mod 0 HRB.

(figure 2-20). It is then struck down (stowed) in a vertical position in the missile magazine beneath the launcher. When handling Tartar missiles, particular care must be taken to avoid damage to the missile radome, to the target detection devices (TDD), and to the missile control surfaces.

When transferring Tartar missiles at sea by STREAM, the procedure is essentially the same as previously discussed in this chapter. Figure

## Chapter 2—MISSILE HANDLING

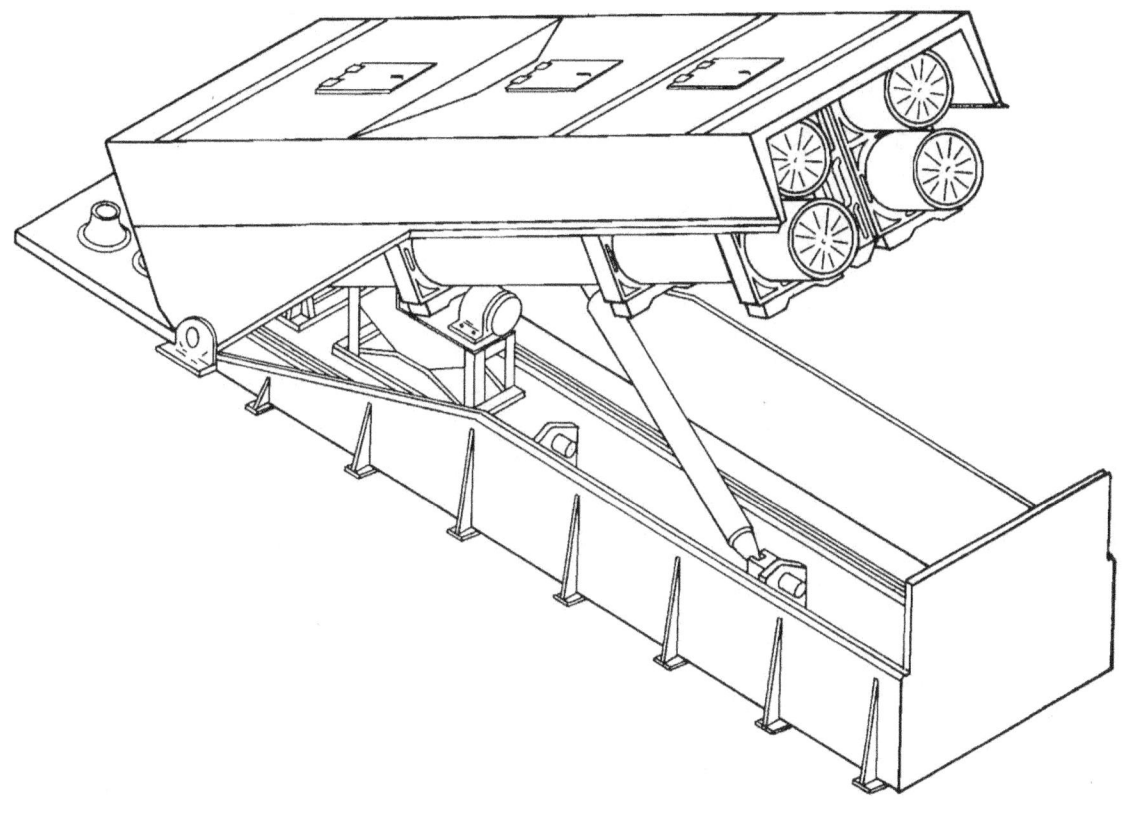

Figure 2-19.—Tomahawk surface launcher (proposed). Armoured Box Launcher (ABL).

94.211

2-2 illustrates the STREAM system attached to a receiving ship's sliding pad eye. Figure 2-3 shows the receiving ship having only a bulkhead kingpost or outrigger available. The Mk 6 Mod 2 or 4 transfer dolly with Tartar rail adapter will be used for dockside/at sea onloading or offloading operations. The Tartar adapter rail is part of the adpater head which is an inclusive assembly of the Mk 6 Mod 2/4 transfer dollies.

### STRIKEDOWN OPERATIONS ON THE MK 13 AND MK 22 GMLSs

For the Mk 13 and Mk 22 launching systems, strikedown operation begins with the crewmemebers attaching the strikedown equipment consisting of a chain drive fixture, a deck control box, and a manual air control valve and air supply lines. The chain drive fixture is attached to the front of the launcher guide arm (discussed in chapter 6), whenever a transfer dolly is used. The purpose of the chain drive fixture is to move a missile from the transfer dolly on to or off of the launcher guide arm. The chain drive fixture is a pneumatically operated unit controlled by a manually operated control valve. A crewmember operates the pneumatic control valve which determines the direction of chain drive movement for either onload or offload operations.

A portable deck control box is plugged into the launcher control system and is operated by

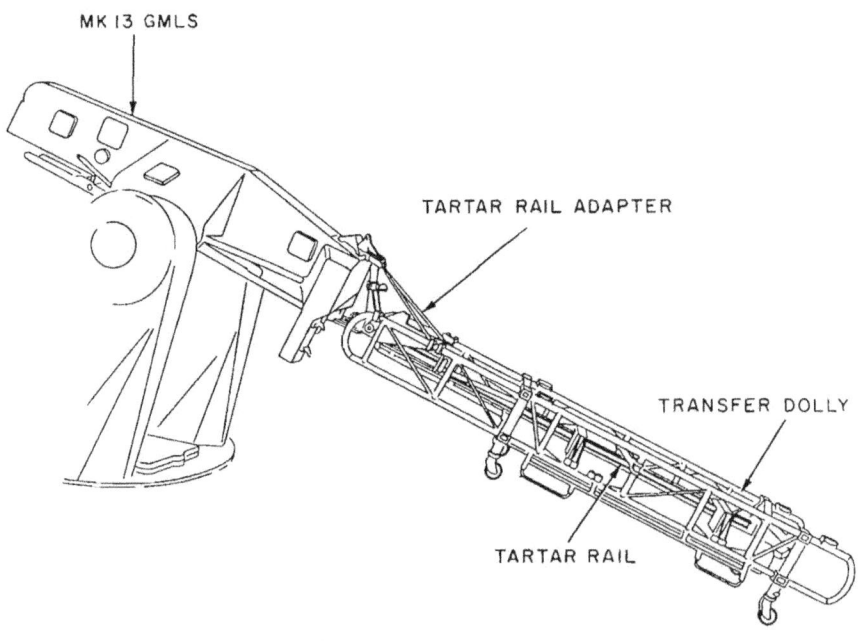

Figure 2-20.—Transfer dolly aligned with Mk 13 GMLS.

personnel on deck (missile handling area) to control the movements of the launcher when mating the launcher with the transfer dolly during strikedown operations. The deck control box is manually operated switching unit contained in a metal box which has two handles, indicating lamps, and toggle switches. A cable attached to one end of the control box is plugged into a receptacle on the launcher stand or bulkhead. For training and elevating the launcher to a strikedown position, an operator used the toggle switches on the deck control box which connect fixed position synchros in the launcher control system to position the launcher to a fixed load position for either port or starboard strikedown operations. For transferring the missile between the launcher rail and missile magazine, an operator at the EP2 panel (launcher system control panel) operates switches in either the step load or step unload mode of operation on orders from the launcher captain. The launcher captain operates the deck control box and orders missile movement for onload or offload operations whenever the launcher is in a position to transfer a missile between the launcher rail and missile magazine. The launcher captain is in charge of all strikedown operations.

Chapter 6 illustrates and describes how the pneumatic chain drive fixture is used to transfer the missile during strikedown operations.

## TARTAR MISSILE NO-TEST PROGRAM

Current NAVSEA instructions limit shipboard testing of Tartar missiles to periodic external missile inspection. All missiles received aboard ship are now certified as reliable and require no test after being issued by a naval weapons station.

It must be emphasized that the no-test concept places certified missiles in the magazine. They are to be fired, returned to a supply source, or jettisoned; the missiles are not to be taken apart or repaired aboard a combatant ship

## Chapter 2—MISSILE HANDLING

## HANDLING

The dual thrust rocket motor (DTRM) is considered a class B explosive and should be handled accordingly. The DTRM produces an extremely hot exhaust blast and noxious gases. It is relatively safe when handled properly, but a sharp blow could crack the propellant grain resulting in an explosion when the missile is fired. If the DTRM is found to be armed, manually move the arming lever to the SAFE position and request disposition and instructions from NAVSEASYSCOM.

Personnel shall keep clear of the area aft of the missile (DTRM area) at all times.

Because the missile contains electro-explosive devices (EEDs), observation of currently prescribed (HERO) safety precautions during handling is mandatory.

Take all possible steps to protect all missiles from extremes of temperature, humidity, vibrations, electrical or magnetic fields, and radiological exposure. Exposure to any of these conditions, when excessive, may require disposal of the missile.

Upon receipt of the missile, a visual inspection shall be made to ensure that no physical damage has occurred during handling. This inspection is to determine if all sections are free of rust and corrosion, that all covers, plugs, tape, and decals are in place and secure, and that safety devices are in the SAFE position. Then any abnormal conditions are indicated, the defective missile shall not be struck down but shall be returned to the replenishing source.

Interlocks and warning bells are built into the handling system as safety features and are not to be bypassed or disregarded at any time except under emergency conditions. In the event that such devices are disabled or bypassed, adequate warning signs shall be posted to indicate that such a condition exists. Also, all applicable safety precautions shall be posted at each operating station of the handling system. Regular handling drills employing dummy or training missiles shall be held to ensure safe operation and improved individual proficiency. During drills, the officer in charge and leading chief shall carefully observe all operations which might create hazardous conditions and shall take the necessary corrective steps to alleviate them.

## HANDLING AND STOWAGE OF OTHER MISSILES

The magazine, or stowage area, is the location with fixed installations designed for stowage of all the various types of aircraft ammunition.

The delivery assembly area is the location aboard carriers where the various components of ammunition are delivered for assembly into complete weapons for use on aircraft.

Since the assembly, testing, and arming of aircraft launched missiles are the duties of the Aviation Ordnanceman, the GMM assigned to the ship's armory has the primary duties of ensuring that all components of aircraft launched missiles are properly stowed and maintained in a state of readiness at all times. The GMM may also be called upon to supply the various components of aircraft launched missiles to personnel in the delivery assembly area where the missiles are assembled for use or transfer to ready service areas. If you are charged with the responsibility of magazine stowage and the transfer of rocket and missile components, you should thoroughly familiarize yourself with the practical methods of safe handling and stowage of such items. Listed below are some safety precautions to be observed when handling rockets and guided missiles.

Firing temperature limits specified for each missile must be observed for safe operation. If a missile is exposed outside of temperature limits stenciled on the unit, it should be set aside and handled in accordance with current instructions.

Continued exposure to abnormal stowage temperatures may cause the propellant to deteriorate, with attendant hazards of possible explosion when the rocket is fire.

Rough handling or blows may break the propellant grain, thus exposing too much surface to burning and leading to possible excessive pressure in the motor. Excessive pressures may cause the motor to explode when fired.

## STANDARD MISSILES (MR/ER)

The Standard MR missile and its components are shipped and stowed in the same types of containers as the Tartar missile. The handling equipment and procedures for loadout, offload, underway replenishment, and stowage are identical for the Standard MR and Tartar missiles.

The Standard ER missile and its components are shipped and stowed in the same types of containers as the Terrier missile. The handling equipment and procedures for loadout, offload, underway replenishment, and stowage are identical for the Standard ER and Terrier missiles.

Power for the Standard missile is supplied by a squib-activated primary battery. This battery will generate gas, when activated, requiring the following special handling procedures whenever the battery is activated or a misfire occurs:

Allow approximately 4 hours for battery termperature to return to normal.

Inspect the battery vent port (forward of the dorsal fin) to determine if termperature has cooled sufficiently to work safely.

Clean up any vented electrolyte (potassium hydroxide, a caustic alkali). Do not allow the electrolyte to contact body or clothing. If it does, immediately flush the contaminated area with large quantities of vinegar and water or freshwater.

After cooling and cleanup, replace the missile battery vent port plug and return the missile to the magazine for future offloading.

If during missile handling or firing exercises black smoke appears from the battery vent port, the missile should be jettisoned immediately. Black smoke indicates a missile battery fire. The appearance of white steam from the missile vent port is due to the battery venting and should not be mistaken for a battery fire within the missile.

Make sure that the DTRM igniter arming level (Standard MR) and the sustainer arming indicator (Standard ER) are in the SAFE position prior to and during handling operations.

## AIRCRAFT MISSILES

Missile magazines in aircraft carriers generally are located below the waterline and within the ship's armor belt. For ease in the handling of missile components, these magazines contain power operated handling equipment such as electrical, hydraulic, or pneumatic hoists, trolleys, etc. To provide adequate and continuous serveillance in magazines containing certain missiles, and to provide assurance that a specific hazard is not actively present, these magazines are equipped with special detection equipment.

Aboard most aircraft carriers the handling, stowage, and assembly of aircraft launched missiles is the responsibility of personnel in the Aviation Ordnanceman rating. Some carriers split this responsibility and utilize personnel in GMG and GMM ratings for the maintenance of stowage magazines and some missile handling equipment.

The movements of aircraft ammunition and explosives between the magazine areas and aircraft involve specific handling and assembling functions that are controlled by areas designed for a specific purpose. Two of these areas, the magazine or stowage area and the delivery assembly area, are of interest to personnel of the GMM rating assigned to the ship's armory aboard a carrier.

Since the Mk 29 GMLS (NATO Seasparrow) is now installed aboard carriers, the magazine spaces for the Sparrow III RIM-7H will probably be an area of responsibility for GMM pesonnel assigned to the Mk 29 GMLS.

## NUCLEAR WEAPONS HANDLING AND STOWAGE

Nuclear weapons used by the Navy may be bombs, torpedoes, missiles, depth charges, and projectiles. Rules for peacetime operation of nuclear weapons (capable) systems issued by the Chief of Naval Operations, along with official Naval Sea Systems Command special weapons checklists, are mandatory directives which must be followed.

The operation of each type of nuclear weapon is described in the applicable special weapons ordnance publications (Navy SWOPs). Nuclear weapons will be handled and stowed in accordance with Navy SWOP 50-1 and SWOPs of the 20 series. No ammunition assemblies or components shall be disassembled or modified unless authorized by applicable technical instructions. Detailed safety precautions and considerations are prescribed in Navy SWOP 50-1.

## Chapter 2—MISSILE HANDLING

Missiles that have nuclear warheads are stowed in a ready service condition in the same missile magazine as those with conventional warheads and require no special handling or testing. The GMM who deals with any weapon must ensure that a proper stowage condition is maintained. This ensures the reliability of the weapon and also guarantees personnel safety. As a leading GMM, it is essential that you have a thorough knowledge of the hazards concerned and the restrictions imposed on nuclear weapons. For this reason your main concern when dealing with a nuclear weapon is its security and protection.

## SAFEGUARDING NUCLEAR WEAPONS

Nuclear weapons require special protection because of their political and military importance, their destructiveness, and the attendant consequences of an unauthorized nuclear detonation. Procedures and responsibilities for the establishment of effective security measures are set forth in DOD directives and implemented by the using agencies. The Navy's security program is outlined in OPNAVINST C5510.83 series, Criteria and Standards for Safeguarding Nuclear Weapons. This instruction is the basis for determining the minimum necessary requirements for all nuclear weapons in Navy custody. It may be augmented by additional security measures as deemed necessary by local commanders.

This section outlines the basic requirements for safeguarding nuclear weapons in the Navy, and is not intended to include all of the local area commander's requirements. It is your responsibility as a senior petty officer to keep informed of the security requirements of your activity.

### Definitions

The definitions that follow are used throughout the Navy in conjunction with nuclear weapons. Navy SWOP 4-1 is the approved source for definitions other than those in OPNAVINST C5510.83 series.

Access: As applied to nuclear weapons, access means physical proximity in such a manner as to allow the opportunity to cause a nuclear detonation. Whenever the word "access" appears in the nuclear weapons program, only this meaning will apply. Access should not be confused with entrance.

Technical knowledge: That knowledge, however obtained, required to cause a nuclear detonation.

Critical position: One in which the incumbent has (1) technical knowledge of nuclear weapons, and (2) access to nuclear weapons.

Controlled position: One in which the incumbent is performing duties physically associated with nuclear weapons, but does not require technical knowledge of, or access to, nuclear weapons.

Exclusion area: The designated area containing one or more nuclear weapons.

Limited area: The designated area surrounding one or more exclusion area.

Controlled area: The entire ship is considered a controlled area.

### Two-Man Rule

No other single area identifies more with the spirit and intent of the nuclear safety and security program than the two-man rule. All personnel working with nuclear weapons should read and understand this rule, which is explained in *GMM 3 & 2*, NAVEDTRA 10199-C, and OPNAVINST C5510.83 series.

### Entry and Access Control

Entry control to limited and/or exclusion areas is formalized and maintained to ensure positive identification of personnel prior to admission. A badge system, entry control lists, visitor escorts, and a duress system are employed.

Unauthorized actions by persons with approved access to nuclear weapons is one of the threats to nuclear weapons. Therefore, entrance to exclusion areas containing nuclear weapons is restricted to properly cleared personnel who have a positive need for access, or to personnel who have to enter a space containing nuclear weapons during the course of their duties. Only persons authorized by the commanding officer can be admitted to exclusion and limited areas.

When projects in a limited and/or exclusion area require the presence of personnel not cleared for normal entry, such persons are kept under constant escort by the security force or supplementary personnel. Their movements are limited to only those necessary for the performance of assigned tasks. A log of persons entering and leaving exclusion areas is kept and maintained locally for a period of at least 2 years.

In those spaces in which nuclear weapons are stowed and manned by only two individuals, all openings and entrances to those spaces (other than those in use) must be locked and alarmed. The unlocked entrances are guarded by an armed guard who also controls entrance to the spaces.

When transporting nuclear weapons from one area aboard ship to another, an appropriately armed guard accompanies the personnel loading, handling, or transporting the weapon.

Once working hours commence and the exclusion area is entered by two authorized persons, the responsibility for maintaining the two-man rule rests with the senior person present. No one individual is allowed to remain in the exclusion area alone.

## SAFETY REQUIREMENTS

Safety rules are issued for every nuclear weapons system. These rules are to be followed in peacetime and wartime, when possible. All safety rules are applied against these safety standards.

All hands must take positive measures to prevent weapons involved in accidents or incidents, or jettisoned weapons, from producing a nuclear yield.

Take positive measures to prevent deliberate arming, launching, firing, or releasing except on lawful orders.

Positive measures must be taken to prevent inadvertent arming, launching, firing, or releasing, and to provide adequate security.

In GMLSs having nuclear weapons capabilities, all technical operations involving conventional weapons, training weapons, and non-nuclear components will be conducted following identical procedures established for handling nuclear weapons, except those procedures involving PMS.

## HANDLING

The best handling equipment designed is only as good and as safe as the personnel who operate it. With nuclear weapons it is imperative that you know the type of material you are handling and its hazards. Further, you must know the capabilities and limitations of the equipment you are using when handling the weapons.

When using hoisting equipment in handling nuclear weapons, it should never be loaded in excess of its rated capacity. No piece of handling equipment should be used for other than its intended purposes. When elevated loads are moved horizontally on monorail, sudden stops or starts must be avoided. Remove any obstructions from the path of the load. Never raise weapons higher or let them remain suspended longer than is absolutely necessary to complete the required handling operation. Suspended weapons or weapons components and the controls of handling equipment must not be left unattended.

A first aid kit shall be available in the area during weapons handling. Drills shall not be conducted when live ammunition is being handled. Hooks used on hoists to lift nuclear weapons shall be fixed with a safety latch or moused. All handling equipment, slings, hoists, elevators, and other equipment used during handling evolutions shall be inspected just prior to use, then checked periodically during handling evolutions. During handling operations firehoses shall be laid out and charged. If handling equipment malfunctions, all operations must cease, the weapons must be made secure and the equipment placed out of service until the malfunction is corrected.

Slings used for handling explosives shall be of all-wire or all-fiber construction. A combination of wire and fiber shall not be used.

## Chapter 2—MISSILE HANDLING

## TRAINING

All personnel assigned to work with nuclear weapons must receive special training in the handling, stowage, and accounting methods of nuclear weapons. Prior to such training they must possess at least a Secret clearance based on a background investigation. Only properly cleared personnel who have need for access to spaces containing nuclear weapons will be allowed entry to these spaces. Only personnel of demonstrated reliability and stability as outlined in BUPERINST 5510.11 series, Nuclear Weapon Personnel Reliability Program (PRP), will be assigned to this type of duty.

## PERSONNEL RELIABILITY PROGRAM

The personnel reliability program is aimed at all personnel who control, handle, have access to, or control access to nuclear weapons or nuclear weapons systems. The program covers selection, screening and continuous evaluation of the personnel assigned to various nuclear duties. The program seeks to ensure that personnel coming under its purview are mentally and emotionally stable and reliable.

## STOWAGE REQUIREMENTS

Stowage requirements for nuclear weapons aboard ship are governed by applicable SWOPs, OPs, and ODs, and directives issued by higher authority. Nuclear and conventional weapons of a given family, i.e., Terrier, ASROC, Harpoon, and Tomahawk are compatible. In most cases a nuclear weapon can be stowed in a tray or adapter rail that has previously been used to stow a conventional weapon. As previously discussed in this chapter, an ASROC adapter rail has to be modified by installing or removing inserts from the snubbers according to ASROC weapon type desired. When changing locations of nuclear/conventional weapons, it is of paramount importance that proper identification be maintained at each location. The actual locations may be preassigned by the commanding officer for nuclear weapons in the ready service rings and/or magazines.

## ALARM AND WARNING SYSTEMS

Numerous alarm and warning signals are installed on ships with nuclear weapons spaces. Some are audible alarms, such as bells and buzzers; others are warning lights. Some are connected to all parts of the ship, and others, only to certain spaces. The nuclear weapons stowage spaces have warning signals for high temperature and security.

The operation of security alarms and warning signals can be mechanically operated switches or pushbuttons activated by the opening of access doors and/or hatches to nuclear weapons spaces. Alarm panels used for security alarm systems are located in ship's areas that normally are manned at sea and in port such as quarterdeck areas and damage control central. When the alarm panels include entry into a nuclear weapons space, special security forces are alerted to safeguard nuclear weapons and components.

Although alarm circuits are tested and maintained by IC electricians, the switches/alarms associated with the magazine sprinkling systems, booster suppression and missile quenching systems are actually tested by GMMs during performance of PMS. So it is essential that the GMMs have a working knowledge of the associated alarm circuits for those areas. Briefly stated, the GMM rating does not have a "need to know" concerning maintenance of the FZ alarm circuits associated with nuclear weapons spaces.

## VENTILATION IN NUCLEAR WEAPONS SPACES

On most ships with nuclear weapon spaces, the ventilation system for those spaces is not connected to the system that services other parts of the ship. The reason for this arrangement is that, in the event of a nuclear accident, radioactive particles will not be carried from nuclear weapons spaces through the ventilation system into other living or working spaces.

Air conditioning is provided to keep the weapon stowage spaces within the required temperature and humidity levels. Intermittently, as prescribed by ship's policy, the air-conditioning system is secured, the exhaust vents are opened, and the exhaust system is turned on to vent the stale air from the stowage areas. After a period of time, again prescribed by ship's policy, the exhaust system is secured and the supply vent system is opened to resupply fresh air to the stowage space. After resupplying the stowage space with fresh air, the supply vent system is secured and the air-conditioning system reactivated.

## REFERENCES

The following publications contain pertinent information on missile handling and stowing and were consulted in the writing of this chapter.

1. NAVTRA 10200-B, *Gunners Mate M 1 & C.*

2. OP 2888, Tartar Guided Missile, Mk 15 Mods 1 and 2, Volumes 1, 2, and 3.

3. OP 2889, Guided Missile Complete Round, Mk 3 Mod 0 (Terrier HT-3) Description, Volumes 1 and 2.

4. OP 3043, Guided Missile Complete Round RIM-2E, 3, 4, 5 and RIM-2F-1, 2, 3, 4/Terrier HT and HTR/General Description/U/, Volume 1.

5. OP 3726, TER/TAR/STD Guided Missile Warheads and Fuze Components, DESC/OPER/MAINT/U/, Volume 1.

6. OP 4060, NATO Seasparrow Guided Missile RIM-7H-1 and RIM-7H-2, Description, Operation, Handling, Volume 1.

7. OP 4006, Guided Missile Launching System, Mk 29 NATO Seasparrow.

8. OP 4341, Harpoon Missile (RGM-84A) Handling Operations and Maintenance Procedures, Volume 4.

9. NWP 14, Replenishment at Sea.

10. OP 2173, Approved Handling Equipment For Weapons and Explosives; Adapters Thru Jigs, Volume 1.

11. OP 2173, Approved Handling Equipment For Weapons and Explosives; Loaders Thru Trucks, Volume 2.

12. OP 3192, Missile Transfer Dolly Mk 6 Mods 1, 2, 3, and 4; D.O.M. and IPB.

# CHAPTER 3

# PREPARING FOR A FIRING EXERCISE

The last chapter dealt with the parts of the ship's missile system that were used in handling and stowing the weapons. You already know a good deal about launching systems from study and experience. What additional information or skills must you gain to qualify as a GMM1 or Chief? Look at the occupational standards that apply to the missile launcher system.

Notice that most of the factors are placed at the GMM3 level. In the personnel advancement requirements, note that the GMM3 must be able to make operational tests. To advance to GMM2 you had to be able to man any station in the system on your ship, use special test equipment, and interpret, record, and report the test results. You learned to inspect and disassemble, clean, lubricate, and reassemble many of the launching system components and the missile handling and dud-jettisoning equipment.

Now you must be able to train individuals and teams in the operation of the launching systems on your ship. If there is more than one type of system on your ship, you have to train on all of them. You must learn to perform all tests of the equipment and to locate trouble in any part. Overhaul, repair, and adjustment of all mechanical, electrical, electronic, and hydraulic equipment in the missile launching systems are part of the job of the GMM1 and C. As you are aware, that includes an array of complicated equipment.

Not only must you be able to do all of these different kinds of work, but you must be able to teach others, to plan programs for getting the work done, and to conduct classes to carry out the programs. Planning of work and supervision of personnel doing the work will be important parts of your job.

## MANNING THE SYSTEM

Practice sessions in preparing a guided missile launching system (GMLS) for firing are necessary to develop coordination, speed, and skill in carrying out the steps in order. Rotate the crew to different positions so each crewmember can become proficient in the different operations. Shifting individuals to different positions undoubtedly will slow down the team for the time being, but it is much more valuable than to train each person to become an expert at only one position. Each crewmember should be able to take over any position in an emergency.

The types of GMLSs and their major components were described in *Gunner's Mate M 3 & 2*, NAVEDTRA 10199 series. Additional information is given on the care and repair of launching systems in chapter 9 of this GMM 1 & C manual. It is assumed that all checks and tests required after repair or maintenance have been performed and the launching system is ready for firing a missile. Tests and checkoffs to be performed in the process of preparing for a firing or firing exercise are given some attention in this chapter but are described more specifically in chapter 9. Alignment of missile batteries is covered in chapter 8.

## WEAPONS SYSTEM OPERATION

Although weapons systems differ in the details of preparation of the launcher for firing, the general steps in the operation of the system are very similar. For the missile system these steps are:

1. Search radar detection.
2. Fire control radar tracking.

3. Missile launching.
4. Guidance and target intercept.

Your special concern is with the third step, but you need to have some knowledge of how the whole system operates so the work will be coordinated. This knowledge will also help you in operating control stations. Figure 3-1 shows the main components at different stations of a Terrier weapons system on a CG, and the flow of information and instructions among the components.

When a target is detected by the ship's search radars (or sonar), radar information concerning the target's range, azimuth, and height is supplied to the ship's combat information center (CIC). The data is evaluated in CIC and the target (or targets) assigned to the weapons control station (WCS). Further target evaluation is made by WCS and a director radar is assigned to track the target. If the target is considered to be an enemy, general quarters is sounded and all personnel take their stations. Power is turned on to activate the system. The launching system captain activates the EP1 power panel and is stationed at the EP2 (launcher captain's) panel. Decisions as to the type of weapon to use and if, and, or when to fire are made by the antiaircraft warfare commander on the basis of information from CIC and WCS, and the decisions are relayed to the various control stations of the weapons system. The operators of the control panels push buttons to set in operation the mechanisms to carry out the decisions.

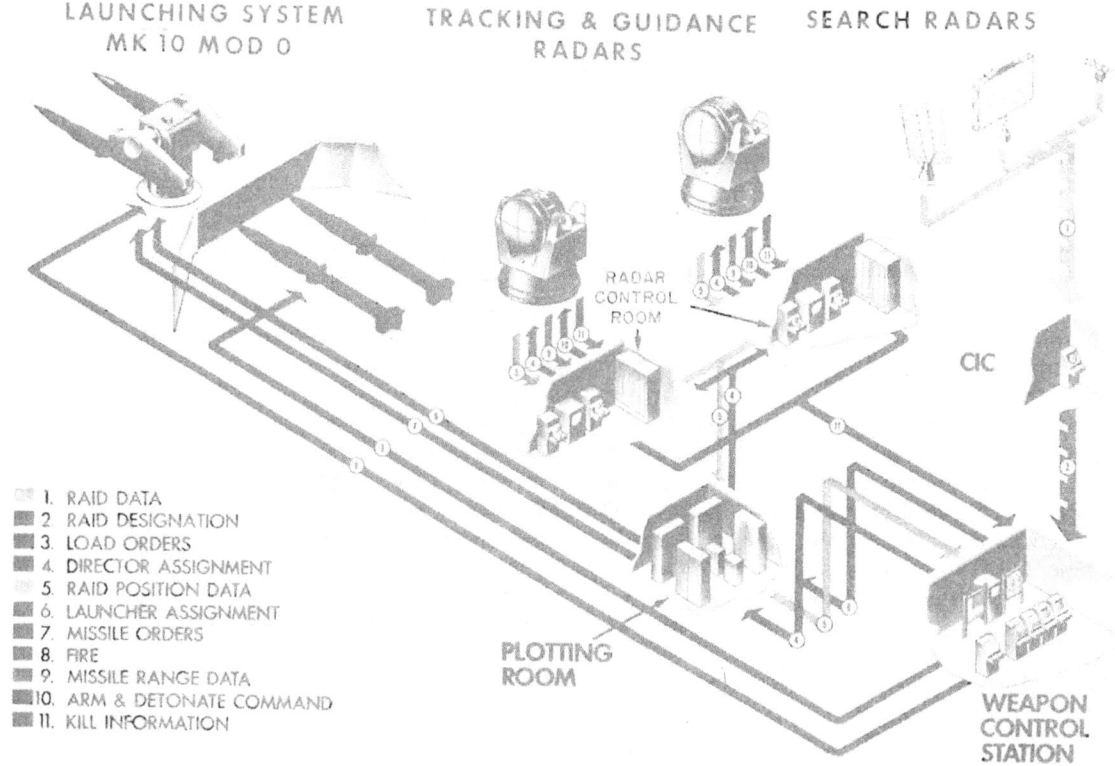

Figure 3-1.—Terrier weapon system on a CG; flow of information and directions.

## Chapter 3—PREPARING FOR A FIRING EXERCISE

## MANNING THE GMLS CONTROL PANELS

As a GMM2 you had to be able to take over the operation of any of the launcher control panels; to advance to GMM1 or Chief you should be able to supervise the panel operators and teach others the panel operation techniques. The number and types of panels vary with the system and the names may differ, but all systems have at least three types—

1. Power panels.

2. System control or launcher captain's panel.

3. Test panels.

Older systems had many small control panels for control of components of the system. In newer systems the tendency is to enclose several panels in two or three large panels. This requires fewer operators but the operators must be alert to many things. The small individual panels are used mostly for unloading and maintenance.

The dud-jettisoning panel is associated with the launching system controls. It is discussed in the next chapter.

### Power Panels

All launching systems have one or more power panels by which the system is connected to ship's power. All of the electrical power is supplied from the ship's electrical system, but the voltage has to be stepped down for many applications. The power panel contains circuit breakers, contactors, and overload relays for the launcher power drives, missile warmup, train and elevation motors, blower motors, and loaders and/or feeders.

In automatic/step operation the power panels are activated from the launcher captain's station.

### Launcher Captain's Panel

Figure 3-2 shows one type of launcher control panel. The launcher control panel will vary with mark and mod of the launching system, the type of shipboard installation and other factors, so it is not possible to tell you just which buttons to push. After you have turned on the power, set the EP2 panel (EP3 on the Mk 11 launching system) on STANDBY and watch and listen for signals that will indicate what to do next. An ALERT signal from the weapons control station will cause a flashing signal on all the launcher system panels and also will give an audible signal. When all the panel operators have set their panels on READY, the signal goes to the launcher captain's panel who sets his panel on READY, which signals WCS that all parts of the launcher system are ready.

Four types of orders are transmitted from the weapons control station to the launching system, and these go through the launcher captain's panel:

1. Missile order—determines the type of round(s) to be loaded.

2. Load select order—distinguishes between simultaneous operation of A and B sides or separate operation of either side.

3. Loading order—distinguishes between hold, single, or continuous loading of the type of missile ordered.

4. Unloading order—distinguishes between unload launcher or unload assembly area.

Pushing the correct buttons on the panels causes the launching system to load automatically the type of round ordered. A few manual operations are required and these differ with the type of round. For example, the Terrier BT-3A(N) requires actuation of the safety switch actuator. All Terrier rounds require installation of booster fins and unfolding the missile's (tail) control surfaces, followed by operation of the safety foot switch by the assembler at each station.

MISSILE ORDER.—The type of round(s) to be loaded will be ordered by the WCS when the system is activated. Although verbal orders may be given in some systems, the order is usually

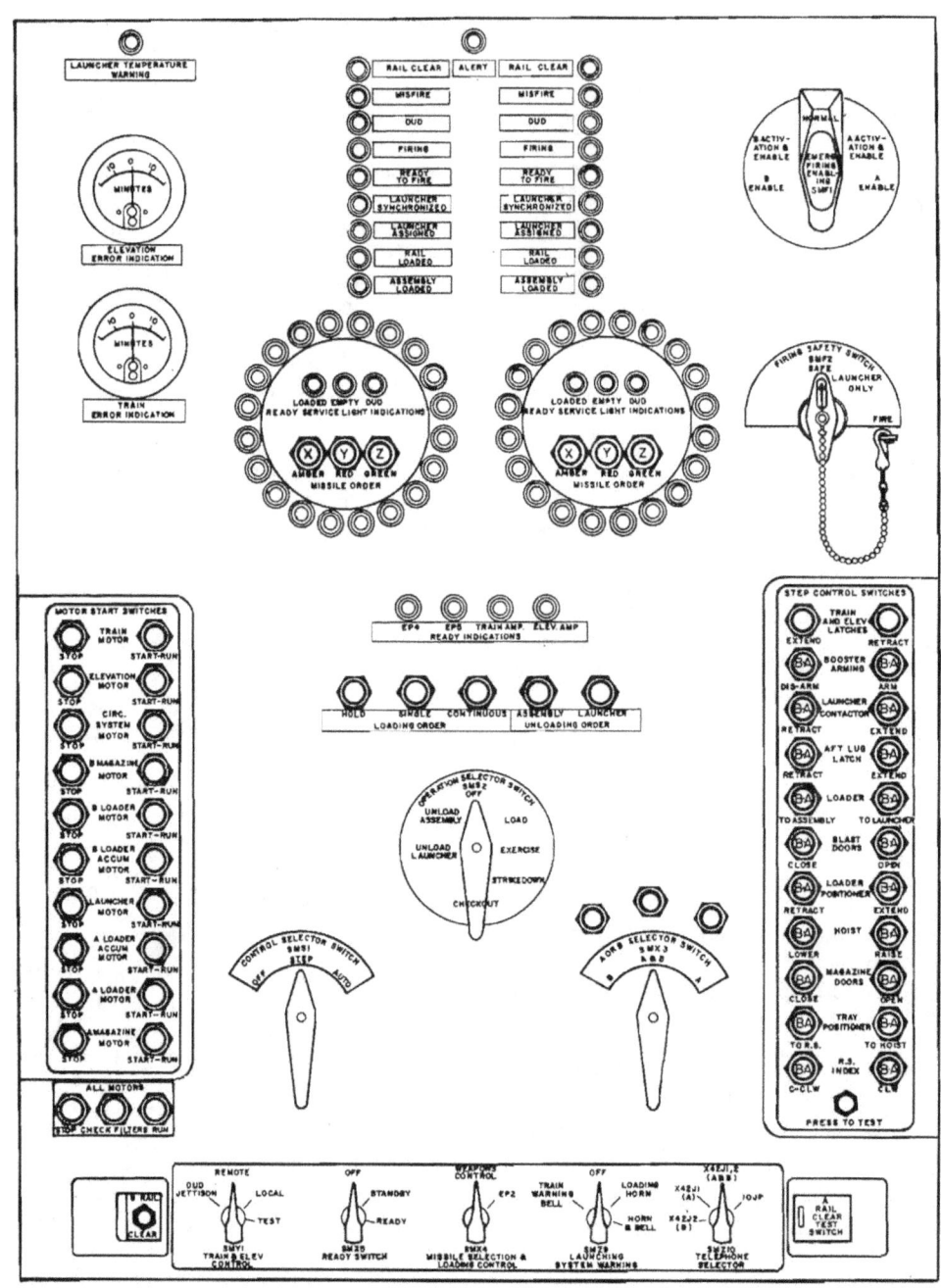

Figure 3-2.—Launcher control panel, EP2 for the Mk 10 GMLS.

## Chapter 3—PREPARING FOR A FIRING EXERCISE

indicated by signal lights on the EP2 panel (IP5 in Mk 11 systems). The operator of the EP2 panel pushes the button that will cause the selected type of round to be indexed to the loading position. If the HOLD order comes, the round is held at the hoist position until the next order is received.

LOAD SELECTOR ORDER.—At the weapons control station the load select switch is set to either SINGLE or CONTINUOUS, the corresponding light signal on the EP2 panel lights up, and the loading operation is started.

LOADING ORDER.—If the loading order is CONTINUOUS, the system will continue to select the same type of round and hoist it to the loader rail each time the empty loader pawl returns to the load position. On SINGLE, one missile will be hoisted and loaded on the launcher. On HOLD, the launching system is held in READY condition but no round is loaded.

UNLOADING ORDER.—When rounds are to be returned to the magazine, WCS will indicate UNLOAD LAUNCHER, or UNLOAD ASSEMBLY, which will cause a corresponding light on the EP2 panel to be illuminated (flashing). The launcher captain will then initiate automatic unloading operations.

The order for cessation of operation of the launching system is transmitted from the weapons control station to the launching system captain via signal lights on the EP2 panel or by sound-powered phone circuits.

Look again at figure 3-2 and note the designations of the pushbuttons and lights. Some panels have more buttons and lights than the model shown. The operator of the launcher captain's panel has to be alert to everything that is taking place in the launching system. The operator needs to know the system and mentally picture what is taking place on the launcher as each signal lights up or when a button must be pushed in response to orders from the weapons control station. In remote operation the launcher slews to position in response to train and elevation orders from the computer in missile plot. (In local operation, train and elevation orders have to be set in at the EP3 test panel (EP5 for Mk 11).) The firing key is on the evaluator console (EC) or the weapons assignment console (WAC) in the weapons control station. The EC or WAC operator does not close the firing key until all applicable status lights for the type of missile to be fired are received at the console. A few examples are launching system ready, warmup power on (BT/HT only), launcher synchronized, ready to fire, etc. Also, depending on the tactical situation, a salvo may be ordered by the EC/WAC operator.

TEST PANELS.—One test panel is used for both port and starboard components of some systems, while other systems have separate test panels for port and starboard. The test panel is used only during launcher test or local control operations and is unmanned during automatic loading operations. The test panel contains switches, synchros, and connections required to perform complete tests on the train and elevation systems. Auxiliary equipment connected to the EP3 panel for testing includes dummy directors, signal generators, and oscillographs. Separate limiter demodulators are used in conjunction with an oscillograph. Launching systems using an error recorder have built in limiter demodulators within the recorder.

The EP3 panel of the Mk 13 launching systems is used during local control of the launcher. At each panel, checklists should be posted for each type of procedure. Figure 3-3 shows part of a checklist posted at the EP2 panel of a Mk 13 Mod 0 launching system. Use the checklist as a verification that all steps are performed in the correct sequence each time the launching system is operated.

OTHER CONTROL PANELS.—Figure 3-4 shows an example of the assembly captain's control panel EP4 on the Mk 10 Mods 0, 5, and 6 GMLSs. Except for A-side and B-side switch and lamp designations, the EP4 (A-side) and EP5 (B-side) panels are identical. These panels are the assembly, strikedown, and checkout control panels for the launching system.

The EP4 and EP5 panels have a key-operated safety switch. When this switch is at the SAFE

## GUNNER'S MATE M 1 & C

At EP1 panel:

1. 440 Volts Power-On lights .................................................. On
2. All circuit breakers ........................................................ ON
3. All switches ............................................................... ON
4. All Power-Available lights .................................................. On
5. All Fuze-Blown lights ....................................................... Off

At EP2 panel:

1. Man station; plug headset into receptacle at right-hand side of panel.
2. SMZ4 (Telephone Selector switch) ......................................... 10JP
3. SMW1 (Missile Warmup Selector switch) .......................... Required position
4. Warmup Status indication light ............................................. On
5. Toggle switch in circular light pattern .................................. NORMAL
6. Report from Safety Observer .......................................... "All Clear"
7. SMS1 (Control Selector switch) ............................................ STEP
8. SMS2 (Operation Selector switch) ........................................... OFF
9. SMF2 (Firing Safety switch) ........................................ FIRE ENABLE
10. SMY1 (Train And Elevation Control switch) ............................... REMOTE
11. SMX4 (Loading Control switch) ................................... WEAPONS CONTROL
12. Ready Indications (3 lights) ................................................. On
13. SMY2 (Launching System Warning switch) ..................................... BELL
14. Open left-hand switch cover
15. Start motors by depressing START-RUN pushbuttons.
    (START-RUN pushbutton lights) ............................................. On
16. All Motors CHECK FILTERS light ............................................ Off
17. All Motors Run light ....................................................... On
18. Open right-hand switch cover
19. Check Step Control Switches. If necessary, use Step pushbuttons
    to obtain following light indications:

    a. Dud Jettison RETRACT light ............................................ On
    b. Launcher Rail EXTEND light ............................................ On
    c. Train Positioner EXTEND light ......................................... On
    d. Elevation Positioner EXTEND light ..................................... On
    e. Arming DISARM light .................................................... On
    f. Fin Opener Cranks RETRACT light ....................................... On
    g. Contactor And Fin Opener Cranks DISENGAGE light ....................... On
    h. Aft Motion Latch RETRACT light ........................................ On
    i. Hoist TO MAGAZINE light ................................................ On
       (Hoist TO INTERMEDIATE and TO LAUNCHER lights off)
    j. Blast Door CLOSE light ................................................. On
    k. Ready Service Index CCW and CW lights ................................. On

20. Close right- and left-hand switch covers.
21. SMS1 (Control Selector switch) ............................................ AUTO
22. SMS2 (Operation Selector switch) .......................................... LOAD
23. SMS3 (Ready switch) ...................................................... READY
24. SMS1 (Control Selector switch) ............................................. OFF
25. SMS3 (Ready switch) .................................................... STANDBY

94.10

Figure 3-3.—Checkoff list, activation procedures, Mk 13 Mod 0 GMLS.

## Chapter 3—PREPARING FOR A FIRING EXERCISE

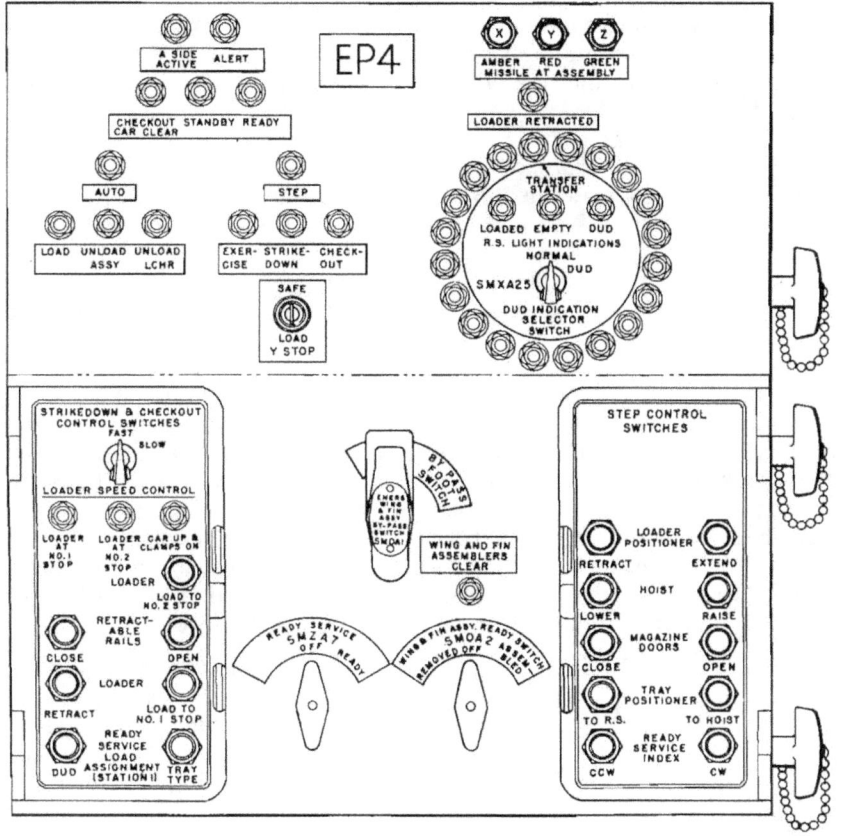

Figure 3-4.—Feeder control panel, EP4 for the Mk 10 GMLS.

position, a Y-type missile cannot be loaded onto the launcher because an interlock prevents the blast doors from opening. Other switches allow the EP4 and EP5 operators to change tray assignments on the ready service rings, to select a fast or slow loader speed, and to indicate the completion of missile assembly or disassembly. In addition, these panels have three safety switches mounted on the right side. These safety switches are for the magazine motor, the loader motor, and the loader accumulator motor. When in the SAFE position these switches prevent the launcher captain from starting the associated motors. After the assembly captain verfies that the assembly area and magazine area are clear, the switches are positioned to RUN to allow starting of the motors. While in the SAFE position, these switches may be pulled out of the panel to prevent inadvertent starting of the motors while personnel are performing maintenance or repair work on the system.

## MANNING OF THE GMLS STATIONS

Because much of the operation of a launching system is automatic, the number of personnel required is small. The number varies with the mark and mod of the system, the type of ship, the type of round, the type of warhead, and the type of operation (automatic, step control, or manual). The mark and mod of the launching system are related to the ship or class of ship on which the system is used. The Mk 10 Mod 0 is placed in DDGs; Mods 1 and 2 are on CGN-9; Mods 3 and 4 are on CVs; Mods 5 and 6 are on CGs and on CGN-25, and the Mk 10 Mods 7 and 8 are placed on CG-26 and later ships. Future changes, revisions, and modifications will assign new marks and mods.

### Checkout Area

Terrier weapons require mating of the missile and booster before stowage, unmating for offloading, and require a minimum of two people in the checkout area. The Tartar and ASROC arrive aboard completely assembled and are not disassembled. In preparation for firing, the checkout area is not manned. The Tartar systems, NATO Seasparrow surface missile system (NSSMS), Harpoon canister/launcher configurations, and the Mk 26 GMLS do not have checkout areas.

### Wing and Fin Assembly Areas

Terrier/Standard ER missiles require eight assemblymen (four for A- and four for B-side). On the Terrier BT/HT and Standard ER missiles, the tail control surfaces are folded (not removed) during stowage and need only to be erected at assembly. The booster fins are installed during assembly.

On Tartar weapons, the fins also are folded. They are erected automatically when the missile is on the launcher in automatic operation, and by pushing the FIN OPENER pushbuttons on the launcher control panel in step operation. When the Tartar missile is to be returned to the magazine, the fins have to be folded manually while the missile is on the launcher.

WARNING: Before folding missile fins, turn the firing safety switch on the the launcher control panel to SAFE, then remove the switch handle and give it to the crewmember going out to the launcher to fold the fins.

When the firing safety switch is set on SAFE, the launching system cannot be activated; removing the switch handle from the panel and placing it in the custody of the person working on the launcher prevents accidental activation.

### Magazines

The missile systems in present use have the mated or assembled rounds stowed in magazines in close proximity to the launchers, so the rounds can be transferred automatically from the magazine to the launcher. The magazines are unmanned in automatic and step operation. In case of power failure, manual operation is necessary. Manual operation is also used for maintenance, checking, and installation purposes. Handcranks are used instead of pushbuttons. When using handcranks, remember that the electric and hydraulic interlocks are ineffective. If the unit seems to bind or is difficult to move, stop cranking and investigate.

The only time personnel are permitted in a Tartar magazine is when it is deactivated, as during maintenance. No assembly is required on the missile and the control panels are outside and adjacent to the missile magazine.

## SAFETY

Safety checks have to be made frequently in all systems, methods, types or modes of operations, and areas of operation. Many safety devices are built into each system to safeguard the personnel working with it, and to prevent damage to the missiles, machinery, or ship. They were placed in the system because they are necessary. Do not deactivate or bypass a safety device for any reason, and do not permit your crew to do it. During preparation for firing a safety observer must be stationed in a position to observe the work, including the personnel and machines or equipment involved.

The checkoff lists posted at each station have the applicable safety precautions listed next to the operational steps to which they apply. The same safety precaution may be given

## Chapter 3—PREPARING FOR A FIRING EXERCISE

several times on one list each time it applies to a step in the operation to be performed. When you prepare a checkoff list, as you will be required to do, insert safety precautions in the same manner next to each step to which they apply. Some of the technical manuals place the safety warnings throughout the text, wherever applicable, and also include a summary listing of all the safety precautions given in the book. Review this safety summary frequently.

Commanding officers may issue additional safety precautions. Observance of the safety precautions is mandatory. If you fail to enforce safety rules, you can be held responsible.

**Types of Danger**

Dangers may be classified according to the type of material or the object causing the danger, such as machinery, explosives, gases, liquids, irritants, pressure, fire, or electricity. A material that is dangerous in itself may be used only under prescribed conditions or circumstances. The caustic electrolyte in missile batteries, for example, will always burn the skin, so the problem is simple—do not let any get on the skin. Safe use of other materials may require compliance with special conditions. Particular conditions necessary for each type of explosive used in missiles are described in *Gunner's Mate M 3 & 2*, NAVEDTRA 10199 series. You should know what types of explosives are used in the different parts of your missiles and the specific precautions for the handling of each type. As all the explosives used in the missiles are enclosed by some form of container, there is little danger of skin contact with explosives that can cause dermatitis. The bulk and weight of the units present more of a problem in safe handling.

MACHINERY.—Before any machinery is set in operation, the area must be checked to be sure no one is in a position to be injured by moving machinery. The launcher captain must sound the warning bell before pushing a button to activate any machinery, and the safety observer must warn away anyone seen in the area. Grisly experience has shown the need for this caution before operating any powered machinery such as a missile launching system.

It is the duty of the safety observer to immediately correct any violation of safety rules. If you are operating the launcher captain's panel, it is up to you to turn off the power if so ordered or when you see a situation that requires quick stoppage of any part of the launcher system.

Checking the launching system equipment and machinery for safe operating condition is part of routine maintenance performed by you and the personnel you supervise. The equipment is cycled without a load; any fault in the operation is corrected before the equipment is used with a load.

EXPLOSIVES.—The "safe-distance" lines are the equivalent of the danger circle painted on the deck around each launcher. Minimum criteria for the danger circle can be obtained from U.S. Navy Ordnance Safety Precautions, (OP 3347). Their purpose is to remind all personnel that the area is covered by movement of the launcher with loaded guide arms. Personnel inside these safety lines are in a danger area where they could be struck by the moving guide arms and/or missiles. Also present is the possibility of accidental ignition during mating, handling, strikedown, or strikeup operations. Remind your personnel, if necessary, never to place themselves where they would be in the path of the blast if the missile or booster were ignited.

The smoking lamp must be out at all times in missile handling and stowage areas.

The detonator in the safety and arming (S&A) device can be set off by a radar beam, static electricity, or a spark. The S&A device is not tested, disassembled, or repaired aboard ship.

NUCLEAR WEAPONS.—With nuclear warheads assembled into missiles, there is always the possibility of nuclear radiation. Probably the chief cause of nuclear incidents and accidents is careless handling, e.g., dropping a weapon with a nuclear warhead. Prevention, therefore, means making sure that the weapon is securely fastened to the hoist, crane, trolley, or other lifting machinery.

## Safety Provisions and Devices

Each launching system has numerous safety devices, some of them entirely automatic in operation. The position of some components or devices is checked on the control panels by means of lights or other signals. Because control panels may be widely separated and on different decks, telephone communications must be established between the panel operators, the safety observer, and the personnel at work in the different areas. A loudspeaker announcement and/or a warning bell should warn people to stay away from the topside loading area before operation begins. Only the persons actually needed for the work are permitted to remain there. (No persons are required within the launching system to operate it, and signs should be posted to keep personnel out of the assembly and checkout areas.) If anyting goes wrong with the launcher or other part of the system and someone has to work on any part to correct the fault, disable the component so no one can start it accidentally. On the power panel, turn off the switch that activates the component, remove the handle of the switch, and give it to the person who is going to do the repairs. Only after repair is accomplished is the handle returned to the switch. The unit can then be activated again and its operation tried out. No one may enter a magazine while a loading or unloading operation is in process. If you are in charge at a control panel, check carefully before you push the button that starts the machinery moving. The operators at their control panels signal and the launcher captain when their respective parts of the system are ready.

The safe-fire switches must remain at the LAUNCHER ONLY (figure 3-2) position throughout all the daily operational checkout. There are safety switches for the magazine, the loader, and the loader accumulator. Each of these switches has two positions, SAFE and RUN. When in the SAFE position, the handle can be removed, to be retained by the person doing the maintenance on the equipment until the work has been accomplished.

After receiving the order to do so, and before activating a launching system, the operator of the EP2 panel must receive the "all clear" report from the safety observer, who is stationed where the whole launcher area can be observed.

Lights on the operating panels indicate various conditions that need to be checked and corrected before proceeding. A light on the EP1 panel, labeled with a warning sign, indicates that there is a blown fuse that deactivates the magnetic door lock. This must be corrected at once. A monitor (ground detector) on the 115-volt power supply triggers an alarm if there is a grounded circuit (Mods 7 and 8 only).

The firing cutout cams are designed for each installation of a launcher so the launcher cannot be fired into any part of the ship. The installation is tested and checked at the shipyard and rechecked and tested after any change or modification of the launcher. The firing cutout cams are also checked during prefiring checks using the applicable maintenance requirement cards (MRCs). The positioning of the launcher in response to train and elevation orders is checked each time the launcher is used during training, preparation for firing, or during and after maintenance work. The angle of train and elevation necessary for target intercept is calculated by the computers in the weapons control station from the data obtained by radar or sonar tracking of the target and the computations made in CIC. In automatic operation the train and elevation synchro signals cause the launcher to slew to the position ordered. In local control the launcher is positioned in train and elevation at the launcher test panel.

Train systems and elevation systems contain similar electric, hydraulic, and mechanical equipment. Each system receives and responds to order signals independently of the other. In normal operation remote orders are supplied by the fire control computer. These signals determine the flight path of the missile during the "boost phase" of its flight. The power drive systems of different launchers are very similar in operation. The principles of operation are the same in all of them.

The voltages used by train and elevation systems are dangerous, and may be fatal if

## Chapter 3—PREPARING FOR A FIRING EXERCISE

contacted. If electircal trouble develops, consider all circuits dangerous until the trouble is located and corrected.

## FIRING EXERCISES

The steps in preparation of a launcher for firing of a particular launching system vary with the mode (surface-to-surface, surface-to-air, surface-to-underwater), the type of ship on which it is installed, the type of missile (conventional warhead, nuclear warhead), the purpose of firing (intercept, destruction, etc.), and other factors. Preparing for an exercise firing may require a number of steps that are different than those required when preparing for a live shot. Considering the different possibilities, a complete and exact list of steps in preparation for firing cannot be made to cover all situations. The checkoff list you prepare must be made to fit your launching system and must be complete in detail.

The following are examples of items that may be listed on your checkoff list.

1. Perform a daily systems operability test (DSOT) within prescribed time limits before the exercise.

2. Check firing cutouts using applicable MRC.

3. Check all accumulator pressures.

4. Test fire dud-jettison units and recharge.

5. Lower lifelines.

6. Ensure dud/misfire procedures are posted near launcher captain's panel.

7. Hold a safety briefing for the launching system crew and safety observer.

## TERRIER GMLSs

The Mk 10 GMLS is the only Terrier system installed on board active duty surface combatants at the present time. Mods of the Mk 10 launching system in current use are Mods 0 through 8. The dual Mods 1 through 6 differ only in the adaptations necessary for use on opposite sides of the ship, or fore and aft location on the ship. Mod 0 is used on DDGs; Mods 1 and 2 on CGN-9, (figure 3-5); Mods 3 and 4 on CVs or CVNs; Mods 5 and 6 on CG 16-class and Mods 7 and 8 on CGs 26-class and later.

### Mk 10 Mods 0 Through 6

In the Mk 10 each magazine contains two ready service rings that holds 20 weapons in each ring. The ready service ring is rotated automatically to bring the selected weapon to the loading or No. 1 position, from which it is raised to the assembly area by the hoist. The loader-positioner rams the missile into engagement with the loader rail, the empty hoist lowers, and the magazine doors close. While the wings and fins are being assembled by the assemblymen and warmup is applied to the missile (BT/HT only), the ready service ring indexes another missile to the No. 1 position.

When the wings and fins are assembled, the eight operators move to a safe area and depress foot switches indicating that assembly is completed. This illuminates a light on the assembly captain's panel; the assembly captain, in turn, operates a switch which indicates to the launcher captain the assembly is completed. On the Mods 5 and 7 launching systems, the tilting rail will raise upon completion of assembly. If the missile is the correct one, it is positioned on the launcher guide arm by the loader. If the missile is not the one ordered, a flashing indication appears on the EP4 or EP5 panel; this situation must be corrected immediately. Then the launcher aft-shoe latches extends, thereby transferring the missile from the loader pawls to the aft-shoe latches. The launcher contactors extend and continue warmup of the missile after the loader pawls retract. When the booster contactors are fully extended, the arming tools extend. After the loader pawls have retracted clear of the spanning rails, the spanning rails retract and the blast doors close. When both blast doors are closed, the train and elevation latches retract and the launcher synchronizes

# GUNNER'S MATE M 1 & C

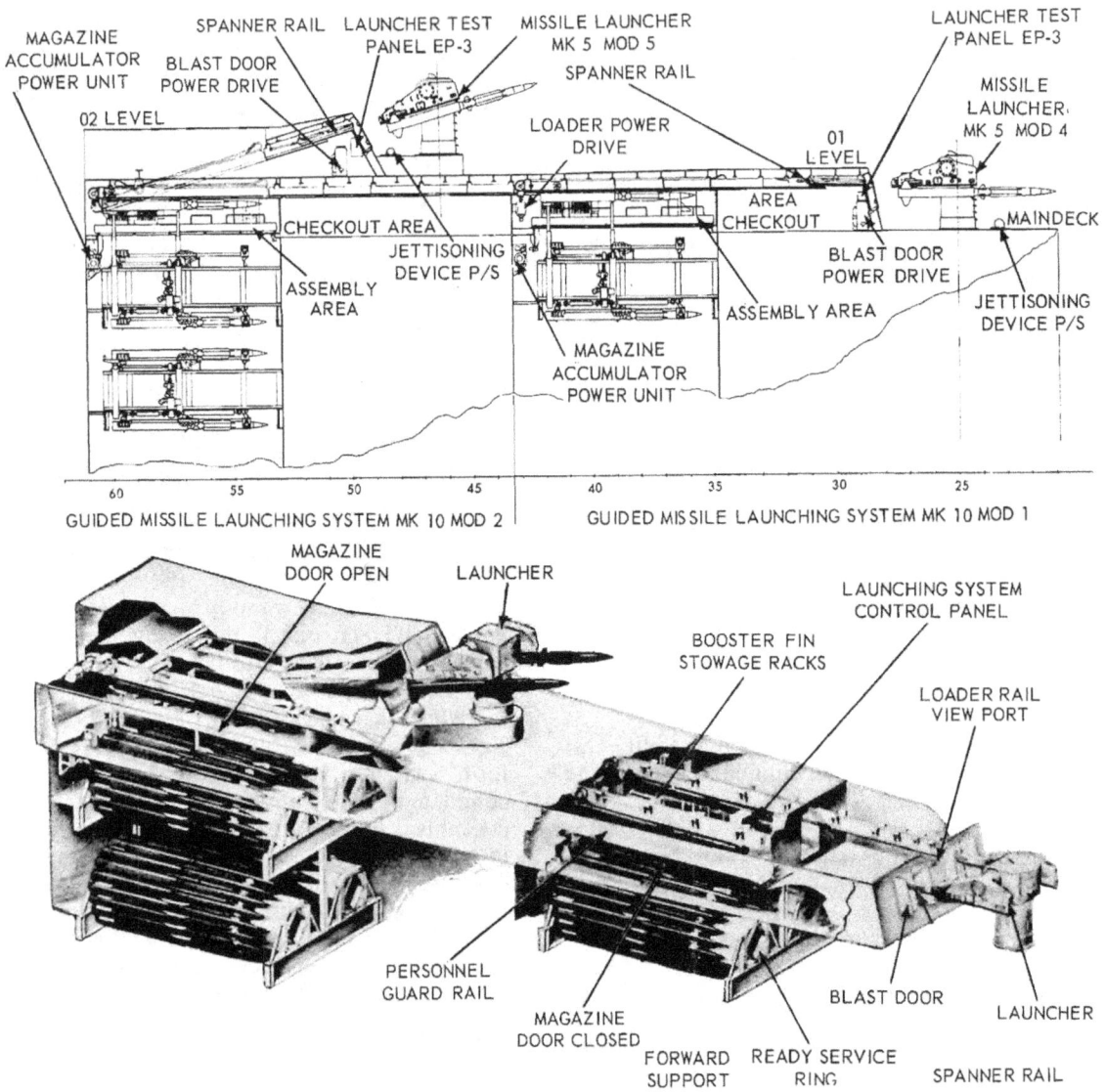

Figure 3-5.—Mk 10 Mods 1 and 2 GMLSs.

with the remote order from the assigned director. The loader pawls and loader positioners move back to position, ready to receive the next weapon from the hoist. Before the launchers slews to the ordered train and elevation positions, the panel operator must receive an "all clear" from the safety observer to be sure that no one is in the path of the launcher.

The firing safety switch is on the launcher captain's panel, and is operated as ordered from the weapons control station. If the weapons

## Chapter 3—PREPARING FOR A FIRING EXERCISE

control operator has the load-selector switch on CONTINUOUS, the weapons are hoisted each time the loader returns to the assembly area, and the loading and firing sequence is repeated. If the load switch is on SINGLE, only one weapon will be loaded until further orders from the weapons control station.

**Mk 10 Mods 7 and 8 Launching Systems**

The outstanding innovation in the Mods 7 and 8 is provision for stowing ASROC weapons alternately with Terrier missiles in the Terrier ready service magazine. *Gunner's Mate M 3 & 2*, NAVEDTRA 10199-C has several pictures and a description of this arrangement. OP 3114 PMS/SMS is the OP for the Mods 7 and 8 systems.

The Mk 10 Mod 8 guided missile launching system is an aft installation aboard a CGN 35-class ship. The loading and firing operations are identical with the Mod 7 system. However, the Mod 8 has no tilting rail in its feeder system. The increased length of its loader rail causes the load and unload cycles to be somewhat longer than on a Mod 7 system. The absence of the tilting rail also affects step operation. There are no pushbutton switches for tilting rail raise and lower on the EP2 panel; but there is an 8-second delay before the blast doors open, which gives time for the gases to clear after missile firing. The ready service mechanisms are identical to those of the Mod 7, having two upper ones, each designed to hold 20 weapons, alternating Terrier and ASROC weapons, and one lower ring that holds Terrier weapons only.

**TARTAR GMLSs**

Launching systems used with Tartar missiles are the Mk 11 Mods 0, 1, and 2; Mk 13 Mods 0, 1, 2, 3, and 4, and Mk 22 Mod 0. In each of these systems the magazine is a compact metal structure and the launcher is placed on top of it. As the missiles are stowed completely assembled, there is no need for an assembly or a transfer room.

The following is a list of the Tartar GMLSs and mods that are placed on surface combatants:

1. Mk 11 GMLS

   Mod 0—DDG-2 thru 14

   Mods 1 and 2—CG 10-class (tentatively scheduled for decommissioning)

2. Mk 13 GMLS

   Mod 0—DDG-15 thru 24

   Mod 1—DDG-31 thru 34

   Mod 2—(Ships decommissioned)

   Mod 3—CGN-36 and 37

   Mod 4—FFG 7-class

3. Mk 22 GMLS

   FFG-1 thru 6

4. Mk 26 GMLS

   Mods 0 and 1—CGN 38-class and AVM-1

The Mk 13 launching system has a single-arm launcher in which the dud-jettison unit is integral with the launcher arm, a missile magazine, and missile launching system control equipment. The Mk 11 is a twin-rail launcher designed to handle two missiles simultaneously.

In automatic control the launching system control initiates and controls the loading cycle, but the launcher is positioned and the missile is fired by the ship's fire control system.

**Mk 11 GMLS**

The Mod 0 is in use on DDG 2-class ships, and Mods 1 and 2 are installed on CG 10-class ships. Panel designations on the Mk 11 system are different than in other systems and this may lead to confusion if you have become accustomed to other systems. On the Mk 11 the

GUNNER'S MATE M 1 & C

EP2 panel is the loader control panel and the EP3 is the launcher control panel. The Mk 205 Mod 3 (RC1), Mod 4 (RC2), and Mod 5 (RC3) relay control panels are used in the Mk 11 Mod 0 GMLS.

If the launcher is to be trained or elevated by local control, the operator of the launcher uses the Mk 211 local control panel (EP5). The EP5 panel contains train and elevation synchros for local operation of the launcher. On the panel are mounted two error (lag) meters (for train and elevation), two synchro control turning knobs for local control of the launcher in train and elevation, coarse and fine indicator dials for train and elevation positions, and an alarm switch to warn personnel that the launcher is to be operated.

The Mk 11 system has other special indicating panels—the IP1, IP2, and the IP5.

The Mk 209 Mod 0 magazine loading indicator panel (IP1), located above the loader control panel (EP2), is an illuminated replica of the launcher, magazine, and magazine cover. It indicates the launcher position, the magazine cover, and blast door positions in relation to the launcher, and indicates whether cells are loaded or unloaded (figure 3-6). The blast doors are

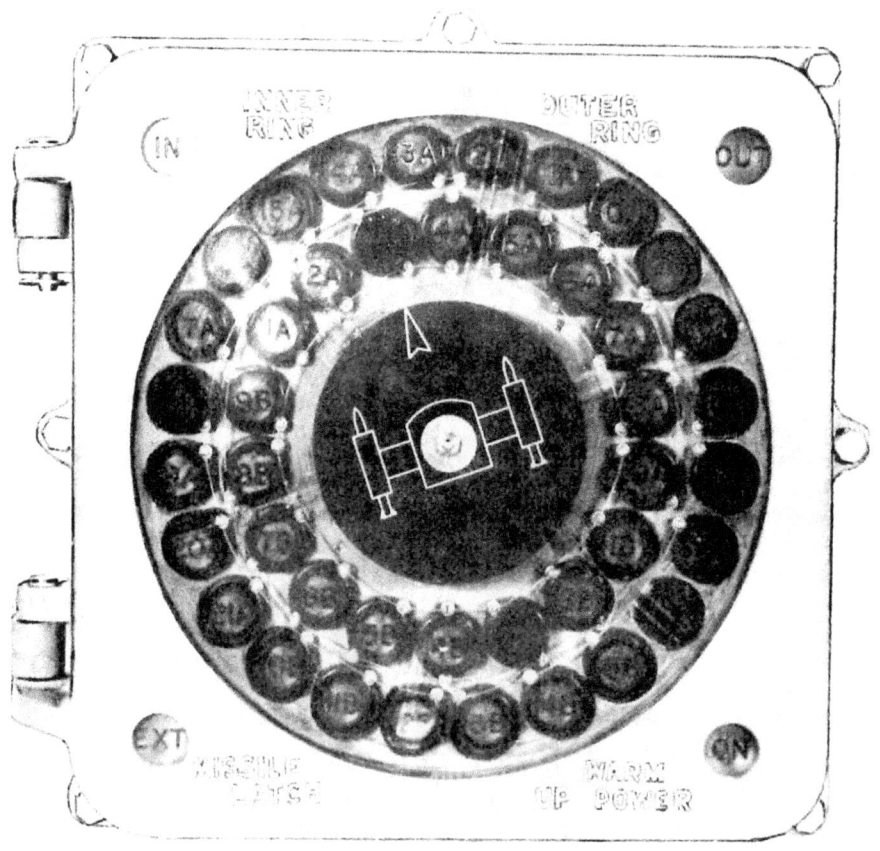

Figure 3-6.—Magazine loading indicator panel (IP1) for Mk 11 GMLS.

94.16

# Chapter 3—PREPARING FOR A FIRING EXERCISE

represented by four red colored discs. The numbered discs in the illustration are the indicating lights for the individual missile cells. The two upper corner lights indicate whether loading operation is from the inner or outer ring section of the magazine. The two lower corner lights indicate if the warmup power is on, and whether the missiles are latched or unlatched.

The missile latch indicator panel (IP2) is located in the open center of the missile magazine. It is used chiefly during missile replenishment and readying for sea. As the missile latches cannot be checked visually to be certain the missiles are secured in the magazine, the unlatched missiles are indicated on the panel. This is an unsafe condition and must be corrected, as the missiles could slide (vertically) in the magazine. Of course, the launching system must be deactivated while anyone is down inside the magazine.

Magazine mode and guide type indicator panel (IP5) displays the following.

1. Magazine mode.

2. Type missile on guide arms.

3. T-SAM test mode.

4. T-SAM on guide arms.

5. Missile type selected not available.

6. Mixed load warning.

## Mk 13 GMLS

The Mk 13 Mod 0 launching system follows the trend in combining several control panels into one. This sytem has three control panels—EP1, power control; EP2, launcher control; and EP3, test panel. The local control panel is made a part of the EP3 panel. In automatic operation the launcher captain operates the EP2 panel; EP1 and EP3 are unmanned. The safety observer watches the launcher area and keeps the launcher captain informed on all phases of launcher operation.

The compact unit construction of this launching system makes it usable on a variety of ships. It may be mounted entirely abovedecks, or the magazine may be placed below deck level. The most noticeable difference between the Mk 11 and Mk 13 systems is the difference in the launchers—Mk 11 has two launcher arms and the Mk 13 has only one (figure 3-7). This, of course, eliminates all operational steps that involve loading, unloading, or jettisoning for a second side. The Mk 13 is an extremely high speed system. The steps in operation are very similar for the Mk 11 and Mk 13 systems. In automatic control there are four steps:

1. Type select. Warmup (IT/ITR only).

2. Loading.

3. Assignment.

4. Firing.

Steps 3 and 4 are performed by remote control from the weapons control station.

WARNING: Do not energize the launching system until communications have been established between the safety observer and the launcher captain, and the safety observer has reported that the launcher area is all clear.

As soon as the launching system is activated after receiving the order from the weapons control station, warmup power is applied to four missiles, IT/ITR, in the magazine of the Mk 11 system, and in the Mk 13 system, to three missiles in sequence, not simultaneously (figure 3-7). Missile warmup selector switch SMW1 can be positioned to AUTO 1, AUTO 2, or AUTO 3. The position of SMW1 determines how many missiles are put on warmup in the magazine. For normal automatic operation, AUTO 3 is selected. As missiles are loaded, warmup is applied to succeeding missiles in the magazine. If for any reason the first missile is not loaded within 14 minutes after start of warmup, this missile is automatically removed from warmup and another is placed on warmup. The warmup status light on the launcher control panel will turn red after 15 minutes. The circuit for the

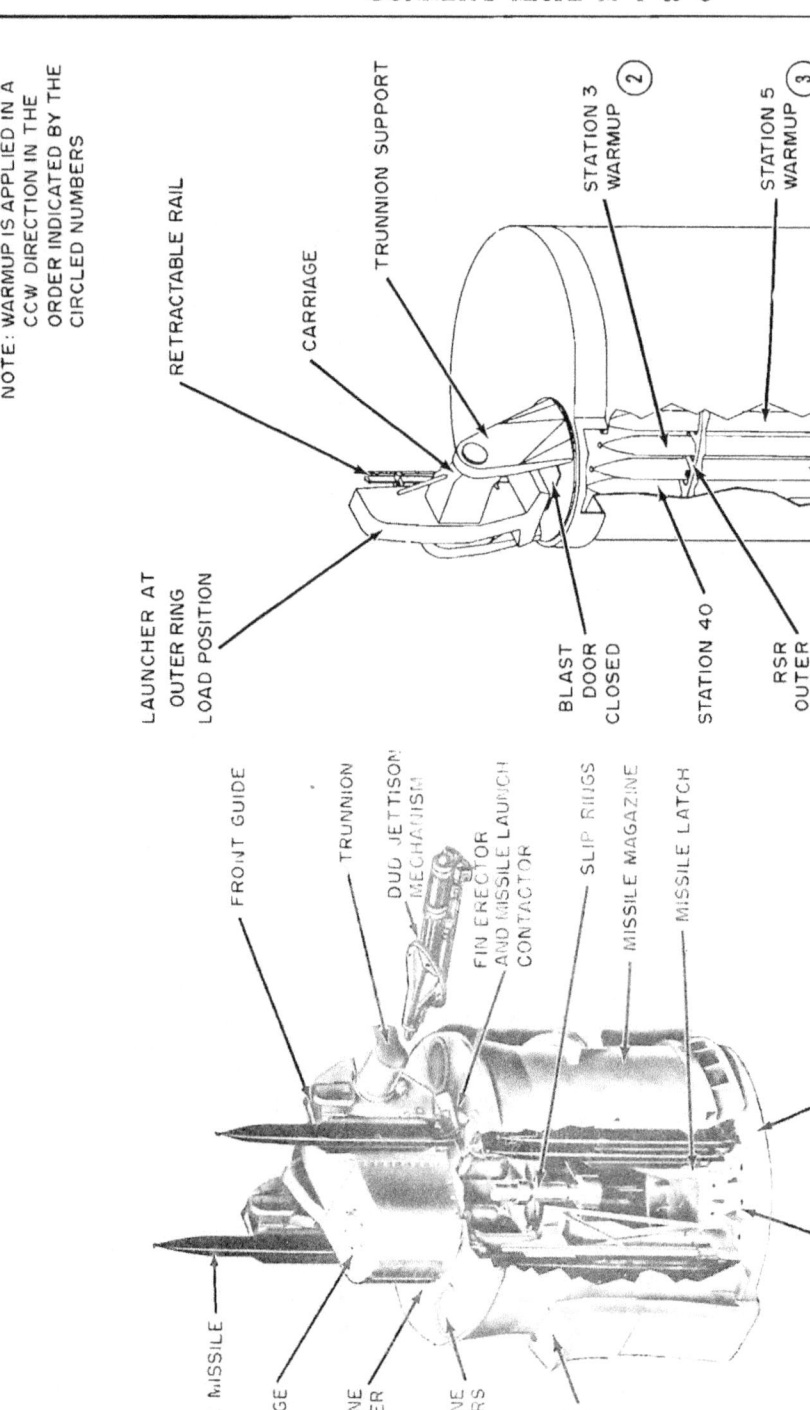

Figure 3-7.—Types of Tartar GMLSs: A. Mk 11 (two launcher arms); B. Mk 13 (one launcher arm).

## Chapter 3—PREPARING FOR A FIRING EXERCISE

application of warmup power is established during strikedown; a warmup contactor enters a socket in each missile as it is placed in a cell in the ready service ring.

NOTE: Standard MR missiles do not require warmup.

The operators at the control panels push the buttons in sequence according to the chart posted at each station and in response to orders from the weapons control station. In the Mk 13 system the EP2 panel operator takes care of all the pushbuttons, but on the Mk 11 system the work must be coordinated between the EP2 and EP3 operators. The safety observer keeps in contact with the panel operators so any part of the system can be stopped quickly if necessary.

If the load order (from the weapons control station) is for continuous loading, the launching system will continue to load missiles until the magazine is empty (of the type ordered), beginning with the outer ring of missiles (if the outer ring was initially selected) and, when those cells are empty, loading missiles from the inner ring. If the inner ring was initially selected, loading will automatically be shifted to the outer ring when the inner ring is depleted. No ITR missiles will be stowed in the outer magazine ring of the Mk 13 GMLSs after completion of surface to surface Standard missile (SSSM) ARM ORDALT 8217. Also, the Mk 13 GMLSs will receive SSSM Active capability upon completion of ORDALT 8239. The Mk 11 GMLSs will receive only the SSSM Active capability upon completion of ORDALT 8204. ARM missiles will receive warmup by segments in the Mk 13. Warmup is applied to the missiles automatically at the successive stations in the outer ring (figure 3-7), and in the inner ring (or in the oppostie order if the inner ring was selected first). This assumes that every cell is loaded and there are no dud missiles among them. If the load order is for one missile, the launching system will stop after one missile has been loaded. With the Mk 11 system a two-missile salvo may be ordered.

As soon as the panel operator receives the load order, the magazine cover rotates to the missile selected, and the launcher is synchronized to the loading position for inner ring or outer ring position (Mk 11 only). If the missile selected is not at the hoist position the ready service ring rotates clockwise to bring the missile to that position (Mk 13 only). After the minimum warmup time has been applied to the missile, hydraulic power is transferred to the hoist and the hoist rises to the intermediate position. As the missile is moved from station at the hoist position the warmup contactor at the base of the missile breaks contact.

At the intermediate level the hoist pawl engages the missile aft shoe and the magazine retractable rail extends to complete the missile track to the spanning rail, which is attached to the blast door. Then the blast door opens extending the spanning rail, the elevation positioner extends into the open blast door, and the hoist with the missile raises to the launcher. When the loader hoist completes its raise cycle, the launcher aft motion latch secures the missile to the guide arm and the hoist returns to the magazine. The launcher warmup is again applied. The fin openers engage the fins on the missile for opening. The train and elevation positioners retract and the blast door closes, retracting the spanning rail.

When the missile is in position on the launcher, the missile aft shoe contacts the forward motion latch and at the same time actuates the rail-loaded indicator plunger. This plunger actuates the launcher rail-loaded switch which lights up an indicating light in the weapons control station and the launcher captain's panel. The weapons control station assigns a target to the launcher, at which time the missile fins are automatically extended. The launcher slews to the train and elevation positions ordered. The automatic tracking cutout system prevents the launcher from pointing into certain areas where a fired missile would be hazardous to the ship's structure. The cutout system opens the firing circuit when the launcher points into an area unsafe for missile firing (nonpointing zones). The launcher synchronized light on the launcher control panel and an indicator in the weapons control station show when the launcher guide and carriage are positioned so the missile can be launched in the proper flight attitude (azimuth and elevation). The blast door must remain closed and the fins on the missile must be unfolded before the ready-to-fire signal is given. Unfolding of the fins can be going on while the launcher is moving in train and elevation to the ordered and corrected position. The fins unfolded light on the launcher panel goes on when the fins are unfolded.

All of these actions of the launching system components should have taken place in less than 6 seconds from the time minimum warmup elapsed in the magazine to launcher synchronization. A prefiring evaluation is made by the launcher captain and the weapons control officer. The firing safety switches must be closed at the launcher control panel and at the safety observer's position. The firing zone clear light must be on (launcher control panel). Launcher warmup must have been applied after the fin opener was engaged and the launcher was assigned. Time delay relays close after the minimum time has elapsed. The launcher warmup switch on the power panel must be on for the minimum number of seconds. The code set in the missile must correspond with the code of the assigned fire control system. The CODE SELECTED light on the launcher panel will go on if the codes match.

Only if all conditions are met, will the missile firing be ordered. The ready-to-fire light will go on in the weapons control station and on the launcher control panel. The fin opener and contactor must be engaged at the time the firing key is pressed in the weapons control station. The circuit to the hot gas generator squibs (in the missile) leads through the launcher contactor. After the hot gas generator squibs have fired, the contactor and the fin opener cranks retract and all circuits through the contactor are broken.

Beneath the closed blast door the hoist has been lowered, and the ready service ring has indexed another missile to the hoist position. Warmup has been applied and the missile is ready to be loaded on the launcher. If only one missile was ordered, no further loading takes place.

This assumes that every valve, switch, etc., works perfectly. If any part fails to perform as expected, repairs must be made by the GMMs. As most of the smaller ships have only one launcher, a failure can be a critical matter. You need to become thoroughly familiar with the system on board so you can locate trouble quickly and remedy it. It is expected that application of the planned maintenance system will reduce such failures to a minimum.

A number of changes have been made to improve the performances of the Mk 13 Mods 1, 2, and 3 launching systems. The base structure of the magazine is completely redesigned. The water injectors (see chapter 7) have been extended below the bottom plates of the base structure. The missile restraint rings now have vertical mounting brackets and are made of heavier material. The magazine rail assembly in each cell now has a latch lock on the magazine rail latch to prevent the latch from being jarred open. The hoist assembly has changes in the hoist pawl unit, the curved track assembly, the retractable rails, and the retractable rail valve blocks. A hand-operated, nitrogen-booster pump has been added to boost the ship-supply pressure for charging the jettison accumulator. It is mounted inside the stand assembly just below the center hatch (figure 3-8).

Pressure-cutout switch assemblies and their associated valves and orifices have been relocated from the safety relief valve to the tank cover of the header tank for the train and elevation drives, and the header tank of the magazine power supply.

The launcher guide, too, has some changes. These are changes in the forward motion latch and lock, the igniters, the fin openers, and the fin opener and contactor assemblies. A key-operated lock on the arming device permits the launcher captain to lock the arming device as a safety precaution during checkout of missiles on the guide arm. The forward motion stop latch has been changed from a movable stop to a fixed one. The igniters have been modified so they contact the missile only when the missile is armed. The changes in the fin opener and contactor assemblies are minor and do not

## Chapter 3—PREPARING FOR A FIRING EXERCISE

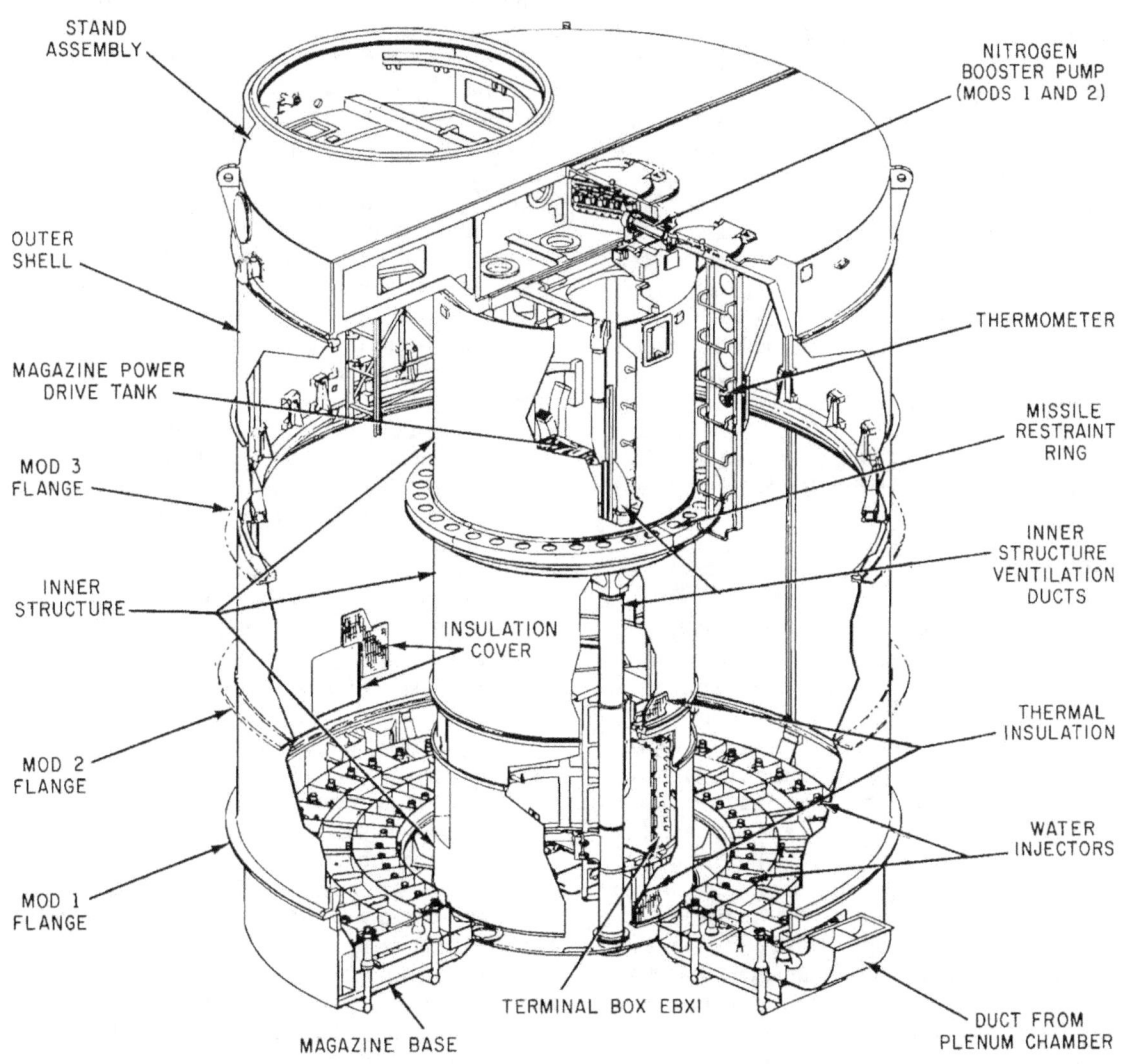

Figure 3-8.—Magazine structure, Mk 13 Mods 1, 2, and 3.

change the operation of the assemblies. The fin opener housing is slightly smaller and shaped slightly differently than that on the Mod 0.

The principal change in the train and elevation systems is the redesign of the electronic servocontrol units. There are also some changes in the train and elevation drive motors, the servo and supercharge hydraulic systems, and in the receiver regulators.

Some of the changes were necessary because of the larger size of the Standard MR missile. The magazine has been modified to allow stowage of seven different types of missiles at a time instead of X-Y-Z which no longer exists.

The missile identification system allows identification of up to 64 different types of missiles, however as stated before only seven types can be loaded into the magazine at a time. The same holds true for the GMLS Mk 11. The missile identification box (MIB) has a two digit display that indicates the numerical code of the missile and several light emitting diodes (LED) to indicate idenfification status. To assign the missiles to their stations in the magazine, the EP2 operator unlocks the missile type assignment switch cover and assigns each empty station to the missiles to be onloaded. Loaded stations are not changed by the operator.

The differences between the Mk 13 Mod 4 and the Mk 13 Mods 0 through 3 GMLSs are covered in chapter 5 of *Gunner's Mate M 3 & 2*, NAVEDTRA 10199-C and will not be discussed in this text.

**Mk 22 GMLS**

The Mk 22 guided missile launching system is installed on small ships (FFGs 1 through 6) where space, weight, and other considerations require a smaller and lighter system than the Mk 13. The Mk 22 is an extremely compact single-arm launching system designed to stow, load, and fire Tartar missiles, and may be adapted for handling, loading, and firing other missiles. It is attached to the ship's structure with a single mounting ring like that of the 5"/54 Mk 42 gun mount. The missiles are stowed vertically in a single ready service ring, which is nonrotatable. The launcher rotates to the loading position over the selected cell. Figure 3-9 shows structural elements of the system in a cutaway view. The train/hoist and elevation power drives and their associated receiver-regulators and miscellaneous controls are supported on the launcher's center column. The control panels are remotely located.

The launcher is bearing-mounted to the upper magazine section, and its base ring forms the top of the magazine. The launcher arm assembly provides the guide rail and latches which support and secure the missile, the fin erectors which unfold the missile tail fins, the launcher-to-missile electrical connector which feeds the prefiring intelligence to the weapon, mechanical input to arm the rocket motor, and firing contacts to ignite the rocket motor. The rail guides the missile for the first 20 inches of travel, then retracts, moving away from the flight path. This gives extra clearance so that the missile will not strike the foward end of the guide under severe ship roll, wind, and other conditions. A dud-jettisoning device is provided in the guide arm to boost a faulty missile overboard if necessary.

Many of the components are the same as on the Mk 13 launching system. The guide arm in its entirety is interchangeable. A major difference is in the power drive. Both loading and training are tied to one power unit. The two operations cannot take place simultaneously. The elevation power drive is a separate unit.

The operational characteristics and controls are similar to the Mk 13, and personnel training does not present unique problems. At general quarters three crewmembers are required—an operator at the main control panel, a safety observer, and an emergency repair tenchnician.

Operation of the system is normally automatic, and the crew merely monitors the system. At LOAD order, the launcher automatically trains and elevates to the selected loading position, the magazine blast door opens, the hoist chain engages the missile from below and pushes it into position on the launcher. The fin erectors engage the missile fins (opening them if the launcher has been assigned a target) and at the same time the contactor makes electrical contact with the missile. The hoist chain is then retracted and the blast door closes. If the launcher has been assigned, it synchronizes in train and elevation with the computer signal. The missile may be fired any time after synchronization. An automatic warmup system ensures that enough missiles are kept on warmup to permit firing continuously but that any missile approaching a condition of excessive warmup will be taken out of sequence and allowed to cool. Indicators provide continuous information on orders received, status of launching system operations, number of missiles in the magazine, and missile warmup in the magazine.

Step control is used for system maintenance, exercise, strikedown, and missile checkout. Safety interlocks, firefighting installations, vents to limit magazine pressure, a plenum chamber

## Chapter 3—PREPARING FOR A FIRING EXERCISE

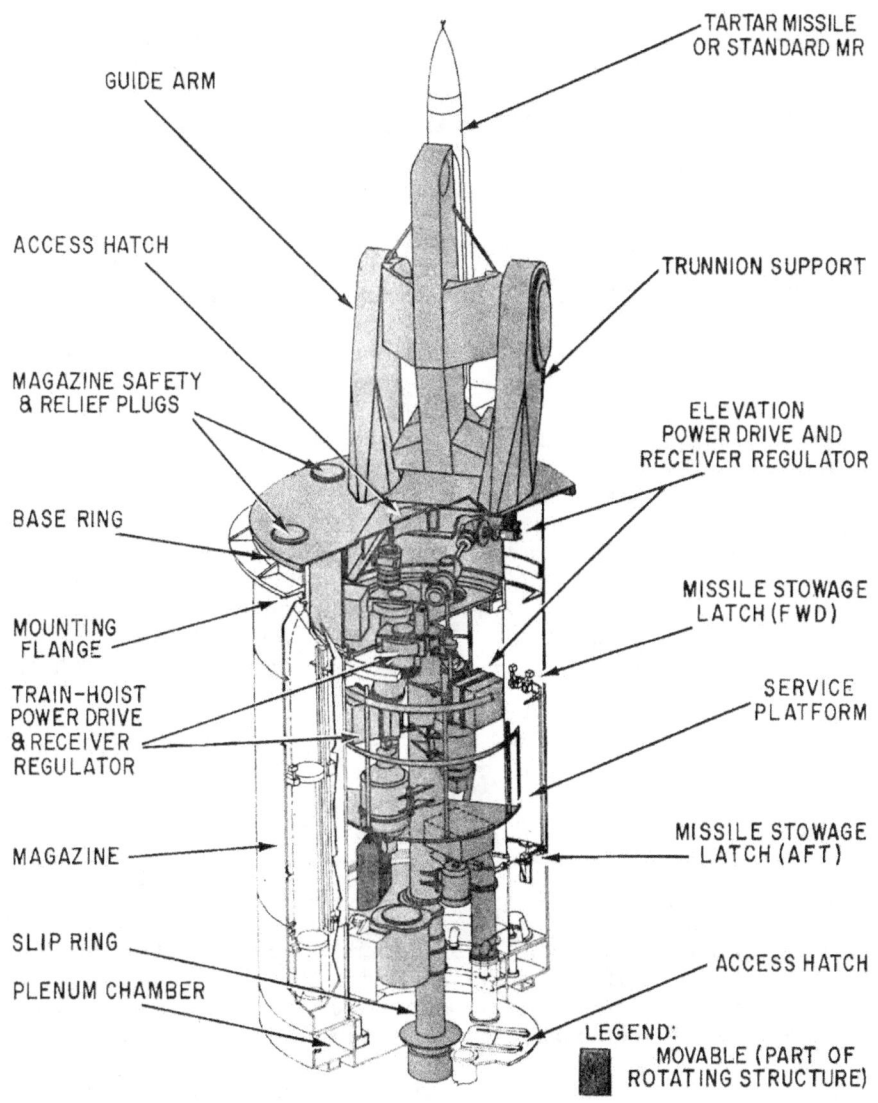

Figure 3-9.—Mk 22 Mod 0 GMLS.

under the missiles, and a water injection system are very similar to those in the Mk 13 Mod 0 system. If a missile should accidentally ignite in the magazine, the plenum chamber receives the exhaust gases and conducts them to an elbow-shaped duct at the edge of the chamber, where the gases escape to the atmosphere.

### STANDARD GMLSs

The already installed Terrier and Tartar launching systems will be used to launch the Standard missiles SM-1 ER/MR and SM-2 ER/MR when they are placed in service on ships. Modifications will be made to the existing launching systems to accommodate the two types of Standard missiles SM-1, the medium range (MR), and the extended range (ER).

To make the Standard missiles compatible with existing shipboard systems, some minor modifications must be made. Actually, two comparisons must be made between (1) Terrier missile systems and Standard (ER) missile systems; and (2) Tartar missile systems and Standard (MR) systems. On some ships either the Terrier or the Standard (ER) missile can be used, and on some Tartar ships either the Tartar or the Standard (MR) missiles can be used.

Relatively minor changes to launching systems include: (1) missile identification circuits, (2) warmup time delay bypass circuits, (Standard missiles need no warmup), (3) circuits to delay missile firing until its one shot batteries are ready for use (stabilized) and (4) a signal comparison network to identify the Standard missile illuminator frequencies.

As the Standard missiles SM-2 (ER/MR) are incorporated into the Standard missile SM-1 (ER/MR) capable launching systems, there will be no major modifications to these launching systems. Simply stated, the Standard missiles SM-2 (ER/MR) are more technically advanced than the SM-1s.

The Mk 26 GMLS is covered in *Gunner's Mate M 3 & 2*, NAVEDTRA 10199-C, chapter 5. Preliminary tests and inspections, required prior to a firing exercise on the Mk 26 GMLS, are similar to those required on the Mks 10, 11, 13, or 22 GMLSs. For example, the DSOT must be performed within a prescribed time before firing, the firing cutouts and accumulator pressures must be checked.

After an AAW or ASW weapon is loaded onto the Mk 26 GMLS rail, the contactor transmits the appropriate electrical orders to the weapon. When weapons control orders firing, the launcher will synchronize to the ordered signal and all other necessary interlock conditions will be satisfied.

The AAW weapons are armed, and the fire-thru latch for ASW weapons is unlocked. For AAW weapons, the contactor retracts when the appropriate ready signal is received from the weapon. The ASW contactor remains extended until the weapon fires and begins to move. The weapon's dual thrust rocket motor (DTRM) is ignited through igniter blades on the guide arm which are in contact with the AAW weapon. ASW ignition is accomplished through the ASW contactor.

As thrust builds, sufficient force is developed to overcome the restraining force of the fire-thru latch and the weapon leaves the rail. For operation with AAW weapons, the retractable rail retracts as soon as the weapon shoes are free of the launcher rail. The aft-shoe latch retracts; the arming tool actuator resets. For continuous system operation, another loading cycle is initiated as soon as the arming tool actuator resets and the launcher returns to load position.

NOTE: For quick reaction mode of operation, SM-2 missiles can be activated as soon as the contactor is extended and the firing key closed. When the launcher is synchronized, final inputs are fed to the missile. As soon as the missile is ready the contactor retracts and the missile is fired.

## SPARROW III, MK 29 GMLS

The Mk 29 launching system does not employ electrohydraulic power drive mechanisms, as do the other launching systems that are maintained by GMMs. This launching system utilizes digital solid-state devices, and the launcher is driven in elevation by a 1 horsepower d.c. drive motor and in train by two, 1 horsepower d.c. drive motors connected in tandem. Since the launcher is loaded all the time, the reaction time is considerably less than with other systems. Once a GMM positions the mode select switch on the maintenance inter-connecting panel to REMOTE, the launcher stow locks will automatically retract and the launcher power drives automatically energize to position the launcher at the air-ready position and then shut down awaiting assignment.

NOTE: Ensure that the rocket motor safe arm mechanisms are in the ARMED position prior to firing, in accordance with ship's policy.

# Chapter 3—PREPARING FOR A FIRING EXERCISE

With the mode select switch still positioned to REMOTE, the GMMs will monitor the launcher and be available in case of a launching system malfunction.

When a hostile threat is imminent, the missiles to be fired are assigned to one of the two radars (dual radar systems) by the operator of the firing officer's console (FOC).

The FOC operator then commands the missile to tune to the assigned radars. After the missiles indicate a tuned condition, they go into a standby state and can be maintained in that state indefinitely. This feature enhances the readiness of the entire system.

At this time the launcher and missiles are ready to automatically accept a firing order from the fire control system.

When a fire control system radar begins to track a target the fire control system computer determines whether the target is in the range of engagement. As the target comes within missile range the computer orders the GMLS to apply runup power to the missile, and the launcher power drives energize to position the launcher to the ordered position. As soon as the launcher synchronizes, the firing command can be ordered automatically by the fire control computer, or manually by the FOC operator if the console is in the semiautomatic mode.

A significant development on the Mk 132 launcher for the Mk 29 GMLS involves use of printed circuit cards instead of mechanical cams in the firing cutout system. The printed circuit cards are engineered to sense launcher position and inhibit firing when the launcher is pointed into a nonfiring zone. Another feature of the launcher is the infrared sensor located in each cell that will sense an inadvertent firing of a missile. If an inadvertent firing occurs the sensor sends a voltage signal to fire the squibs on a nitrogen cylinder. The nitrogen pressure sets the launcher brakes and prevents the launcher from pinwheeling. This emergency brake system is necessary because the Mk 132 launcher does not have brakes, as other GMLS launchers have, that set automatically when in a power off condition.

## HARPOON GMLSs

The only Harpoon GMLSs presently in use are the canister-launcher configurations as previously discussed in chapter 2. However, the Harpoon weapon is compatible with the Mk 11 and Mk 13 GMLSs and it is anticipated that it will be used in the Mk 26 GMLS.

Heater power is required for the Harpoon missiles when the temperature is less than 10°C and is applied by activation of a switch on the weapons control indicator panel. The missile heater operation is then controlled by the thermostats in the missile. All eight missiles can be provided heater power at the same time, and up to four missiles can be provided warmup and electronic power at the same time. This feature allows for ripple firing of all missiles on the same launcher.

The Harpoon ship command launch control set (HSCLCS) uses target data from the existing combatant ship's fire control system, and own ship motion data for preparation and launch of the Harpoon missile. Capability for manual entry of known target data, or selection of bearing-only launch (BOL) when range is unknown is also provided. The HSCLCS computes the fire control solution, initializes the missile, monitors missile status, provides cell selection, and controls the launch sequence.

The existing fire control system provides target data to the HSCLCS in digital or synchro form, and the ship provides synchro motion data from its navigation system (heading, pitch, roll, and speed), and primary power.

## TOMAHAWK GMLSs

The Tomahawk missile is undergoing acceptance tests, as is the launcher system at this time; thus, there are no firm criteria as to the equipment prefiring preparations and tests.

## REFERENCES

1. NAVEDTRA 10200-B, *Gunner's Mate M 1 & C*

2. OP 2593, Guided Missile Launching System Mk 11 Mods 0, 1, and 2, volume 1

3. OP 2665, Guided Missile Launching System Mk 13 Mods 0, 1, 2, and 3, volume 1

4. OP 2676, Guided Missile Launching System Mk 10 Mods 3 and 4, volume 1

5. OP 3114, Guided Missile Launching System Mk 10 Mods 7 and 8, volume 1

6. OP 3115, Guided Missile Launching System Mk 22 Mod 0, volume 1

7. OP 4006, Guided Missile Launching System Mk 29 Mod 0, volume 1

8. OP 4104, Guided Missile Launching System Mk 26 Mods 0, 1, and 2, volume 1

9. OP 4341, Harpoon Missile (RGM-84A) Handling Operations and Maintenance Procedures

10. OP 4093, Terrier BT and HT, Standard (ER) Missiles and Training Rounds

11. OP 2351, Guided Missile Launching System Mk 10 Mods 0, 5, and 6, volume 1

# CHAPTER 4

# LOADING, UNLOADING, AND DUD JETTISONING

This chapter emphasizes the role of the GMM1 and Chief in loading and unloading missiles and discusses dud-jettisoning equipment used on Navy ships. The occupational standards for dudjettisoning are assigned to lower rated personnel, but you have the responsibility for supervising the activity.

Most of the work of loading and unloading is done automatically by the launching system, and lower rated personnel do most of the assembling and disassembling. You, as a GMM1 or Chief, may operate a control panel, supervise the work of the assembly team, act as a safety observer, troubleshoot the equipment, and make the more difficult repairs, including overhaul and adjustment of equipment.

You must become familiar with the system or systems you have on board, but you must also learn about the other types of systems.

## TERRIER GMLSs

Each side of the Terrier launcher is serviced by a complete and independent loading system. Each of these systems is serviced by a corresponding handling system. Except for minor differences, the operation of the two sides is identical. The location of the loaders and ready service rings for the Mk 10 Mods 7 and 8 are shown in chapter 6 of this manual.

## LOADING

Loading is the process of removing a round from the magazine, unfolding missile fins, attaching booster fins, and placing the round on the launcher.

The sequence of steps in moving a round from the magazine to the launcher was given in chapter 3 for the Mk 10 GMLS. If you have had duty on a ship with Terrier capability, these steps are familiar to you. If your experience has been with other missile systems, you will recognize the similarities. The only manual labor involved consists of unfolding missile control surfaces and installing booster fins in the assembly room. If any part of the system fails to act automatically on signal, you need to know how to find the trouble and correct it.

The launcher captain monitors the launching system functions by watching the indicating lights on the EP2 panel during automatic operation. (The step control lights and switches on the EP2 panel are covered and are not in operation during automatic procedure.) The launcher captain reports any malfunction of the equipment by telephone to the feeder system captain and the operator of the weapons assignment console (WAC) or evaluator console (EC) in the weapons control station (WCS). Under emergency conditions, or during any malfunction, the launcher captain stops the launcher movement with the train and elevation motor switches.

When acting as assembly captain, do not allow the assemblers to remove booster fins from the racks until the missile has stopped in position in the assembly area. Wait until all the assemblymen have completed their wing and fin assembly, stepped back to the clear area, and pressed their safety switches. Then signal the launcher captain that the assembly area is ready. If a safety switch is inoperative, or malfunctions in any way, check to see that all the assemblymen are in the clear area after completing assembly; then signal the launcher captain CLEAR BYPASS. The safety switch used by the assemblymen is a foot switch; each person has one at each station in the assembly area.

## UNLOADING

Unloading is the process of removing the round from the launcher, removing and folding the fins, and returning the round to the magazine.

With the exception of the Mk 11 GMLS (Tartar), all GMLSs can be unloaded in an automatic mode of control. The unload order is sent by WCS and will indicate the side—A, B, or both—by causing a blinking light to appear on the launcher captain's panel. The missile may be in the assembly area or on the launcher when the unload order is given. The launcher captain positions the switches on the EP2 panel to conform to the unload orders, and this initiates the automatic unloading. The launcher synchronizes to load position and proceeds through the unloading steps.

When the weapon reaches the assembly area, the missile control surfaces are refolded and the booster fins are removed and returned to the racks. During the unload operation, visually inspect to ensure that the missile control surfaces are folded, the booster fins are removed, the booster is unarmed, and the missile sustainer S&A device is in the SAFE position before returning the weapon to the magazine area. If the S&A device is found in the ARMED position, the missile must be offloaded at the earliest possible time. When working with a BTN, ensure the safety switch actuator is in the SAFE position. A dud or misfired booster being returned to the magazine must not be removed from the wing and fin assembly area until the feeder system captain is notified that the booster is in the UNARMED position and the missile sustainer S&A device is in the SAFE position.

The assembly area is the most dangerous section of the entire launching system during loading and unloading operations. It is the responsibility of the assembly captain (usually a GMM 2 or 1) to ensure strict compliance with all safety instructions. Although the assemblymen are responsible for their own safety, you must give them frequent reminders regarding safety precautions and be on constant watch to see that they are observed. When the feeder system is in operations, the assemblymen remain on the station with their foot switches depressed, except during actual assembly or disassembly of the wings and fins. The assembly captain must not give the READY signal until all personnel have stepped back to the safe area and have depressed their safety switches. There is an emergency wing and fin assembly bypass switch on the panel, but this must never be used except in case of a malfunctioning foot switch and during equipment checkout when personnel are clear of the assembly area.

During continuous firing, weapons will be in the assembly area as well as on the launchers. Before the weapon to be unloaded from the launcher can be moved, the weapon in the assembly area of that side has to be returned to the magazine. In automatic unloading, the launcher captain positions SMS2 to UNLOAD LAUNCHER; the assemblymen remove the booster fins and fold the missile fins; the assembly captain positions the assembly-ready switch on the panel to REMOVED; and the weapon is moved by the system mechanisms back to the magazine or ready service ring. At this time, the weapon on the launcher is moved by the system to the assembly area and the aforementioned actions repeated.

NOTE: If there are weapons in the assembly area and none on the launcher at the end of firing, the UNLOAD ASSEMBLY mode of operation is used.

An unloading cycle is necessary after every firing of an ASROC weapon from a Terrier system because the adapter must be returned to the magazine tray.

## STEP CONTROL

For checking, maintenance, or emergency purposes, the unloading operation may be carried out in step control. Step control is always used when moving the missile-booster combination from the ready service ring to the checkout area for routine care and maintenance. Step control must also be used for

# Chapter 4—LOADING, UNLOADING, AND DUD JETTISONING

exercise and strikedown. The steps are initiated one at a time by use of the pushbuttons on the launchers captain's panel. The launching equipment is always started in step control. The OP for your launching system contains procedural steps and designations of the switches to be activated. The lights and switches on the control panels are plainly numbered and labeled, but operation will be more efficient if you have become familiar with the panels ahead of time.

The indicating lights on the launcher captain's panel show switch actuation. Each pushbutton contains two light bulbs, separated by a center divider. One bulb (or one-half of the push-button) corresponds to the A-side and the other to the B-side of the launcher. To check for a faulty bulb, push the PRESS TO TEST pushbutton (figure 4-1) at the bottom of the EP2 step control panel. If the bulb tests "good," but still no light shows, investigate for the source of trouble.

Before beginning the unloading procedure, the launcher must be latched in the load position. The arming tool must be unwound before the blast doors are opened and it must be retracted before the aft lug latch is retracted. Sound the loader warning horn to warn everyone away from the loader area. After the loader has moved the missile from the launcher back to the assembly area, close the blast doors. Personnel in the assembly area remove the booster fins and stow them in the racks provided for them. Other assemblymen fold the fins on the missile.

NOTE: There are no safety interlocks to prevent the lowering of a round into the magazine with missile fins unfolded and booster fins installed; thus, it is essential that the utmost care is taken to ensure that this does not happen.

## DUD JETTISONING

Jettisoning of a missile may become necessary under the following hazardous conditions:

1. There is a fire on deck or in the vicinity of the launcher.

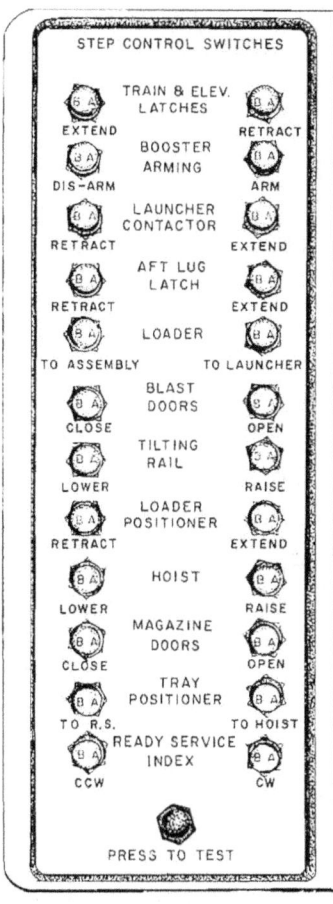

94.120
Figure 4-1.—Step control switches, EP2 panel, Mk 10 Mod 7 GMLS.

2. The weapon is damaged by enemy action.

3. The weapon failed to fire and circumstances do not permit returning it to the magazine or checkout area. Do not jettison a missile with a nuclear warhead. The decision to jettison comes from the commanding officer via the weapons control station.

A dud-jettisoning unit (figure 4-2) is associated with each launcher to permit the ejection of missile rounds from the launcher. Each unit consists of two ejectors and a control panel (figure 4-3). The dud-jettisoning units are

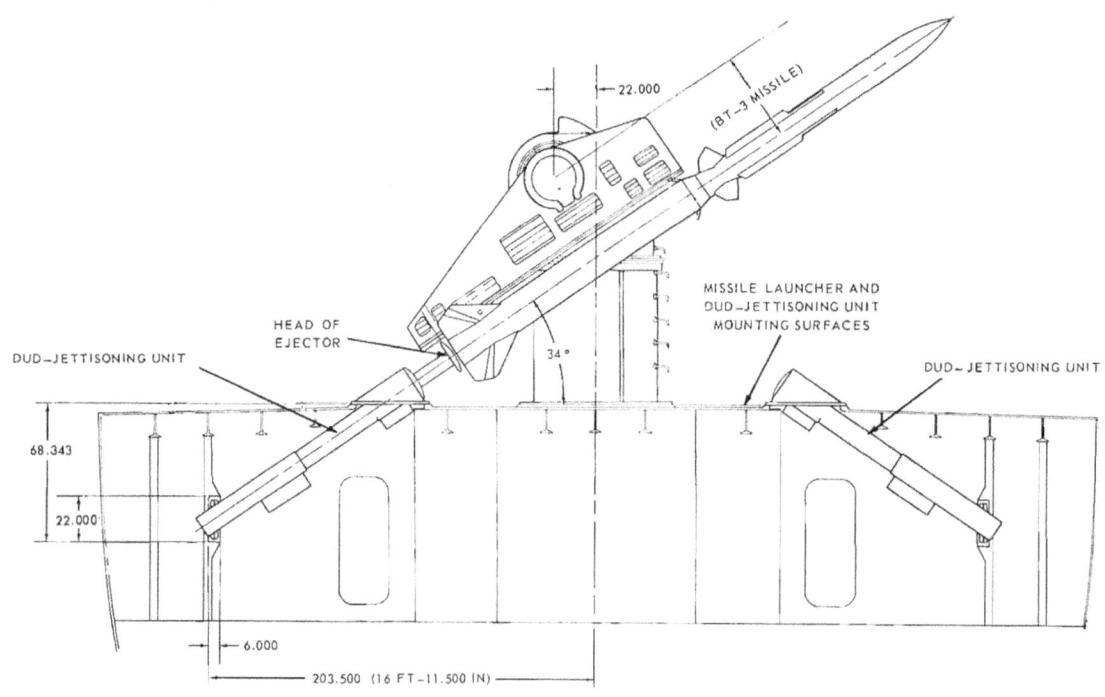

Figure 4-2.—Mk 108 dud-jettisoning unit; cross section view. Mk 10 GMLS installation on a CG.

designed to permit weapon ejection at train angles of approximately 95° or 275° and 34° of elevation. (The train angles may be different for each system Mod and class of ship.) The launcher automatically trains and elevates to bring the after end of the round in line with the dud-jettisoning ejector, and a pneumatic mechanism in the ejector elevates a piston in line with the round. The piston extends slowly, under low pressure air, until its mushroom-headed piston mates with the after end of the round, then it extends rapidly with a short, powerful pneumatic stroke (3500 psi), forcing the round off the launcher and over the side.

The control panel for the dud-jettisoning unit is mounted in the deckhouse, and is operated by the launcher captain (or the port side assembler captain) upon orders from the WCS by sound-powered telephone. In the Mk 10 Mod 8 system, the control panel is adjacent to the A-side blast doors, within the aft compartment. When the ship's roll exceeds 20°, jettisoning must be performed only on the downroll. A standard bubble type inclinometer with a 45° index scale is mounted next to the dud-jettison control panel to indicate ship's roll.

The launcher captain initiates jettisoning by positioning the A or B selector switch (SMX3) to the desired position and the train and elevation control switch (SMY1) to the DUDJETTISON position at the EP2 panel, which casues the launcher to synchronize automatically to the DUDJETTISON position.

The dud-jettisoning procedure may be applied to a dud missile, a misfired booster, or any other condition which necessitates a decision to jettison a weapon.

### Operation

Whenever the firing key is depressed, the dud and misfire lamps light momentarily, until

Chapter 4—LOADING, UNLOADING, AND DUD JETTISONING

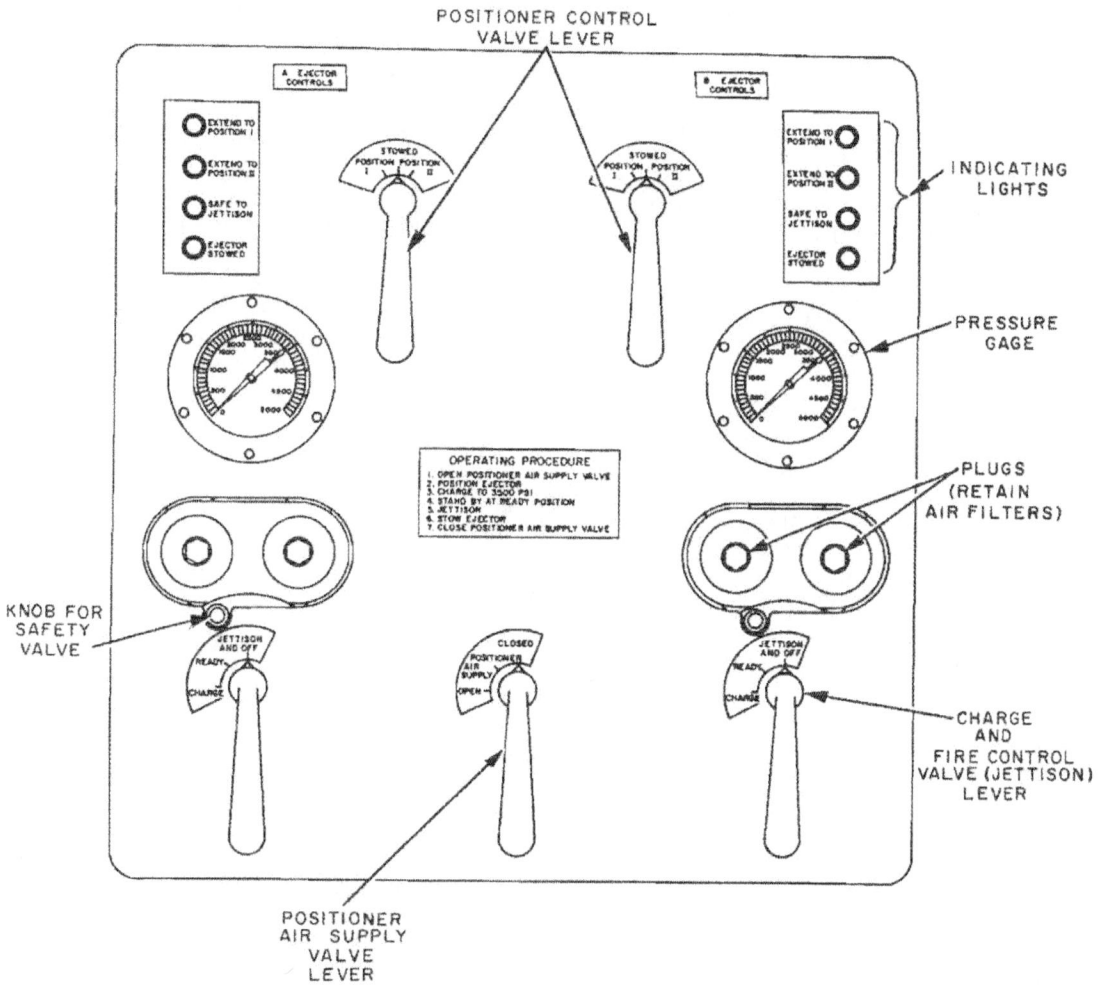

Figure 4-3.—Mk 10 Mod 0 GMLS dud-jettisoning panel.

the missile has cleared the rails. However, if the missile is a dud, the dud lamp continues to be lighted and the contactor fails to retract. The operator may try to fire the missile by placing the emergency firing enabling switch (SMF1) to the ACTIVATE and ENABLE position. If this succeeds, the contactor retracts and the dud lamp, the rail lamp, the warmup timer rail lamp, and the ready to fire lamp all go out. When the arming tool unwinds, the ready lamp also goes out.

If attempts to fire the missile are unsuccessful, it may be returned to the magazine for later inspection and possible repair; its location is marked on the control panels. In some situations (emergency or combat), it may be necessary to jettison a dud missile, but wait for the order to do so.

A dud indication will also occur if the firing key is released too quickly (before 1.5 seconds have elapsed). As a result, the booster firing

83

relay is not energized and a dud missile is left on the launcher.

PANEL OPERATION.—Suppose you are the assembly captain of a Mk 10 launching system when the jettisoning of a missile is ordered. First, establish sound-powered telephone communications. Then open the positioner air supply valve (at the control panel [figure 4-3]) which connects to low pressure air. Next, rotate the A- or B-side positioner control lever to POSITION I for a BT-3, HT3 or the Standard ER missiles. (NOTE: POSITION II is no longer utilized.) When the ejector is in the raised position, rotate the jettison to CHARGE and hold the lever in this position until the air pressure meter reads 3500 psi. (Pressure requirements for your installation may be different.) A SAFE TO JETTISON light on the control panel indicates when the ejector is fully raised to the firing position.

WARNING: Do not cycle below the designated operating pressure.

Rotate the jettison lever to READY and wait for the command. Upon receiving the command to jettison, check that the air pressure meter reads 3500 psi and rotate the jettison lever to JETTISON. The head of the ejector is forced against the booster base by air pressure from the firing valve, and the weapon is forced overboard from the launcher. Note that the air pressure drops rapidly. Lastly, rotate the positioner control lever to the STOW position. The dud-jettison unit must be lowered all the way before the launcher can be trained and elevated for reloading.

There are differences in control panel switches and nomenclature, but the principles of operation are very similar. The steps in operation of the dud-jettison panel on your ship should be posted beside the panel. On some mods, a metal instruction plate is permanently fastened to the dud-jettison panel directly in front of the operator. After the jettison operations are completed, the dud-jettison panel operator moves the lever of the positioner air supply valve to CLOSED, and the launcher captain returns control to the EP2 panel by repositioning the switches to the desired type of operation.

If the round is considered dangerous to the ship, the launcher captain positions the emergency enabling switch to ENABLE upon telephoned order from the WAC/EC operator. The WAC/EC operator then holds down the dud emergency firing switch until the round leaves the rail (rail loaded light goes out). This is dud firing (not jettisoning), and is carried out in WCS. This method of dud firing could disable one side of the launching system. It is used only in case of real danger from the missile on the launcher.

SAFETY RULES.—Under emergency conditions or during any malfunction, the launcher captain must stop the launcher movement with the train and elevation operation selector switch or with the train and elevation motor switches.

In case of booster misfire, do not permit personnel to approach the launcher for at least 30 mintues after the last attempt to fire (unless otherwise prescribed in applicable instructions), and the firing circuits have been disabled prior to approaching a dud/misfire. The time limit is at the discretion of the commanding officer.

Before returning a dud or a misfired weapon, or initiating dud-jettison procedure, rotate the firing safety switch on the EP2 panel to the LAUNCHER ONLY position.

During all operations for disposal of misfires or dud, the launcher captain should remain at the control panel to guard the firing safety switch (SMF2) and to observe and make certain the launcher remains in a safe position.

Do not position the emergency enabling switch to ACTIVATION AND ENABLE during firing unless specifically ordered to do so by weapons control. Use caution as to the proper side and the position ordered.

Before returning a missile to the magazine area, visually inspect to ensure that missile control surfaces are folded, the booster is unarmed and its fins removed, and the missile sustainer is in the SAFE position. A dud or a

## Chapter 4–LOADING, UNLOADING, AND DUD JETTISONING

misfired weapon being returned to the magazine must not be removed from the wing and fin assembly area until the feeder system captain is notified that the booster and the missiles sustainer have been checked and reset to the UNARMED position.

### Dud or Misfire?

Note the difference between a dud and a misfire. If the dud lamp on the WAC lights when the firing key is pressed, nothing happens to the missile—it does not transmit the electrical energy to set off any explosive components. In a misfire, some part or parts of the explosive system were actuated when the missile firing key was depressed, but not enough to send the weapon off the launcher. A misfire presents a dangerous situation. If the misfire lamp lights at any time during the firing cycle, there are three alternatives: (1) emergency firing procedures may be used; (2) the missile may be aimed in a safe direction for the required waiting period and then be returned to the magazine for later examination and repair; or (3) the round may be jettisoned.

A missile is considered to have misfired when its booster fails to fire after the missile's electrical and hydraulic systems have been activated and the booster firing relay has been energized. When the firing key is depressed, the dud lamp lights and the misfire lamp flashes. When the missile fails to clear the rail, the misfire lamp continues to flash and the dud lamp should go out when KCFA(B)2 energizes. You cannot tell whether the explosive train inside the missile will sputter and go out, if it will burn and explode on the launcher, or if it will fire in a short time. All factors of the situation—known, calculated, and surmised—have to be considered in deciding whether to wait and see what happens or to jettison the weapon. In a battle situation, it may be necessary to fire a missile from the other side of the launcher while leaving the misfire on the first side. Several attempts may be made to fire the missile by means of the emergency firing key. When the emergency enabling switch is at NORMAL, the emergency firing key can be held down as long as desired in an attempt to activate and fire the missile. If it is placed at ACTIVATION AND ENABLE, the booster firing transformer is energized through an alternate circuit and many of the normal firing relay contacts are bypassed. If the firing is successful, it shows the relays were at fault on the first try. If the missile cannot be fired by this method, it will probably have to be jettisoned. The decision must be made in WCS.

The two emergency firing circuits in the Mk 10 Mod 7 launching system, emergency activation and enable, and emergency enable, are used only in the Terrier mode.

### Maintenance

The dud-jettisoning unit has been designed to provide maximum service with a minimum of maintenance. A major difficulty with dud-jettisoning units is that ice may form in the valve passages. This is caused by rapid expansion of moist, compressed air. Any mositure traps in air lines should be drained regularly. Deicing lines port heated fluid to the cover door sections to prevent formation of ice during cold weather, permitting operation of the jettisoning unit in the most adverse weather conditions. This anti-icing system also protects the launcher from icing.

Since the dud-jettisoning unit is intended for emergency use, it must be kept in operating condition, ready for instant use when needed. Check out the equipment at regular intervals by exercising each dud ejector (no weapons on launcher rails). Replace indicator lamp bulbs on the control panels when necessary. The outside of the panel should be cleaned periodically. Usually wiping with a dry cloth is enough; a damp, soapy cloth may be needed to remove grease spots or fingerprints. Wipe dry. The dud-jettison unit does not require lubrication. In particular, take care NOT to lubricate the firing piston head or stem.

WARNING: If it is necessary to disassemble any of the air lines, be sure the valve in the ship's high pressure line (4500 psi) and the nearest shutoff valve in the 100-psi ship supply line are closed. Bleeder valves in the ejector unit accumulators should be open. As you work, tag all valves. Protect any open ends of pressure lines with suitable caps or plugs to prevent entry of dirt, moisture, or other foreign matter.

It may be necessary to replace a gasket or a defective limit switch on an ejector unit. The need for a new gasket may be discovered when checking the air-charging chamber of the ejector for moisture. To make the check, remove the drain plug from the lower end of the ejector assembly. If there is any drainage, wipe the drain port clean, check the gasket and plug for signs of deterioration, and replace if necessary. Wait at least 2 hours after a unit has been cycled before making the draining check. Further disassembly of ejector units is not necessary, short of battle damage.

There are six air filters and six air breathers in the Terrier jettisoning equipment which require regular inspection to see if cleaning is needed. Cleaning is done by washing the filter or breather in solvent, rinsing in clear water, and drying with a stream of compressed air. Never direct compressed air at yourself or other; it can be fatal.

Before unscrewing a plug that holds a filter, be sure the manual shutoff valve on the jettison panel is closed and that the pneumatic lines leading to the ejector are vented (JETTISON AND OFF on panel). (See figure 4-3.) There are four of these plugs (and filters) on the face of the panel.

The sensitive switch assemblies, solenoid assembly, and dud-jettison synchro-transmitters all need periodic inspection and adjustment or replacement as required. The adjustment is determined at installation and is not changed, but units can be brought back into adjustment if they vary from it. Two sensitive switch assemblies are located on each ejector. The four solenoid assemblies are all located in the jettison control panel. Any malfunctioning parts are replaced. The synchros are located within the EP2 panel. The synchro control transformers are either adjusted or replaced. Instructions for this are given in the applicable corrective maintenance cards (CRCs).

Manual switches are not repaired but are replaced if they do not function. These manual switches include pushbutton switch assemblies, toggle switch assemblies, and rotary switch assemblies used on control panels.

Maintenance of the electrical cables includes periodic checking of the cables, connectors, or other associated components. Measure the insulation resistance of power supply cables with a megger. A ground-detection indicator on the EP1 panel continuously monitors the control supply circuit. Disconnect this indicator before making a megger test of a cable in the system. If an insulation breakdown is indicated, track it down and correct it, then test again. If a cable is damaged so it requires replacement, get a spare cable of the same kind from spare parts stock and install it. Do not splice a cable except in an emergency. (Umbilical cables are always replaced, not repaired.) Attach identification markers to all cables. All terminal lugs should be crimped to their connectors.

## LOADING ASROC WEAPONS

Loading the ASROC weapon from the magazine of the Mk 10 Mod 7 Terrier GMLS is similar to loading Terrier missiles but a few of the steps are different. If the decision is made to use an ASROC weapon (torpedo or depth charge configuration), the ASROC mode switch on the EP2 panel is pushed upon orders from weapons control. This changes the launching system to the ASROC mode of operation. Automatic control is used except for exercise, testing, or in an emergency. The indicating lights on the EP2 panel show the steps taking place in the loading operation.

The ASROC weapon is stowed with motor fins attached. The snubbers on the ASROC adapter rails have to be closed and secured after firing. The ready service ring tray does not shift from hoist to ring position after bringing a weapon to the assembly area, as it does with Terrier, because the adapter tray must first be returned. This is done after the ASROC is launched.

The type of ASROC weapon must be visually identified when it arrives in the

# Chapter 4—LOADING, UNLOADING, AND DUD JETTISONING

assembly area to confirm the identification. If it is not the one ordered, it must be returned to the magazine and the correct one brought up. The circuit to identify the weapon in the adapter is energized through the loader pawl warmup contactor. This causes the identification light to blink on the EP4 or EP5 panel and the operator can notify the WCS and the EP2 operator.

When the ASROC is at assembly, the ASROC arming tool on the launcher is automatically rotated into position. (There is another arming tool for Terrier.) If the weapon is a Y-type, the Y stop keylock switch must be positioned to LOAD, or the blast doors will not open to permit loading the launcher. When the weapon is on the launcher, the arming tool winds and retracts, and this opens the snubbers on the adapter rail.

Although the ASROC does not have an auxiliary hot-gas power generation system (APS), the activation indication is supplied to the ASROC relay transmitter. This indication is required by ASROC fire control before it supplies the ASROC ready indication on the panel. However, an ASROC failure on the launcher is very unlikely.

When enough thrust has developed, the ASROC travels the length of the adapter rail into ballistic flight. After the weapon is fired, the adapter rail is returned to the magazine tray in an unload sequence. When it arrives in the assembly area on its return trip, the snubbers must be closed manually (with the aid of special tools). A new umbilical cable must be inserted in the adapter, and this is done in the checkout area. This may be done later, depending on tactical circumstances.

### Care of Cable Assemblies

The Mk 10 umbilical cable provides the necessary electrical connection between the adapter and the ASROC weapon. Each time an ASROC weapon is launched from an adapter, the umbilical cable must be removed and replaced with a new cable. The cables are enclosed in a dust cover (figure 4-4); do not remove the dust cover until just before installing the cable.

Remove the cable cover by loosening the locking studs (figure 4-4, view A) with a snubber cam wrench. Lift the forward end of the cover upward, and slip the after end of the cover free of the cover retainer bar. Remove the expended cable by disconnecting the cable connectors at the after end and disengaging the missile retractor-connector from its support. The location of the cable is shown in figure 4-4, view B, on the after handling shoe support. The new cable is placed in the trough of the after handling shoe support and the cable connectors are attached to the adapter wiring connectors.

Cable assemblies frequently are damaged while being connected or disconnected. The keyways must be properly aligned; proceed carefully when connecting the cable plug to the ignition separation assembly (ISA) receptacle to avoid breaking or bending the receptacle pins. Lubricate the rubber ring on the receptacle and inspect the seating surfaces. Consult the OP or the SWOP for the correct lubricant. If the rubber ring has raised out of the groove, it prevents proper plug latching or positive electrical connection of the cable assembly. This work is done in the checkout area when loading an ASROC weapon into the adapter before loading it into the magazine.

### Depth Charge Configuration

Remove the four depth charge streamer tapes during the receipt inspection prior to loading the depth charge into the adapter rail. The tapes, with lead foil barrier, are placed over the hydrostatic ports of the depth charge fuze to prevent entry of dirt, etc. The tape is not a safety device. The red streamers attached to the tape are merely reminders to remove the tapes.

When the ASROC is fired, the arming tool winds to trigger the release of the snubbers. It requires approximately 2000 pounds of thrust to cause the forward restraining latch to release. A positive stop on the launcher guide arm holds the adapter rail so it is not fired with the weapon. The blast doors will not open for the Y-type weapon until the Y stop keylock switch

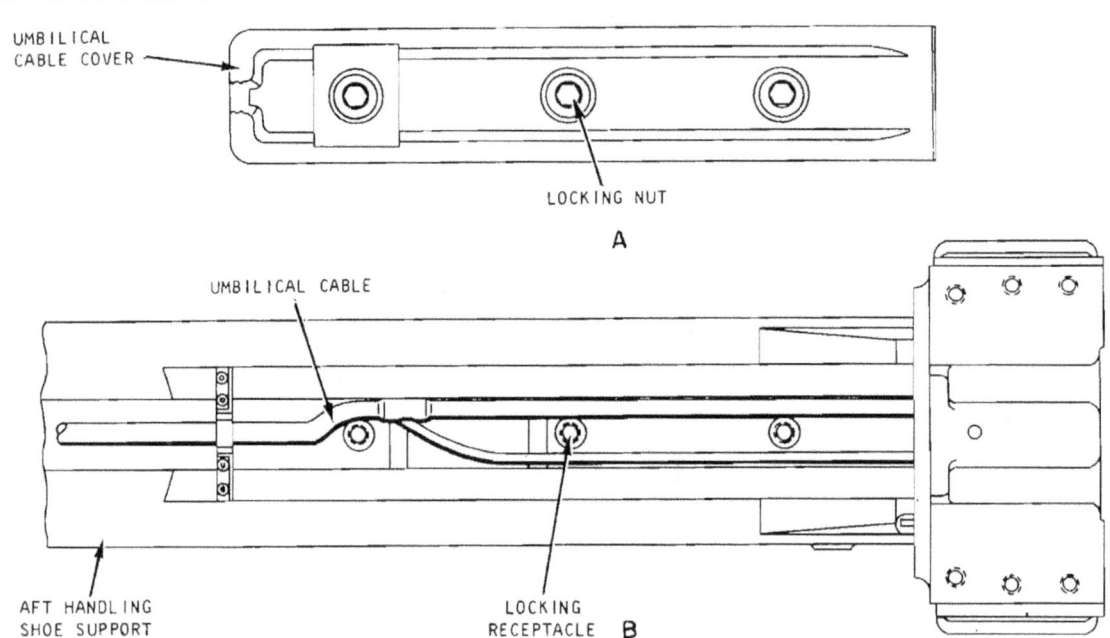

Figure 4-4.—Umbilical cable for ASROC weapon: A. Umbilical cable cover; B. Location of umbilical cable on aft handling shoe support.

is positioned to LOAD. When the weapon is on the launcher and the type indications appear on the EP2 and attack console panels, the operator of the attack console panel immediately checks out the weapon.

Because the depth charge is larger in diameter than the torpedo configuration of ASROC, inserts are not needed in the adapter rail when the depth charge is to be loaded. Side and bottom snubbers in the adapter prevent lateral movement of the weapon; a missile-restraining mechanism prevents fore-and-aft motion.

The steps in loading an ASROC into the adapter must be followed exactly. Be sure to stand clear of the snubbers when they are being unlocked before placing the weapon on the adapter. In a firing cycle, the initial action for opening the snubbers is accomplished when the ASROC arming tool on the launcher extends and winds. At that point snubbers are opened by torsion bars and their acceleration, constant velocity, and deceleration are controlled by hydraulics within the adapter rail; however, they have to be released or closed with a special wrench when the ASROC is being loaded (or unloaded) into the adapter.

## UNLOADING ASROC WEAPONS

To return the ASROC to the magazine assembly area, automatic unloading may be used. After each firing of an ASROC, the adapter must be unloaded before another weapon can be brought up and placed on the launcher. If time permits, the umbilical cable should be replaced on the empty adapter before it is returned to the magazine. When the assembly area is clear, the EP2 operator can return the weapon (or the adapter) that is on the launcher guide arm.

The ASROC with a torpedo warhead uses a torpedo exploder, which must be in the SAFE

# Chapter 4—LOADING, UNLOADING, AND DUD JETTISONING

position when the missile is in the magazine or assembly area. Figure 4-5 shows one type of exploder. Follow the instructions in the OP for the exploder in your weapon.

## Offloading the ASROC Depth Charge or Torpedo

When preparing to offload the depth charge or torpedo, the thrust neutralizer must be installed (figure 4-6, view A) on the weapon. Note that a special wrench is used. Do not tighten the screws too much or the pins in the nozzle plate may be sheared off. The torque requirement is 100-125 foot pounds.

The power supply (figure 4-6, view B) for the depth charge is not installed in a depth charge in peacetime, but stored in a special safe. The blanking plate and seal will remain in place on the weapon as shown in figure 4-6, view C.

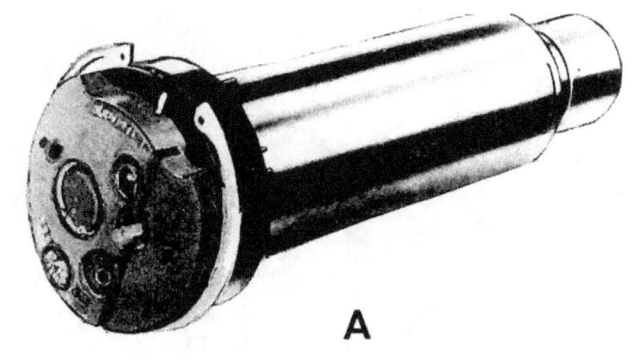
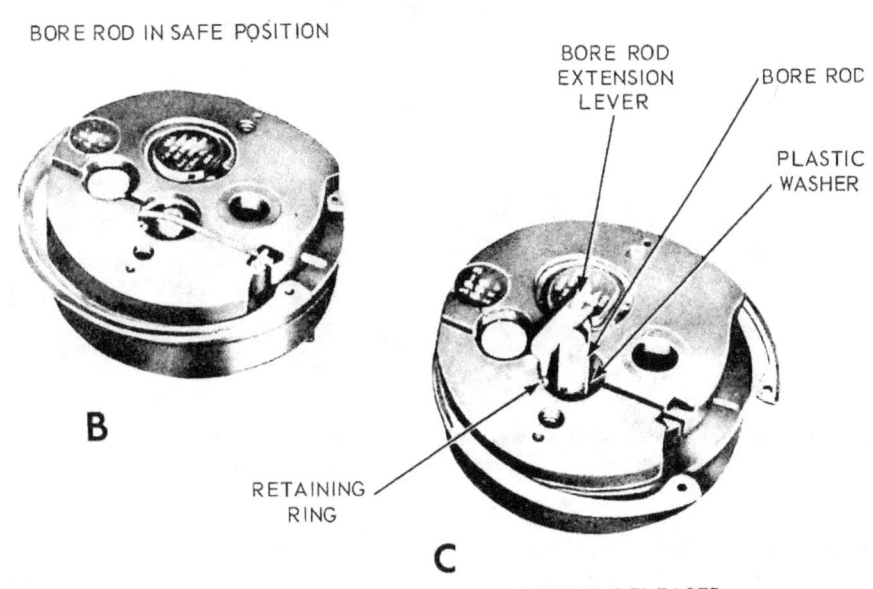

Figure 4-5.—Mk 19 torpedo exploder: A. Assembled exploder; B. Safe position of bore rod; C. Unsafe position of bore rod.

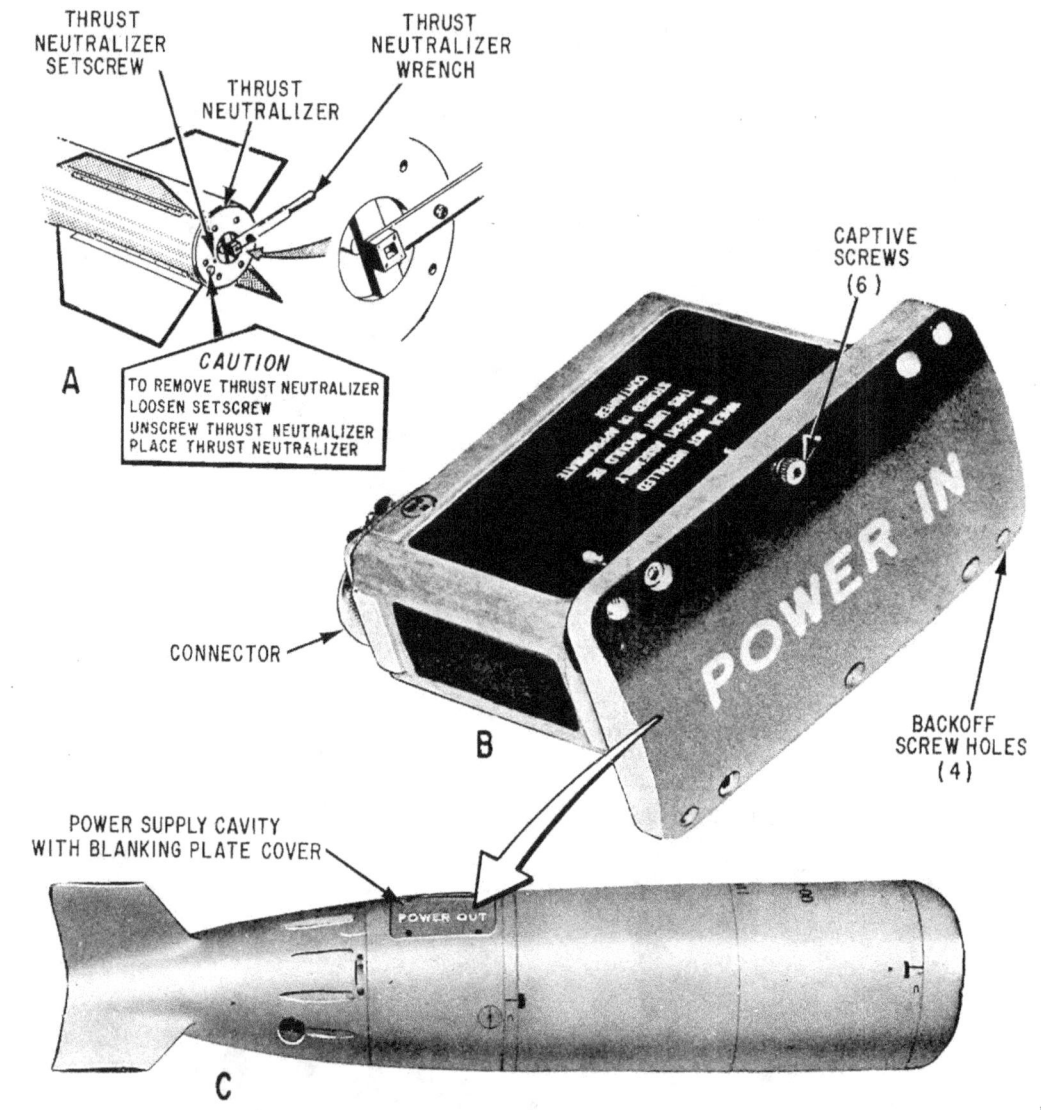

Figure 4-6.—Safing devices restored to ASROC weapon when unloaded: A. Thrust neutralizer; B. Power supply with cover; C. ASROC depth charge; showing blanking plate cover placed over power supply cavity.

When the weapon is to be offloaded, have the container placed in position on deck so the weapon can be lowered into it with the trolley hoist without striking or bumping the weapon. Attach the container ground wire to the thrust neutralizer (figure 4-7) before disconnecting the hoist from the weapon. The ground strap receptacle to which the wire is to be attached is on the neutralizer and should NOT be removed at any time. Secure the weapon in its container so it cannot shift.

## ASROC DUDS AND MISFIRES

With the extensive weapon tests and circuitry checkout required for ASROC, it is not

# Chapter 4—LOADING, UNLOADING, AND DUD JETTISONING

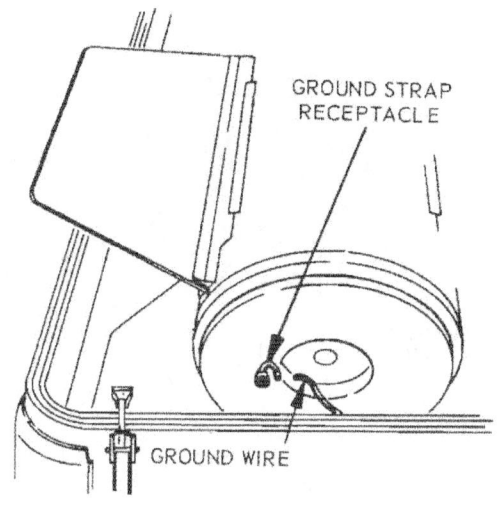

Figure 4-7.—Thrust neutralizer ground connection to container.

94.32

likely that an ASROC weapon will have to be jettisoned, as duds will be discovered before the weapon is placed on the launcher. If it is necessary to jettison an ASROC, the adapter rail is jettisoned with it. It is jettisoned in the same manner as a Terrier missile. If a dud results from loss of synchronization, it should be handled according to ship's doctrine.

### Depth Charge Configuration

The weapon configuration that carries the depth charge payload is designated as the Mk 2 rocket-thrown depth charge (RTDC). If a dud or restrained firing occurs, notify the ASW officer. The ASW officer will decide whether to return the weapon to the magazine or to package it into a container and return it to a depot for repair. You will check both the rocket and the Mk 17 warhead. The weapons in the other trays of the magazine may be used to continue the firing exercise. If a misfire signal shows on the launcher captain's control panel and/or the attack console, you must safe the depth charge at once, before it is removed from the launcher. The nuclear weapons officer must decide, in accordance with rules established for these weapons, what to do with the depth charge.

After the weapon has been safed, proceed with unloading according to the checklist, observing all the safety precautions. If the weapon is to be off-loaded, you will need a Mk 183 container and two Mk 42 or Mk 45 trucks. If your ship does not carry these, you have to stow the weapon. Order the Mk 183 container, the handtrucks, and other needed material, and when they arrive, offload the weapon into the container. Thrust neutralizers and ISA shorting plugs are also needed; these are carried on the ship. Check the supply system and order replacements if necessary. The thrust neutralizer and shorting plug (figure 4-6) were removed when the missile was stowed during strikedown.

Few ships have GMTs aboard; consequently, you need to know how to safe the depth charge. Depth charge safing (disarming) consists of resetting the arm/safe switch to the SAFE position and removing the power supply (figure 4-6, view B). Use the checkoff list from SWOP W44.34.1 and follow it precisely.

### Torpedo Configuration

The weapon configuration that carries the torpedo payload is designated the Mk 3 rocket thrown torpedo (RTT). There is no provision for jettisoning the RTT unless the tactical situation warrants that decision. If, upon attempting to fire the torpedo, the DUD light goes on, the weapon can be returned to the magazine after the established waiting period. It might be offloaded into a container and returned to a depot for repair.

To remove the dud or misfired weapon to a shipping container, follow the checkoff list for unloading.

### Safing the ASROC Torpedo for Unloading

ASROC Mk 44 torpedoes must be returned to a tender or depot every 36 months for maintenance. When a warshot torpedo is to be offloaded, the position of the Mk 19 exploder (figure 4-5, view A) must be checked before the torpedo can be moved. (Torpedo exploders may also be installed in exercise heads to give an

# GUNNER'S MATE M 1 & C

electrical "hit" signal.) The exploder bore rod must be in the cocked depressed position (figure 4-5, view B); if it has moved (figure 4-5, view C), it must be sterilized by turning the sterilizing switch (figure 4-8). To reach the switch, break the foil seal in the top of the exploder, then turn the switch 90° in either direction by using a screwdriver. This short circuits the exploder power supply. (The arming device is the part of the exploder that contains the explosive, and must be handled with great care. Spares are packaged and shipped separately, not assembled in the exploder.) All exploders must be considered armed if the bore rod has released. If you have to offload a warshot torpedo and the bore rod on the exploder is released, sterilize the exploder before moving the torpedo. Once the sterilizing switch has been used, the exploder is useless for firing until a new sterilizing switch is installed (at overhaul).

Details of the different mods of the Mk 19 exploder vary. For example, the Mk 19 Mod 12 exploder does not have the bore rod extension lever shown in figure 4-5, view C; the Mod 12 has a double-acting bore rod with a latch to lock the rod in the cocked position.

On dummy training missiles, inspection of the exploder bore rod is not necessary, but operation of the depth charge arm/safe switch is required for training purposes. Removal and replacement of the thrust neutralizer is also practiced on dummy training missiles. (NOTE: Mk 44 torpedoes are being phased out at the present time.)

The Mk 20 exploder is installed in the ASROC Mk 46 torpedo. The armed condition of this exploder is indicated through a window. If the letter A appears in the window (the letter need not be centered in the window), assume that the exploder is fully armed. If the letter S appears in the window (the letter need not be centered in the window provided the background is completely green), assume the arming rotor and rotor lock are in the SAFE position.

The sterilization switch for the Mk 20 exploder is under a plastic cover. To safe this

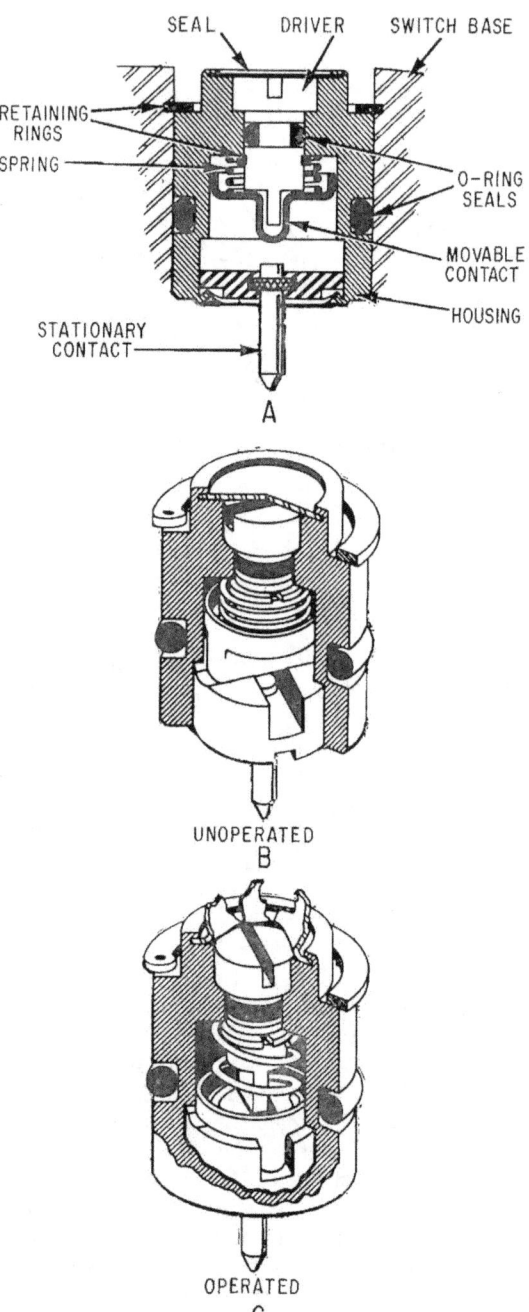

94.33
Figure 4-8.—Mk 1 Mod 0 sterilizing switch: A. Sectional view; B. Unoperated; C. Operated (foil seal broken; switch turned).

Chapter 4—LOADING, UNLOADING, AND DUD JETTISONING

exploder, remove the plastic cover and rotate the sterilization switch slot 90° with a nonferrous screwdriver in either direction until the slot is aligned with the notch pointing to RECOVER on the top of the exploder. The firing capacitors are now grounded and the exploder is safe. Further information can be obtained from OP 3112.

## TARTAR GMLSs

The steps in the operation of a Tartar launcher in bringing a missile from the magazine to the launcher arm were described in chapter 3. In automatic loading, no one is permitted in the magazine and no manual operations are needed in the magazine. This is true of all Tartar systems. There are no wings or fins to be assembled; the Tartar fins are erected automatically by launcher equipment and are folded manually. Figure 3-3 lists the activation procedures for the Mk 13 Mod 0 launching system; warmup of the missiles is shown in figure 3-7. Warmup is not required for the Standard MR missile.

## LOADING

The operational sequence in automatic loading with a Mk 13 GMLS is as follows. The launcher guide arm is empty and the launcher is at the load position.

1. Missile warmup is applied automatically for a minimum of 24 seconds to the selected number of missiles (1, 2, or 3).

2. The ready service ring inner and outer magazine latches retract, the ready service ring positioner retracts, and the ready service ring indexes clockwise (CW) to place a missile at the hoist. The ready service ring positioner extends, and missile warmup is applied for a minimum of 24 seconds.

3. After the warmup period, hydraulic control is transferred from the ready service ring to the hoist.

4. The raise latch retracts and the hoist raises to the intermediate position, where the hoist pawl contacts the missile aft shoe.

5. When the hoist is at the intermediate position, the magazine retractable rail extends to align the magazine rail with the carriage fixed rail.

6. The blast door opens and a spanning rail (an integral part of the blast door) completes the missile track from the magazine to the launcher rail.

7. The elevation positioner extends into the open blast door to secure and align the launcher in elevation (90°) during a load or unload cycle.

8. The hoist intermediate raise latch retracts and the hoist raises a missile to the launcher.

9. When the loaded hoist completes its raise cycle, the launcher aft motion latch extends to secure the missile on the guide arm. The warmup contactor on the launcher engages the missile (figure 4-9) and warmup power is applied for a minimum of 1.8 seconds. The fin openers engage the fins for unfolding. Mods 1, 2, 3, and 4 have minor differences in the fin opener and housing.

10. The hoist lowers to the magazine position. When the hoist is clear of the launcher, the train positioner retracts, freeing the launcher in train. The elevation positioner retracts into the launcher guide arm, clear of the blast door. (See figure 4-10.)

11. The blast door closes and provides a flameproof seal to the magazine.

With the launcher loaded and the hoist down, a new load cycle commences (see automatic loading steps 1 and 2) if continuous loading has been ordered by weapons control. However, before this second cycle can proceed further than step 2, the missile on the launcher must be fired. The following four missile firing steps set up the conditions for continuation of the second load cycle.

1. With a missile loaded on the launcher rail, the launcher free to train and elevate (positioners retracted), and the launcher assigned to a director, the launcher trains and depresses to the ordered position.

2. When the loaded launcher is within 1° of the ordered position, a firing order may be given by weapons control.

93

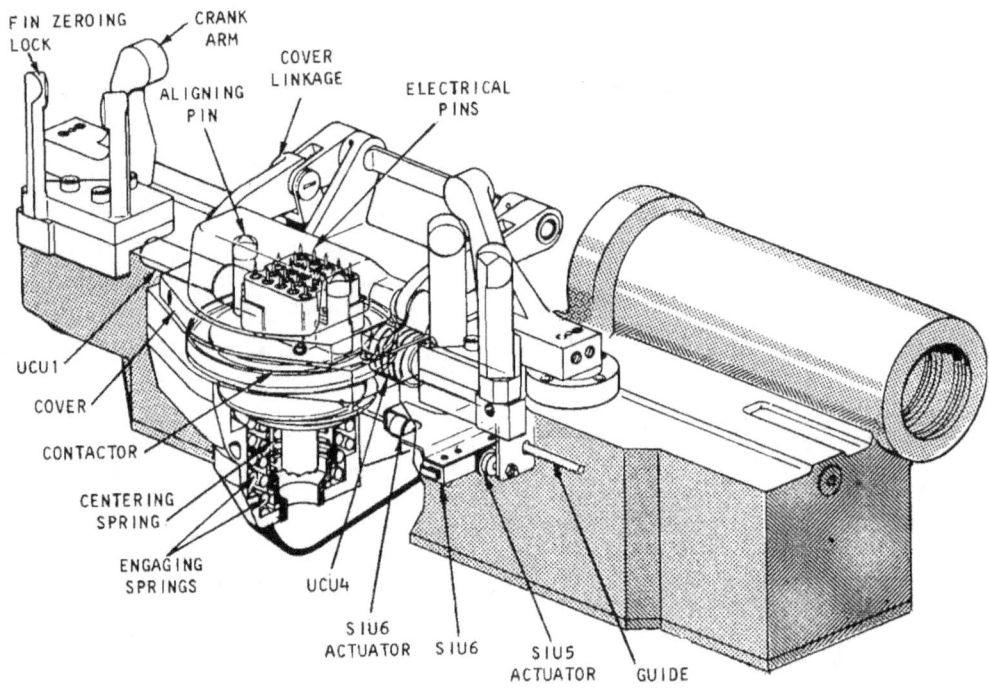

Figure 4-9.—Mk 13 GMLS fin opener and contactor.

3. The firing order begins a sequence of APS firing, arming and unlocking of the forward motion latch, disengagement of the contactor and fin-opener cranks (missile fins are erected), and ignition of the rocket motor.

4. After the missile leaves the guide arm, the guide arm components are set up to receive the next missile. At the same time, the launcher trains and elevates to the load position, permitting the second load cycle to proceed.

## UNLOADING

Unloading may be ordered if the tactical situation changes, or if the missile is a dud or a misfire and WCS decides to stow it for future offload. The steps in unloading depend on the location of the missile at the time the decision is made to stow the missile. In the first situation, the missile may be either on the launcher or on its way to the launcher. If it is on its way to the launcher, it would continue to finish the load cycle in normal operation. In the second and third situations it is on the launcher but the conditions are not the same. In the misfire situation, the missile must be disarmed by the arming device and then retracted, but the contactor and the fin-opener cranks do not have to be retracted, as they are already disengaged. A dud missile (third situation) must be disarmed by the arming device, then retracted; however, the contactor and the fin-opener cranks have to be retracted. In all situations the fins are manually folded after the fin cranks are disengaged. Folding the fins after the launcher has trained and elevated to the load position may be difficult, but sometimes it is necessary. Remember the warning about danger from launcher movements. Place the firing safety switch on the EP2 panel at SAFE (which breaks the powerlines to the motors in the train,

# Chapter 4—LOADING, UNLOADING, AND DUD JETTISONING

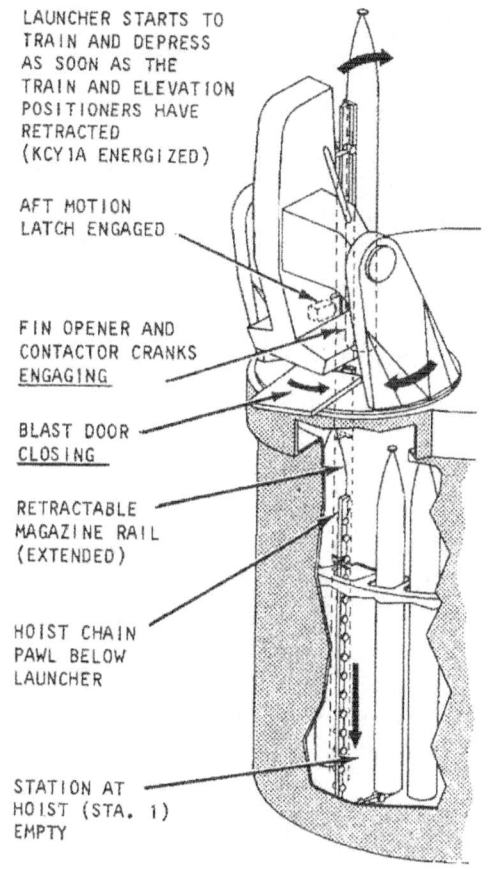

**Figure 4-10.—Mk 13 GMLS: Missile on launcher, hoist lowered below launcher, and blast door closing.**

elevation, and launcher power units); remove the switch lever; manually fold the fins of the missile on the launcher; return the switch lever to the panel and reposition the switch to close the motor circuits; then restart the motors. Depress the fins-manually-folded switch, and automatic unloading be be resumed.

The missile has to be returned to the same ready service ring from which it was taken. In automatic unloading the chain shifter will automatically shift to the proper ring. The ready service ring then rotates counterclockwise (CCW) to the empty cell position. Hydraulic control then shifts to the hoist, the blast door opens, and the hoist rises to the intermediate position. The retractable rail extends when the hoist leaves the magazine. When the hoist reaches the launcher, the hoist pawl engages the missile aft shoe. Then the aft motion latch retracts and the hoist (with the missile) lowers to the intermediate position. When the hoist lowered the missile to the intermediate position, the pawl disengaged from the missile shoe. Disengagement of the pawl now actuates a switch that allows the hoist chain to lower to the magazine position.

During automatic load and unload, the associated step control circuits are required to be open. All Tartar systems except the Mk 11 GMLS can be unloaded in the automatic mode.

## STEP CONTROL

During step operations, the control selector switch on the EP2 panel is in the STEP position. This breaks the automatic load and unload circuits, and prevents feedback into the automatic circuit. Step operation is used if the automatic mode malfunctions, and for all exercise operations. (Parts of exercise circuits and step control circuits are not the same.)

Strikedown, offloading, and checkout procedures are also done in step control. The same equipment is used in both strikedown and offloading, but the procedures are reversed.

Figure 4-11 is a schematic of step operation of fin openers and contactors on the Mk 13 launching system. The step control switches on the EP2 panel are manually actuated after the system is placed in step control. The broken lines in the drawing represent unload and exercise circuits; the solid lines are load circuits.

## DUD JETTISONING

The dud-jettisoning equipment for the Mk 11 and Mk 13 Tartar launching systems differs in a number of important details, and will be treated separately.

## MK 11 GMLS

The dud-jettisoning unit, figure 4-12, jettisons defective missiles from the guide arms

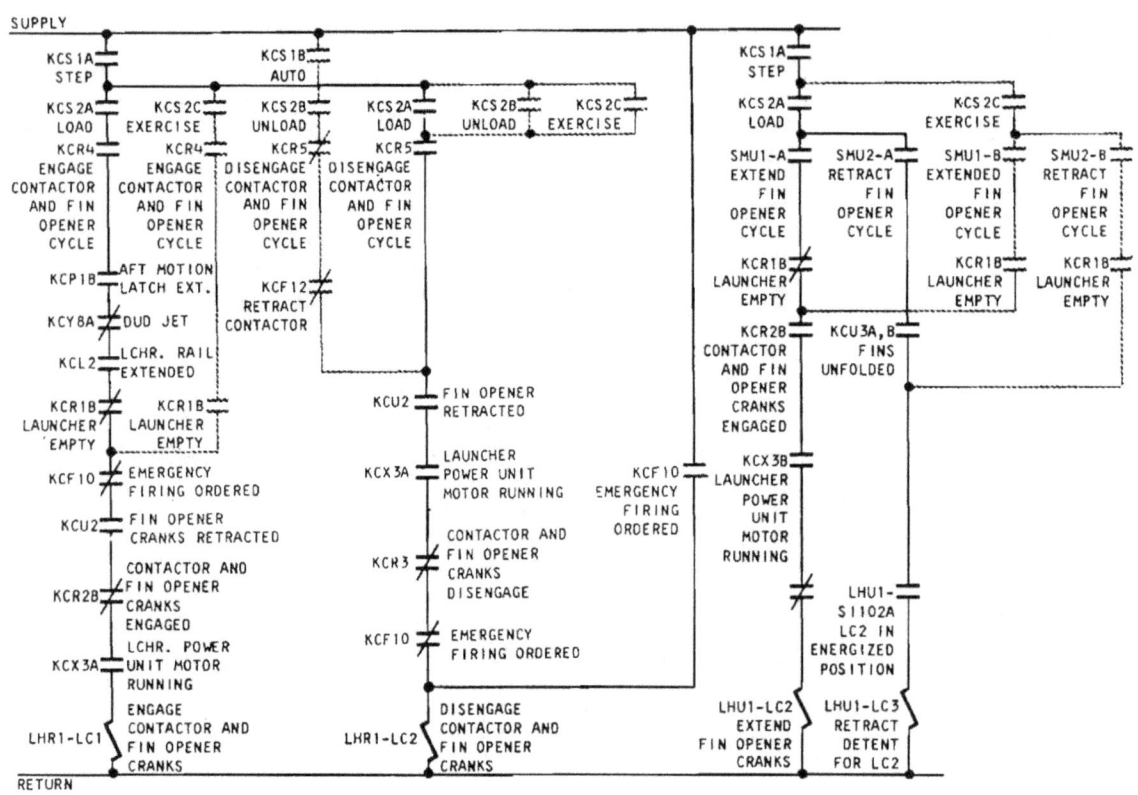

Figure 4-11.—Step operation of fin openers and contactor, load circuits on Mk 13 Mod 0 GMLS.

94.26

when the tactical situation requires it or if the missile is unsafe for return to the magazine. The dud-jettisoning unit consists of two ejectors (one for each guide arm), and a dud-jettisoning control valve panel. The ejectors align with the aft end of the missile on the launcher when the launcher is moved to either dud-jettison position. The ejector spud extends hydraulically to contact the missile. Air pressure from the accumulator, in conjunction with hydraulic pressure, acts on the ejector piston to move the spud forward and force the missile off the guide arm and overboard. The dud-jettisoning unit can be operated either automatically or manually.

To jettison a missile, the launcher is trained and elevated to either the A or B dud-jettison position, where the applicable guide arm aligns with a dud-ejector unit. A spud attached to the booster piston, extends to contact the missile, then ejects the missile at a sufficient velocity to clear the ship structure. The spud then retracts, and the launcher returns to a load position if missile firing is to be continued.

The dud-ejector unit is controlled and operated by a combination of compressed air and hydraulic pressure. Pressurized hydraulic fluid controlled by pneumatic-hydraulic accumulators generates the fluid pressure. Low pressure air (100 psi) generates the hydraulic pressure to extend and retract the spud; high pressure air (2100 psi) generates the hydraulic pressure to eject the missile from the guide arm.

Chapter 4—LOADING, UNLOADING, AND DUD JETTISONING

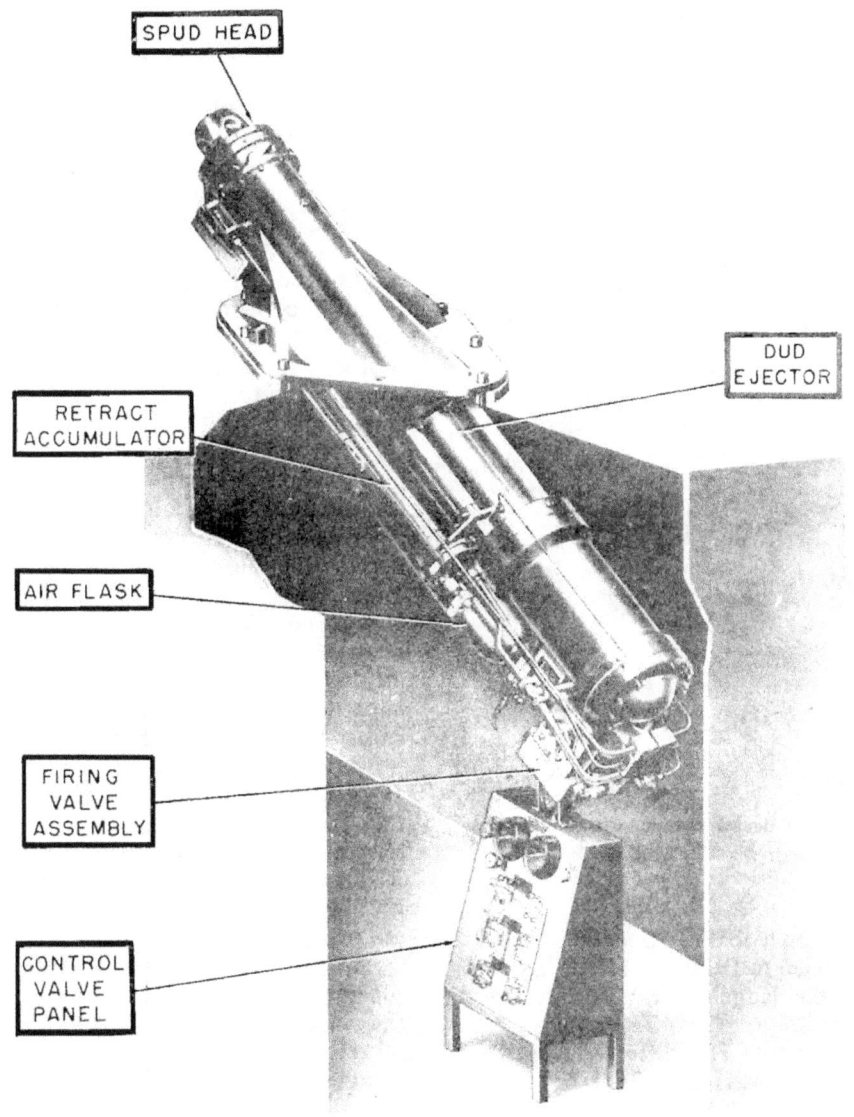

Figure 4-12.—Mk 11 GMLS dud-jettison unit in position to ram.

## Mk 13 GMLS

The jettison devices are identical in the Mk 13 and Mk 22 GMLSs. The dud-jettisoning device in the Mk 13 GMLS is located in the launcher arm as shown in figure 4-13. It is a nitrogen-actuated piston that applies force to the aft face of the forward missile shoe. The piston is hydraulically retracted after jettisoning. It is controlled locally from the launcher control

# GUNNER'S MATE M 1 & C

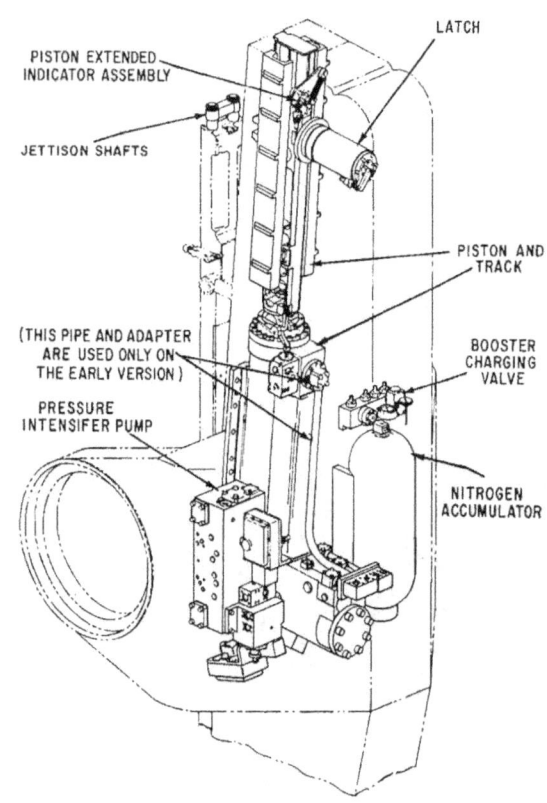

Figure 4-13.—Jettison device components in launcher arm on Mk 13 GMLS.

94.28

panel. When switch SMY1 (located on EP2 panel) is positioned to DUD JETTISON LOCAL or REMOTE, the launcher will slew to fixed dud-jettison positions (36°40′ elevation and 090° or 270° Train). The difference between DUD JETTISON LOCAL and DUD JETTISON REMOTE is that the FCS gyro-compass (stable element) is introduced into the launcher elevation control system to compensate for the ship's pitch and roll when SMY1 is positioned to DUD JETTISON REMOTE. When the DUD JETTISON pushbutton is pressed, the jettison piston extends to jettison a missile.

The following safety precautions apply during dud jettisoning:

1. Obtain permission from the weapons control station before jettisoning.

2. Make sure that communications have been established between the safety observer and the launcher captain before jettisoning.

3. Do not attempt to jettison with less than 2000 psi nitrogen pressure.

4. When the ship is rolling excessively, jettison on the downroll. In local control, observe the inclinometer to determine the degree and direction of roll.

At the EP2 panel when the train and elevation control switch (SMY1) is positioned to the DUD JETTISON REMOTE or LOCAL position, the jettison relay energizes, the contactor retracts, the arming device extends and unlocks the forward motion latch, and the launcher rail retracts automatically. Observe the inclinometer near the EP2 panel and do not jettison until the ship is on the downroll.

When the launcher synchornized light is on, depress the dud-jettison JETTISON pushbutton on the EP2 panel. Since nitrogen pressure is always present inside the jettison piston, the piston creeps forward. As the piston creeps, the aft shoe of the missile forces the forward motion latch out of the way, and the dud-jettison pawls engage the missile forward shoe.

After the piston jettisons the missile, the JETTISON pushbutton light goes on, showing that the piston is extended. The next step is to retract the piston by depressing the dud jettison RETRACT pushbutton, and the launcher rail EXTEND button. After that, loading operations can be resumed.

When the jettison device is exercised (operated when the guide is empty), it operates the same as when jettisoning except that the rate of travel of the piston is retarded while it is extending. If it were not retarded when not loaded with a missile, damage to the equipment would result. The throttle valve and the main check valve control the speed by restricting the passage of hydraulic fluid from the front side to the back of the jettisoning piston land. During jettisoning, the main check valve lifts, permitting the hydraulic fluid to flow to the back of the piston land and accelerating the piston to eject the missile. Two seals near the forward end of the piston spud prevent leakage of the nitrogen pressure, and thus prevent mixing of the hydraulic fluid and the nitrogen.

## Chapter 4—LOADING, UNLOADING, AND DUD JETTISONING

The OP for the equipment contains schematics, circuit diagrams, and detailed illustrations of the parts of the jettisoning equipment. In order to be able to make repairs and adjustments, you need to have a grasp of what happens inside the equipment when you push a certain button on the control panel. Study the OP for the system you have aboard. On the schematics, trace through the actions as they are described in the OP.

The nitrogen booster pump used with the jettison device aids in charging the jettison accumulator. This pump is mounted to the top of the inner structure inside the magazine assembly. It is a manually operated pump (figure 4-14) that boosts the pressure of the nitrogen supply system. When the jettison tank is to be recharged and the pressure is found to be low, run a temporary line from the discharge port of the nitrogen booster pump to the nitrogen-charging valve block for the jettison accumulator tank. The nitrogen line that connects to the adapter block (supply connection) is a permanent one. Opening the supply valve when the handle of the booster pump is in stow position permits nitrogen to flow until pressure stabilizes. If the pressure then is less than that required by the jettison device, turn the pump handle to the pressure (PRESS.) position. This permits nitrogen from the supply area to push against the piston head, forcing back and permitting nitrogen to flow to the jettison accumulator and increase the

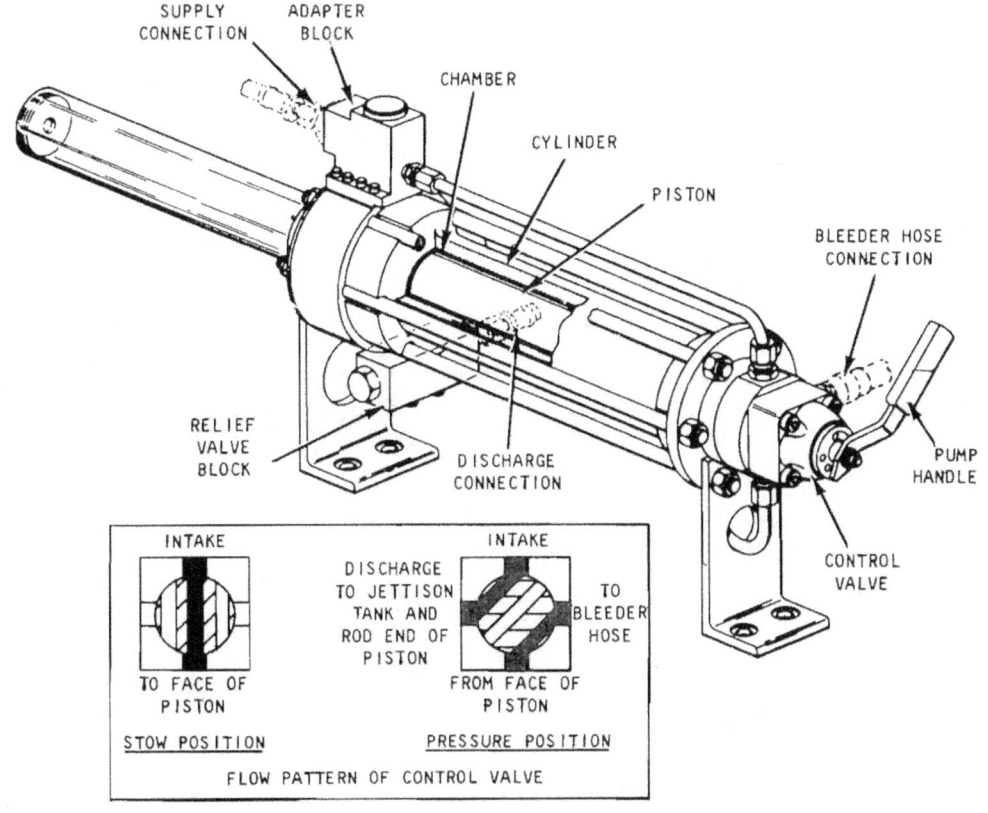

94.123

Figure 4-14.—Nitrogen booster pump assembly for jettison device.

pressure. When the piston is fully extended, move the pump handle to the STOW position. In this position, nitrogen on the face of the piston escapes to the atmosphere and nitrogen pressure on the back of the piston forces it to its former position. Continue stroking the piston by moving the handle as above until the required pressure is reached. A safety valve prevents excessive buildup of pressure.

MAINTENANCE AND REPAIR OF JETTISON DEVICE.—The nitrogen pressure in the jettison accumulator tank should be checked every week, and the tank recharged as necessary. The nitrogen charging assembly, located in the right-hand yoke of the launcher guide, is reached by opening the hinged access door with the special tool provided.

WARNING: Before doing any work on the launcher, remove the firing safety switch handle from the EP2 panel so the launcher cannot be activated inadvertently.

The required pressure varies with the temperature and depends on whether the jettison piston is retracted or extended. There should be a data plate posted on the inside of the access door. The nitrogen pressure tolerance of the dud jettison unit is plus or minus 5% of the pressue indicated on the data plate. Be sure all the valves are positioned for system operation and the nitrogen plug is firmly secured. Then close and secure the access door and return the firing safety switch handle to the EP2 panel. However, if the pressure is not within limits, proceed to charge or bleed the tank to the required pressure. Follow the procedures of MRCs W-1 and R-1 for charging the jettison unit with nitrogen.

In addition to weekly checking of the pressure, once every 2 weeks cycle the jettison device to keep the lubricants distributed and to detect (and correct) any malfunction. At the same time check the jettison device for signs of corrosion.

WARNING: If the nitrogen tank must be disconnected for repair work, first vent nitrogen to the atmosphere by opening the SV4, SV3, SV2, and SV1 valves (figure 4-15).

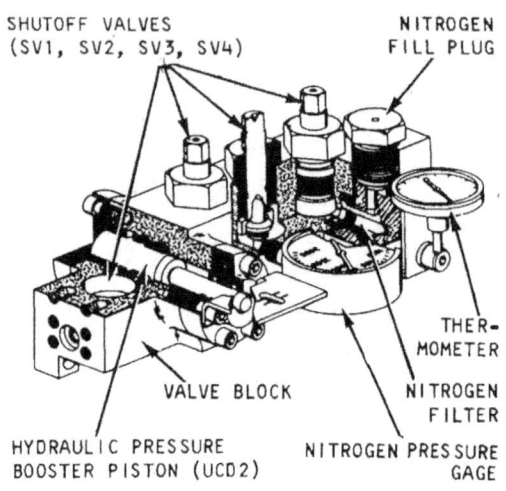

VIEW A

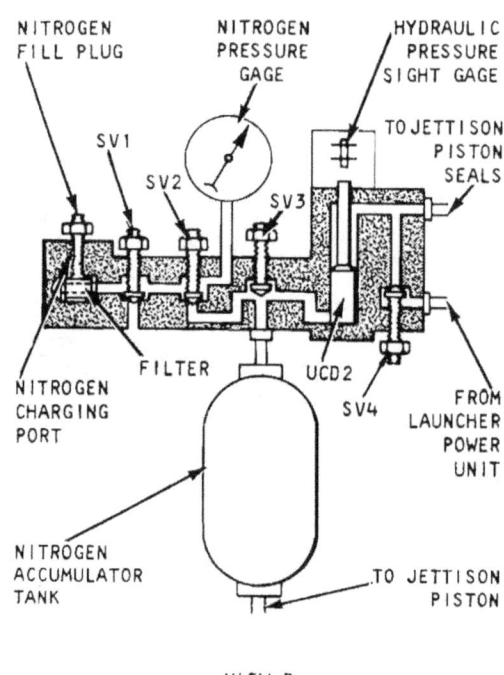

VIEW B

94.29
Figure 4-15.—Nitrogen-charging valve block and booster for jettison device on Mk 13 GMLS.

## Chapter 4—LOADING, UNLOADING, AND DUD JETTISONING

The hydraulic fluid level in the jettison booster piston is a weekly PMS check. This cannot be done with the launcher system inactivated, so you do not remove the safety switch handle. Instead, station safety personnel at the EP2 panel to make sure the train and/or elevation power drives are not started. The hydraulic booster assembly is located in the nitrogen-charging valve block (figure 4-15, view A), and is reached by the same access door in the launcher guide. The hydraulic pressure sight gage (figure 4-15, view B) is marked with red lines to indicate the recharging area. If the indicator (rod end of the booster piston) can be seen in the marked area, the fluid level is unsatisfactory and recharging is necessary. Follow the instructions in MRCs W-1 and R-1 for the launching system. When preparing a checksheet, you need the correct designation for each pushbutton, switch, and valve to be used. In figure 4-15, valves SV1, SV2, SV3, and SV4 are shown for the Mk 13 Mod 0 GMLS.

MISFIRE.—If the rocket motor fails to ignite when the missile-firing relay is energized, the firing relay remains energized and gives a misfire indication on the EP2 panel and in weapons control. The misfire light remains on until the missile is cleared from the rail, whether by dudjettisoning, emergency firing, or unloading. The course of action will be determined by ship's doctrine and the situation. Under combat conditions the dud/emergency firing key may be used to clear the rail quickly and with some chance of a tactical launching resulting.

The dud/emergency firing key is pressed in weapons control. If the missile still does not fire, another circuit energizes and starts contactor retraction and missile arming. After a brief delay, the rocket squibs fire regardless of the contactor position.

The daily operation of the equipment, using a training missile, includes checking of normal firing and misfire, and normal firing resulting in a dud. Firing of a dud and emergency firing, using the dud/emergency firing key, are tested weekly. In any testing of circuits, be sure that switch SMW2 on the EP1 panel is on EXERCISE, so that no missiles in the magazine will be put on warmup. The Tartar Mk 1 Mod 0 dud jettisoning slug is used to test the operability of the dud-jettison device. Emergency firing is tested with a training missile in three phases: for normal firing conditions to simulate an attempted firing; with the training missile set for dud; and with the training missile set for dud but with the electrical contactor extended. The operator of the EP2 panel notes the sequence of action in each case by the lights on the panel. The correct sequence is given in the OP.

Timing tests are made at 3-month intervals. To test the timing of the jettison device, use a stopwatch and record the time it takes for the jettison piston to retract. The allowable time is 20 seconds after depressing the dud jettison RETRACT pushbutton. No missile is on the guide arm for this test. The conditions of the test are simple; they are given in the OP.

Once a year the jettison device is serviced by draining any seepage of hydraulic fluid. The drain plug is at the base of the jettison cylinder. Before attempting to drain the seepage, the nitrogen in the accumulator tank must be vented to the atmosphere. This nitrogen is under 2400 psi pressure. Open SV1 and SV2 valves (figure 4-15) on the nitrogen-charging valve assembly. After cleaning up the seepage and replacing the drain plug, the accumulator must be recharged, the same as in daily maintenance procedures.

WARNING: Use only compressed nitrogen gas to charge accumulator flask bladders. NEVER charge with oxygen or compressed air. A mixture of oxygen and hydraulic fluid is extremely explosive. Since nitrogen and oxygen are both furnished in metal cylinders, use extreme caution to avoid taking the wrong cylinder by mistake. An oxygen cylinder is colored green; a nitrogen cylinder is colored gray with one or two black bands near the top.

### Mk 22 GMLS

The dud-jettison device in the Mk 22 launching system is in its guide arm (it has only one). The complete description of its operation is given in OP 3115, volume 2. If normal firing was ordered on an auto-load cycle and the missile does not leave the rail, a light on the EP2 panel gives a dud indication. If the missile firing relay energizes but the missile does not fire, a misfire indication appears on the EP2 panel. WCS must then decide whether to jettison the missile or return it to the magazine for later rework. If the decision is to jettison, the EP2 operator positions SMY1 to DUD JETTISON REMOTE. The remote-jettison sequence automatically disengages the contactor and the fin-opener cranks, extends the arming device, and unlocks the forward motion latch if these actions did not take place during the auto-load sequence as they should have. The retract-launcher-rail cycle then starts. When the launcher synchronized light goes on, indicating that the launcher has trained and elevated to the jettison position, the EP2 operator pushes the JETTISON pushbutton (SMD1) and the jettison piston ejects the missile overboard. The aft motion latch retracts, and when the jettison light (DSD1) turns steady, the EP2 operator pushes the jettison RETRACT pushbutton (SMD2). While the jettison piston retracts, the operator returns switch SMY1 to REMOTE.

Loss of launcher synchronization breaks the firing circuit. If the period of loss is short, the firing sequence resumes from the point of interruption, but if synchronization is lost for some time, the missile may have to be fired as a dud or misfire. The internal power supply of the missile is reduced rapidly once it is actived, so if the missile is not fired quickly, its range may be greatly reduced. In each case, weapons control will decide how to dispose of the missile. If possible, the missile will be returned to the magazine, marked as a dud, to be repaired later. The emergency firing circuit has a power supply independent of the normal firing circuit; it is resorted to in a tactical situation when a dud or misfire missile must be disposed of. The dud-emergency key in the weapons control station is closed in a second attempt to activate the APS squibs in the missile. If this does not clear the rail, the dud firing circuit switches to the emergency firing circuit. If the failure was in the contactor retract circuit, and the contactor did not retract, this method of firing would damage the contactor pad; therefore, it is used only as a last resort if the missile endangers the ship or its personnel.

Time intervals mentioned actually are very short. From the moment the dud/emergency firing key is pressed until the rocket motor ignites and launches the missile is less than 2 seconds.

## MK 26 GMLS

Two jettison devices are mounted on the launcher platform at the missile A- and B-side jettison positions. These devices jettison defective missiles from the launcher guide rails. The two jettison devices are similar. Each solenoid-controlled hydraulic-mechanical jettison device uses a gas generator to fire an expendable piston.

In preparation for jettisoning, a hydraulic motor extends the expendable piston to near contact with the aft end of the missile. Engergizing an electrical circuit fires the gas generator to propel the piston. Both the piston and the missile are jettisoned overboard. After the jettison device is fired, the expendable piston and the gas generator must be replaced before the device can be used again.

Principal components of the jettison device (figure 4-16) are:

- Jettison mounting bracket
- Jettison housing
- Motor mounting housing
- Extender mechanism
- Expendable piston assembly
- Hydraulic motor
- Control valve assembly
- Gas generator

The jettison mounting bracket, exposed on the launcher platform, supports and encloses the upper end of the jettison device. The mounting bracket, along with the expendable piston cap, protects the jettison device from weather and blast. To prevent ice buildup, heated anti-icing fluid is circulated through a passage in the

## Chapter 4—LOADING, UNLOADING, AND DUD JETTISONING

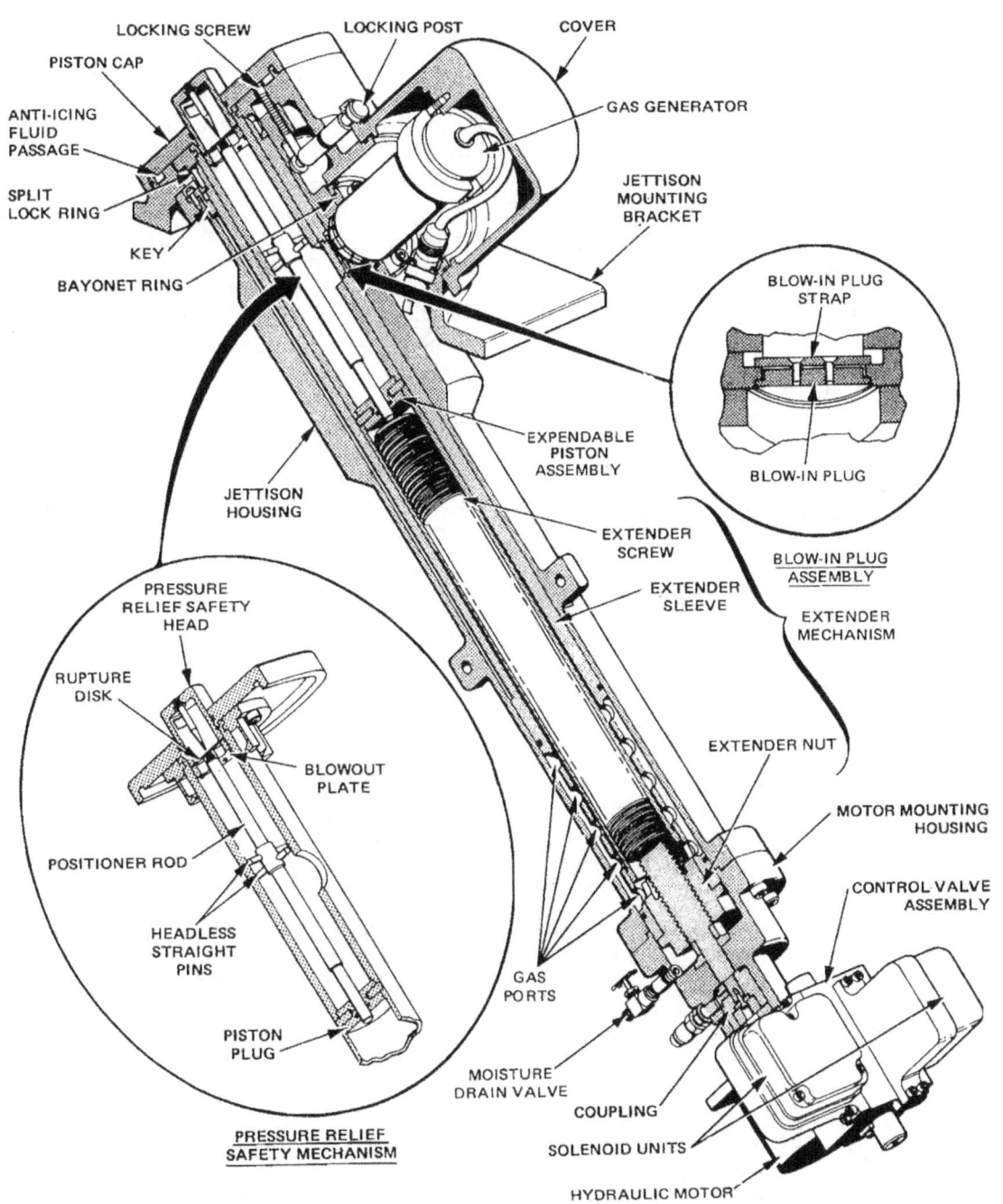

Figure 4-16.—Jettison device on Mk 26 GMLS.

bracket near the piston cap. The covered gas generator is secured to the mounting bracket.

The jettison housing is a cylinder about 4 feet long. The housing contains the extender mechanism and the expendable piston. The upper end of the housing is thicker than the lower end to withstand the explosive force of the gas generator. The housing bolts to the mounting bracket and to a support bracket secured to the launcher platform weldment. A gas escape tube on the housing vents the housing chamber to the atmosphere when the expendable piston is extended.

The motor mounting housing is bolted to the lower end of the jettison housing. It serves as an end cap for the jettison housing, a support for the extender mechanism, and a mount for the hydraulic motor. A moisture drain valve at the lower end of the housing drains the jettison housing.

The extended mechanism, driven by the hydraulic motor, extends and retracts the expendable piston assembly. The mechanism has an extender screw, extender nut, and extender sleeve. The shaft end of the extender screw is bearing-mounted in the motor mounting bracket and coupled to the hydraulic motor.

The extender nut is threaded onto the extender screw and bolted to the extender sleeve. A positive stop on the extender nut keeps the nut from jamming against the motor mounting housing.

The extender sleeve serves as the barrel for the expendable piston assembly. Gas ports in the extender sleeve let expanding gas from the gas generator into the sleeve. The expanding gas propels the expendable piston from the sleeve when the sleeve is extended and the generator is fired.

A blow-in plug is in the wall of the extender sleeve. If the gas generator accidentally fires when the sleeve is retracted, the blow-in plug assembly collapses into the sleeve. This lets gas escape harmlessly to the atmosphere through the expendable piston. A key and keyway keep the extender sleeve from turning when extending and retracting.

The expendable piston assembly fits inside the extender sleeve. The 75-pound assembly is propelled from the sleeve when a missile is jettisoned. The piston assembly consists of a piston sleeve, a piston cap, a piston plug, and a pressure relief safety mechanism.

A split lockring is bolted to the cap. The lockring attaches the cap to the piston sleeve and attaches the piston assembly to the extender sleeve. The piston plug inside the piston sleeve is a guide for the positioner rod of the safety mechanism and a header for gas pressure.

During jettison operation, gas pressure acts against the piston plug to propel the piston from the extender sleeve. If the gas generator accidentally fires with the piston retracted, expanding gases (through the blow-in plug in the extender sleeve) act against the piston plug to keep the piston in the sleeve.

A pressure relief safety mechanism is inside the expendable piston sleeve. This mechanism serves as a moisture seal, as a blow-in plug capture cage, and as a gas vent if the gas generator accidentally fires. The mechanism consists of a positioner rod, headless straight pins, a blowout plate and rupture disk, and a pressure relief safety head. All the pressure relief mechanism components are assembled to the positioner rod. The piston plug and the piston cap hold the rod in the center of the piston sleeve. When the expendable piston is retracted, the positioner rod forces the extender screw to extend the pressure relief safety head from the piston cap.

The rupture disk (when intact) keeps moisture out of the jettison device. The disk breaks if the gas generator accidentally fires. Two straight pins are staked at right angles at the center of the positioner rod. The pins serve as the capture cage for the blow-in plug.

If the gas generator accidentally fires, the capture cage keeps the pieces of the blow-in plug in the piston sleeve. At the same time, the taper pin in the safety head breaks the rupture disk. Gas then escapes harmlessly to atmosphere through the opening in the pressure relief safety head.

The hydraulic motor is bolted to the motor mounting housing. The 8-horsepower motor

## Chapter 4—LOADING, UNLOADING, AND DUD JETTISONING

turns the extender screw clockwise or counterclockwise to extend or retract the expendable piston assembly.

The control valve assembly, mounted on the hydraulic motor, has components that extend and retract the expendable piston assembly. Two solenoid units are mounted on the control valve assembly. They start the hydraulic actions for positioning the expendable piston in preparation for jettisoning.

The gas generator contains an electrically ignited explosive charge. As a result of the charge, gas pressure builds up and propels the expendable piston. The gas generator must be replaced each time the jettison device is fired. A bayonet ring attaches the gas generator to the housing. A locking post and locking screw under the piston cap attach the cover to the housing.

### REFERENCES

The following publications were used in the development of this chapter.

1. OP 3112, Exploder Mechanism Mk 20
2. OP 2236, Torpedo Exploder Mk 19 All Mods
3. NAVTRA 10200-B, *Gunner's Mate (M) 1 & C*
4. OP 2593, Guided Missile Launching System Mk 11 Mods 0, 1, and 2, volume 1
5. OP 2665, Guided Missile Launching System Mk 13 Mods 0, 1, 2, and 3, volume 1
6. OP 3114, Guided Missile Launching System Mk 10 Mods 7 and 8, volume 1
7. OP 3115, Guided Missile Launching System Mk 22 Mod 0, volume 1
8. OP 4104, Guided Missile Launching System Mk 26 Mods 0, 1, and 2, volume 1

# CHAPTER 5

# ELECTRICITY AND ELECTRONICS

All the electrical and electronic components used to operate and test the GMLS are part of the GMM's responsibility. A review of the occupational standards and personnel advancement requirements (PARs) shows that the GMM third and second class are required to obtain the knowledge factors in these fields (electricity and electronics) and to perform the more routine tests and repairs, while the GMM first and chief are expected to perform the more sophisticated troubleshooting, tests, adjustments, and repairs.

The principles of electricity and electronics are covered in the *Navy Electricity and Electronics Training Series (NEETS)*. The rate training manual (RTM) for *GMM 3 & 2* covered the operation and use of many of the electrical and electronic devices used in GMLSs. It also described a typical firing circuit and power motor control circuit. The action of each component in these circuits was explained so that the circuits could be traced. Troubleshooting techniques and charts also were presented and explained. This chapter shows in more detail how the devices described in the RTM for *GMM 3 & 2* are used in GMLS circuits. It also goes into more depth in the areas of circuit analysis and troubleshooting. The main portion of this chapter, however, is directed to the circuits used in the newer GMLSs: solid-state, digital, and logic circuits.

## CONTROL SYSTEM

The GMLS's control system includes equipment that controls, sequences, and interlocks system operations, including the programming of the missile on the launcher rail or guide. Electrical control modes include remote, local, and test. The power distribution equipment receives 440-volts a.c. at 60 hertz from the ship's supply, and 115-volts a.c. at 400 hertz from the fire control switchboard. In general, the power distribution equipment converts the a.c. to d.c., filters it, and distributes it to the GMLS power control and indication circuits.

### POWER AND CONTROL PANELS

Marks 10, 11, 13, 22, 26, and 29 GMLSs have one main power distribution panel. For the Marks 10, 11, 13, and 22 GMLSs, the power distribution panel is designated EP1. On the Mk 26 GMLS, the power distribution panel is known as the power distribution center, and on the Mk 29 GMLS it is called the maintenance interconnecting cabinet. Regardless of the nomenclature, these panels are used basically for the same functions.

The launching system control panel (EP2) on the Mk 10, Mk 13, and Mk 22 GMLSs is the operational control panel for the GMLS. This panel contains the switches and relays required to select and control the type of operation desired; the lamps to indicate the sequence of operation; and the synchros for launcher load, jettison, and strikedown positions (Mk 13 and Mk 22 GMLSs only). Also, the amplifier and error meters for train and elevation are located within these panels. The EP2 panel operator monitors the weapons control orders, which normally will be transmitted by steady or blinking lights. On the Mk 26 and Mk 29 GMLSs, the launching system control panel is

## Chapter 5—ELECTRICITY AND ELECTRONICS

referred to as the main control console and the guided missile control cabinet, respectively. It should be noted that, on the Mk 29 GMLS, local or remote operation of the launcher is selected at the maintenance interconnecting cabinet, and not at the guided missile control cabinet. Solid-state electronics are used extensively in the Mk 13 Mod 4 and the Mk 26 and Mk 29 GMLSs. As with all electrical and electronic equipment, you must have a thorough understanding of the basic circuits, whether conventional or solid-state.

## CONTROL CIRCUITS

The control circuits for each launching system are used to control sequence and interlock systems operation. The control circuit voltages vary according to the system and the application. Some control voltages are obtained from stepdown transformers within the main power distribution panels; some of these voltages are then rectified. Solid-state devices use a variety of d.c. voltages, most of which are supplied by d.c. power supplies. Weapons control information, synchro supply, and the firing circuits are distributed to the GMLS from the fire control (FC) switchboard.

You must know which circuits can be turned off at your main power distribution panel, and which ones are controlled at the FC switchboard. All the GMLSs have a certain commonality, and knowing one system will aid you in understanding another.

### Conventional Type

The conventional type of control circuit shown in figure 5-1 is for the Mk 10 GMLS. This type of circuit was described in the rate training manual for *GMM 3 & 2*. It is included here to provide a reference to compare with the solid-state and logic circuits that will be described in this chapter.

### Solid-State Type

The control and indicating circuits for the newer GMLSs, such as the Mk 13 Mod 4, and the Mk 26 and Mk 29, use transistors as switching elements. These circuits, also known as solid-state control circuits, have not eliminated the use of the relay, but have greatly reduced the number of relays required for a system.

NOTE: As this text was going to press the Navy announced that the Mk 13 Mods 1 through 3 and Mk 22 GMLSs are to be modified to include the new design features and solid-state circuitry used on the Mk 13 Mod 4 GMLS.

The use of transistors as switching elements for control circuits increases the complexity of the circuits but reduces the chance of circuit malfunctions. A relay's switching action involves the opening and closing of mechanical contacts. After a period of time, the contacts become corroded and pitted due to electrical arcing and, eventually, they fail. On the other hand, transistor switching action is nonmechanical and, according to theory, should not fail.

These solid-state control circuits still must use interlock switches. However, most of the switches are of the proximity type, which are described later in this chapter.

Transistors can be used as the switching element in solid-state control circuits because, like a simple switch, they present low resistance to current flow when in a conducting state (switch closed) and a high resistance to current flow when in a nonconducting state (switch open). The terms used to indicate that a switching transistor is conducting are "turned on," "saturated," and "forward biased," while the nonconducting condition is described as being "turned off," "cut off," and "reverse biased."

Both the NPN and PNP types of transistors are used in the GMLS solid-state control circuits. However, the NPN type is used more often. The PNP type, when used, is normally in the output portion of the circuit.

Before proceeding into actual solid-state control circuits used in GMLSs, a short review of how switching transistors operate may be helpful. Figure 5-2 shows functional diagrams and schematic symbols for the NPN and PNP transistors. The (b) and (d) portions of this

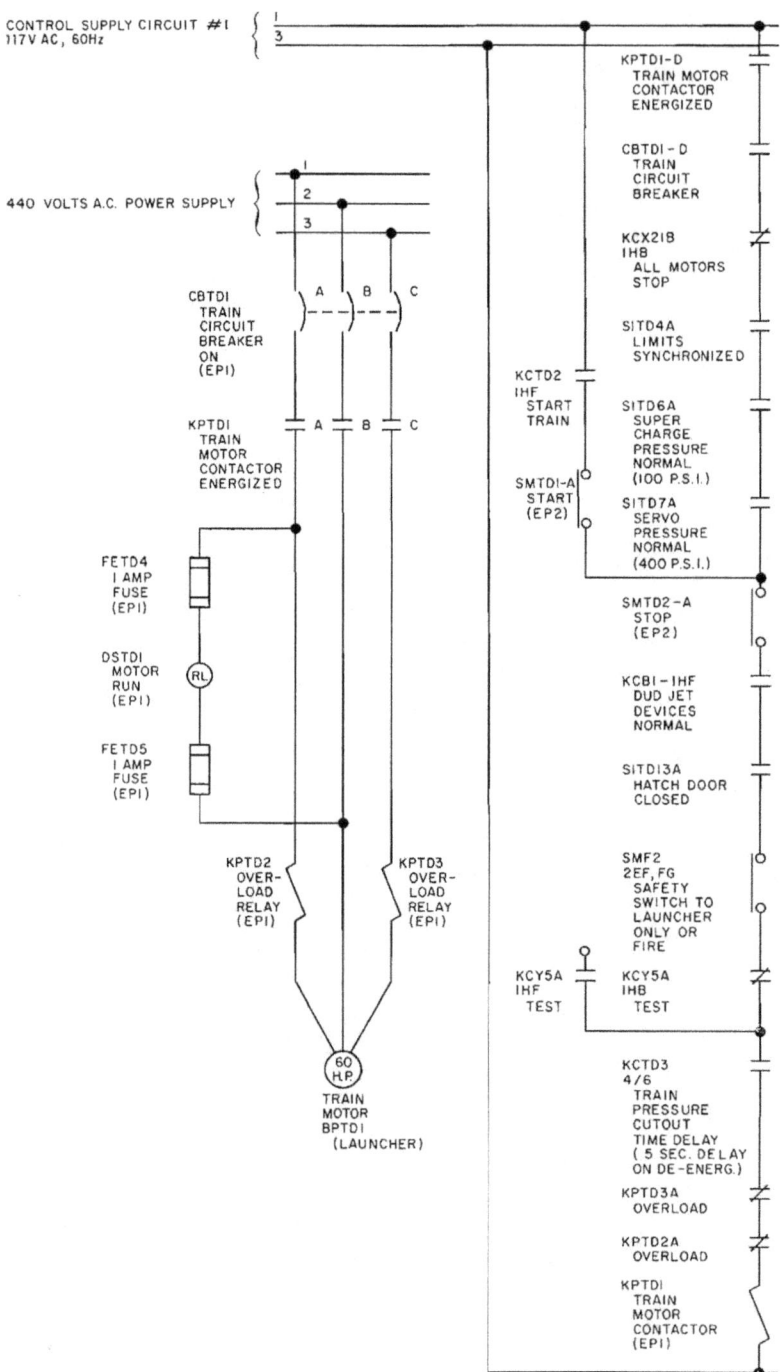

Figure 5-1.—Train start circuit, Mk 10 GMLS.

## Chapter 5—ELECTRICITY AND ELECTRONICS

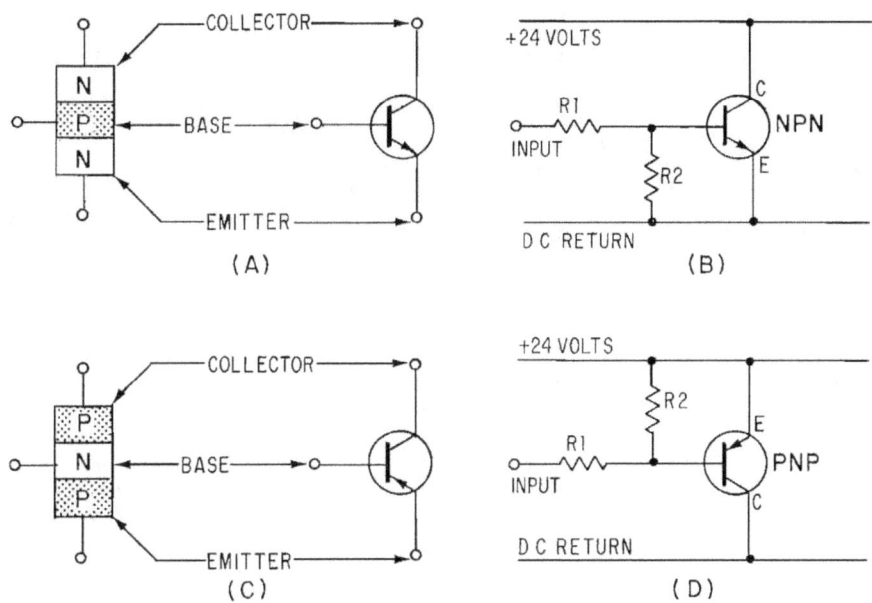

Figure 5-2.—Symbols and functional schematic of the NPN and PNP switching transistors.

94.220

illustration also show how these transistors normally are connected into the switching circuits. Note that the collector of the PNP transistor is connected to d.c. return, while the collector for the NPN transistor is connected to the high voltage side of the circuit. Also, note that the PNP transistor in the (d) portion of figure 5-2 is shown inverted from the way it is shown in the (c) portion of the figure. This inversion merely simplifies the circuit and has no effect on its operation.

The input that is applied to a switching transistor's base normally is called the signal voltage. This voltage is described as being either in a high or low state. The values of the high- and low-signal voltages vary with the circuit in which they are used. The low value is normally at, or near, zero volts (in some circuits, a small negative value is used), while the high value is normally between +5 and +24 volts.

As previously stated, a transistor must be forward biased to conduct. To forward bias an NPN transistor, the base must be made more positive than the emitter, and to forward bias a PNP transistor, the base must be less positive than the emitter. Therefore, the NPN transistor in the (b) portion of figure 5-2 must have a high signal applied to its base to conduct because, with a low signal applied, its base and emitter are both at return (ground) potential.

The PNP transistor in the (d) portion of figure 5-2 operates in a similar manner. If no signal is applied to its base, its emitter and base are both at a +24 volt potential and cannot conduct. However, when a high signal is applied to the PNP transistor's base, current will flow from the input through resistors R1 and R2 to the +24 volt (high) side of the line. The voltage drop across the resistors reduces the bias applied to the base, while the emitter remains biased at +24 volts. Under this condition, the PNP transistor is forward biased and it conducts.

If an NPN and PNP transistor are connected in the manner shown in the (a) portion of figure 5-3, you have a simplified example of an actual switching circuit that will be presented later in

# GUNNER'S MATE M 1 & C

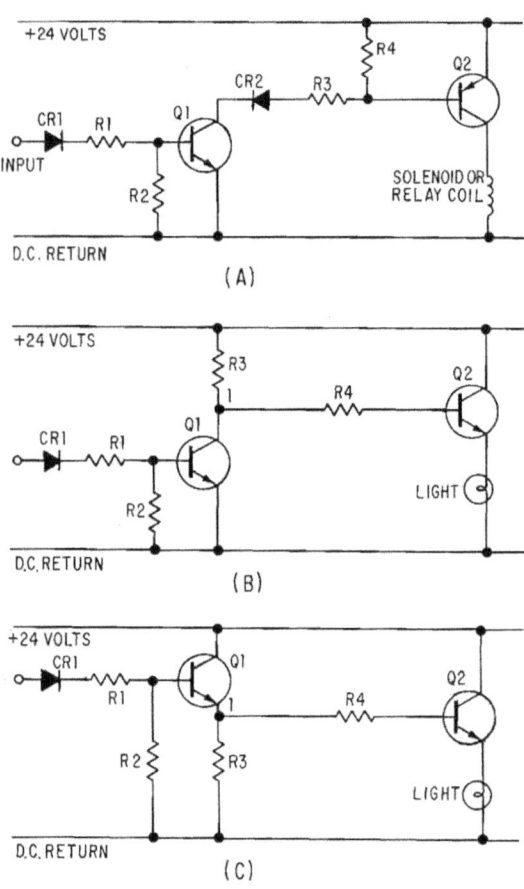

**94.221**
Figure 5-3.—Examples of simplified solid-state switching circuits.

this chapter. To energize the solenoid or relay coil in this circuit, a high signal is applied to the base of Q1. With current flowing from d.c. return through R2, R1, and the input diode CR1, Q1 is forward biased and turns on. Current now flows from d.c. return through Q1, CR2, R3, and R4, to the +24 volt line. This current flow forward biases Q2, causing it to turn on and pick up (energize) the solenoid or relay coil.

The circuit in the (b) portion of figure 5-3 is similar to the circuit in the (a) portion of the figure, except that Q2 is an NPN transistor. Q1

is turned on by applying a high signal at its base. However, with Q1 conducting, point 1 (input to the base of Q2) is at ground potential. Since both the base and emitter of Q2 are at the same potential, Q2 is reverse biased (turned off) and acts like an open switch. If a low signal is applied to Q1, it turns off (acts like an open switch) placing point 1 near +24 volts. Q2 is now forward biased and turns on. Thus, a low input to Q1 is required to turn on the light.

The circuit in the (c) portion of figure 5-3 is designed to function in opposition to the circuit in the (b) portion of the figure. When Q1 is turned on by applying a high signal at its base, point 1 (input to Q2) is shorted to the high side of the line placing a bias of approximately +24 volts on the base of Q2. Since Q2 is an NPN transistor, the +24 volts causes it to become forward biased and it is turned on. Thus, a high input to Q1 turns on the light in this circuit.

The following solid-state circuits are used on the Mk 26 GMLS:

1. Proximity switch.
2. Optical limit switch.
3. Buffer.
4. Inverter-buffer.
5. Diode matrix.
6. Solenoid driver.

Many of these circuits, or variations of these circuits, will be used also in the Mk 13 Mod 4 and Mk 29 GMLSs. Each GMLS has its own peculiar circuits. Information on the Mk 26 GMLS is presented in this chapter.

Figure 5-4 is a block diagram showing how these individual circuits are arranged into a control circuit. It should be understood that inputs to the buffer, inverter-buffer, and diode matrix circuits can come from other sources than those illustrated, and the diode matrix outputs can be applied to other circuits than the one illustrated. Note that only one output from each of the buffer and inverter-buffer circuits is shown connected to the diode matrix circuit. This was done to simplify the circuit. These disconnected outputs normally will be applied to the diode matrix circuit illustrated or some other diode matrix circuit.

Chapter 5—ELECTRICITY AND ELECTRONICS

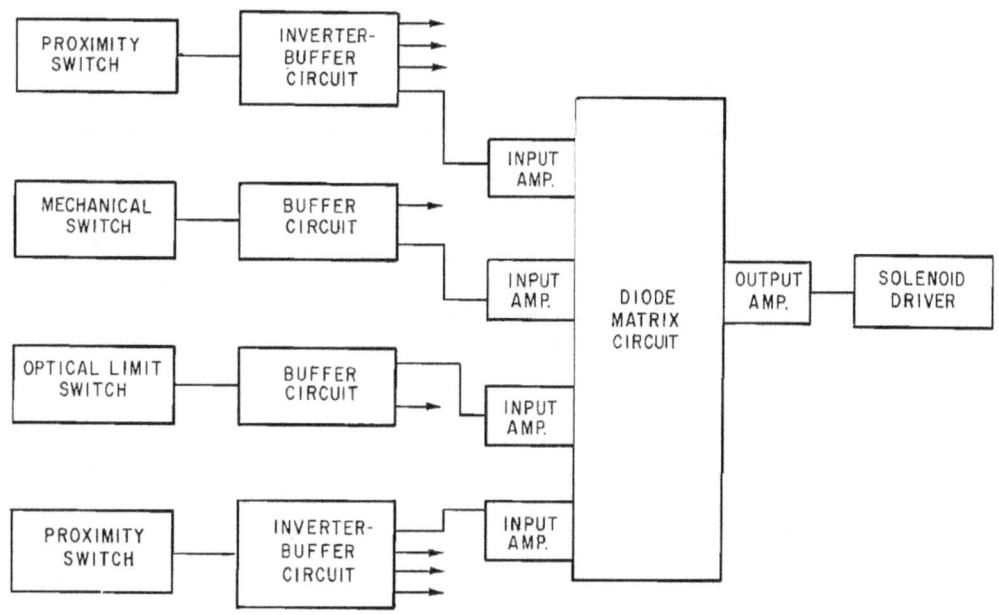

Figure 5-4.—Simplified block diagram of a control circuit for the Mk 26 GMLS.

PROXIMITY SWITCH CIRCUIT.—The proximity switch is a magnetically activated switch which detects launching system component positions. The proximity switch does this by sensing the presence of a permanent magnet mounted on the moving component. When the magnet is in proximity (mechanical alignment) with the switch, the switch activates and produces a high-voltage output.

Internally, the proximity switch uses a Hall-effect generator (figure 5-5) to sense the presence of the magnet. The Hall-effect is basically the generation of a voltage across the opposite edges of an electrical conductor that is carrying current and is exposed to a magnetic field. The voltage output of the Hall-effect generator is proportional to the strength of the magnetic field to which it is exposed.

When the permanent magnet is not near the switch, the Hall-effect generator applies a low voltage to the amplifier. The amplifier, in turn, provides a low voltage to the switch circuit.

Since the voltage input to the switch circuit is below its switching threshold level, the switch circuit is considered open. In this condition, the proximity switch is deactivated and its output voltage is a nominal +5.5 volts d.c.

When the magnet field comes near the switch, the Hall-effect generator applies a high voltage to the amplifier. The amplifier, in turn, provides this high voltage to the switch circuit.

Since the voltage input to the switch circuit is now above its switching threshold level, the switch circuit closes. When this happens, the resistor is effectively bypassed or shorted out. In this condition, the proximity switch is activated and its output voltage is between +14 and +24 volts d.c. The proximity switch output is applied to an inverter-buffer circuit in the GMLS circuitry.

OPTICAL LIMIT SWITCH CIRCUIT.—The optical limit switch electronic circuitry consists of a solid-state transmitter and a solid-state detector (figure 5-6). When d.c. power (24 volts)

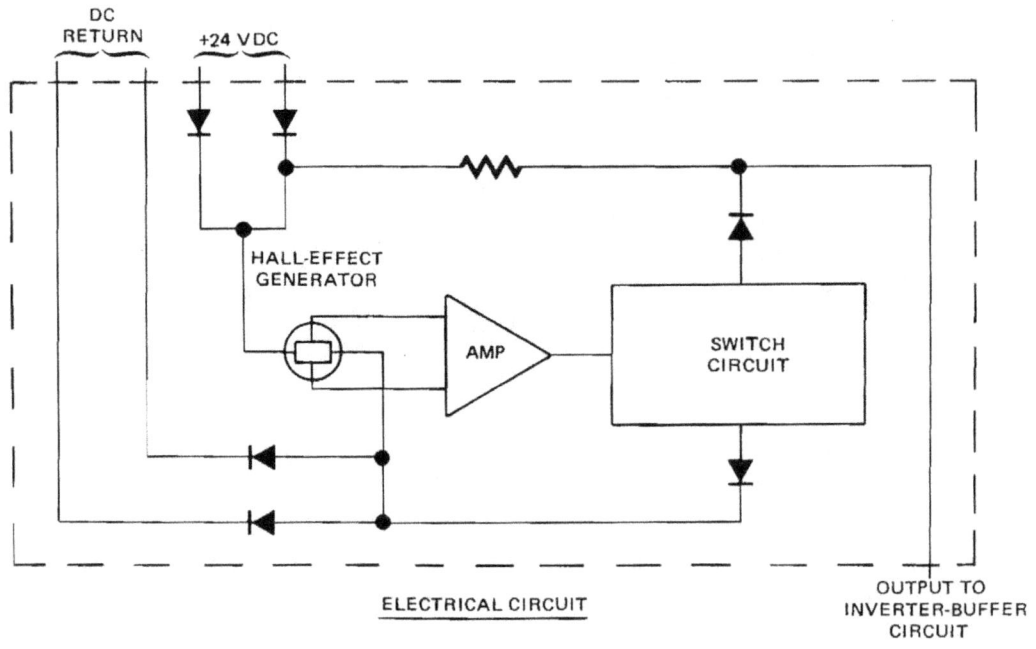

Figure 5-5.—Proximity switch circuit, Mk 26 GMLS.

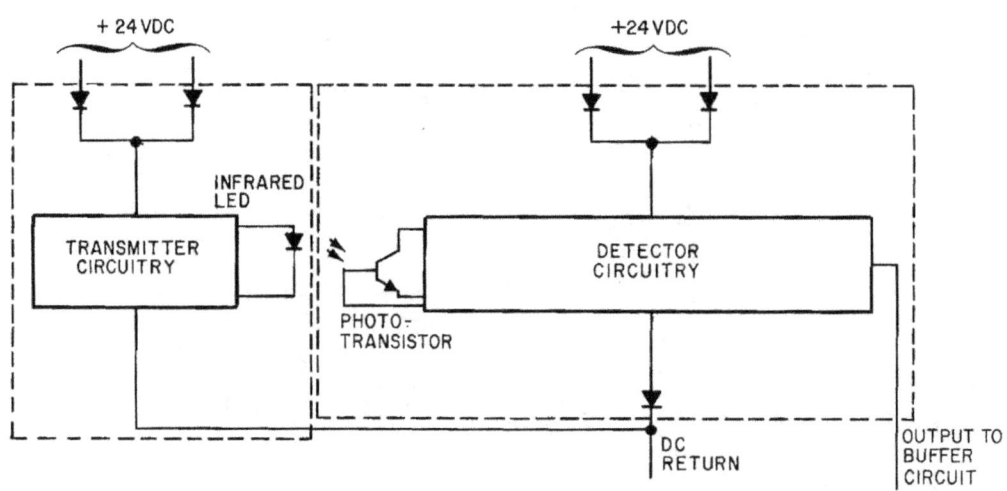

Figure 5-6.—Optical limit switch circuit.

## Chapter 5—ELECTRICITY AND ELECTRONICS

is applied to the transmitter, the transmitter circuitry pulses an infrared light-emitting diode (LED) at 500 hertz. The infrared beam generated by the LED is focused by the transmitter lens and aimed at an external reflector. The reflector then reflects the light beam back to the detector lens which focuses the beam on the phototransistor connected to the detector circuitry.

When the phototransistor is activated by the light beam, it provides a positive input to the detector circuitry that, in turn, provides a high output to the buffer circuit.

When the light beam is interrupted by a missile or moving equipment, the phototransistor deactivates. This causes the detector circuitry to provide a low input to the buffer circuit.

BUFFER CIRCUIT.—The buffer circuit (figure 5-7) is an electronic circuit which acts as an interface (buffer) between an optical limit switch, a mechanical switch, or a diode matrix circuit, and the rest of the launching system circuitry. This circuit provides the inverted and noninverted signal outputs needed by the launching system circuitry. Although the buffer circuit has two input lines, only one input is used. In some systems, this circuit is called an inverter circuit.

When an optical limit switch is deactivated, a low signal is applied to the buffer circuit. This low-voltage potential is applied across a resistor to the base of transistor Q1, turning it off. Because Q1 is not conducting, a high-voltage potential is sensed at the bottom of the resistor above it and is applied to the buffer circuit's left-hand output as a high signal. This high-voltage potential also is applied across a resistor to the base of transistor Q2, turning it on. With Q2 conducting, d.c. return is sensed at the bottom of the resistor above it and is applied

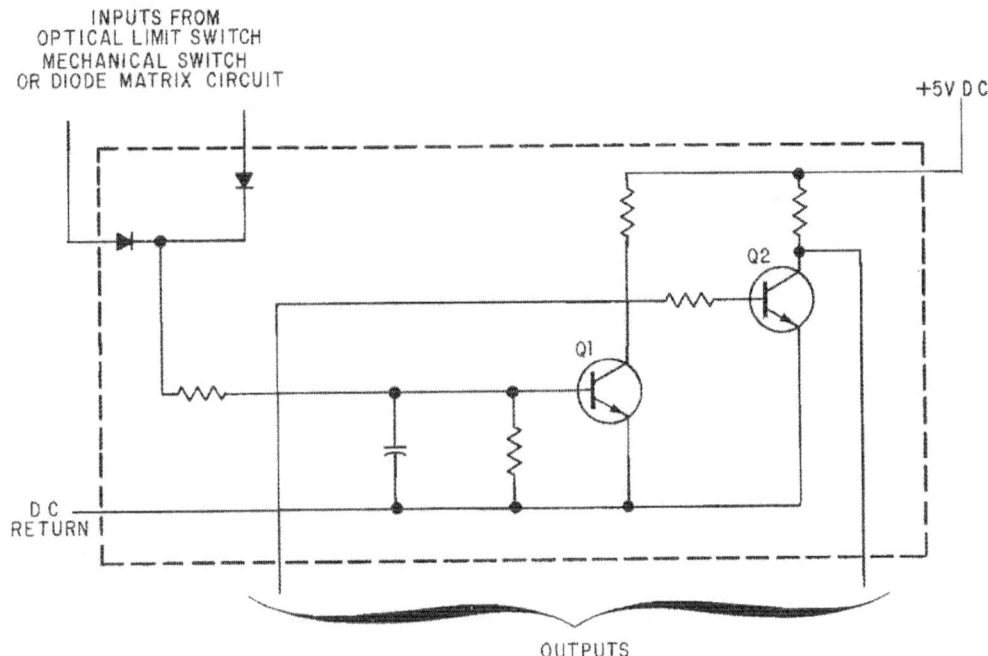

Figure 5-7.—Buffer circuit.

to the buffer circuit's right-hand output as a low signal.

When an optical limit switch is activated, a high signal is applied to the buffer circuit. This high-voltage potential is applied across a resistor to the base of transistor Q1, turning it on. With Q1 conducting, d.c. return is sensed at the bottom of the resistor above it. This is applied to the buffer circuit's left-hand output as a low signal.

The low-voltage potential also is applied across a resistor to the base of transistor Q2, turning it off. Because Q2 is not conducting, a high-voltage potential is sensed at the bottom of the resistor above it. This is applied to the buffer circuit's right-hand output as a high signal.

INVERTER-BUFFER CIRCUIT.—The inverter-buffer circuit (figure 5-8) is an impedance-matching electronic circuit which receives the voltage output from a proximity switch. This circuit provides the inverted and noninverted signal outputs needed by other launching system circuitry.

When the proximity switch is deactived, a +5.5 volt d.c. input voltage is applied to the inverter-buffer circuit. This voltage is applied across resistor R34 and capacitor C1. The voltage drop developed across resistors R2, R3, and R4 is not enough to break down zener diode CR2, which has a breakdown level of 8.2-volts d.c. Therefore, a low-voltage potential is sensed at the anode of CR2. This low-voltage potential is applied across resistors R1 and R8 to the base of transistors Q1 and Q4, turning them off.

Because Q4 is not conducting, a high-voltage potential is sensed at the bottom of resistor R9 and is applied to the inverter-buffer flagged 1 output as a high signal. This high-voltage potential also is applied across resistor R10 to the base of transistor Q5, turning it on. With Q5 conducting, d.c. return is sensed at the bottom

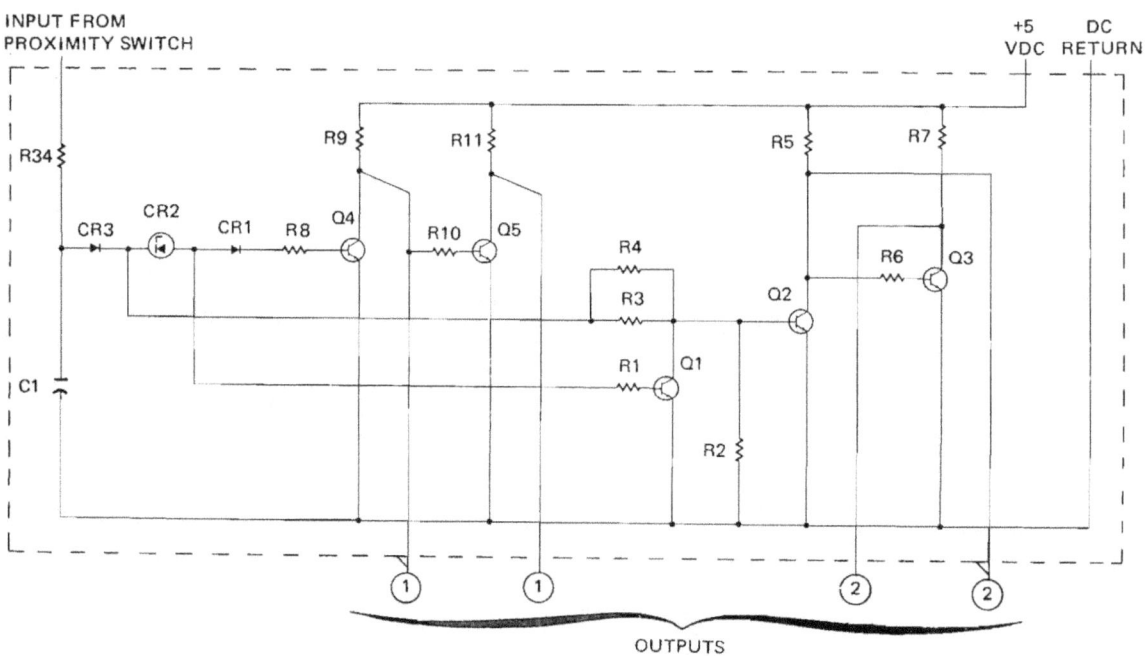

Figure 5-8.—Inverter-buffer circuit.

94.226

# Chapter 5—ELECTRICITY AND ELECTRONICS

of resistor R11 and is applied to the inverter-buffer unflagged 1 output as a low signal.

Because Q1 is not conducting, a high-voltage potential is applied to the base of transistor Q2, turning it on. With Q2 conducting, d.c. return is sensed at the bottom of resistor R5 and is applied to the inverter-buffer flagged 2 output as a low signal. This low-voltage potential also is applied across resistor R6 to the base of transistor Q3, turning it off. With Q3 not conducting, a high-voltage potential is sensed at the bottom on resistor R7 and is applied to the inverter-buffer unflagged 2 output as a high signal.

When the proximity switch is activated, a +14 volt d.c. to +24 volt d.c. input voltage is applied to the inverter-buffer circuit. This increase in voltage at the cathode of zener diode CR2 causes it to break down, applying a high-voltage potential through it. This high-voltage potential is applied across resistors R1 and R8 to the base of transistors Q1 and Q4, turning them on.

With Q4 conducting, d.c. return is sensed at the bottom of resistor R9 and is applied to the inverter-buffer flagged 1 output as a low signal. This low-voltage potential also is applied across resistor R10 to the base of transistor Q5, turning it off. Because Q5 is not conducting, a high-voltage potential is sensed at the bottom of resistor R11 and is applied to the inverter-buffer unflagged 1 output as a high signal.

With Q1 conducting, d.c. return is applied to the base of transistor Q2, turning it off. Because Q2 is not conducting, a high-voltage potential is sensed at the bottom of resistor R5 and is applied to the inverter-buffer flagged 2 output as a high signal. This high-voltage potential is also applied across resistor R6 to the base of transistor Q3, turning it on. With Q3 conducting, d.c. return is sensed at the bottom of resistor R7 and is applied to the inverter-buffer unflagged 2 output as a low signal.

The inverter-buffer circuit also uses fail-safe provisions if it receives no input voltage. This would occur if the input line to the circuit were either open or shorted to ground. The lack of enough voltage to break down zener diode CR2 causes transistors Q1 and Q4 to turn off, and transistor Q5 to turn on. Under this condition, a high signal is applied as the inverter-buffer flagged 1 output and a low signal is applied as the unflagged 1 output.

Since there is no voltage drop across resistor R2, transistor Q2 turns off and transistor Q3 turns on. Under this condition, a high signal is applied as the inverter-buffer flagged 2 output and a low signal is applied as the unflagged 2 output. In these states, the outputs of the inverter-buffer circuit deactivate all of the launching system components that they control, and all are in a safe condition.

DIODE MATRIX CIRCUIT.—A diode matrix circuit is an electronic circuit which receives input signals from either the launching system circuitry or another diode matrix circuit. The diode matrix circuit combines these signals and generates an output signal which is applied to a solenoid driver, a lamp driver, relay driver, or another diode matrix circuit. This solid-state circuit is the equivalent of a relay circuit such as the train motor start and run circuit shown in figure 5-1. However, the solid-state circuit is more reliable, uses less power, and takes up less space. A diode matrix circuit consists of an input amplifier, a diode matrix, and an output amplifier.

Input Amplifier.—The input amplifier (figure 5-9) is an impedance-matching electronic circuit. It gets its input from either the launching system circuitry or another diode matrix circuit. This circuit acts as an interface between the source of the input and the diode matrix.

When the input source applies a low signal to the input amplifier, a low-voltage potential is applied across a resistor to the base of transistor Q1, turning it off. Because Q1 is not conducting, a high-voltage potential is sensed at the bottom of the resistor above it. This high-voltage potential is applied to the base of transistor Q2, turning it on. With Q2 conducting, d.c. return is sensed at the bottom of the resistor above it and is applied to one row of the diode matrix as a low signal.

115

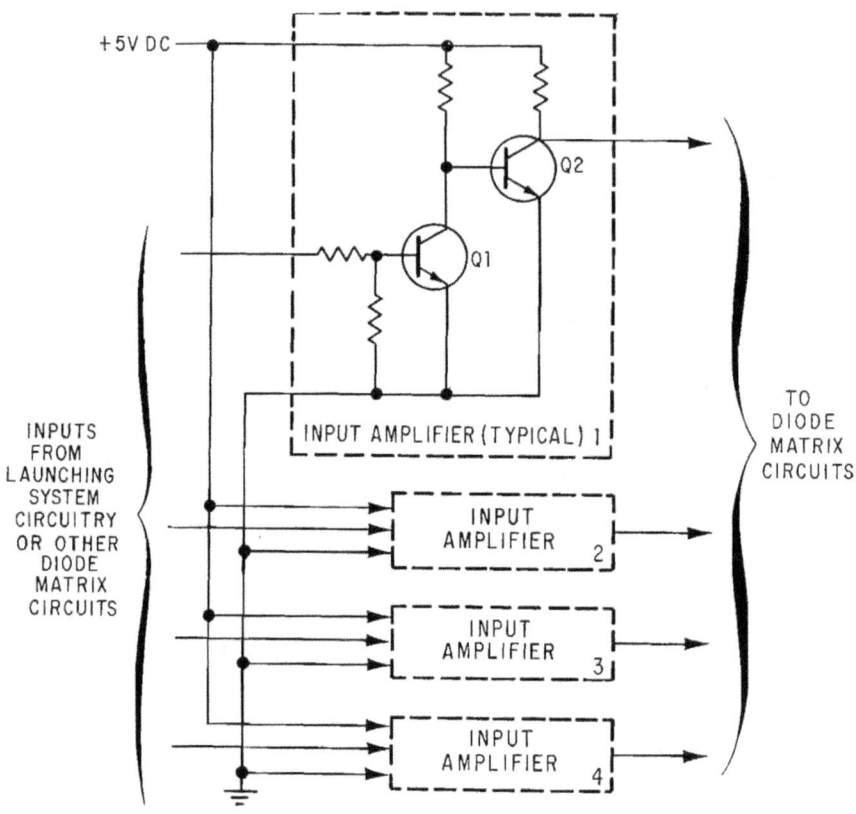

Figure 5-9.—Diode matrix input amplifier circuit.

94.227

When the input source applies a high signal to the input amplifier, a high-voltage potential is applied across a resistor to the base of transistor Q1, turning it on. With Q1 conducting, d.c. return is sensed at the bottom of the resistor above it. This low-voltage potential is applied to the base of transistor Q2, turning it off. Because Q2 is not conducting, a high-voltage potential is sensed at the bottom of the resistor above it and is applied to the row of the diode matrix as a high signal.

Diode Matrix.—The diode matrix is an arrangement of interacting horizontal and vertical columns of electrical conductors, interconnected by diodes. Each horizontal conductor is connected to either an input amplifier or an output amplifier. The vertical columns of conductors are connected to a voltage source of +5 volts d.c. The diodes that interconnect the horizontal and vertical conductors are arranged so that their cathodes are connected to the horizontal conductors while their anodes are connected to the vertical conductors. Under this setup, the diode can be made to act like a switch when the input amplifier applies a high or low signal to the diode's cathode. With the diode's anode connected to +5 volts d.c., a low signal forward biases the diode and it acts like a closed switch, while a high signal causes the diode to be reverse-biased and it acts like an open switch.

## Chapter 5—ELECTRICITY AND ELECTRONICS

These diodes, along with the horizontal and vertical conductors, form the series and parallel portions of a control circuit. Figure 5-10 is a straight-lined schematic of the diode matrix for the Mk 26 GMLS train motor start and run circuit. Note that the symbol used to represent a diode is the same as the one used to represent switch or relay contacts. In some systems, the diode matrix circuit is called a summary circuit. In fact, some drawings used in the Mk 26 GMLS refer to the diode matrix circuit as a summary circuit.

Output Amplifier.—The output amplifier (figure 5-11) is an electronic circuit which receives its input from the diode matrix. This circuit acts as an interface between the diode matrix and a solenoid driver, a lamp driver, or another diode matrix circuit. Note that in figure 5-10 the output amplifier is identified as the output driver.

To function, the output amplifier must first be enabled. The circuit has a separate enable input specifically for this purpose. The enable input may come either directly as a voltage from the +5 volt d.c. power supply or from a logic signal generated by the launching system circuitry.

When a low signal is applied to the enable input, a low-voltage potential is applied across a resistor to the base of transistor Q4, turning it off. Because Q4 is not conducting, a high-voltage potential is always sensed at the base of transistor Q5, turning it off. Because Q5 is not conducting, a low-voltage potential is sensed at the bottom of the resistor below it and is applied to the circuit output as a low signal. In this condition, therefore, the output amplifier is disabled and its output is always a low signal.

When a high signal is applied to the enable input, a high-voltage potential is applied across a resistor to the base of transistor Q4, turning it on. With Q4 conducting, the voltage potential sensed at the base of transistor Q5 depends on whether transistor Q3 is turned on or off by the signal input from the diode matrix. In this condition, therefore, the output amplifier is enabled and its output can be either a low or a high signal.

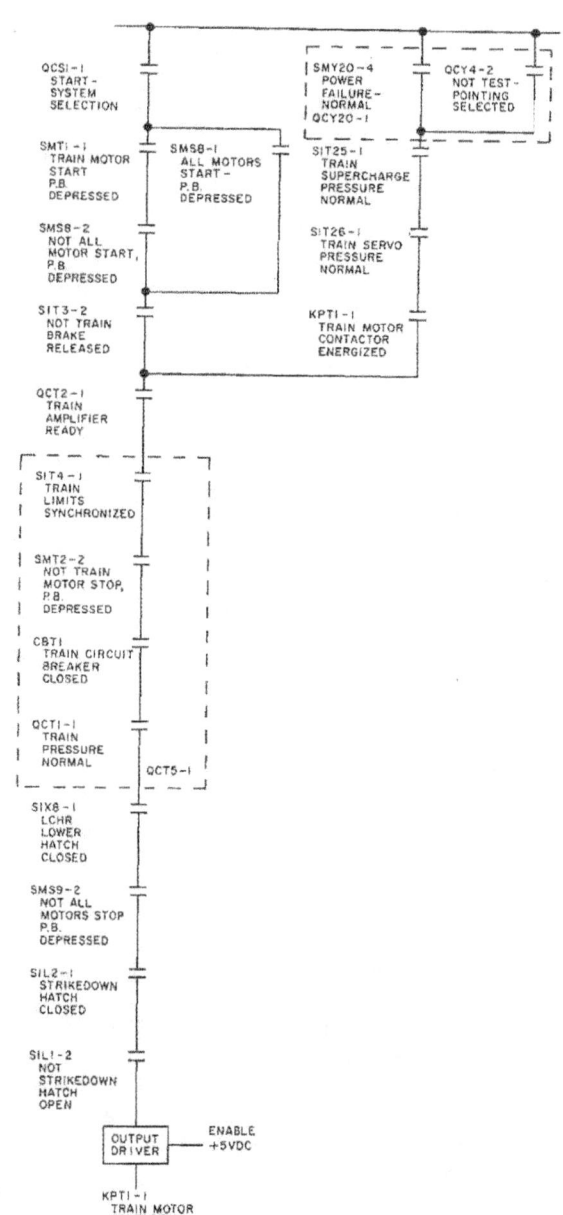

94.228
Figure 5-10.—Diode matrix logic circuit, Mk 26 GMLS.

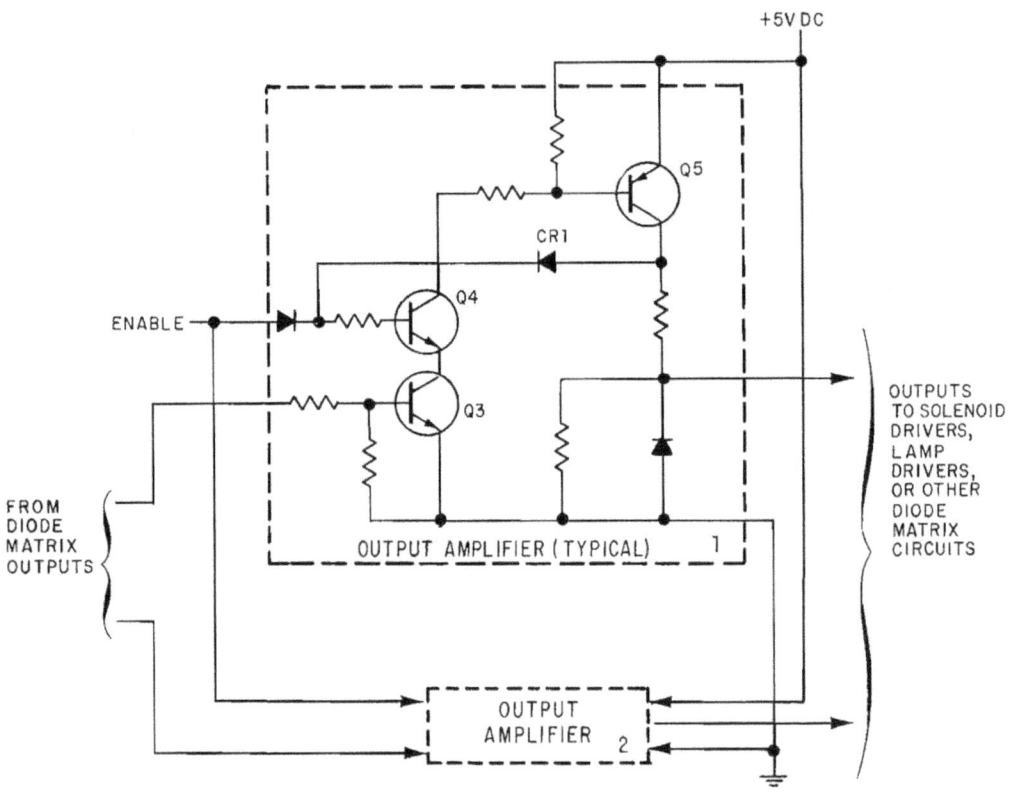

Figure 5-11.—Diode matrix output amplifier circuit.

When the output amplifier is enabled and a low-signal input is received from the diode matrix, a low-voltage potential is applied across a resistor to the base of transistor Q3, turning it off. Because Q3 is not conducting, a high-voltage potential is sensed at the base of transistor Q5, turning it off. Because Q5 is not conducting, a low-voltage potential is sensed at the bottom of the resistor below it and is applied to the circuit output as a low signal.

When the output amplifier is enabled and a high-signal input is received from the diode matrix, a high-voltage potential is applied across a resistor to the base of transistor Q3, turning it on. With Q3 conducting, d.c. return is applied through transistor Q4 to the resistor above it. This causes a low-voltage potential to be sensed at the base of transistor Q5, turning it on. With Q5 conducting, +5 volts d.c. is sensed at the top of the resistor below it. This causes a high-voltage potential to be sensed at the bottom of this resistor and to be applied to the circuit output as a high signal.

OUTPUT DRIVER CIRCUITS.—A solenoid driver circuit is shown in figure 5-12. This circuit, like the proximity switch circuit, is provided +24 volts d.c. from the GMLS power supply. When a high signal (+5 volts d.c.) is applied to the base of Q1, it is forward biased by the current flow in its base circuit (voltage drop across R2) and turns on. With Q1 conducting, the voltage drop across R3 foward biases Q2 and it will turn on, completing the solenoid coil circuit. A low input to the solenoid driver circuit

## Chapter 5—ELECTRICITY AND ELECTRONICS

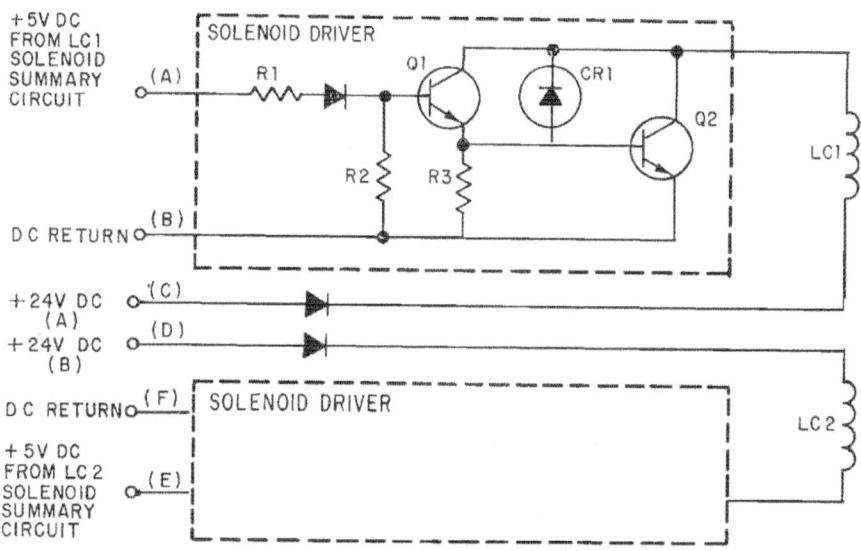

Figure 5-12.—Solenoid driver circuit.

94.230

cuts Q1 and Q2 off and deenergizes the solenoid. Note in figure 5-12 that the input to the solenoid driver is called a summary circuit. Summary circuit is another name for the diode matrix circuit.

The light driver circuit, like the solenoid driver circuit, has its own +24 volt d.c. supply. However, unlike the solenoid driver, the light driver consists of a single transistor connected in series with the light. When a high input is applied to the base of this NPN transistor it turns on, and turns the light on.

An electronic relay is a solid-state switching device that is activated by d.c. circuits. These relays are used when switching relatively large current-carrying conductors, both a.c. and d.c. is required. Figure 5-13 shows an example of an electronic relay driver used in the Mk 26 GMLS. This is the driver circuit for the train power motor start and run diode matrix shown in figure 5-10. Note that Q1 in this driver circuit is an NPN transistor, while Q2 is a PNP transistor. Q1 is turned on by applying a high signal to its base. With Q1 conducting, the base of Q2

becomes less positive than the emitter, and Q2 turns on. If the other conditions of the driver circuit are in their normal run setup, the electronic relay picks up and energizes the coil of the train motor contactor KPT1. With the coil energized, the A, B, and C contacts of KPT1 (shown in figure 5-14) close and complete the 440-volt power circuit to the train motor.

SUMMARY.—The solid-state control circuit can be compared to the conventional control circuit by studying the train motor start and run circuit for the Mk 26 GMLS in figures 5-10, 5-13, and 5-14, and the same circuits for the Mk 10 GMLS in figure 5-1. Regardless of the type of circuit, the object is to pick up the train motor contactor KPT1 in the solid-state circuit, or contactor KPTD1 in the conventional circuit. Certain conditions have to be met in either circuit before the contactors pick up. Once all the conditions have been met in the solid-state diode matrix circuit (figure 5-10), the output driver (also called output amplifier) sends a signal to the electronic relay driver circuit (figure 5-13). When the conditions are met in the electronic relay drive circuit, the electronic relay

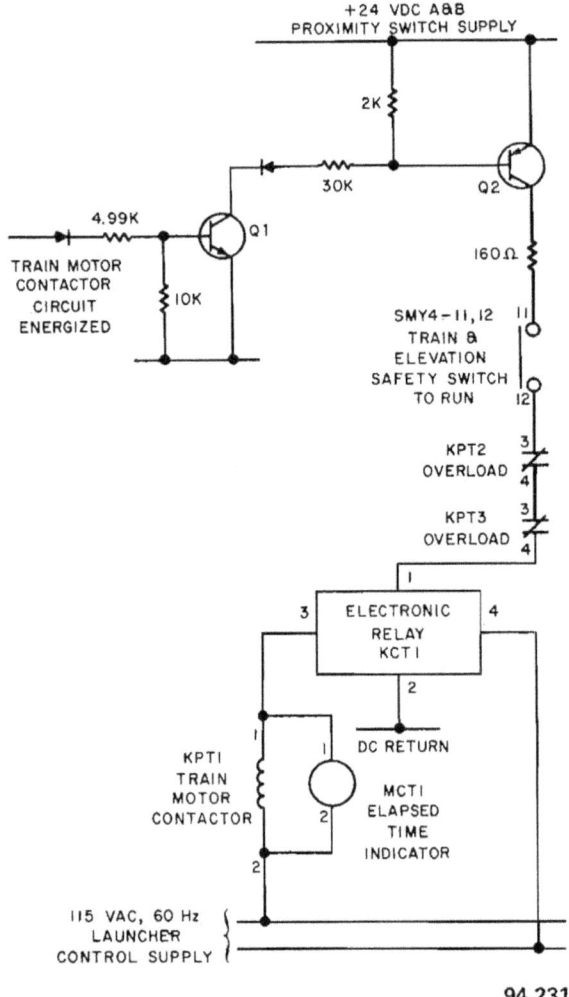

Figure 5-13.—Train start circuit, Mk 26 GMLS.

KCT1 picks up and completes the 115-volt circuit to the train motor contactor KPT1. With 115-volts a.c. applied to KPT1, it picks up and closes its A, B, and C contacts in the train motor circuit (figure 5-14).

### Logic Circuits

In this portion of the chapter we have included logic circuits, along with conventional circuits and solid-state control circuits under the heading "Control Circuits." This arrangement may be misleading because logic circuits are not a type of control circuit, but merely a way of presenting any type of circuit (electric, electronic, digital, mechanical, or hydraulic) whose components can be described as operating as a two-state device (on-off, high-low, open-closed, energized-deenergized, etc.). Presently, conventional control circuits are not shown as logic circuits. However, solid-state control circuits used in the Mk 13 Mod 4, Mk 26, and Mk 29 GMLSs all include logic circuits. Also, the missile identification circuits for the Mk 13 Mods 1 through 3, and the Mk 22 GMLSs are shown as logic circuits. All digital circuits are shown as logic circuits. However, the digital logic circuits differ considerably from the non-digital logic circuits. The digital circuits are covered later in this chapter.

Logic circuits use functional symbols instead of component symbols to represent circuit functions or sequential events in circuit operation. The skill in using logic circuits depends, to a degree, on the ability to classify components as two-state devices. For example, a transistor is either turned on or turned off; an input to a circuit may be either high or low; a switch or relay contact is either open or closed; a light is either on or off; and a solenoid or relay coil is either energized or deenergized.

The standardization of logic symbology has been an ongoing program since the early 1960s. The current standard for logic symbols (ANSI Y23.14-1973/IEEE Std 91-1973) was adopted by the Department of Defense for its use. To please everyone, the standard was made fairly broad, and its interpretation varies with the company that receives the contract to write the OP on a GMLS. Therefore, there are variations in the manner of presenting logic circuits, as will be pointed out later in this chapter.

This coverage of logic control circuits is fairly basic and will act as a lead-in to the digital circuit portion of this chapter.

LOGIC SYMBOLS.—There are three basic logic symbols (gates) used in most logic control circuits. They are the inverter gate, the AND gate, and the OR gate. The logic signals applied to these gates are of two signal states: either a

# Chapter 5—ELECTRICITY AND ELECTRONICS

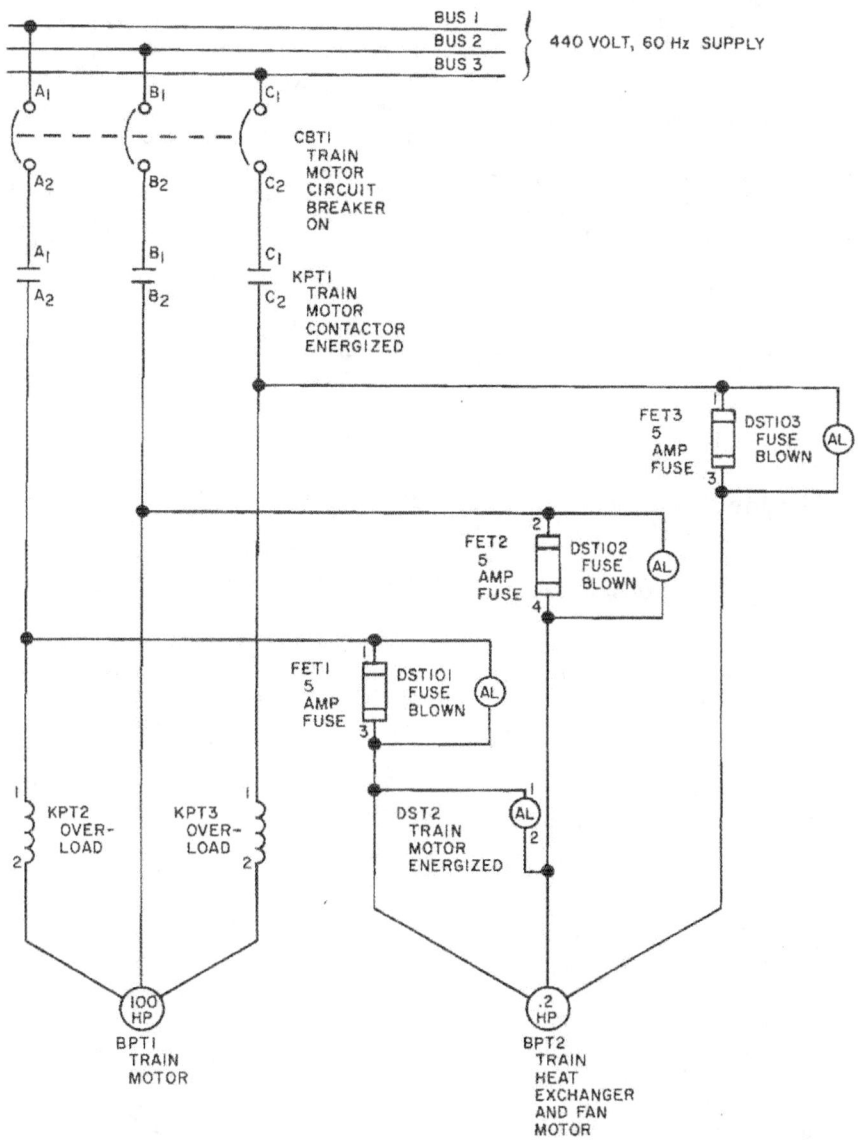

Figure 5-14.—Train motor circuits, Mk 26 GMLS.

high signal (the more positive voltage level) or a low signal (the less positive voltage level). The gate circuit recognizes the different signals at their input, performs an operation, and provides an output. The state (high or low) of the ouput signal from a gate depends upon how that component operates.

The operation of a gate circuit is described by defining the condition of the output for each set of input conditions. Each gate circuit described in this chapter is provided with a symbol and a truth table. The truth table summarizes the output of a gate when a set of inputs is present.

A logic circuit may be documented in either of two ways; (1) the negative indicator system used in the Mk 29 GMLS and (2) the polarity indicator system (also known as mixed logic) used in the Mk 26 GMLS. Since this portion of the text is primarily written around the Mk 26 GMLS, the polarity indicator system is covered first.

In the Mk 26 GMLS, the state of the inputs and output of a logic symbol is described in this GMLS's OP as being either at a high- (more positive voltage) signal level or at a low- (less positive voltage) signal level. The signal-level indicators in the polarity indicator system are known as polarity indicators (△); in the OP for the GMLS, they are called unfilled flags. A flag at the input of a logic symbol indicates that a low-level signal, rather than the normal high-level signal, is required to activate the logic circuit. Additionally, a flag at the output of a logic symbol indicates that an activated logic circuit produces a low output, rather than the normal high output. The absence of flags at the inputs and the output of a logic symbol indicates that its circuit operates in the normal logic manner; a high input produces a high output.

NOTE: The current standards on graphic symbols for logic diagrams state that regardless of the logic system (negation indicator or polarity indicator) the level of the inputs and output should be described in terms of 0-state and 1-state. Since the OP for the Mk 26 GMLS describes the inputs and output for logic symbols as being at a low- or high-signal level, this text will present them in the same manner.

In the Mk 29 GMLS, the state of the inputs and output of a logic symbol is described as being at a 0-state or at a 1-state. The signal level indicators in the negation indicator system are called negation indicators (O). In this logic system the drawings will indicate which type of logic (positive or negative) is used. In positive logic the 0-state is the less positive level and the 1-state is the more positive level. While in negative logic, the states are reversed; the 0-state is the more positive level and the 1-state is the less positive level. With the exception of the negative and positive logic consideration, the negation indicator is used in the same manner as the flags in the logic circuits for the Mk 26 GMLS. In some logic circuits the negation indicator is called a NOT symbol. However, this usage is not in accordance with the current standards on graphic symbols for logic diagrams.

The logic circuits in this portion of the chapter will first be explained in the basic logic manner; a high input produces a high output.

Inverter Gate.—The inverter gate symbol (figure 5-15) is used to represent a logic circuit that changes (inverts) the logic signal. If the signal is low (L), it will be changed to a high (H) level; if the signal is high, it will be changed to a low level. Note how the flags are used in figure 5-15 to indicate the function of the inverter. The inverter in the (a) portion of the figure has a flag at its input; therefore, its function is to change a low input to a high output. The opposite condition is depicted in the (b) portion of the figure. With a flag at its output, it functions to change a high input to a low output.

Figure 5-16 is an example of an inverter circuit that often is used in solid-state control

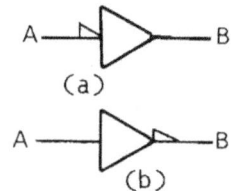

| INPUT A | OUTPUT B |
|---|---|
| L | H |
| H | L |

94.233

Figure 5-15.—Inverter circuit.

## Chapter 5—ELECTRICITY AND ELECTRONICS

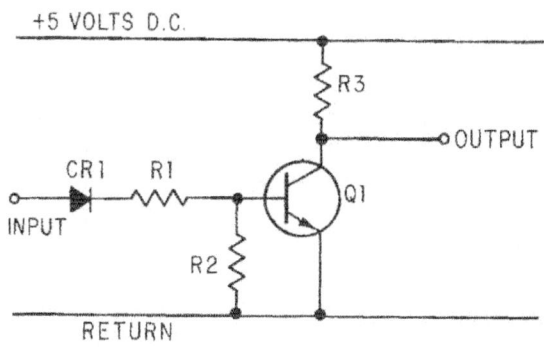

Figure 5-16.—Inverter circuit.

circuitry. A high-signal input to the base of the NPN transistor turns it on. However, this action, in effect, connects the inverter's output to return, and the output signal is low. With a low-signal input at the base of the transistor, it is cut off and the output signal is high.

AND Gate.—The basic AND gate symbol (a) and an equivalent control circuit (b) are shown in figure 5-17. The basic AND gate is any device having two or more inputs and a single output, and the output of which is high only when all inputs are high and is low when any one of the inputs is low. Thus, all three inputs (A, B, and C) in figure 5-17(a) must be high to have a high at the output (D) and, if any one of the three inputs (A, B, or C) is low, the output is low. The same condition is depicted by the simple series circuit in the (b) portion of figure 5-17. The inputs A, B, and C could be switches or relay contacts, as in a conventional control circuit; transistors, as in one type of solid-state control circuit; or diodes, as in the diode matrix circuit for the Mk 26 GMLS. A high output in the depicted control circuit indicates that the coil is energized. However, the AND gate normally represents only a portion of a series circuit and, under this condition, only indicates that all inputs are high.

Figure 5-18 depicts the six types of AND symbols, along with their truth tables, used in the logic circuits for the Mk 26 GMLS. Although only two inputs for each AND symbol are depicted, many symbols in the actual circuits contain three or four inputs. The symbol in the (a) portion of the figure is the basic AND symbol described in the preceding paragraph.

The symbol in the (b) portion of figure 5-18 uses flags at all inputs and at the output. This symbol depicts a logic circuit in which all inputs must be low to produce a low at the circuit's output. If any one or more of the inputs are high, the output will be high.

The symbol in the (c) portion of figure 5-18 uses a flag at its output. It depicts a logic circuit in which all inputs must be high to produce a low output. If any one or more of the inputs are low, the output is high. Note that this symbol is the basic AND gate with an inverted output. The circuit this symbol represents is used frequently, and the symbol is sometimes called a NAND.

The symbol in the (d) portion of figure 5-18 uses flags on all its inputs but none at the output. It depicts a logic circuit in which all inputs must be low to produce a high output. If any one or more of the inputs are high, the output will be low.

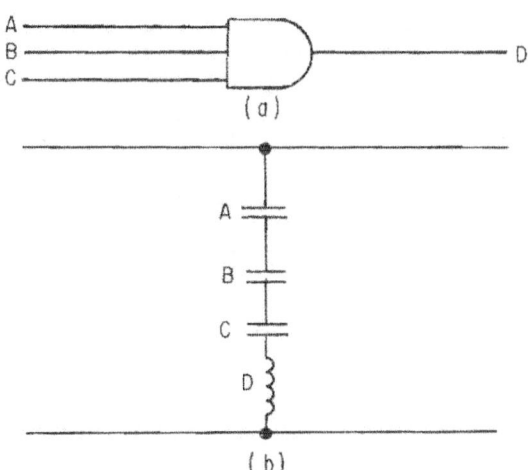

Figure 5-17.—AND symbol and equivalent circuit.

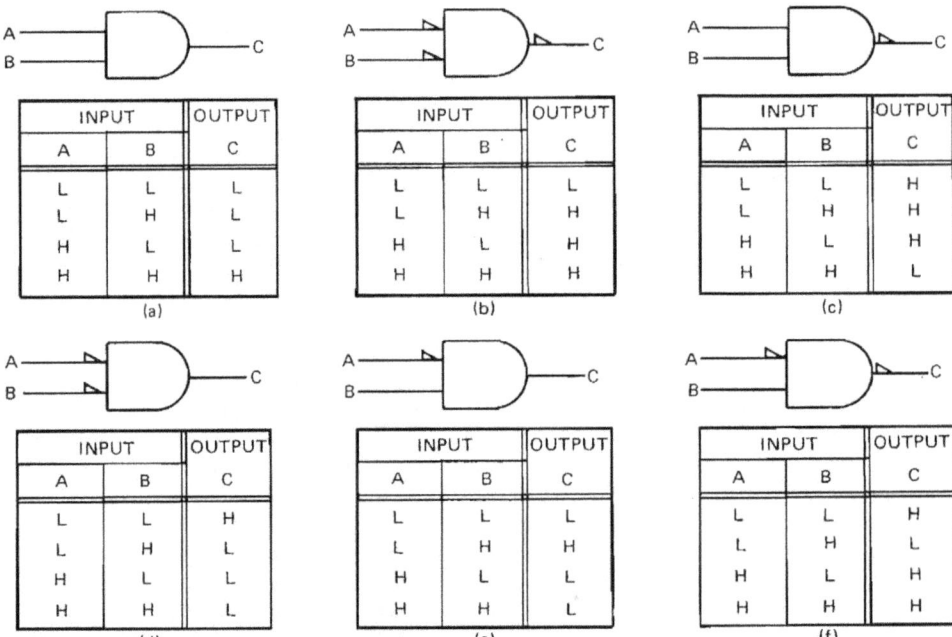

Figure 5-18.—AND gates.

The symbol in the (e) portion of figure 5-18 is a variation of the symbol in the (d) portion of the figure. The output of this logic circuit will be high only when the flagged input(s) is/are low and the unflagged input(s) is/are high.

The symbol in the (f) portion of figure 5-18 is a variation of the symbol in the (b) portion of the figure. The output of the logic circuit will be low only when the flagged input(s) is/are low and the unflagged input(s) is/are high.

OR Gate.—The basic OR gate symbol (a) and an equivalent control circuit (b) are shown in figure 5-19. The basic OR gate is any device having two or more inputs and a single output, the output of which is high when any of the inputs is high, and is low when all inputs are low. Thus, if any one or more of the three inputs (A, B, or C) are high, the output (D) is high. All three inputs must be low to produce a low output. The same condition is depicted by the parallel circuit in the (b) portion of figure

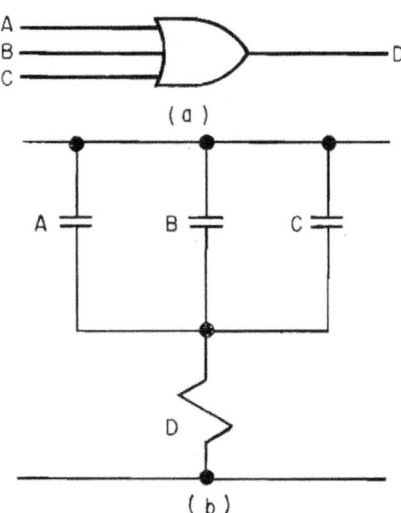

Figure 5-19.—OR symbol and equivalent circuit.

5-19. The output of the simple parallel circuit is depicted as a coil of a solenoid or relay. However, the OR gate normally will represent parallel branches of a more complicated circuit.

Figure 5-20 depicts six types of OR symbols, along with their truth tables. Although only two inputs for each symbol are shown, some symbols in the actual GMLS circuits contain three or four inputs. The symbol in the (a) portion of the figure is the basic OR gate described in the preceding paragraph.

The symbol in the (b) portion of figure 5-20 uses flags at all inputs and at the output. This symbol depicts a logic circuit whose output is low when any input is low, and high when all inputs are high. Note that the truth table for this symbol and the truth table for the basic AND symbol, shown in figure 5-18(a), indicates that the logic circuits for these symbols produce the identical output for a given input. For each OR gate there is an AND gate with this characteristic.

The symbol in the (c) portion of figure 5-20 uses a flag at the gate's output. This symbol depicts a logic circuit whose output is low when any input is high and is high when all inputs are low. Note that this symbol is the basic OR gate with an inverted output. The logic circuit this symbol represents is frequently used, and is sometimes referred to as a NOR gate.

The symbol in the (d) portion of figure 5-20 uses flags in all its inputs but none at the output. This symbol depicts a logic circuit that is high when any of the inputs is low, and low when all inputs are high.

The symbol in the (e) portion of figure 5-20 is a variation of the symbol in the (d) portion of the figure. The output of the logic circuit this symbol represents is high when any of the flagged inputs is low, or when any of the unflagged inputs is high. The output is low when all flagged inputs are high and all unflagged inputs are low.

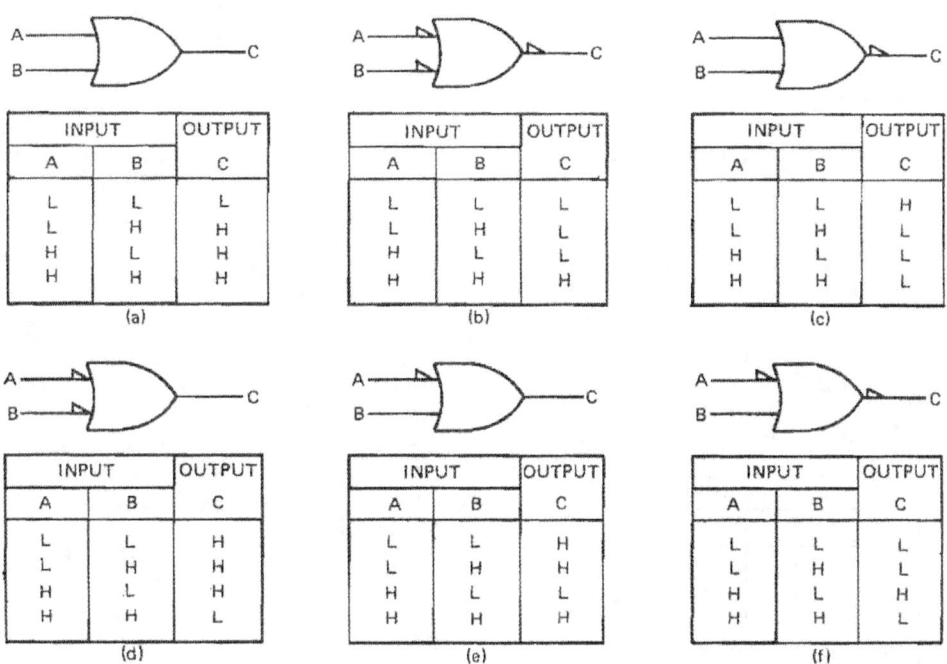

Figure 5-20.—OR gates.

The symbol in the (f) portion of figure 5-20 is a variation of the symbol in the (b) portion of the figure. This symbol depicts a logic circuit that has a low output when any of the flagged inputs is low or when any of the unflagged inputs is high. When all the flagged inputs are high and all the unflagged inputs are low, the output is high.

Exclusive OR Gate.—The symbols and truth table for the exclusive OR gate circuit are shown in figure 5-21. These symbols depict a logic circuit with only two inputs. The output of the logic circuit this symbol depicts is high when either, but not both, of the inputs is high. If both inputs are either high or low, the output is low. OP 4104 (Mk 26 GMLS) indicates that any one of the three symbols in figure 5-21 may be used to depict an exclusive OR gate. However, the current standard on graphic symbols for logic diagrams recognizes only the symbol in the upper left-hand portion of the figure.

Distributed Connection (DOT-OR) Gate.—The symbol and truth table for the DOT-OR gate are shown in figure 5-22. This symbol depicts a circuit consisting of two or more elements whose outputs are joined (wired) together to achieve the effect of an OR logic

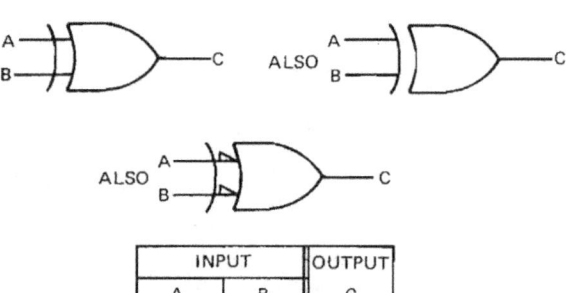

| INPUT | | OUTPUT |
|---|---|---|
| A | B | C |
| L | L | L |
| L | H | H |
| H | L | H |
| H | H | L |

94.239

Figure 5-21.—Exclusive OR gate.

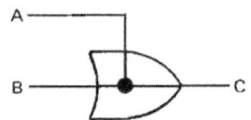

| INPUT | | OUTPUT |
|---|---|---|
| A | B | C |
| L | L | L |
| L | H | H |
| H | L | H |
| H | H | H |

94.240

Figure 5-22.—DOT-OR gate.

circuit. The output of this circuit is high when one or more of the inputs are high, and low when all inputs are low. OP 4104 (Mk 26 GMLS) calls this circuit a wired OR gate and depicts the symbol as the same size as the general AND and OR symbols. However, the current standard on graphic symbols for logic diagrams recommends that it be one-half the size (or smaller) of the general symbols. There is also a DOT-AND gate circuit. However, it normally is not used in control circuits.

LOGIC CIRCUIT OPERATION.—The operation of a logic circuit will be covered by comparing a simple motor start and run circuit—first as it would be presented in a nonlogic manner, then as it is presented in the Mk 26 GMLS. The start and run circuit discussed is for the guide arm exercise and emergency accumulator system that provides an alternate source of hydraulic fluid for operation of the Mk 26 GMLS guide arm, guide arm components, and the elevation positioner latch.

Nonlogic Circuit.—Figure 5-23 shows the start and run guide arm emergency motor circuit as it would be presented in a nonlogic manner. The circuit, when complete, energizes the emergency motor contactor KPX7. When energized, KPX7's contacts complete the circuit

## Chapter 5—ELECTRICITY AND ELECTRONICS

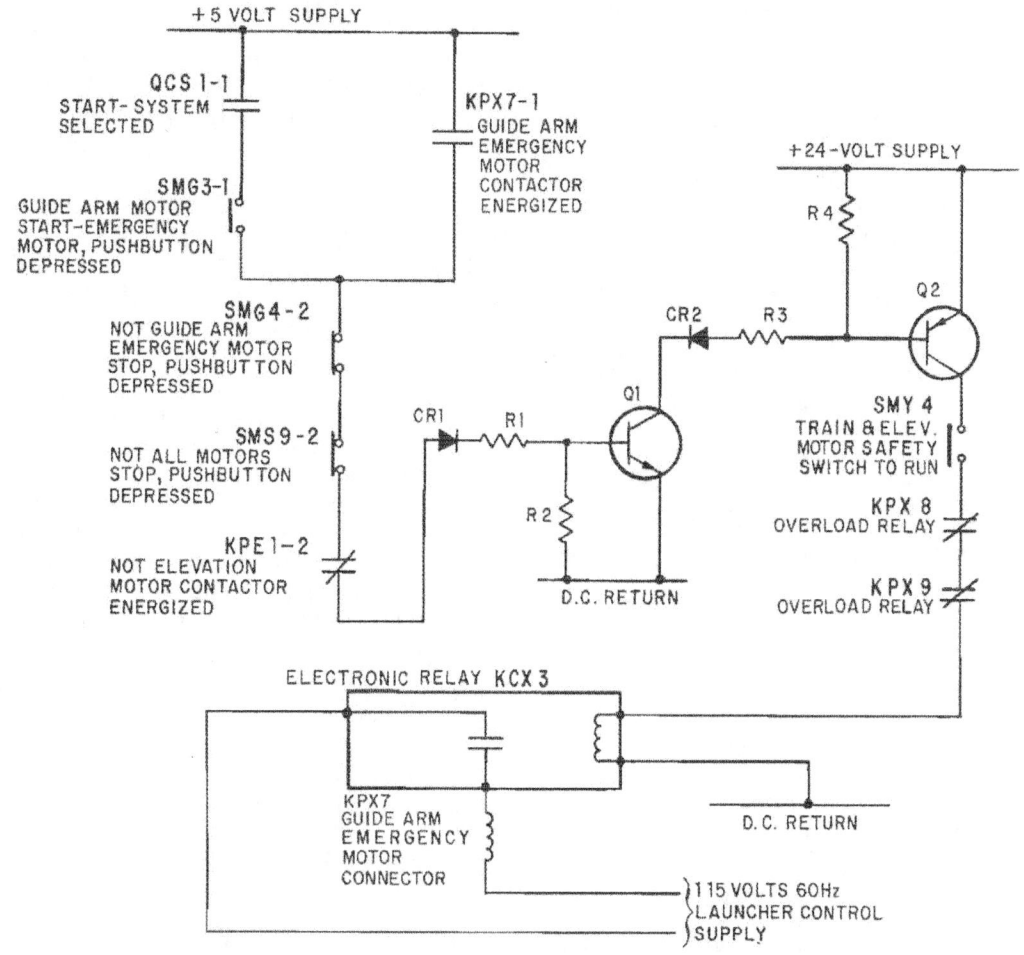

Figure 5-23.—Start-run nonlogic circuit for the guide arm emergency motor.

94.241

to apply 440 volts to the electric motor for the guide arm emergency accumulator system. One contact of KPX7 (KPX7-1) is applied to the run (hold) portion of the start and run circuit (figure 5-23).

The circuit (figure 5-23) to energize KPX7 can be divided into four circuits; a 5-volt circuit, two 24-volt circuits, and one 115-volt circuit.

When the 5-volt circuit is complete, current flow is from d.c. return through R2, R1, CR1, KPE1-2, SMS9-2, SMG4-2, and one of the parallel branches of the circuit to the 5-volt supply line. The current flow through R1 and R2 produces a voltage drop that forward biases (turns on) transistor Q1. A circuit is now complete from d.c. return through Q1, CR2, R3, and R4 to the 24-volt supply line. With Q1 conducting, the voltage drop across R3 and R4 forward biases (turns on) Q2. Under normal conditions, a circuit is complete from d.c. return through KCX3, KPX9, KPX8, SMY4, and Q2 to the 24-volt supply line. With current flowing through the coil of KCX3, its contact closes and

127

completes the 115-volt circuit through the coil of KPX7.

Under normal conditions, SMG3-1 is depressed to energize KPX7 and to start the motor for exercising or emergency operation of the launcher guide arms. As soon as KPX7 energizes, one of its contacts (KPX7-1) completes a holding circuit in one leg of the parallel portion of the start and run circuit. This contact of KPX7 permits SMG3-1 to be released, and KPX7 to remain energized. To stop the motor, the normal method is to depress STOP pushbutton SMG4-2. The motor also can be stopped by depressing the ALL MOTORS STOP (SMG9-2) pushbutton or by positioning SMY4 to SAFE.

A brief description of the function of the other interlock components in this circuit follows.

1. QCS1-1 is a transistor circuit that prevents starting the motor unless the control system start mode is selected.
2. KPE1-2 is a contact of the motor contactor for the main elevation (guide arm) accumulator system. This interlock prevents the emergency and the main systems from being activated at the same time.
3. SMY4 is the safety switch for both the train and elevation power drives.
4. KPX8 and KPX9 are contacts of the overload relay coils in the 440-volt supply lines to the electric motor for the guide arm emergency accumulator system.

Logic Circuit.—Figure 5-24 shows the guide arm emergency motor logic start and run circuit as it is depicted in the OP for the Mk 26 GMLS. Since the function of this circuit and its interlocks was covered in the nonlogic circuit (figure 5-23), this coverage is limited to the portion of figure 5-24 that is shown in logic.

NOTE: The logic symbols in the Mk 26 GMLS are identified by the number of the printed circuit board on which the gate is mounted, while the inputs and output of the logic symbols are identified by the pin numbers on the printed circuit board jack.

When inputs QCS1-1 and SMG3-1 are high (start system is selected and emergency motor pushbutton is depressed), the output of AND gate 42-19 is high. This high is applied through a DOT-OR gate to AND gate 41-47. (In the OP for the Mk 26 GMLS, the DOT-OR gate is called a wired OR gate.) If the other three inputs to AND gate 41 are high, this high will cause the output of gate 41-51 to go high. The high output of AND gate 41-51 is applied to the 24-volt relay circuit to turn on transistor Q1.

The high output from AND gate 41-51 is applied also to AND gate 60-7 to permit the gate to establish a hold (run) circuit after input KPX7-1 goes high. When the output of AND gate 60-19 is high, the high is applied through the DOT-OR gate to AND gate 41-47. The high output of gate 60-19 keeps the output of gate 41-51 high after SMG3-1 goes low (is released). After the hold circuit is established, the output of the logic circuit normally stays high until SMG4-2 or SMS9-2 goes low (is depressed).

If this is your first encounter with logic diagrams and you have acquired skill in reading nonlogic control circuits, you have reached the conclusion that a logic diagram is not necessary for circuits as simple as the one covered in figure 5-24. This conclusion is correct. Logic circuits are intended to simplify the interpretation of complicated control circuits.

## DIGITAL CIRCUITS AND DEVICES

Digital equipment based on digital techniques is becoming increasingly common in the newer guided missile launching systems; i.e., Mk 13 Mod 4, Mk 26, and Mk 29. Further, as this text was going to press, the Navy announced that the Mk 13 Mods 0 through 3 and Mk 22 GMLSs are to be updated to the solid-state electronic controls and digital circuitry incorporated in the Mk 13 Mod 4 GMLS. Looking at today's developments in the weapons system field, we see more and more applications for digital equipment. Consequently, Gunner's Mates will find a collection of electronic systems which use digital equipment as integral parts of the guided missile launching systems and the

## Chapter 5—ELECTRICITY AND ELECTRONICS

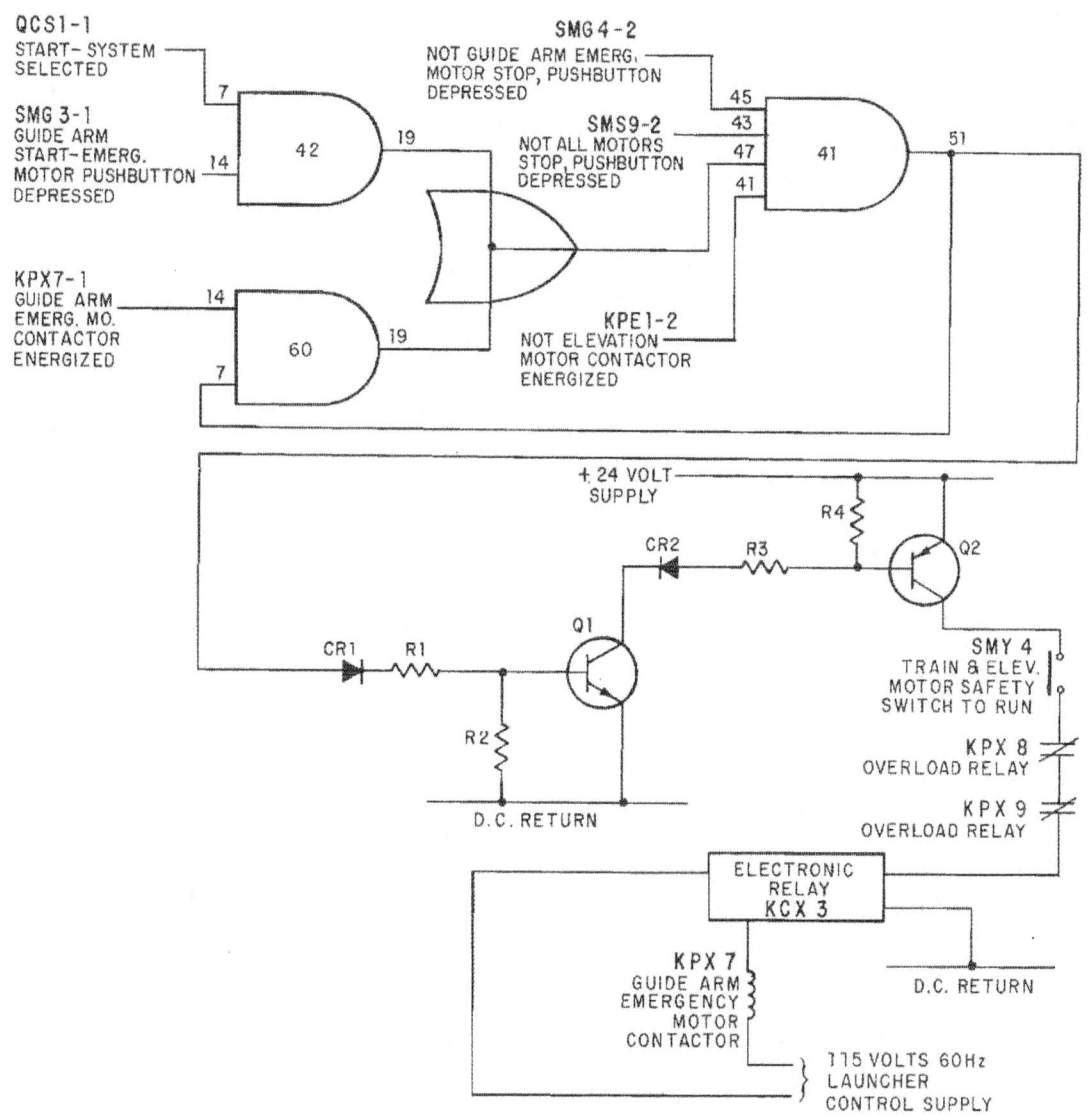

Figure 5-24.—Start-run logic circuit for the guide arm emergency motor.

overall weapons control system. To understand the operation of the system in general, and the launching system portion in particular, it is necessary to have a basic understanding of digital functions, circuitry, terminology, and applications. Module 13 of the *Navy Electricity and Electronics Training Series (NEETS)*, and rate training manuals *Mathematics, Vol. 3*, NAVPERS 10073 series, and *Fire Control Technician M 3 & 2*, NAVEDTRA 10209 series, present basic coverage of the purpose, operation, and theory of digital equipment.

## Computers

A computer is any device capable of accepting information, applying mathematical operations to that information, and obtaining results of these operations. Therefore, a computer must have an input, processing, and output section.

Before any problem can be solved by a computer, the quantities involved must be expressed in terms of common units; that is, digital computation is the process in which digits alone are used to solve the problem.

Basically, the purpose of a digital computer is the same as that of any other computational aid. However, it is more accurate within practical limits of speed. Computers can perform thousands of repetitive computations involving hundreds of thousands of digits without making an error. Further, they can store millions of items of information for future use.

Digital computers also have limitations in that a continuous variable cannot be processed. Simple and explicit instructions (programming) must be provided for each operation that is to be performed. Instructions as to where information is stored, how to use it, what to do with it, and what step is required next are sometimes more complicated than the problems the computer is meant to solve. Approximations and rounding off are also problems encountered with computers.

A digital computer performs its calculations by counting and comparing. With this simple capability it can add, subtract, multiply, divide, and make logical decisions. Therefore, to perform computations with a digital computer, all that is needed is a method which will enable the machine to count, compare, and transfer a digit from one place to another. To understand the operation of computer devices, it is necessary to learn the language of the computers.

While there are many different numbering systems, this section will deal with two: the base ten (decimal), and the base two (binary) systems. The binary system will be compared with the decimal system because all computers use the binary system. The bistable nature of simple electronic components such as switches (open and closed), and tubes and transistors (conducting or cutoff) makes the binary numbering system a convenient tool.

In the decimal system, ten digits (0 through 9) are used, and in the binary system, two digits (0 and 1) are used. The symbols or digits used are repeated as many times as desired or needed. Both systems use the place value of positional notation concept. This means that a numeral has a specific value of its position.

## Position Notation

An example of place value is shown in the two numbers 230 and 203. The 3 has a different value in each number by virtue of its position.

In this example, when the number 230 is expressed as two hundred thirty, some assumptions are made which must be understood to compare the decimal and binary systems.

The first assumption the reader makes upon looking at the number 230 is that the number is a decimal number. When no base symbol is given, the decimal system is assumed. If the number had been a base four number, it would have been written as $230_{(4)}$ with the 4 indicating the base of the numbering system. Some examples and their numbering base symbols are as follows:

| Number | Base |
|---|---|
| 398 | ten (decimal) |
| $342_{(5)}$ | five (quinary) |
| $101_{(2)}$ | two (binary) |

When a number is written in base ten, no base indicator is given. If the number is written in a base other than ten, the base indicator is used.

Another assumption made is that the number 230 is two hundred thirty; that is, the digit 2 indicates hundreds, the digit 3 indicates

## Chapter 5—ELECTRICITY AND ELECTRONICS

tens, and the digit 0 indicates units. In base ten this is true. The digits have specific values according to their position. Place value or positional notation for the number 378924 in the decimal system is as follows:

| Place value | $10^5$ | $10^4$ | $10^3$ | $10^2$ | $10^1$ | $10^0$ |
|---|---|---|---|---|---|---|
|  | 100000 | 10000 | 1000 | 100 | 10 | 1 |
| Number | 3 | 7 | 8 | 9 | 2 | 4 |

Notice that the place value column on the right is $10^0$ or 1. The increasing place values, moving to the left, are indicated by the base (10) raised to one greater power. The second place value column is, then, $10^1$, the third is $10^2$, etc.

The number 378924, then, is 3(100000) + 7(10000) + 8(1000) + 9(100) + 2(10) + 4(1).

The place value columns for the binary (base two) system follow the same pattern as the base ten system except the base two is used. For example, the number $10110_{(2)}$ is indicated in the place value chart for the base two as follows:

| Place value | $2^4$ | $2^3$ | $2^2$ | $2^1$ | $2^0$ |
|---|---|---|---|---|---|
|  | 16 | 8 | 4 | 2 | 1 |
| Number | 1 | 0 | 1 | 1 | 0 |

The number $10110_{(2)}$, then, is 1(16) + 0(8) + 1(4) + 1(2) + 0(1) = $22_{(10)}$.

Notice that place value in base two has a different value than the same place in base ten.

To obtain further computer information, refer to *Digital Computer Basics*, NAVPERS 10088 series.

### Binary Numbering System

In base ten, when counting from 0, the numbers are 0, 1, 2, 3, 4, 5, 6, 7, 8, 9. The next number to be counted is ten, but there is no single symbol in base ten to indicate the next number. Therefore, a combination of symbols is used by repeating the 0 and indicating a 1 in the tens column. This means one group of tens and no groups of units.

When counting in base two, there are only two digits or symbols, 1 and 0. The same mechanical process is used in base two that was used in base ten. Starting with 0, in base two, the count is 0, 1.

There are no other symbols to indicate the next number, so a 1 is written in the next place value column to the left and a 0 is written in the first place value column; that is, 0, 1, $10_{(2)}$.

This last number, then, indicates one group of two and no groups of units. The next number is $11_{(2)}$, which means one group of twos and one group of units. A number such as $1101_{(2)}$ would indicate one group of eights, one group of fours, no group of twos, and one group of units, and in this case is the same as decimal number 13.

Counting comparisons between the decimal and binary systems are made as follows:

| Decimal | Binary |
|---|---|
| 0 | 0 |
| 1 | 1 |
| 2 | 10 |
| 3 | 11 |
| 4 | 100 |
| 5 | 101 |
| 6 | 110 |
| 7 | 111 |
| 8 | 1000 |
| 9 | 1001 |
| 10 | 1010 |
| 11 | 1011 |
| 12 | 1100 |
| 13 | 1101 |
| 14 | 1110 |
| 15 | 1111 |
| 16 | 10000 |
| 17 | 10001 |

Notice that when counting objects, the decimal 17 and binary 10001 refer to the same object; therefore, decimal 17 must be equivalent to binary 10001.

## Data Transmission

To avoid confusion between GMLSs using digital equipment, data transmission for the Mk 26 GMLS will be presented. The control system consists of components that process orders and responses. The components that process orders make up the order data network; components that process responses make up the response data network. The information processed, networks, and equipment of the control system are:

1. Orders.
2. Responses.
3. Order data network.
4. Response data network.
5. Digital serial transceiver (DST).
6. Local control module (LCM).
7. Right- and left-hand circuit card housings (RCH and LCH, respectively).
8. Digital interface module (DIM).
9. Electronic servocontrol unit (ESCU).
10. Relay transmitter.

The following information obtained from OP 4104 is not complete; but will give you an overview of these complex functions and equipment. Just the main points will be discussed here.

The orders to the control system can be either from a remote, local, or test source. The order source depends on the mode selection made by the main control console (MCC) operator.

During remote mode, weapons control (WC) and the digital serial transceiver (DST) use control line signals (table 5-1) to start each data transfer. Then WC sends the orders in the form of four digital words (figure 5-25), one word at a time. Words are sent at 128 words per second. The first word has train information; the second, elevation information; the third, rail-A and system information; and the fourth, rail-B information.

Each word consists of 32 bits of order information. These bits, together with launching system conditions, cause the launching system to cycle through its operations. Each bit of the order is one binary digit with three equal segments (phases) of 333 nanoseconds each—one-third of a microsecond. The first and third phases serve as timing signals for the DST, digital interface module (DIM), and local control module (LCM). The second phase contains the order information.

When phase 2 is high (positive level), it shows that the particular bit is active and contains an order (information). If the bit is low (negative level), it shows no order except for bit 26 of words 1 and 2. At its low level, bit 26 shows clockwise train motion in word 1 or depress the launcher in word 2.

The first 26 bits of each order word can be identical in the total number and placement of bits. However, bits 27, 28, and 29 always appear in a definite combination to positively identify each order word.

Table 5-1.—Weapons Control and Digital Serial Transceiver (WC/DST) Interface Control Signals

| Signal Name | Origin | Purpose |
|---|---|---|
| **Orders** | | |
| Output data request | DST | Notifies WC that DST is ready to accept order words. In response, WC sends first order word. Each order word must be requested separately |
| Output acknowledge | WC | Notifies DST that WC sent an order word. Permits DST to accept the order word |
| **Responses** | | |
| Input data request | DST | Notifies WC that DST is transmitting a response word |
| Input acknowledge | WC | Notifies DST that WC has accepted the response word. Now WC is ready for the next response word |

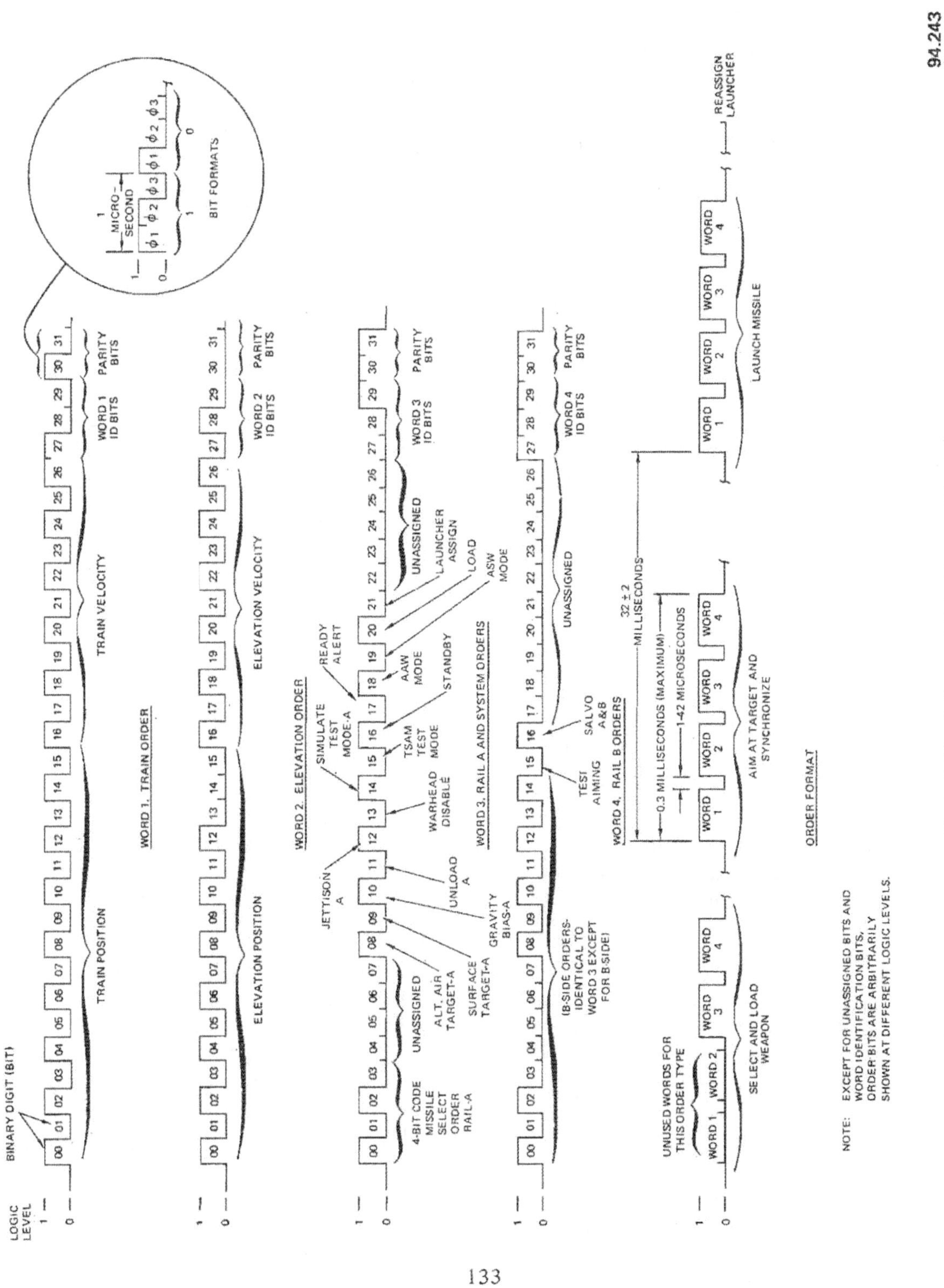

## Chapter 5—ELECTRICITY AND ELECTRONICS

Figure 5-25.—Order word digital format, Mk 26 GMLS.

Parity bits 30 and 31 at the end of each word can also be identical from word to word, or within a word. These bits give an odd-parity indication if the word is valid. Bits 30 and 31 are added to the word after the rest of the order information is determined. When the order word is ready for transmitting, the computer at WC sums up bits 00 through 15 to generate bit 30. It sums up bits 16 through 29 to generate bit 31.

The control system provides responses in remote, local, and test modes of operation. The responses are in both +28 volts d.c. hardwired and digital form.

The hardwired responses are operator signals sent to WC. Where applicable, they are sent to nearby gun mount control. These responses tell WC that the GMLS is available, the launcher is empty, and the blast doors are closed. The digital responses shows the status of the launching system.

The digital responses are in binary coded form. The responses consist of four digital words. Each word (figure 5-26) contains 32 bits of information. Each bit, or combination of bits, shows launching system status.

Each bit has three phases. These phases are identical to the order bit phases in time and phase placement within that bit. By comparing the response bits with the order bits, WC can decide if the GMLS has correctly responded to an order.

The response bits are generated by the circuit card housings for system, rail-A, and rail-B responses. These responses are combined in the LCM with train and elevation position responses.

The order data network processes the electrical signals for the operation of the hydraulic-mechanical components of the GMLS. This network also provides the circuit pathways for programming the missile(s).

The response data network provides electrical response signals to WC and indicators on the MCC. These response signals show the status of the GMLS and the missile(s).

In all modes, the responses are supplied to the WC and MCC. These responses come from switches and synchros located in the hydraulic-mechanical equipment.

The DST is a parallel-to-serial and serial-to-parallel converter for information which is sent between WC and the GMLS. The DST gets digital order words from weapons control in parallel form, changes them to serial form, and applies them to the DIM and LCH.

At the same time, the DST receives the digital response words from the LCM in serial form, changes them to parallel form, and applies them to WC. Thus, the DST acts as an interface between weapons control and the Mk 26 GMLS.

The right- and left-hand circuit card housings contain switch and summary circuits. These circuits receive order and response signals and then generate the output signals that sequence launching system operations. The right- and left-hand circuit card housings are on either side of the MCC. The circuit card housings contain printed circuit (PC) cards, switches, and indicators which are used during all launching system functions.

**Converters**

The DIM is a digital-to-analog converter for train and elevation information which is sent from WC to the GMLS. The DIM receives the WC digital order words 1 and 2 from the DST in the remote mode, and the LCM while in local control.

The electronic servocontrol unit (ESCU) is a servoamplifier for the train and elevation orders sent from WC to the GMLS through the DIM. In local control, the LCM supplies the signals to the GMLS through the DIM.

The ESCU compares actual position with desired position in either remote or local control.

The relay transmitter is an interface module that provides an electrical path for programming AAW missiles on the launcher. Through electronically controlled relays, circuits in the relay transmitter can be switched to either a simulate test mode or normal operating mode. Selecting a simulate test mode prevents the relay transmitter from applying signals to the missile.

Chapter 5—ELECTRICITY AND ELECTRONICS

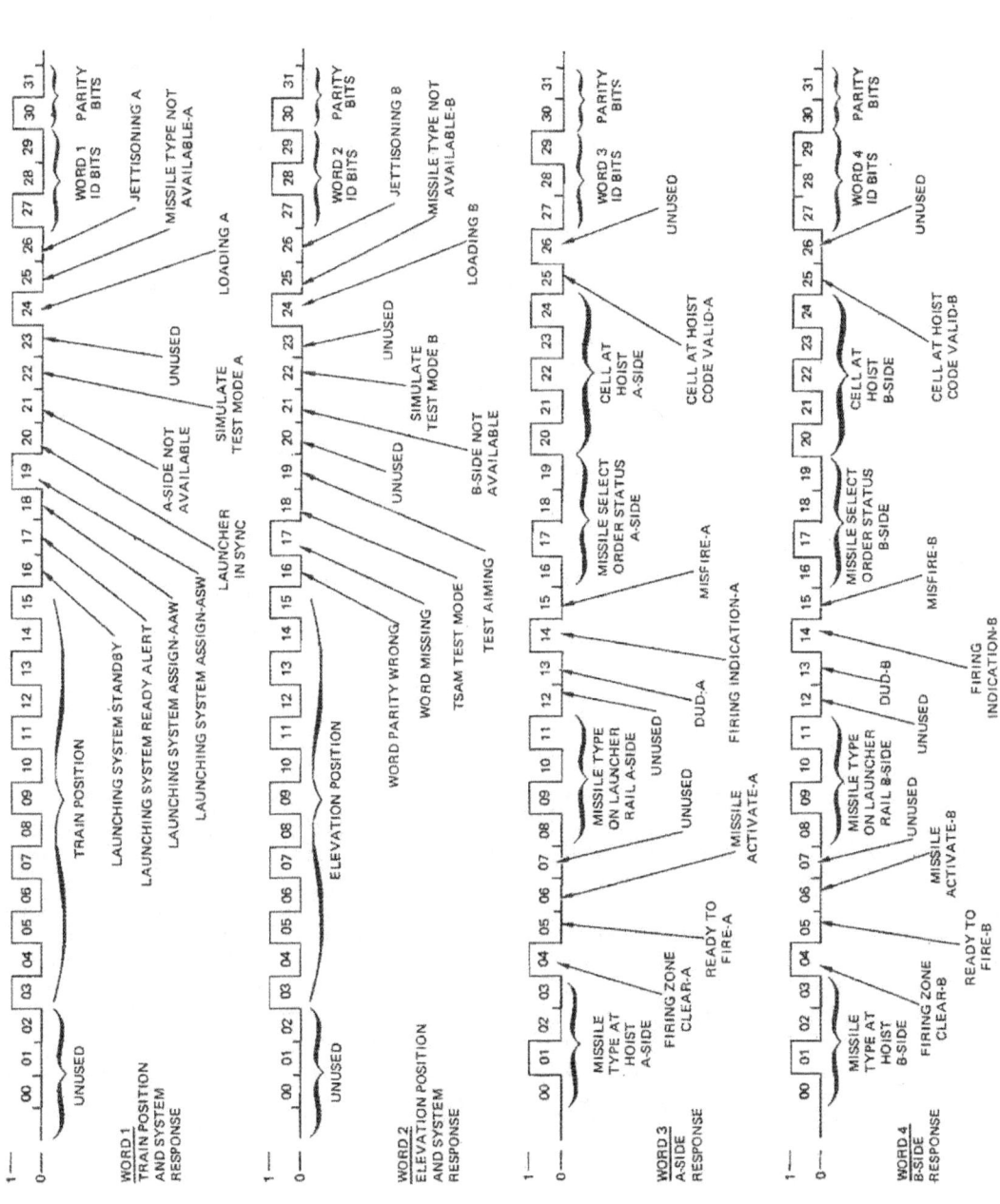

Figure 5-26.—Response word digital format, Mk 26 GMLS.

In a normal tactical operation, the relay transmitter receives orders from two sources: the right- and left-hand circuit card housings and an AAW fire control system. The circuit card housings provide the signals for programming, activating, and firing AAW missiles. The AAW fire control system provides the signals for programming the electronic section of the missile.

## TROUBLESHOOTING CONTROL AND POWER CIRCUITS

*Gunner's Mate M 3 & 2,* NAVEDTRA 10199 series contains a chapter on the use of meters for testing, making electrical measurements, and troubleshooting circuits. Review any parts about which you are uncertain. A solid understanding of the underlying principles is necessary before trying to understand complicated variations.

NOTE: The testing of ship's weapons control circuit wiring makes use of 500- to 1000-volt meggers. Repeated high potential tests (over 300 volts peak) can damage synchros and other small rotating components. High potential tests involving these components should be limited to those required for qualification and acceptance at the time of manufacture. Synchros, servomotors, resolvers, tachometer generators, etc., should be disconnected from the circuit when megger or ground tests are being conducted.

Missile system installations greatly increased the requirements for 400-hertz power supplies having varying degrees of voltage and frequency regulation. Missile ships have had to install 400-hertz generating plants to satisfy the demand. All missile ships have three separate 400-hertz power systems, each consisting of two or more motor generators. One is used for the ship's service system; another supplies the continuous wave illuminators used with guidance radars. The third system is the most closely regulated (voltage and frequency) and is used for the missile system(s).

Launcher electric motors are started and run under the power of a 440-volt, 60-hertz ship's power supply. The slipring assembly on the launcher stand and carriage provides continuous interconnection between on-launcher and off-launcher electrical connections while allowing unlimited train motion of the launcher. On the Terrier launcher, the slipring consists basically of a 440-volt collector ring assembly, a 115-volt collector ring assembly, and a fluid slipjoint. Each collector ring assembly has a rotating and nonrotating section. The rotating sections mount collector brushes that are connected by cabling to circuits of on-launcher equipment. The nonrotating sections mount collector rings which are connected by cabling to the circuits of the off-launcher power and control components. The rings are engaged by the brushes of the rotating sections to complete the electrical circuits. The four brushes contained in each brush ring are electrically connected to a terminal on the outer surface of the ring. The launcher cabling connects to the terminals of the assembled brush rings.

Close voltage and frequency regulation are necessary for use in the missile system. Voltage and frequency regulated equipment can now be provided in 30-, 60-, 100-, 200-, and 300-kilowatt sizes, with voltage balance regulators supplied when necessary. Supplying the power needed for the missile system is the responsibility of the ship's engineering department.

While depending on the engineering department to supply the power in the voltage and frequency desired, you have tested circuits and tubes and have used schematic and block diagrams, and voltage, resistance, and troublshooting charts. Experience and study will help you improve your ability to interpret the results of the tests and trace a malfunction. It is possible to track down a malfunction by checking each part or component in the circuit, following the circuit diagram until you come to the defective part. But that may take hours of tedious work. A study of the problem may reveal a shortcut that will locate the trouble in much less time.

### Preliminary Isolation

With the enormous amount of wiring and electrical components required in a weapons

## Chapter 5—ELECTRICITY AND ELECTRONICS

system, it is not surprising that a high proportion of the failures are in the electrical system. The control and power relays of the Mk 10 Terrier GMLS, for example, consist of more than 400 miniature rotary relays, 6 subminiature relays, and 6 small-size rotary relays. These relays are in the EP1, EP2, EP4, and EP5 panels.

Conscientious application of the 3-M Systems is intended to reduce the incidence of failure. The MRCs give step-by-step detail of what to do for routine matinenance, but when any part of the equipment fails to perform as it should, you have to turn to the OPs for aid in troubleshooting.

POWER CIRCUITS.—Let us concentrate on the EP1 panel, which is the basic distribution panel for all electrical power to the launching system. It contains switches, circuit breakers, fuses, relays, and contactors for the power and control circuits. The launcher captain turns on the various circuits before manning the EP2 (launcher captain's) panel. In figure 5-27, the items are sectioned off by number. Lights in section 1 indicate that the 440-volt power has been turned on and is available on the panel. As the motors in the launching system are energized, lights in section 2 come on: (a) B-side magazine motor; (b) train motor; (c) elevation motor; and (d) A-side magazine motor. The circuit breakers for these motors are in section 3. Lights in section 4 indicate that the following motors are energized: (a) B-side loader motor; (b) launcher rails motor; (c) circulating system motor; and (d) A-side loader motor. Section 5 has the circuit breakers for these motors. The two lights in section 6 are for A- and B-side loader accumulator motors, and the circuit breakers for these are in section 7, with a third circuit breaker for the control system. When you activate the panel, you turn on all these switches and circuit breakers unless only one side of the launcher is to be used, and then you turn on only the circuit breakers and switches for that side.

Lights in section 11 (figure 5-27) indicate that power is available for the 120-volt warmup circuits and the light in section 13 indicates power available in the 115-volt control circuits. The ON/OFF switches for warmup supply circuits and control supply circuits are in section 14 and the fuses are in section 15. Each fuse block has two fuses and two fuse-blown indicating lights. Two extra fuses are in section 16, with screw-on watertight caps.

If the light in section 8 (figure 5-27) is on, it indicates that the door interlock on the panel is inoperative. A magnetic latch on the door prevents its opening while the power is on. Before the door can be opened to make repairs, etc., the 440-volt power must be turned off and then the door handle (9) can be turned to open the door. Number 10 is an emergency release for the magnetic door latch. Number 12 is a ground detection indicator. It monitors the 117-volt control supply circuits and triggers an alarm if there is a grounded circuit. Figure 5-28 shows in outline the EP1 functions.

Let us assume that you have turned on all the switches and circuit breakers on the EP1 panel to activate the system. You notice that a fuse-blown light for switch d in section 14 (figure 5-27) is on. This means that control supply circuit (3) for the B-side feeder is disabled in some way. You have to find where the trouble is. Check the fuse-blown light first. You will need to look inside the panel. Before you can do that, you must disconnect the power supply; the panel door will not open while the power is on. Remember safety rules for working with electrical equipment: wear no rings, wristwatches, bracelets, or similar metal objects. Do not work with wet hands or wet clothing. Wear no loose or flapping clothing. Discharge any capacitors before touching them—they retain a charge after they are disconnected from their power source.

You may see the cause of the failure as soon as you look behind the panel door, but more than likely you will need to get the electrical drawings and trace the wiring until you find the trouble. The power distribution cables are numbered 0 to 99, and the wires are numbered 0 to 999. Loading control cables for the A-side are numbered 200 to 299, and for the B-side the numbers are 300 to 399. Wire and cable

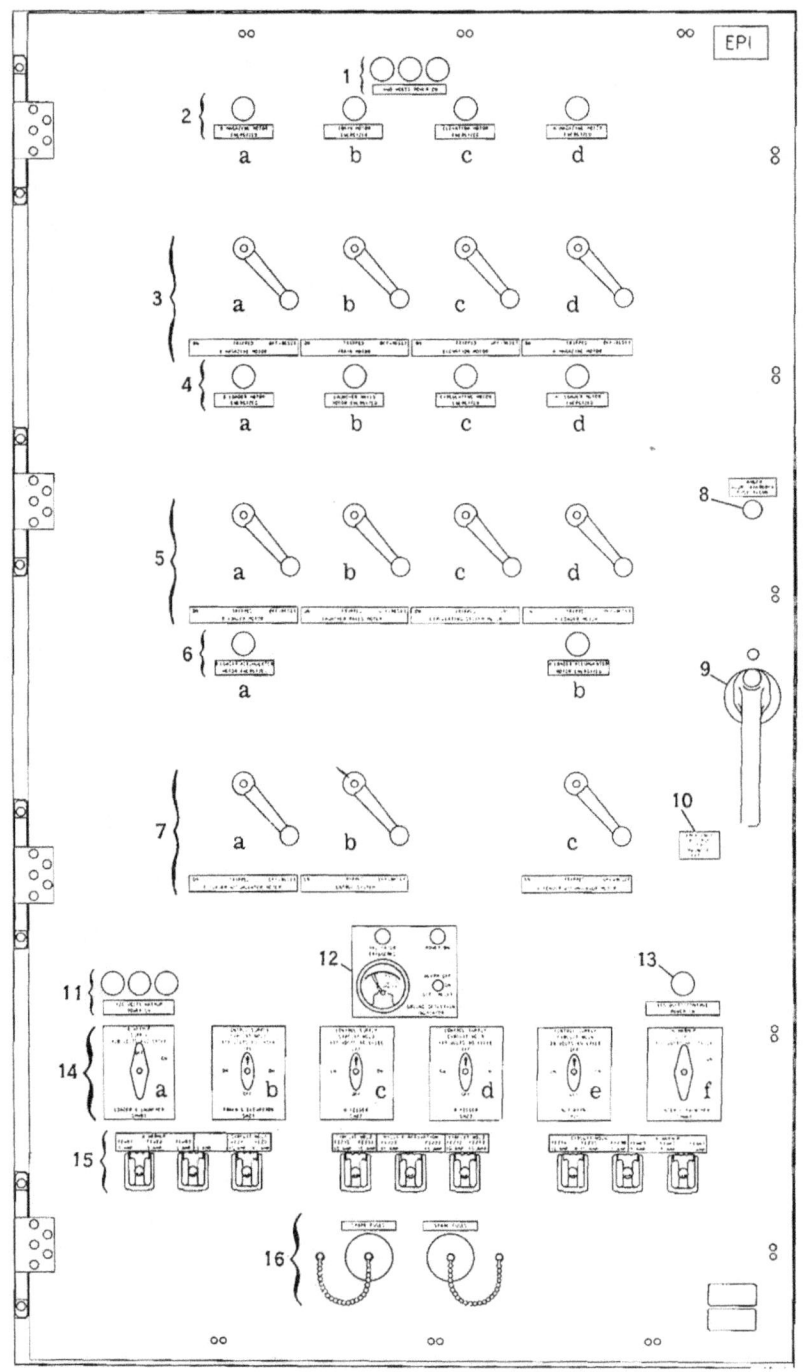

Figure 5-27.—EP1 panel, Mk 10 Mod 7 GMLS.

## Chapter 5—ELECTRICITY AND ELECTRONICS

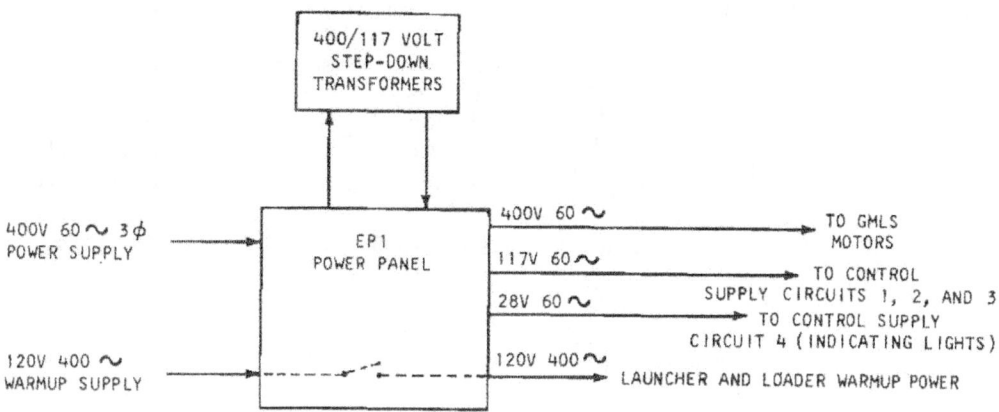

Figure 5-28.—EP1 Control panel functions.

94.12

numbers are assigned in groups, with A or B added to indicate the side served by the wire or cable. For example, WSA2022 means wire, single conductor, number 2022 of the A-side loaded circuitry. The cabling schematic also identifies the type and size of wire used in each application. The drawing explains the component type designations used, such as WS above, and the major assembly designations, such as LB for loader, B-side, or BA for dud jettison, A-side. All electrical and hydraulic components are identified by a combination of letters and numbers that indicate the kind of device or component, the identification of the major assembly of which it is a part, and identification of the specific component. These reference designations do not replace drawing, part, or stock numbers. They identify the part on the schematic. For example, KCLA1-1AB can be interpreted as follows:

KC—relay, control

LA—Loader, A-side

1—No. 1 among the relays associated with the A-side of the loader.

1AB—the A and B contacts on the first wafer or section of the relay. It also indicates that the A and B contacts on the first section of the relay are wired in that circuit application.

To return to the EP1 panel and your problem—if the trouble is only a faulty fuse, replace it. However, remove and replace fuses only when the associated circuit is completely deenergized. Use a fuse puller made of insulating material. Use a fuse of the same rated voltage and amperage capacity. After you have replaced the fuse, replace the fuse cover (if it has one), then energize the circuit. A fuse may explode when the circuit is energized.

When you have located the trouble that caused the fuse to blow, and have repaired it, reactivate the panel to check the work you have done.

Since the EP1 panel is connected directly to the ship's electrical system for its power supply, you need to work with the ship's electricians when there is a failure in any of the lines connected to the ship's power supply.

CONTROL CIRCUITS.—Assume that you have turned on all the connections at the EP1 panel and power is available for all the circuits. You are now ready to take your position at the EP2 panel. You receive orders from weapons control regarding the mode of operation, the type of missiles to be used, single loading or continuous loading, and whether A-side or B-side, or both, are to be used. You are ready to activate the EP2 panel, through which electrical power is supplied to the different units of the launching system.

The magazine, which consists of the ready service ring, the load status recorder, the hoist mechanism, and the magazine doors, is operated by hydraulic power from the Mk 64 power drive. One power drive is located on the A-side and the other on the B-side. Individual controls for the units are on the EP2 panel. Circuit #2 (figure 5-29) for control supply furnishes the 117-volt a.c. electricity to energize the coil of KPXA1 magazine motor contactor. The start circuit for the magaine accumulator motor is controlled from the EP2 panel. When the contactor (KPXA1 in figure 5-29) is energized, it closes contacts which complete the 440-volt supply to the magazine accumulator motor (BPXA1).

Normally there will be no trouble starting the magazine accumulator motor by depressing the START-RUN pushbutton switch SMXA16A (figure 5-29). However, a malfunction may occur at any time in such complex equipment. It is important, therefore, to understand the motor start circuit and the relay elements it includes.

To complete the start circuit (figure 5-29), you position SMS1 (control selector switch) at STEP, SMS2 (operations selector switch) at OFF, and SMX3 (A- or B-side selector switch) at A or A AND B for A-side operation. To start the motors, control selector switch SMS1 must be positioned at STEP during activation, and switch SMS2 must be at OFF during that time to prevent system operation until activation is completed.

With these manual switches (figure 5-29) positioned, it is time to position the switches or relays for the components powered by the accumulator unit. The positioning latches, both the clockwise and counterclockwise, for the ready service ring must be extended so the ready service ring will not start indexing before the system is ready. Both tray shift solenoids (LHDA1-LC1 and LHDA1-LC2) must be deenergized and the associated solenoid rocker arm must be at neutral to prevent indexing ahead of readiness. The normally closed (n/c) contacts of these switches are wired into the start circuit, so the switch elements are closed when not actuated. Hoist solenoid switches (LHHA1-SI101 and LHHA1-SI102) and magazine door solenoid switches (LHGA1-SI101 and LHGA1-SI102) perform the same function—they prevent premature activation of the associated parts of the launching system.

Relay elements KCHA1B and KCHA2B (figure 5-29) ensure that the hoist is either up or down when starting BPXA1. The magazine door solenoid switches (LHGA1-SI101 and LHGA1-SI102) remain deenergized at this time so the doors will not open. Both overload relay elements (KPXA2) and (KPXA3) are closed because there is no overload in the 440-volt power supply to magazine motor BPXA1. The remaining elements between SMXA16A and the KPXA1 coil remain closed during the motor-state procedure. The magazine motor STOP switch (SMXA17) is spring-held in the closed position unless it is depressed to stop the motor. Also closed is LHXA1-SI101A, the solenoid switch to dump magazine accumulator pressure if it becomes necessary. The solenoid LC1 will not energize until the motor has been stopped.

Now, with all the manual switches properly positioned and the associated interlocks closed, you are ready to press the magazine-motor START-RUN button, SMXA16. This completes the 117-volt circuit to the coil of the motor contactor KPXA1.

When the contactor coil is energized, it closes contacts A, B, and C of relay KPXA1 in the 440-volt motor power circuit and contact D in the motor-run circuit. The motor should start and begin driving the parallel piston pump.

Suppose the motor does not run after you have pressed the start button. Maybe somebody forgot to position the magazine safety switch (SMZA12) to RUN. This manual switch on the EP4 panel must be positioned. If that is not the cause of the nonoperation, you will need to get the drawings for the system to trace the cause of the failure. The schematic helps you picture the layout of the system, but you will need the electrical diagrams to make the proper corrections. Review the checklist to make sure you did not omit any step in the activation. The checklists posted at the panel should be used every time the panel is activated.

## Chapter 5—ELECTRICITY AND ELECTRONICS

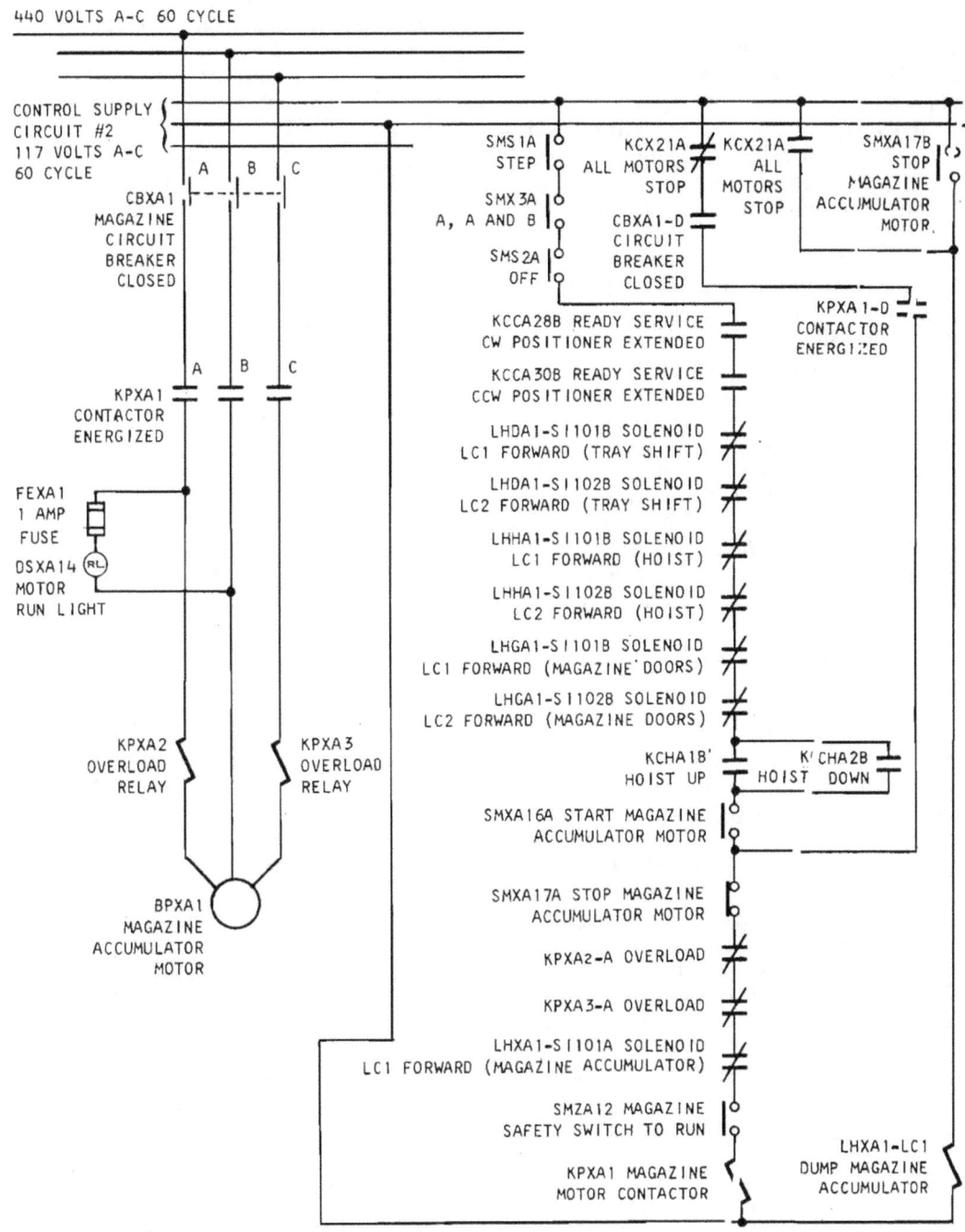

Figure 5-29.—Start and run circuits for the magazine accumulator system motor.

## Component Isolation

Once the source of trouble has been isolated to a particular circuit, several aids and shortcuts are available for isolating the defective component. Four probable sources of trouble in circuits are an open relay coil, a fused relay contact, an open diode, or a shorted diode. When isolating troubles, first determine which coils of the relays are energized when a pushbutton is pressed. The drawing or the maintenance manual may have a listing of the coils of the relays for each circuit. Check each circuit systematically for opens, and for shorts. There is little likelihood of a shorted relay coil, but a diode wired across the coil of the relay may be shorted, and that would cause a fuse to blow as soon as the circuit to the relay is completed. Shorted diodes in other circuits may cause no such giveaway reaction but may permit current to pass through other diodes. Those are more difficult to locate. When the shorted diode is isolated from the associated circuitry, do not assume it is bad; its forward and backward resistance should be checked.

Since the launchers must be trained and elevated every day as part of routine training and maintenance, any defects or failures in the servomechanisms of those systems will be evident. Servomechanisms are used in connection with so many parts of a missile launching system, no one application can be considered typical. Their use in the training and elevating system is one of their most extensive applications. The receiver-regulators are described in the next chapter. The emphasis there is on the hydraulics of the system. Following are some suggestions for troubleshooting the electrical parts. First, review the four steps:

Step 1—Observe the equipment's operation.

Step 2—Make an internal visual check.

Step 3—Isolate the trouble to the faulty parts, using meters, electrical prints, and maintenance publications.

Step 4—Replace or repair the defective part; test the system's operation afterward.

There are several publications containing lists of symbols, and from past experience you can probably identify many of them. As a first class or chief you must enlarge your knowledge in this field beyond the basics required for the third class.

For the most part symbols are standard, but there are variations. For all their variations, symbols are really simplified sketches of the devices they stand for. If you are reasonably familiar with the devices they represent, you should have lttle trouble identifying the symbol in the schematics. Uncommon symbols are explained on the drawing.

As there are tricks in all trades, there is one in circuit tracing.

Wiring diagrams and schematics are often a complicated maze of many circuits, accomplishing many functions. You must acquire the ability to disregard all circuits that are unnecessary to the one you are attempting to trace. The resulting circuit, depicted on one drawing, will show only the circuits necessary for one particular function. This important feature of circuit tracing is called straight lining.

RELAYS.—Relays suspected of faulty action may be checked with the relay test equipment mounted on the inner side of the EP2 panel front door (figure 5-30), next to the door latch. Before testing relays, the pins should be examined to be sure that they are not bent. To straighten bent pins, firmly seat the relay in the pin straightener mounted in the top of the test panel (figure 5-30). Terminal pins on a plug-in type of relay are shown in figure 5-31. After any necessary straightening, insert the terminal pins into the test socket (figure 5-30). The toggle switch SMZ19 applies (or removes) power to (from) the coil of the relay being tested. SMZ19 also switches the circuitry of the test socket to permit testing of the normally open or normally closed internal circuits of the relay as desired. Selector switch SMZ18 permits checking the individual internal circuits of the relay, normally open or normally closed, as determined by the position of SMZ19. As each internal circuit of the relay is tested by positioning SMZ18, indicator light DSZ13 indicates whether the relay is operating properly.

## Chapter 5—ELECTRICITY AND ELECTRONICS

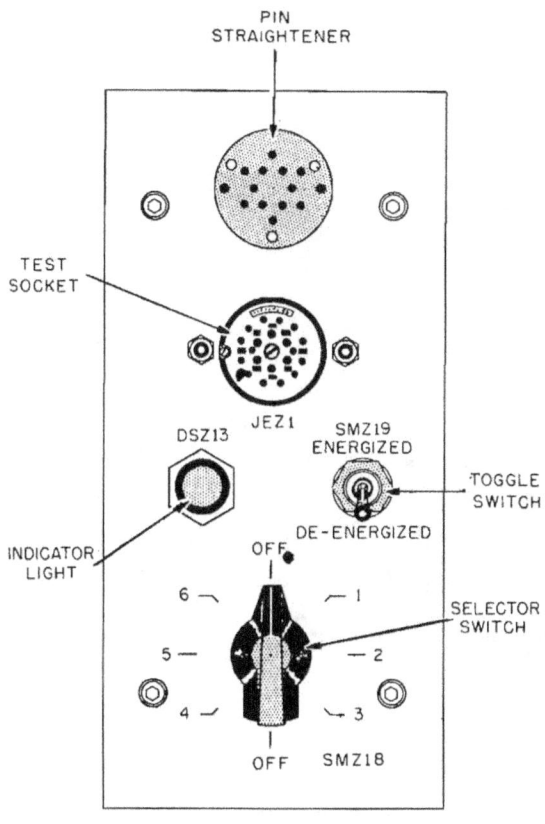

Figure 5-30.—Relay test panel on the EP2 panel, Mk 10 GMLS.

INTERLOCK SWITCHES.—The switches on the control panels are chiefly manual switches of pushbutton, rotary, or toggle types. Numerous interlock switches are used throughout the launching system. They are actuated by mechanical motion or hydraulic pressure, and are used to monitor equipment functions. The design varies with the application, but usually consists of one or more switch elements mounted to an actuating device. They assume that related equipment is at a certain position or has performed a certain function, so that operation will be in sequence. For example, the hoist cannot raise a missile to the loader if the magazine doors are closed. The circuit which controls hoist operation contains an interlock that does not allow circuit completion until the magazine doors are open and latched. This interlock is a relay wired into the solenoid circuit. When the relay is energized, the interlock is closed. The relay energizes when the associated interlock switch, mounted to the magazine door equipment, is actuated. This switch actuates when the magazine doors have fully opened and the door lock latch is engaged. Other interlock switches in the circuit assure that the loader is in position (retracted) above the magazine doors and that the tray shift on the ready service ring is positioned to hoist. Even the motor start circuit includes interlocks. They require all equipment to be in a certain position/condition in order to start the motors.

When interlock switches malfunction, the entire switch assembly is removed and a replacement unit is installed. Before the replacement unit is installed, it should be checked electrically with the switch test device (special tool 1614018) to be sure that it functions properly.

The interlock switches of the Mk 10 Mod 0 launching system control are of two types. The majority of the switches are sensitive switch assemblies, and the rest are microswitches mounted in the solenoid housings and in the load status recorder assembly. The OP for the system has a listing of all the sensitive switches, the location of each, its function, the reference drawings, and instructions for adjustment, with an additional listing of solenoid interlock switches mounted on brackets and secured to the supporting frames of the primary solenoids in the switch housing (figure 5-32). The assemblies are of right-hand and left-hand configuration, so when you are replacing one, be sure to get the correct one.

Before disconnecting any switch for replacement, be sure to mark down or note the connection of each lead so you can connect the leads of the replacement in exactly the same way. Use a soldering iron to remove the leads, and when attaching the new leads, solder them in place, after placing the switch assembly in position and securing it lightly. Adjust the air gap (figure 5-32) with the solenoids deenergized, according to the corrective maintenance cards (CMCs). Tighten the locknuts after making adjustments.

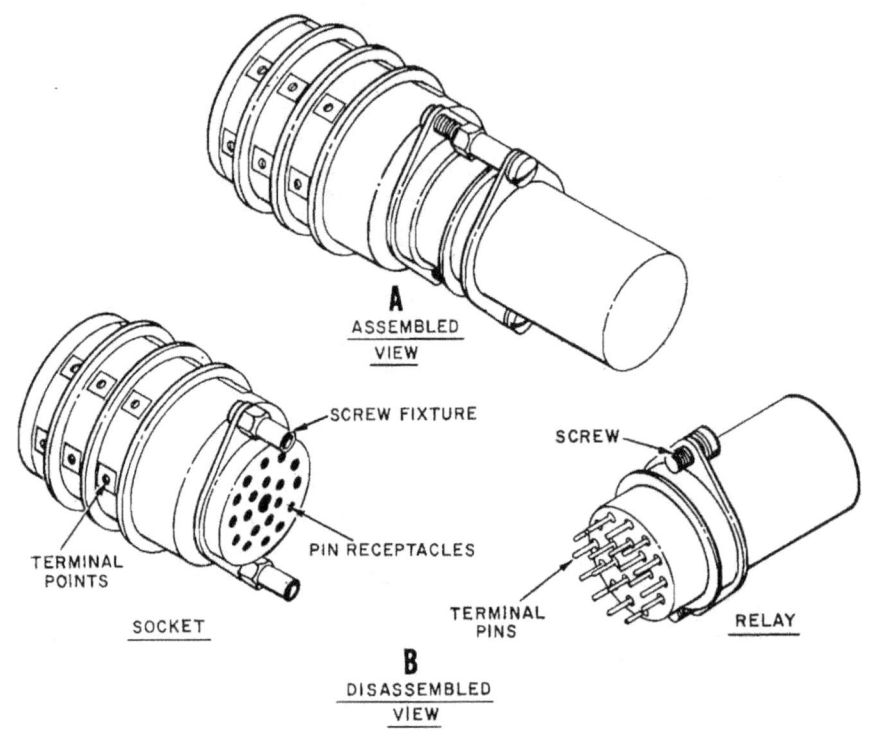

Figure 5-31.—Plug-in relay and socket assembly: A. Assembled view; B. Disassembled view.

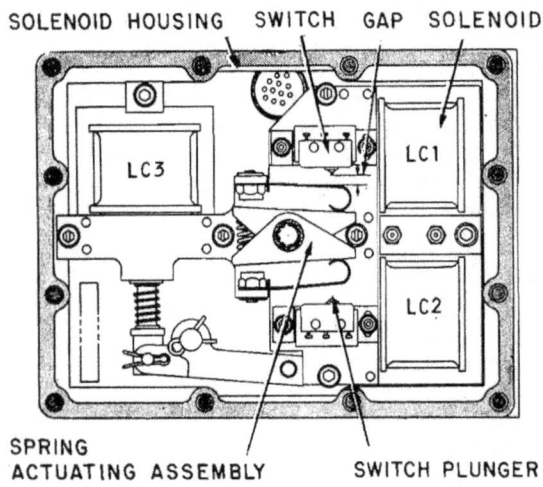

Figure 5-32.—Typical solenoid switch assembly.

*Gunner's Mate M 3 & 2*, NAVEDTRA 10199 series, traced you a typical power control circuit and a typical firing circuit, and showed you how interlocking worked in the circuits, and how parts of the circuit operated in a definite sequence. When you are tracing a circuit to locate a casualty, remember to include the interlocking switches that can prevent activation along any part of the circuit. If a faulty switch is found, it should be replaced or adjusted. Be absolutely certain that a switch is faulty before replacing it. It may only need adjustment to operate properly. If a switch is replaced, it must be adjusted within the equipment. Refer to the applicable corrective maintenance card (CMC) for the adjustment procedures.

Interlock switches must be checked periodically to be sure they are actuating and deactuating properly. Check them electrically to

## Chapter 5—ELECTRICITY AND ELECTRONICS

make sure that they are making and breaking as required. When an interlock switch malfunctions because of mechanical wear or damage, replace the entire switch. The Mk 10 Mod 7 Terrier GMLS uses eight types of interlock switches: (1) sensitive switch assembly—used throughout the system; (2) micro-sensitive switch—two are used in the EP1 panel; (3) 2PB switch assembly—used within the loader-control cam housing; (4) type-A rotary switch—used in the ASROC adapter rail; (5) single switch assembly—used within the dud-jettison solenoid housings and within the load status recorders; (6) paired switch elements—used within the solenoid housing, loader-control cam housing, contactors, and magnetic circuit breakers; (7) paired switch-element assemblies—used throughout the system in standard solenoid assemblies and in loader-control solenoid assemblies; and (8) triple switch-element assembly—used in the load status recorder.

SOLENOIDS.—Solenoids do not require planned maintenance, but the associated switches in a solenoid housing may, at various times, require adjustment. Your corrective maintenance cards (CMCs) will have the switch adjustments. Problems with the solenoid coils will be nominal, but occasionally they will short out or open. The lifespan for solenoids, as with switches, is very long. Troubleshooting solenoids is relatively simple, but you must be familiar with the complete sequence of operation for your particular system. The resistance of solenoids will vary between launching systems because there are various manufacturers involved. Once you have established the resistance for your system's solenoids, however, there will be only a slight variation between the coils. A solenoid coil that reads zero ohms is shorted; an infinity reading would indicate an open circuit.

SOLID-STATE DEVICES.—Solid-state devices, as you have already observed, are being used extensively in the newer systems. Increased reliability and decreased power consumption are the major benefits derived from solid-state devices. Printed circuit boards will have test points to aid in troubleshooting. The operating voltages for solid-state devices, transistors, gates, flip-flops, etc., are usually below 40-volts d.c. You will need a highly sensitive voltmeter. When you discover a faulty component, you will replace either the individual component or the printed circuit board. The trend is to replace the printed circuit board. The faulty board should be returned to the supply department, which will send it to a repair facility. If there are no spare cards, and the individual components are available, you may have to make the repairs yourself. Before attempting the repair, you should review the techniques of soldering printed circuit boards. It is very easy to ruin an entire board by using a high-voltage soldering iron. Further, failing to use heat sinks where applicable will ruin a printed circuit board component.

LOAD STATUS RECORDER.—We include the load status recorder here as an example of a complex electromechanical assembly (figure 5-33). The load status recorder is used exclusively on the Mk 10 GMLS Mods 0 thru 8. One is located on each ready service ring, mounted on the outboard side of the truss. Its two basic sections are relay board assembly and a switch and cam actuator assembly. It monitors the missile type and condition at all 20 stations in the ready service ring and sends this information to the control panels (EP2, EP4, and/or to the EP5) in the form of interlock switch and visual light indications. The ring of lights on the EP2 panel shows what the recorder tells it; i.e., what is in the ready service ring at each station. Shipboard correction or adjustment of the electrical components should not be attempted. Remove the defective unit, such as a triple switch or single switch element, and install a new unit. If the load status recorder malfunctions mechanically, order a replacement from the supply system.

The proper operation of the load status recorder can be checked during the daily exercise of the launching system. Each time the ready service ring is indexed to another station, notice if the lights representing the stations in the ready service ring rotate in the same direction. If there are empty trays or trays with dud missiles, the EMPTY and DUD indications

# GUNNER'S MATE M 1 & C

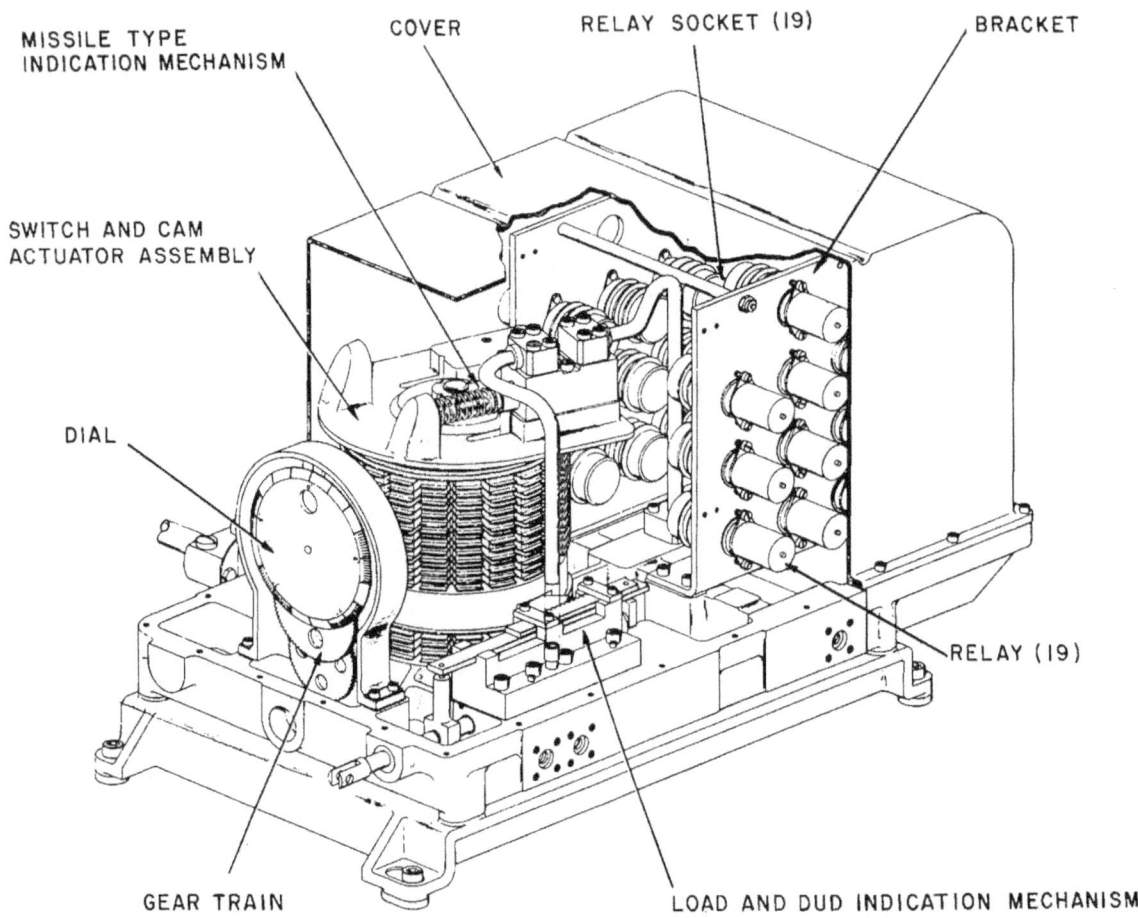

Figure 5-33.—Load status recorder, Mk 10 GMLS.

can be checked. The checking must be done in the step mode, operating from the EP2 panel. For unload assembly, unload launcher, checkout, or strikedown, the EP4 (or EP5) panel must be used to rotate the ready service ring. The loading pattern was set into the load status recorder at the time the missiles were loaded, and if the recorder is operating properly, the lights on the control panels should read back the same as the loading pattern. The color of the light indicates the type of tray or round. Three amber lights inside each circle of lights (representing the ready service rings) indicate the meaning of the lights in the circular pattern as DUD, EMPTY, or LOADED. If you push the DUD button, the lights should go on for all the trays that hold dud missiles. If you push the EMPTY button, the lights representing trays that are empty should come on. If you activate the LOADED button, the lights for all the trays containing missiles should come on, the color indicating the type of missile in each. In each case, the light indications should agree with the loading pattern established at the time of loading, unless the tray assignment has been changed, or the missile has been unloaded.

## Chapter 5—ELECTRICITY AND ELECTRONICS

At the present time one ship, the USS *Truxtun* (CGN-35), has solid-state load status recorders. Although much of the mechanical gearing and switches have been replaced by solid-state devices, the function and associated light indications have not changed.

## SAFETY

As a prospective PO1 or chief, you will have the responsibility for impressing on your subordinates the need for safety rules, and you must be firm in enforcing them.

Volumes have been written about electrical and electronic safety, yet accidents continue to happen. Frequently, the victim of such an accident is a person, such as an electrician, who should have known the dangers of electricity. Certainly an electrician is aware of the dangers of electricity, so why does he or she fall a victim to them. Perhaps the answer is in the old saying, "Familiarity breeds contempt." People become careless when they work with electricity every day.

What can you do about this attitude? You can keep reminding the individuals involved of the safety rules that must be observed in each operation. Watch to see that they observe them. Check to be sure that all electrical tools are in good repair and that the end of the ground wire within the tool is connected to the tool's metal housing. The other end must be connected to ground. The grounded type plugs and receptacles, which must be used, automatically make this connection.

Missile components must be protected against stray voltages by adequate grounding during all phases of handling, assembling, disassembling, and testing. Attach the ground straps where indicated on the containers.

The following general safety precautions are applicable during all phases of maintenance and operation:

1. Remove all power from the equipment when conducting operations requiring no power. Under certain conditions, capacitors may retain a high-voltage charge after the equipment is turned off. To avoid the possibility of electrical shock, discharge circuits before touching them.

2. All electrical test equipment using 115-volt a.c. line power is provided with a means of grounding the chassis systems through the power cable. Be sure the proper power cable is used. Do not use damaged power cables or test leads. Damaged test leads should be replaced, not repaired.

3. Under no circumstances should personnel perform servicing or maintenance of the equipment without the immediate presence of another person capable of rendering aid.

4. Personnel working with or near high voltages should be familiar with the methods of artificial respiration. *Standard First Aid Training Course*, NAVTRA 10081 series, describes the accepted method of mouth-to-mouth resuscitation. Charts illustrating and describing the method should be posted at a number of places on the ship. Everyone needs to know it. There is no time to look it up after there has been an accident. Resuscitation should be started in seconds after an accident. After 3 minutes, the chances of revival decrease rapidly.

Investigations have proved that nearly every accident could have been prevented if the individuals involved had observed the safety precautions and procedures contained in the applicable equipment OPs.

Although a number of protective devices are built into modern electronic equipment, personnel can still receive severe burns and lethal shocks under certain conditions. One source of danger sometimes ignored by repair personnel is the multiple inputs of electronic equipment. Turn off ALL sources of power, including power from other equipment such as synchros and remote control circuits. Moreover, the rescue of a victim shocked by the power input from a remote source often is hampered because the power source must be found before it can be turned off.

Another source of trouble occurs when a unit is removed from its usual location and energized while it is outside its normal enclosure, thus removing the protection given by built-in safety features. In such cases, special precautions are necessary.

Personnel working on electronic equipment and circuits should take the time required to make the operation safe. Schematics and wiring diagrams of the system should be studied carefully in advance and notice taken of all circuits, in addition to the main power supply, which must be deenergized. Electronic equipment usually has more than one source of power; ALL sources must be deenergized before equipment is serviced. Unless it is absolutely necessary, do not allow anyone to work on a circuit with the primary power applied. In those cases where such a procedure is necessary, the individual involved should stand on approved rubber matting, keeping one hand free of the equipment at all times.

Many accidents have occurred from 115-volt power because people ignore safety measures when using it; 115-volt power is not a low, harmless voltage.

The following commonsense safety precautions should be observed at all times.

1. Use one hand when turning switches on or off. Keep the doors to switch and fuse boxes closed. Use a fuse puller to remove cartridge fuses.

2. Unless it is absolutely necessary, do not work on energized circuits. Always take the time to lock out, or block out, the switch and tag it. Locks for this purpose should be readily available. If a lock cannot be obtained, remove the fuse and tag it.

3. All supply switches or cutout switches from which power can be fed should be secured in the open position and tagged. The tag should read:

DANGER – SHOCK HAZARD

Do Not Change Position of Switch Except by Direction of (person making, or directly in charge of, repairs)

4. Never short out, tamper with, or block open an interlock switch. If your ship has a tag-out system in operation, you will have to ensure proper entries are recorded in the tag-out log and the tags are properly filled in and secured at the proper location. Shipwide tag-out systems are the outgrowth of the engineering department's Propulsion Engineering Boards (PEBs), which are held annually and at the completion of overhaul.

5. Inform remote stations of the circuit on which work is being performed.

6. Keep clear of exposed equipment; if it is necessary to work on it, use only one hand and keep the other clear of the equipment. Use insulated tools and insulated flashlights of the molded type when working on exposed parts.

7. Keep clothing, hands, and feet dry if at all possible. If work must be done in wet or damp locations, use a dry platform or wooden stool to sit or stand on, and place an approved rubber mat or other nonconductuve material on top of the wood.

8. Do not remove hot tubes from their sockets with your bare hands. Use a tube puller.

9. Use a shorting stick to discharge all high-voltage capacitors.

10. Be aware of nearby high-voltage lines or circuits. Use rubber gloves, when applicable, and stand on approved rubber matting (MIL-M-15562). Not all so-called rubber mats are good insulators.

11. Do not work on high-voltage equipment alone. A safety observer, qualified in first aid for electrical shock, should be present at all times. The safety observer should know the circuits and switches controlling the equipment and should be instructed to pull the switch immediately in case of accident.

12. Check circuits with a meter, never with bare fingers, and avoid touching any of the metallic surfaces of the test prods. When measuring voltages over 600 volts, do not hold the test prods.

13. Turn off the power before connecting alligator clips to any circuit.

14. Make certain that the equipment is properly grounded. Ground all test equipment to the equipment under test.

15. Solvents should be used to the minimum extent possible for routine cleaning. Solvents should not be used on hot equipment because of the increased fire or toxicity hazard. If cleaning of the equipment with solvents is authorized, the applicable MRC for the equipment will list the type of solvent to use.

In case of electrical shock, prompt rescue is essential to survival. As the rescuer, you must first shut off the voltage or, if this is not immediately possible, you must observe the

## Chapter 5—ELECTRICITY AND ELECTRONICS

following precaution in freeing the victim from the live conductor. While standing on dry insulating material, use a dry board, belt, clothing or other nonconducting material to free the victim.

Two types of injuries require prompt first aid. In the case of severe electrical shock, the victim's breathing may have become paralyzed and the heartbeat stopped. Immediate artificial respiration and external heart massage may be necessary.

An accident with tools may cause serious bleeding which must be controlled at once. The proper first aid measures to take are described in publications issued by the Bureau of Medicine and Surgery.

All personnel concerned with electronic equipment should be familiar with external heart massage and mouth-to-mouth resuscitation techniques. The personnel qualification standard (PQS) for division damage control petty officer has a section on first aid to which you should refer.

## SERVOMECHANISMS

An apparatus that includes a servomotor (or servo for short) is often called a servomechanism. And what is a servomotor? It is a power-driven mechanism, commonly an electric motor, which supplements a primary control operated by a comparatively feeble force. The primary control may be a simple lever, an automatic device such as an optical limit switch, or a meter for measuring position, speed, voltage, etc., to whose variations the motor responds, so that it is used as a correctional or compensating device. A servo is a control device, a power amplifier, and a closed-loop system. *Gunner's Mate M 3 & 2*, NAVEDTRA 10199 series, described and illustrated the fundamentals of servomechanisms. They are used in all the power drives, and the principles apply to all of them—only the details of application vary in the different launching systems. Servos may be electrical, mechanical, electronic, hydraulic, or combinations of these, but all use the feedback principle. One or more power amplifiers are part of any servosystem. There must be an input and an output, and between these, an error detector and an error reducer. Each of these essential components may have many parts, so that even a simple schematic may seem like a complicated maze. Remembering the essential parts of a servo and the direction of the signals is helpful in tracing through the schematic.

## SERVO TEMINOLOGY

In addition to those already mentioned, a number of specialized terms are used in connection with servosystems. The most common terms are defined here:

OPEN-CYCLE (OPEN-LOOP) CONTROL of a servosystem means actuation of the servo solely by means of the input data, the feedback device being either removed or disabled. It should be understood that open loop control is not employed in launcher power drive servos; but in testing certain servo characteristics, an open-cycle control is often useful. Under such conditions the elements involved are frequently referred to as an open servo loop.

CLOSED-CYCLE (CLOSED-LOOP) CONTROL refers to normal actuation of the system by the difference between input and output data, with the feedback device operative.

CONTINUOUS CONTROL is used to describe uninterrupted operation of the servosystem on its load, regardless of the smallness of the error.

DEVIATION or error of a servo, is the difference betwen input and output.

ERROR SIGNAL or error voltage is the corrective signal developed in the system by a difference between input and output.

INSTRUMENT SERVOS and POWER SERVOS are designations used to classify servomechanisms according to their power output. An instrument servo is one rated at less than 100 watts maximum continuous output; a servo whose rating exceeds this amount is a power servo.

## TYPES OF SERVOS

Servosystems can be divided into two basic types—open-loop and closed-loop systems. The essential features of each are indicated by the block diagrams in figure 5-34.

In both systems, an input signal must be applied which represents in some way the desired condition of the load.

In the open-loop system shown in figure 5-34, view A, the input signal is applied to a controller. The controller positions the load in accordance with the input. The characteristic property of open-loop operation is that the action of the controller is entirely independent of the output.

The operation of the closed-loop system (figure 5-34, view B) involves the use of followup. The output, as well as the input, determines the action of the controller. The system contains the open-loop components plus two elements which are added to provide the followup function. The output position is measured and a followup signal proportional to the output is fed back for comparison with the input value. The resultant is a signal which is proportional to the difference between input and output. Thus, the system operation is dependent on input and output rather than on input alone.

Of the two basic types, closed-loop control (also called followup control) is by far the more widely used, particulary in applications where speed and precision of control are required. The superior accuracy of the closed-loop system results from the followup function which is not present in open-loop systems. The closed-loop device goes into operation automatically to correct any discrepancy between the desired output and the actual load position, responding to random disturbances of the load as well as to changes in the input signal.

A convenient classification of servosystems can be made in accordance with their use. The most common of these servosystems are position servos and velocity servos. The position servo is used to control the position of its load and is designed so that its output moves the load to the position indicated by the input. The velocity servo is used to move its load at a speed detemined by the input to the system.

Many servosystems cannot be fitted into either category. For example, a third type of servo is used to control the acceleration rather than the velocity of its load. Special applications of the different types are used for calculating purposes—the servo making a desired computation from mechanical or electrical information and delivering the answer in the form of mechanical motion, an electrical signal, or both.

## SERVOS USING SYNCHROS

A servo, servosystem, or servomechanism (the three terms mean the same thing) is an automatic control device widely used in the Navy and distinguished by several special characteristics. There are many different types of servosystems, and not all of them use synchros. The purpose of servosystems in which control synchros are used is to supply larger amounts of power and a greater degree of accuracy than is possible with synchros alone. Another equally important characteristic of the servo is the ability to supply this power automatically, at the proper time, and to the degree regulated by the need at each particular

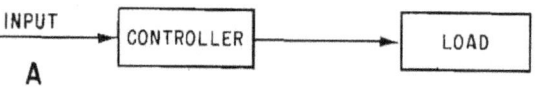

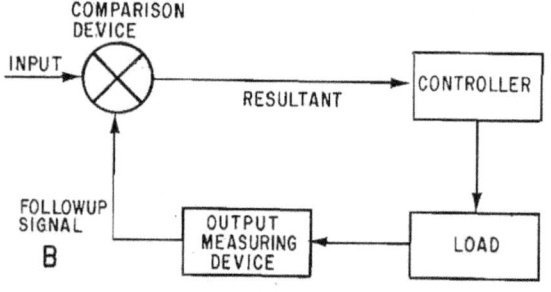

33.61
Figure 5-34.—Basic types of servosystems: A. Open-loop; B. Closed-loop.

moment. All that the system requires to perform the specific task for which it is designed is an order specifying the desired results. When such an order is received, the servo compares the desired results with the existing conditions, determines the requirements, and applies power accordingly, automatically correcting for any tendency toward error which may occur during the process.

To function, a servosystem must meet five basic requirements:

1. It must be able to accept an input order defining the desired result, and translate this order into usable form.
2. It must feed back, from its output, data concerning the existing conditions over which it exercises control.
3. It must compare this data with the desired result expressed by the input order and generate an error signal proportional to any difference which this comparison shows.
4. It must, in response to such an error signal, issue the proper correcting order to change existing conditions to those required.
5. It must adequately carry out its own correcting order.

In functional terms, the components normally found in a servosystem using synchros are a synchro system, an amplifier, a power control device, a drive motor, and a feedback device.

The remainder of this chapter covers the synchro system and amplifiers, while the next chapter covers the hydraulic and mechanical components of the power drives, which include the control, drive, and feedback devices.

## SYNCHRO SYSTEMS

In a launcher power drive servosystem, the synchro system provides the input. The control transformers (CTs) of the synchro system also act as the comparison devices shown in the B portion of figure 5-34. The CT receives the order signal from the controlling station, compares it to the followup signal (also known as B-end response or position signal) and, from these two inputs generates the error signal that is applied to the amplifier.

There are two basic types of synchro systems—torque and control. The main difference between the systems is in their output. A torque system produces a mechanical output, while the control system's output is electrical. The transmitters for these systems are functionally and physically similar. However, in a torque synchro system, the transmitter is designated a torque transmitter (TX), while the transmitter in the control system is designated a control transmitter (CX). The receivers in the two synchro systems are not only functionally and physically different, they are designated differently. The receiver in the torque system is designated a torque receiver (TR), while the receiver in the control system is designated a control transformer (CT).

Except for two TXs in each launcher receiver-regulator that transmit launcher position to the fire control system, the synchros in the launcher power drive's servosystems are of the control type (CXs and CTs). Therefore, the main emphasis in this portion of the text is directed toward the control synchro system.

## SYNCHRO SYSTEM ALIGNMENT

If synchros are to work together properly in a system, it is essential that they be correctly connected and aligned in respect to each other and to the other devices, such as directors and launchers with which they are used. The best of ordnance equipment would be ineffective if the synchros in the data transmission circuits were misaligned electrically or mechanically. Since synchros are the heart of the transmission systems, it stands to reason that they must be properly connected and aligned before any satisfactory firing can be expected.

Electrical zero is the reference point for alignment of all synchro units. The mechanical reference point for the units connected to the synchros depends upon the particular application of the synchro system. As a GMM on board ship, your primary concern with the mechanical reference points will be the centerline of the ship for launcher train and the standard reference plane for launcher elevation.

Remember that whatever the system, the electrical and mechanical reference points must be aligned with each other.

There are two ways in which this alignment can be accomplished. The most difficult way is to have two individuals, one at the transmitter and one at the receiver or control transformer, adjust the synchros while talking over sound-powered telephones or some other communication device. The better way is to align all synchros to electrical zero. Units may be zeroed individually, and only one person is required to do this work. Another advantage of using electrical zero is that trouble in the system always shows up in the same way. For example, in a properly zeroed TX-TR system, a short circuit from S2 to S3 causes all receiver dials to stop at 60° or 240°.

In summary, zeroing a synchro means adjusting it mechanically so that it will work properly in a system in which all other synchros are zeroed. This mechanical adjustment is accomplished normally by physically turning the synchro rotor or stator. Module 15, *Principles of Synchros, Servos, and Gyros,* NAVEDTRA 172-15-00-79, of the Navy Electricity and Electronics Training Series (NEETS), describes standard mounting hardware and gives simple methods for physically adjusting synchros to electrical zero. Additional information about synchros may also be obtained from OP 1303 and Military Standardization Handbook MIL-HDBK-225(AS), Synchros, Description and Operation.

### Zeroing TXs, TRs, and CXs, Using a Synchro Tester

Synchro testers of the type shown in figure 5-35 are used primarily for locating a defective synchro. They also provide a fairly accurate method of setting transmitters and torque receivers on electrical zero.

Control transmitters, torque transmitters, and torque recievers are functionally and physically similar. Therefore, they are zeroed in the same manner.

To zero a synchro with the tester, connect the units as shown in figure 5-35 and turn the synchro until the tester dial reads zero degrees. This is the approximate electrical zero position. Momentarily short S1 to S3, as shown. If either the synchro or tester dial moves, the synchro is not accurately zeroed, and should be shifted slightly until there is no movement when S1 and S3 are shorted.

NOTE: By exercising proper caution it is possible to perform all the preceding zeroing procedures using 115 volts where a source of 78 volts is not available. If 115 volts is applied instead of 78 volts, do not leave the synchro connected for more than 2 minutes or it will overheat and may be permanently damaged.

### Zeroing TXs, TRs, and CXs, Using a Voltmeter

Using a voltmeter, the zeroing procedure is broken down into steps as follows:

1. Carefully set the equipment, whose position the synchro transmits, to its zero or mechanical reference position.
2. Deenergize the synchro circuit and disconnect the stator leads. Set the voltmeter to its 0- to 250-volt scale and connect it into the synchro circuit as shown in figure 5-36, view (A). Many synchro systems are energized by individual switches; therefore, be sure that the synchro power is off before working on the connections.
3. Energize the synchro circuit and turn the stator or rotor until the meter reads about 37 volts (15 volts for 26-volt synchros). This is the course setting and places the synchro approximately on electrical zero.
4. Deenergize the synchro circuit and connect the meter as shown in figure 5-36(B), using the 0- to 5-volt scale.
5. Reenergize the synchro circuit and adjust the rotor or stator for a null (minimum voltage) reading. This is the electrical zero position.

A simple check on the common electrical zero position of a TX-TR synchro system can be done with a jumper. Put the transmitter and receiver on zero and intermittently jumper S1 and S3 at the receiver. The receiver should not

## Chapter 5—ELECTRICITY AND ELECTRONICS

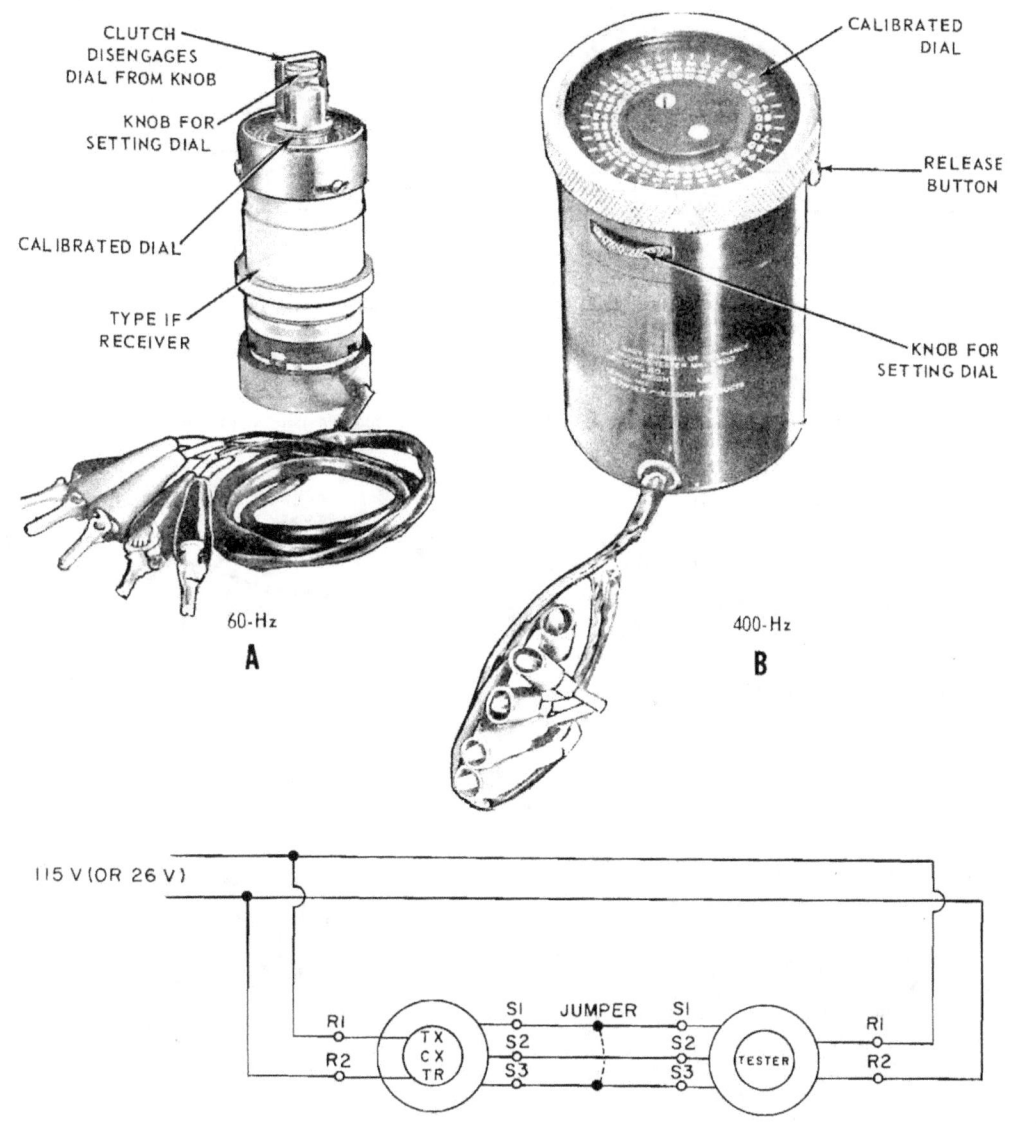

Figure 5-35.—Zeroing a TX, CX, or TR, using a synchro tester.

move. If it does, the transmitter is not on zero and should be rechecked.

### Zeroing a CT, Using a Voltmeter

A synchro control transformer is zeroed if its rotor voltage is minimum when electrical zero voltages are applied to its stator. Turning the CT's shaft slightly counterclockwise will produce a voltage between R1 and R2, which is in phase with the voltage between R1 and R2 of the CX supplying excitation to the CT stator.

The procedure used for zeroing depends upon the facilities and tools available and how

153

# GUNNER'S MATE M 1 & C

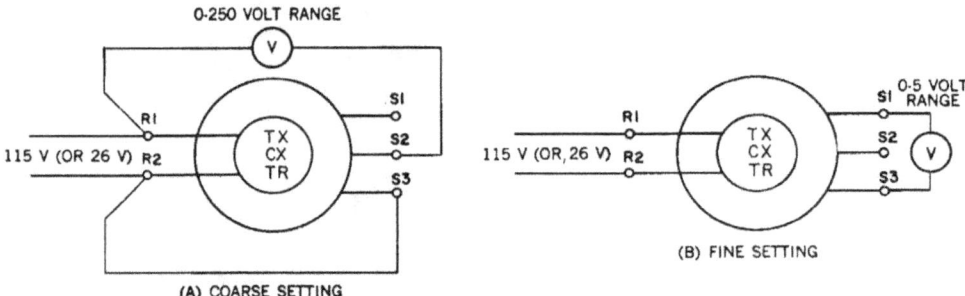

Figure 5-36.—Zeroing a transmitter or a torque receiver using a voltmeter.

the synchros are connected in the system. CTs may be zeroed by use of only a voltmeter, synchro testers, or other synchros in the system.

Regardless of the method used, there are two major steps in each zeroing procedure: first, the coarse (or approximate) setting; second, the fine setting. Many units are marked in such a manner that the coarse setting may be approximated physically on standard units; an arrow is stamped on the frame and a line is marked on the shaft extension, as shown in figure 5-37.

Using a voltmeter with a 0- to 250- and 0- to 5-volt scale, CTs may be zeroed as follows:

1. Remove connections from the CT and reconnect as shown in figure 5-38, view A.
2. Turn the rotor or stator to obtain minimum voltage reading.
3. Reconnect meter as shown in figure 5-38, view B, and adjust rotor or stator for minimum reading.
4. Clamp the CT in position and reconnect all leads for normal use.

### Alignment Summary

The described zeroing methods apply to all standard synchros and prestandard Navy synchros.

Before testing a new installation and before hunting trouble in an existing system, first be certain all units are zeroed. Also, be sure the

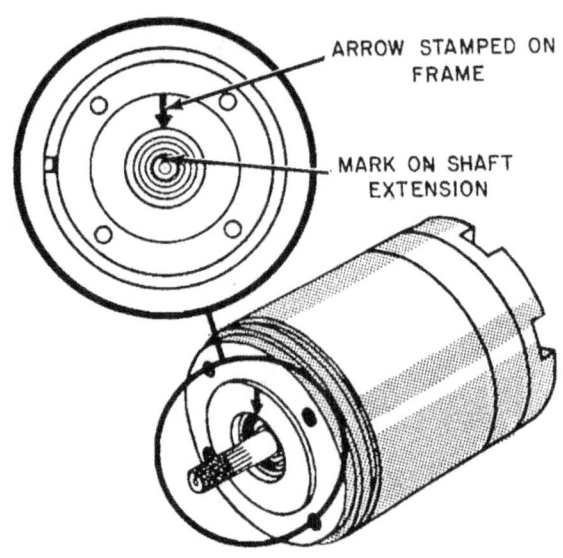

Figure 5-37.—CT coarse electrical zero markings.

device's mechanical position corresponding to electrical zero position is known before trying to zero the synchros. The mechanical reference position corresponding to electrical zero varies; therefore, it is suggested that the instruction books and other pertinent information be carefully read before attempting to zero a particular synchro system. The MRCs and the OP for the system should be studied, as there are

## Chapter 5—ELECTRICITY AND ELECTRONICS

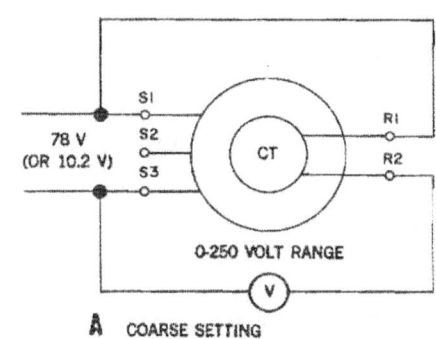

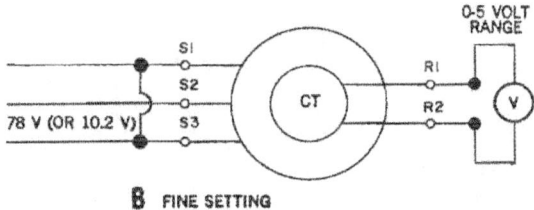

Figure 5-38.—Zeroing a CT, using a voltmeter.

55.36

likely to be some differences from the general instructions given in OP 1303 or module 16 of NEETS. For example, OP 2665, volume 3, Guided Missile Launching System Mark 13 Mod 0, gives step-by-step instructions for replacement and adjustment procedures for train and elevation regulator CTs. If an operational check indicates that a CT in the regulator is not operating properly, replace it and zero the new CT.

### MAINTAINING A SYNCHRO SYSTEM

Synchro units require careful handling at all times. Never force a synchro unit into place. Never drill holes in its frame. Never use pliers on the threaded shaft. Never use force to mount a gear or dial on its shaft.

Two basic rules apply to synchro units:

1. If it works, leave it alone.
2. If it goes bad, replace it.

Synchros are no longer considered to be repairable items. Replaced synchros should be disposed of in accordance with current instructions. Unless it is an emergency and no replacement is available, you must never take a unit apart or try to lubricate it. The gearing (figure 5-39) should be lubricated, using an atomizer, any time the cover of the reciever-regulator is removed, but do not lubricate switches or the tachometer. Follow the instructions in the MRCs for maintaining the synchro system.

### TROUBLESHOOTING SYNCHRO SYSTEMS

Shipboard synchro system troubleshooting is limited to determining whether the trouble is in the synchro or in the system connections.

All synchro casualties are not electrical, however, and do not require special equipment to uncover. One fairly common trouble affecting synchro operation is friction. Bearings must be especially clean, allowing the synchro rotor to turn freely. The slightest sticking will cause an error in route position, because there is little torque on the rotor when it is nearly in agreement with the incoming signal. Friction may also be caused by bent shafts and improper mounting of the synchro in the equipment. Early consideration should be given to the possibility of friction when troubleshooting faulty synchro operation. The synchros are not tested individually but are checked in the shipboard performance tests. If the test does not meet the standard requirements, a search is made for the faulty component.

While adjustments are a vital part of maintenance, they are too numerous to be covered here. Instead, a word of caution: At the time of installation, your control equipment was adjusted by well-qualified personnel using special tools and equipment. For this reason, adjustments should be undertaken only after qualified personnel have verified that an adjustment is necessary. A good habit to cultivate when making adjustments is to scribe gears at their original point of mesh, and count threads or teeth to the position of the new adjustment. These measures will prove most

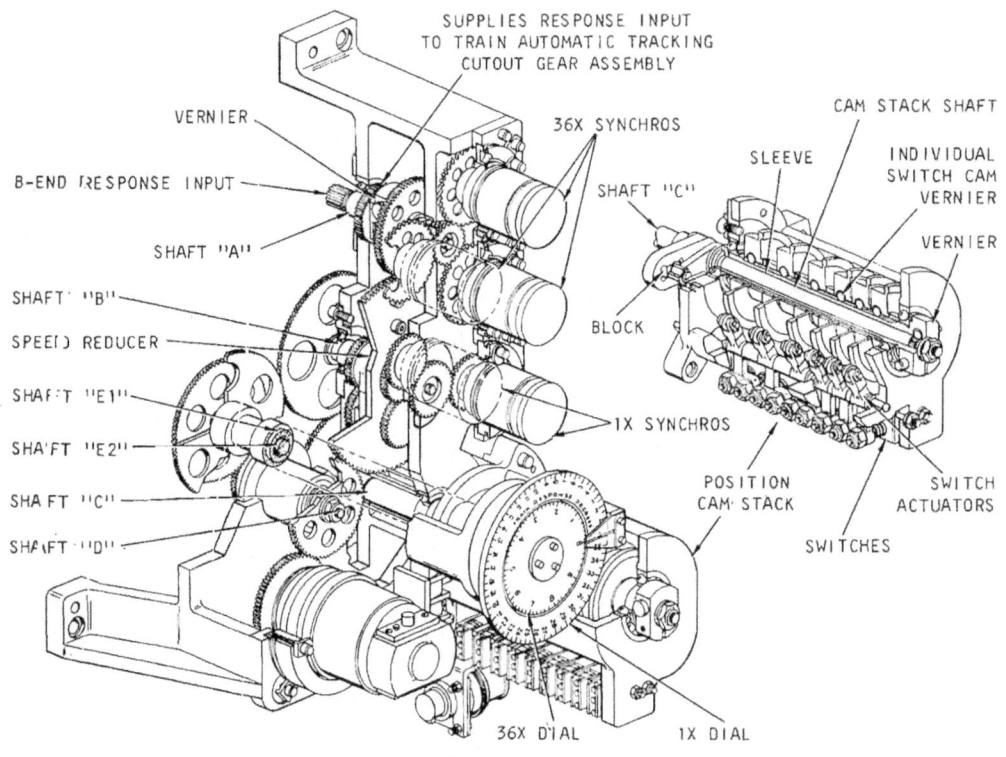

Figure 5-39.—Train synchro gear assembly.

valuable when an adjustment is later found to be incorrect or unnecessary.

In a newly installed system, the trouble probably is the result of improper zeroing or wrong connections. Make certain all units are zeroed correctly, then check the wiring. Do not trust the color coding of the wires—check them out with an ohmmeter. A major source of trouble is improper excitation. Remember, the entire system must be energized from the power source for proper operation.

In systems which have been working, the most common trouble sources are:

Switches—shorts, opens, grounds, corrosion, wrong connections.

Nearby equipment—water or oil leaking into the synchro from other devices. If this is the trouble, correct it before installing a new synchro.

Terminal boards—loose lugs, frayed wires, correction, and wrong connections.

Zeroing—units improperly zeroed.

Wrong connections and improper zeroing in any system are usually the result of careless work or inadequate information. Do not rely on memory when removing or installing units. Refer to the applicable instruction book or standard plan. Tag unmarked leads or make a record of the connections. Someone else may need the information.

It is an excellent idea to check individual conductors for legible markings on the vinyl (spaghetti) sleeving. Repair activities have the capability of preparing new sleeving for you.

## Chapter 5—ELECTRICITY AND ELECTRONICS

### SERVOAMPLIFIER

The purpose of a servoamplifier is to control an output in a manner dictated by an input. Normally, the servosystem's signal input is at a low energy level and must be increased greatly to perform an appreciable amount of work. This is the job of the servoamplifier. The amplifier controls a large power source which is activated by a low-powered error signal.

There are just about as many different amplifiers as there are jobs for amplifiers to do. Each part of the amplifier is selected to do a particular part of the total job. You cannot just look at a circuit and understand why everything is there. The best way to analyze an amplifier is to divide it into stages, coupling circuits, decoupling circuits, and biasing networks. *NEETS* modules 6, 7, and 8 introduce you to vacuum tube and solid-state amplifiers and power supplies. However, as of now, *NEETS* does not cover magnetic amplifiers.

Servoamplifiers can be divided broadly into functional stages. You have learned how the error signal is selected and modulated, or deomodulated, to suit the individual amplifier. The first stage or stages of amplification increase the voltage of the error signal. When the signal voltage is amplified a sufficient amount, it is used as the input to the power stage. Here the primary concern is current delivered at a steady voltage under load conditions. The push-pull type of amplifier is used extensively in missile servosystems. A push-pull amplifier is preceded by a phase inverter or paraphase amplifier. The power stage may be one or more stages, depending on the power output needed.

In general, the higher the gain of the amplifier, the tighter the control and the more accurate the servosystem. An increase in the system gain will reduce the system velocity errors and increase the speed of response to inputs. An increase in system gain also reduces those steady-state errors resulting from restraining torques on the servo load. However, to obtain these advantages, the servosystem must pay a price in the form of a greater tendency toward instability. A linear servosystem is said to be stable if the response of the system to any discontinuous input does not exhibit sustained or growing oscillations. The highest gain that can be used is limited by consideration of stability.

### LAUNCHER POWER DRIVE SERVOAMPLIFIERS

*Gunner's Mate M 3 & 2*, NAVEDTRA 10199 series, described and illustrated servosystems (with amplifiers) used to control error signals in launcher power drives. Amplifiers associated with ordnance actually do more than amplify. Some power drive amplifiers change the incoming a.c. synchro signal to a d.c. signal that can be used to control a servomotor. In amplifiers associated with ordnance equipment, the power supply normally is built into and, therefore, is physically a part of the amplifier.

Examples of other amplifier functions include stabilizing, synchronizing, speed limiting, position limiting, and current limiting. Amplifiers associated with ordnance equipment are nearly always classed as power amplifiers. A voltage amplifying stage is used only if it is necessary to increase an input voltage. The number and type of amplifier functions is determined to some extent by the type of output controlled by the amplifier.

### Mk 10 GMLS Launcher Servoamplifiers

The amplification of the train and elevation signals for the Mk 5 Terrier launcher is an example of the use of servoamplifiers.

The small electrical input signals must be amplified into usable signals of sufficient magnitude to operate the electrohydraulic servo valves of the receiver-regulators. The amplification system is common to both power drives and consists of a dual channel magnetic amplifier, made up of four magnetic amplifier stages mounted on a common chassis, and a power supply. One channel of the amplifier services the train power drive and the other channel services the elevation power drive. In each channel, one magnetic amplifier stage is the primary servosystem amplifier and the other is the velocity system servoamplifier.

The primary system servoamplifier receives a position error voltage signal from the 1- and 36-speed CTs in the receiver-regulator. The amplifier also receives an unfiltered velocity signal from the rate generators in the remote, local, or dummy director. It mixes and amplifies these signals and uses the resultant output to operate the primary electrohydraulic servo valve. The input circuit of the primary amplifier limits the voltages to the magnetic amplifier stage control windings and provides automatic changeover from the 1-speed signal control to the 36-speed signal control when the launcher position error reduces to less than 5 degrees of correspondence with the order signal. It also receives an amplifier load supply voltage and a synchro offset voltage from the power supply. The train primary amplifier input circuit applies the offset voltage to the output of the 1-speed synchro control transformer for stick-off purposes. The offset voltage is not applied to the elevation primary amplifier.

The velocity system servoamplifier receives a filtered velocity signal from the rate generators in the remote, local, or dummy director. The amplifier also receives an electrical feedback signal from the velocity and integration potentiometers of the receiver-regulator. The velocity amplifier mixes and amplifies these signals and uses the resulting output to operate the velocity electrohydraulic servo valve. The input circuit of the velocity amplifier provides the gain control for the velocity input and voltage controls for the potentiometers; it mixes the velocity signal input with the potentiometer signals, and applies the resulting signal to the control windings of the magnetic amplifier stage.

The potentiometer voltage supply circuit provides a frequency-sensitive, regulated, and filtered voltage for the velocity and integration potentiometers of the receiver-regulator. The regulated voltage supply prevents fluctuation of the integration and velocity system outputs and compensates for the varying line frequencies to stabilize the electric drive motor and B-end error of the power drive.

MAGNETIC AMPLIFIER.—The magnetic amplifier is an important device in electrical and electronic equipment. Amplifiers of this type have many features which are desirable in missile systems. The advantages include (1) high efficiency (90%); (2) reliability (long life, freedom from maintenance, reduction of spare parts inventory); (3) ruggedness (shock and vibration resistance, overload capability, freedom from the effects of moisture); (4) stability; and (5) no warmup time. The magnetic amplifier has no moving parts and can be hermetically sealed within a case similar to the conventional dry type of transformer.

The magnetic amplifier has a few disadvantages. For example, it cannot handle low-level signals (except for special applications); it is not useful at high frequencies; it has a time delay associated with magnetic effects; and the output waveform is not an exact reproduction of the input waveform.

The term "amplification" generally refers to the process of increasing the amplitude of the voltage, current, or power. The term "amplification factor" is the ratio of the output to the input. The input is the signal that controls the amount of available power delivered to the output.

Until comparatively recent times, magnetic control has had little application in missile electronic equipment since existing units were slow in response and were of excessive size and weight. But with the development of new and improved magnetic materials, there has been a parallel development of magnetic circuits for tubeless amplification, and many of these units are now used in automatic pilots, static a.c. voltage regulators, and in associated test equipment

Magnetic amplifiers are devices which control the degree of magnetization in the core of a coil to control the current and voltage at the load or output. One of the oldest forms of magnetic amplifiers, the saturable reactor, contains at least two coils (see figure 5-40) wound on a common core made of magnetic material. A d.c. control voltage is applied to one of the coils, and the resulting current serves to modify the inductive reactance of the second

## Chapter 5—ELECTRICITY AND ELECTRONICS

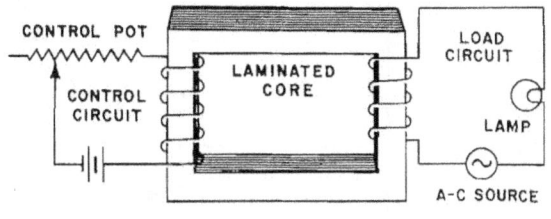

BE 389

Figure 5-40.—A basic magnetic amplifier.

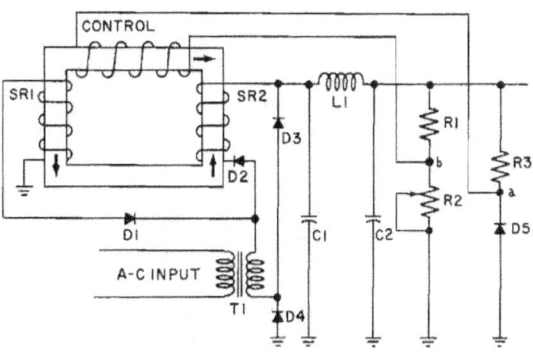

12.213

Figure 5-41.—Power supply using magnetic amplifier voltage regulator.

winding by causing magnetic saturation of the common core. The second coil is a series element in the a.c. load circuit so that current variations take place in the load in accordance with those made in the control voltage. In more complex magnetic amplifiers the input, or control signal, may be either a d.c. or a properly phased a.c. voltage.

To understand the theory of magnetic amplifiers, it is necessary that you understand the theory of magnetism and magnetic circuits. This information may be found in module 1 of *NEETS*. OP 2350 also provides a very good description of the basic theory of operation of a magnetic amplifier.

SERVOSYSTEM POWER SUPPLY.—The equipment power supplies of missile systems must meet certain basic requirements which include ruggedness, long life, and freedom from excessive maintenance problems. To meet these requirements, the development of power supply equipment has resulted, in many cases, in the elimination of the electron tube as the chief cause of failure. The magnetic amplifier has been used to replace the complex arrangements usually necessary for good voltage regulations; and the solid-state power diode is often used instead of the fragile vacuum tube. An example of a circuit with these components is shown in figure 5-41.

The circuit is a conventional full-wave bridge rectifier using a magnetic amplifier to control the output and a zener diode as a part of the regulating system. The zener diode element is a solid-state equivalent of the gaseous regulator tube and maintains a constant voltage across the terminals regardless of variations of the current it conducts, within the specified operating range. In the schematic shown (figure 5-41), the connection of the zener diode is the reverse of that of an ordinary rectifying diode since, in this example, it is the inverse breakdown voltage characteristic which is used for regulation.

Current flow (figure 5-41) during one half cycle is through the load, choke L1, diode D3, the secondary of T1, and diode D1, then returning to ground through SR1 of the reactor. During the other half cycle, the current flows through the load, L1, SR2, D2, the secondary of T1, and D4 to ground. In addition to the load current, there is conduction through D5 and R3 and through R2 and R1.

The control winding of the magnetic amplifier is energized by the voltage between the junction of R1 and R2 and the upper terminal of the zener diode, D5. When the output voltage is of the proper value, the potential across the control winding (and, therefore, the current through it) sets the magnetic bias of the reactors at the operating point, which is well up on the magnetization curve to obtain a high percentage of the source voltage

If the output voltage tends to rise, the voltage at point A remains constant due to the action of the zener diode; but the voltage at point B increases. This causes a change in the

159

current flowing in the control winding so that the bias point is shifted to the value that results in lower conduction in the load coils. As a result, the voltages across SR1 and SR2 increase, and the output voltage decreases.

When the output voltage tends to decrease, the potential at point B falls with respect to that at point A and the control current changes the bias to a point of higher conduction. This lowers the voltage drops across the a.c. coils of the reactors and increases the value of the output. Capacitors C1 and C2, together with L1, are connected to form a filter which smooths the output to give a nearly pure d.c. voltage. Resistor R2 is adjustable, being set to the value for optimum operating voltage in normal use. It also provides a means for making adjustments to compensate for any changes that occur in the circuit components.

Avalanche breakdown is sometimes called zener effect, after the American physicist Clarence Zener, who made theoretical investigations of the problem of electrical breakdown of insulators. The breakdown mechanism in PN transistor junctions is not the same as in insulators but, in spite of this, the term "zener voltage" is often given to breakdown voltage of junctions. The reverse voltage at which the current suddenly begins to make its sharp descent is called zener breakdown voltage. The use of the word "breakdown" does not mean that the diode is destroyed; but rather that the normal negative reverse current increases suddenly and sharply. A typical zener diode curve is shown in figure 5-42.

Zener diodes are used chiefly as regulation and reference elements. When a reverse voltage is applied, no current will be passed until there is a breakdown in the covalent bond of the atoms, causing a sharp increase in current flow in the reverse direction. If this happened in a regular PN junction diode, it would be considered defective, but zener diodes are designed to be self-healing and can be used repeatedly without damage. The point of breakdown or avalanche is built into the diode and can be made to occur at various voltages. In figure 5-42, approximately 20 volts is applied.

The zener effect and zener diodes are covered extensively in module 7 of *NEETS*.

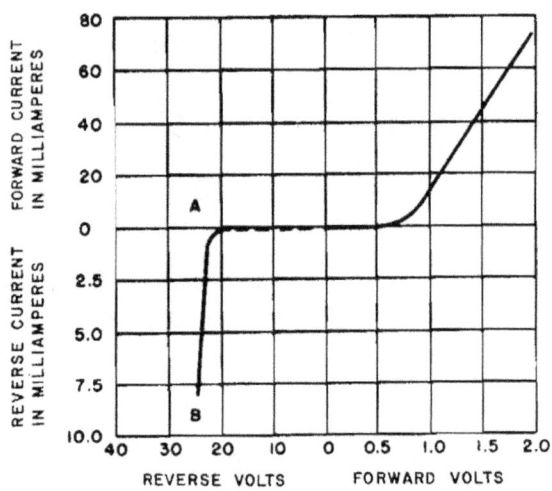

Figure 5-42.—Zener diode characteristic curve.

### Mk 13 GMLS Launcher Servoamplifiers

The train and elevation amplifiers for the Mk 13 Mods 1 through 3 GMLSs are identical, and share a common power supply in a transistorized electronic servocontrol unit which is mounted in the EP2 control panel. (The Mk 13 Mod 0 GMLS power drive amplifiers use both tubes and transistors.) Thirteen printed circuit cards in a rack on top of the main chassis plug into 13 female receptacles in the back of the compartment. Each of the amplifiers (train and elevation) requires an identical set of six printed circuits, one card for each of six primary stages in the functioning of the amplifier.

Figure 5-43 is a functional diagram of the servoamplifier with each stage and its printed circuit board identified. The CTs in the receiver-regulator are the error detectors for the system. They compare the remote order signal from the fire control computer to the actual launcher position (B-end response) and generate the error signals. The error signals (1- and 36-speed) are then applied to the first stage (PC1) of the servoamplifier. The amplifier selects the appropriate error signal (1- or

## Chapter 5—ELECTRICITY AND ELECTRONICS

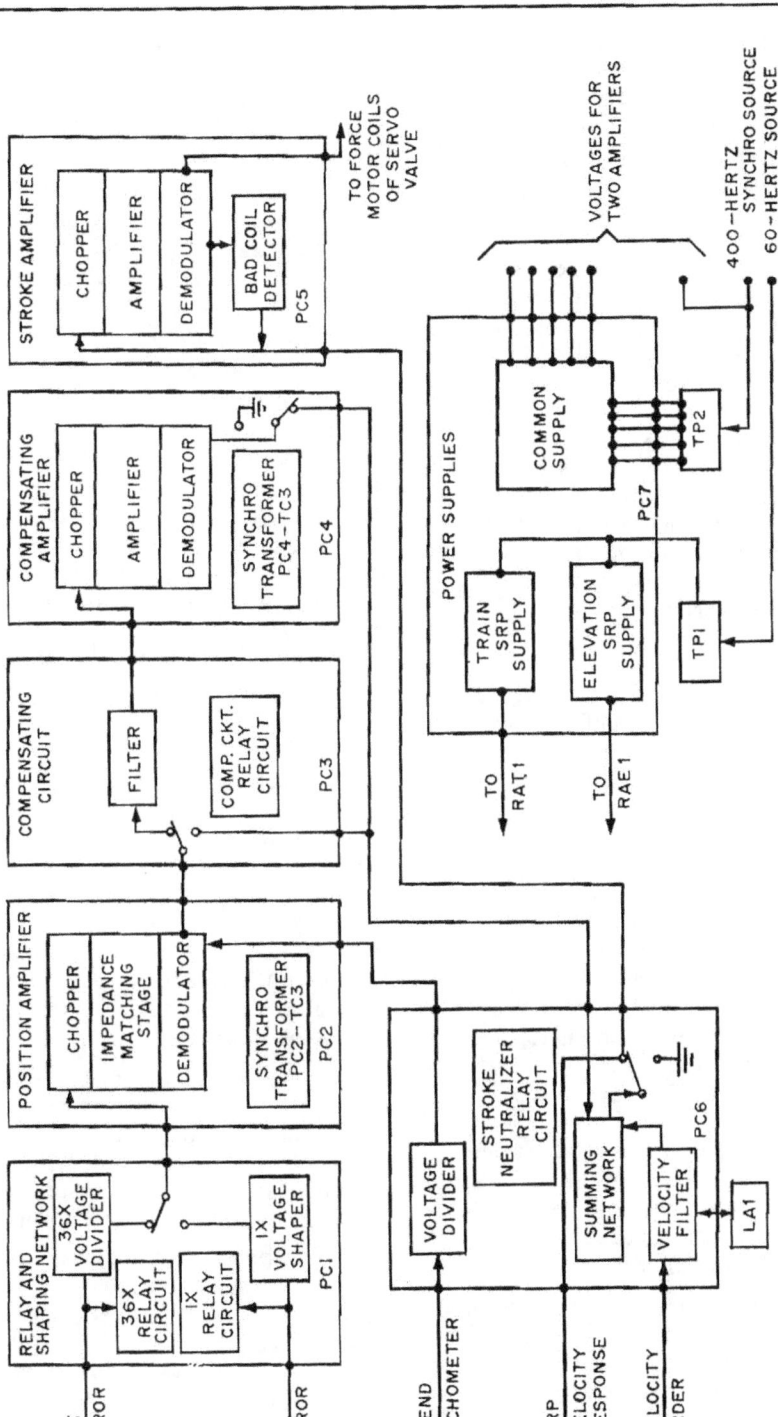

Figure 5-43.—Amplifier circuit cards, functional diagram.

36-speed), mixes the error signal with the velocity signals (B-end response, stroke response, and director velocity), amplifies the error signal, and converts the a.c. input into a d.c. output that operates the force motor coils of the electrohydraulic servo valve.

The velocity signals are applied to PC6 (figure 5-43). These signals help the power drive to prevent overshooting the ordered position and to follow constant velocity and roll signals smoothly. The velocity inputs include a director velocity signal, an A-end stroke response potentiometer signal, and a B-end response tachometer-generator signal. The director velocity and A-end stroke velocity signals are summed and applied to the error signal in PC6, while the B-end velocity signal is routed to PC2 where it is applied to the error signal.

## TROUBLESHOOTING AND ADJUSTMENTS

Unless specifically directed otherwise, defective train and elevation amplifier units are removed as a unit and replaced. The CMCs will give you step-by-step instructions for making adjustments for the train and elevation amplifiers of your particular GMLS.

Some adjustments made at the factory are not changed on shipboard. Hermetically sealed components are always replaced rather then repaired. Before replacing such a unit, double check all associated circuitry (resistors, wiring, etc.). When a defective component is replaced, adjust it and the channel in which it operates, following the instructions for your equipment.

All amplifier channel balance adjustments have been set at the factory. On installation, and thereafter, the balance of both stages of amplification should be checked using the meters installed in the amplifier panel.

Demodulators are balanced at the factory and normally no further adjustment should be necessary except on replacement, or in case the setting at balance adjustment is disturbed.

Rectifiers are very important components of magnetic amplifiers. Series rectifiers may be checked by the use of cathode-ray oscilloscope. Whenever possible, the waveform across a rectifier suspected of being defective should be compared to waveforms observed across other rectifiers in the same circuit.

In many servosystems the gain of the amplifier can be varied by an adjustment. The gain adjustment governs the amplitude or amount of the signal voltage applied to the amplifier or one of its stages. Normally, the highest gain possible, with the servosystem possessing a satisfactory degree of stability, is the most desirable.

In a.c. servosystems another adjustment which can control the sensitivity of the system is the phase adjustment. The phase adjustment is used to shift the phase relationship between the signal voltage and a reference voltage. In an amplifier with phase shift control, the grid signal is shifted in phase with reference to the plate voltage of a tube. The tube's firing point is delayed or advanced, depending upon the phase shift of the grid signal. The phase shift can vary the firing time of the tube over the plate's entire positive alternation.

A phase control is included in some servosystems using a.c. motors. The two windings of the a.c. servosystems using a.c. motors should be energized by a.c. voltages that are 90° apart. This phase adjustment is included in the system to compensate for any phase shift in the amplifier circuit. The adjustment may be located in the control amplifier or, in the case of a split-phase motor, it may be in the uncontrolled winding.

Servosystems using push-pull amplifiers must be balanced so that when there is no signal input to the amplifier, its output will be null, and the servomotor will stand still with no creep. The push-pull amplifier must ensure equal torque in both directions of the servomotor.

Gain, phase, and balance adjustments are often present in one amplifier. These adjustments tend to interact so that when one of them is changed, it may affect the others. Therefore, after making any one adjustment it is a good practice to check the other adjustments.

On the latest GMLSs, transistor amplifiers are used in place of electron tube and magnetic

## Chapter 5—ELECTRICITY AND ELECTRONICS

amplifiers. A transistor amplifier must have three-element (two-junction) semiconductors to amplify a signal, just as a three-element electron tube is needed for amplification. There are also three types of transistor amplifiers, according to which part is grounded: grounded emitter, grounded base, and grounded collector. The theories and operating characteristics of vacuum tubes and of transistors are covered in modules 6 and 7 of *NEETS*. Transistors are designed to perform the same functions as vacuum tubes. As they are solid-state semiconductors, they are much less fragile than vacuum tubes. Of course, failure can be caused by misuse, such as current overloading, or application of too high a voltage. Faults in manufacturing, or flaws in the material can cause mechanical failure. Radiation affects them, so they must be shielded. Most failures are caused by the effects of moisture on the surface. Hermetic sealing of the transistors by manufacturers is now the usual practice. Since transistors are so very small, a speck of dust falling across a junction can completely short-circuit it. A dust-free atmosphere is a practical necessity in a transistor-fabrication plant.

### REFERENCES

The following publications were used in developing this chapter.

1. OP 2665, Guided Missile Launching System Mk 13 Mods 0, 1, 2, and 3
2. OP 3114, Guided Missile Launching System Mk 10 Mods 7 and 8
3. OP 3115, Guided Missile Launching System Mk 22 Mod 0
4. *Navy Electricity and Electronics Training Series (NEETS)*
5. OP 2350, Guided Missile Launcher Mk 5
6. NAVEDTRA 10077 series, *Blueprint Reading and Sketching*
7. NAVEDTRA 10209 series, *Fire Control Technician M 3 & 2*
8. NAVPERS 10073, *Mathematics, Vol. 3*
9. OP 4104, Guided Missile Launching System Mk 26 Mods 0, 1, and 2
10. OP 4006, Guided Missile Launching System Mk 29 Mod 0
11. OP 3000, Weapons Systems Fundamentals
12. NAVPERS 10088 series, *Digital Computer Basics*

# CHAPTER 6

# HYDRAULICS AND PNEUMATICS

The missile launching systems currently maintained by GMMs are operated in part by hydraulic power. The only exception is the Mk 29 GMLS (NATO SEASPARROW), which uses no hydraulics.

Hydraulic systems can actuate mechanisms instantaneously, with almost 100 percent efficiency; however, introduction of foreign matter into the system or leakage can make the whole system inoperative. Daily checks for leakage and constant vigilance against the entry of foreign matter into the system are important factors of hydraulic system maintenance.

The hydaulic network of piping required to carry the oil and the valves which direct, restrict, relieve, or shut off the flow afford numerous places for trouble to develop; therefore, regular inspections and maintenance are necessary to keep the systems operating smoothly.

Hydraulic power drives have been used in the Navy for many years to train and elevate guns. When launchers were needed for rockets and missiles, power drives were adapted for them. The principal advantage of hydraulic power drives is their ability to move large loads smoothly and quickly: A disadvantage is their need for constant maintenance. To retain the power drive's characteristics of power, speed, and control, and to help eliminate extensive repairs and costly replacement of parts, strict adherence to instructions from maintenance publications and maintenance requirement cards (MRCs) is essential. Correct casualty analysis can only be made by someone with a thorough knowledge of how the equipment operates. When trouble develops, reach first for the OP—not for the wrench.

## HYDRAULICS IN GMLSs

Hydraulic power drives are used in missile systems to move missiles from one shipboard station to another, position missile launchers in train and elevation for firing, load and offload missiles from ship to ship or from ship to shore and, in some systems, to jettison unwanted missiles. Hydraulics are also used in the missile to control the missile flight attitude through movement of missile control surfaces (tails). This chapter will discuss hydraulics in launching systems and in missiles. How hydraulic systems control the missile's flight path is explained in *GMM 3 & 2*.

## HANDLING AND STRIKEDOWN EQUIPMENT

The use of handling equipment (handlift trucks and transfer dollies) was described in chapter 2. Hydraulics on this equipment is limited to shock absorbers and brakes.

The hydraulic coverage of strikedown equipment for this chapter will be limited to the strikedown elevator and hatches for the Mk 10 GMLS. The loader accumulator power units are the hydraulic power source for this strikedown equipment.

In the Mk 10 GMLS, the A- and B-side loader accumulator power units are located in the strikedown and checkout area and supply hydraulic power to:

1. The strikedown elevator and hatch (port or starboard).
2. The spanning rails and blast doors.
3. The retractable rails.
4. The floating rails or tracks.
5. The loader positioner.

# Chapter 6—HYDRAULICS AND PNEUMATICS

The strikedown elevators and hatches for the Mk 10 GMLS were developed under the auspices of NAVSHIPS (now NAVSEA), and are contained in **NAVSHIPS 378-0320**. The hydraulic schematics for this equipment use JIC symbols which differ considerably from the more common hydraulic schematics for the Mk 10, Mk 13, Mk 22, or the Mk 26 GMLSs.

Strikedown elevators and hatches are designated as either port or starboard units. On a forward Mk 10 GMLS installation, the port strikedown elevator and hatch are supplied hydraulic power by the A-side loader accumulator, and the starboard side strikedown elevator and hatch are supplied by the B-side loader accumulator. On an after Mk 10 GMLS installation, the port strikedown elevator and hatch are supplied hydraulic power by the B-side loader accumulator, and the starboard strikedown elevator and hatch are supplied by the A-side loader accumulator. The EP4 panel is always the A-side assembly captain's control panel regardless of the installation; the EP5 panel is for the B-side, again regardless of the installation.

NOTE: Before attempting to operate the strikedown elevators and hatches, you must manually remove the mechanical locking mechanisms from the elevator and position the manually operated hatch dogs clear for unobstructed movement of the hatch. Hydraulically operated latches (dogs) are ganged to one actuator shaft. The hydraulic schematic for the strikedown elevator and hatch on the Mk 10 GMLSs will be shown later in this chapter. Although the schematic does not specify port or starboard, it is adaptable to either side since the components on each side have the same designations and perform the same functions regardless of physical location. Valve A should be closed after securing the strikedown elevator and hatch. This prevents hydraulic shock from possibly damaging the equipment, in case SMS2 on the EP2 panel is rotated to strikedown while the loader accumulator is running.

The electrical controls for the strikedown elevator and hatch consist of two pushbutton switches, four solenoids, and two interlock switches. The control station is in the checkout area of the strikedown/checkout compartment.

One pushbutton switch, designated A1, is at the elevator/hatch control station. The other pushbutton switch, B1, is by the strikedown hatch at the 01 level. These switches have three positions: UP, NEUTRAL, and DOWN.

The four solenoids are on valve blocks located near the control station. One set of solenoids, designated A and B, controls a valve designated J; the other set, also designated A and B, controls a valve designated L.

The two interlock switches have designations of K and N. Switch K, actuated by the shaft of the hatch dogs, is on the overhead of the missile house. Switch N, actuated by a hatch cam, is at the aft end of the strikedown hatch.

### Electrical Functions

The following explanation refers to the electrical and hydraulic schematics in figures 6-1 and 6-2, respectively.

With the elevator/hatch system activated and the pressures up, the operators at the elevator/hatch control station raise the elevator by depressing and holding A1 and B1 at UP. With the A1 and B1 contacts closed, the n/c contacts of switch K complete the circuit to solenoid A of valve J. Solenoid A energizes and shifts the spool of valve J to the right. When the hatch dogs fully retract, a shaft actuates switch K.

The n/c contacts of switch K open the circuit to solenoid A of valve J, causing valve J to return to neutral. At the same time, the n/o contacts of switch K complete a circuit to solenoid A of valve L, causing the plunger of valve L to move to the right.

The elevator starts upward when the hatch fully opens. When the elevator reaches its upper limits, the operators release switches A1 and B1 to deenergize solenoid A of valve L. If the

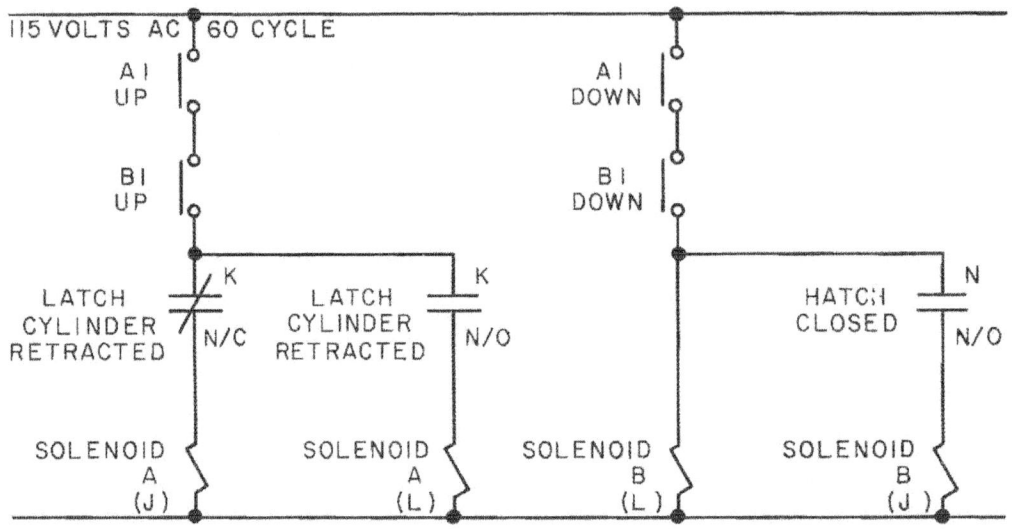

Figure 6-1.—Strikedown elevator and hatch: electrical schematic.

operators release A1 and B1 prematurely, the raise cycle stops prematurely.

To lower the elevator and to close and dog the hatch, the operators depress and hold A1 and B1 at DOWN. This action completes a circuit to energize solenoid B of valve L and thereby initiates the lower cycle. When the elevator is fully down, the hatch starts to close. As the hatch nears the end of its cycle, it actuates switch N; the n/o contacts of switch N, in turn, complete a circuit to solenoid B of valve J. The resulting hydraulic action moves the dogs toward the engaged position.

When the dogs fully engage, the operators release switches A1 and B1 to complete the cycle. The operators can initiate the next cycle when ready.

The following is a step-by-step set of operating instructions for the Mk 10 GMLS strikedown elevator:

NOTE: Valves D and E are set to specifications at installation, then locked. Valve F should be set and locked at the proper position to ensure the correct speed of the strikedown elevator and hatch.

1. Activation:
   a. At the EP2 panel:
      (1) Start the loader accumulator motor.
      (2) Rotate Operation Selector Switch SMS2 to STRIKEDOWN.
   b. Open valves T and U.
   c. Turn valve G to AUTOMATIC.
   d. Turn valve H to STOP.
   e. Open valve A leading to the accumulator.
2. Raise cycle:
   a. Depress and hold switches A1 and B1 at UP.
   b. When the elevator reaches the end of its cycle, release switches A1 and B1.
3. Lower cycle:
   a. Leave the valves in the position they were in for the raise cycle.
   b. Depress and hold switches A1 and B1 at DOWN.
   c. When the elevator reaches the end of its cycle, release switches A1 and B1.

Chapter 6—HYDRAULICS AND PNEUMATICS

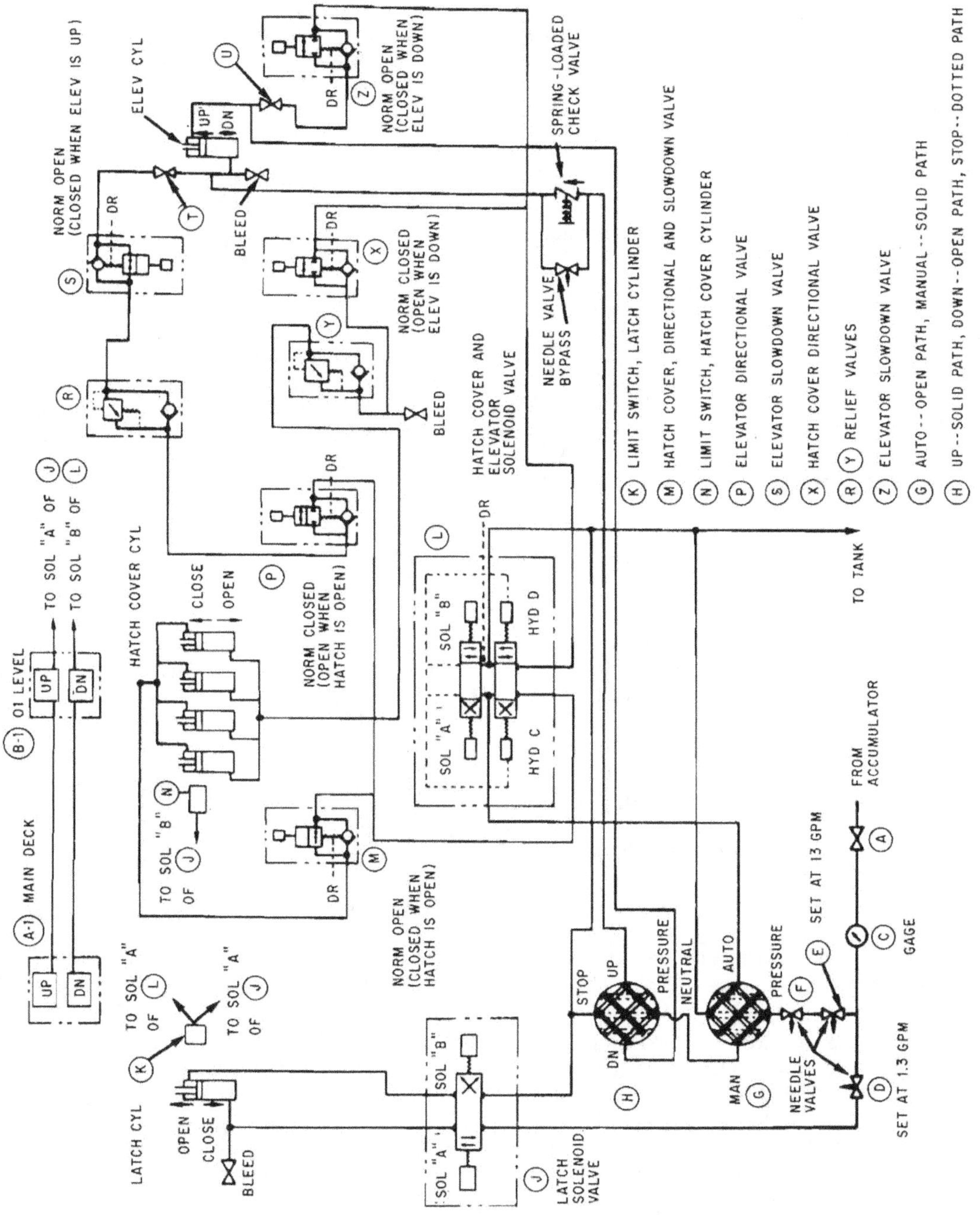

Figure 6-2.—Strikedown elevator and hatch: hydraulic schematic.

## Hydraulic Functions

The following is an explanation of the hydraulic functions in raising and lowering the elevator in automatic control (refer to figure 6-2):

1. Raise cycle:

This explanation starts with the conditions in step 1 (Activation), letter e.

Hydraulic pressure from the accumulator is present at valve J of solenoid A and through valve G (open path) to valve L of solenoid A.

When the operators depress A1 and B1 to UP, latch solenoid A of valve J energizes to position the valve-J plunger to the right and thereby port PA to the large-area side of the latch cylinder. This action raises the cylinder piston (unlocks the hatch dogs).

With the valve J-plunger of solenoid A at the right, the small-area side of the latch cylinder is open to tank.

When the piston of the latch cylinder raises fully, limit-switch K closes, solenoid A of valve J deenergizes, and solenoid A of valve L energizes to shift the valve-L plunger to the right. Valve L now ports PA to the right end of hydraulically operated valve D, and shifts this valve to the left. Also, PA at valve L goes through valve D to valve M, the hatch-cover directional and slowdown valve. From valve M, PA goes to the small-area sides of the four hatch-cover cylinders. This action opens the hatch cover by moving the cylinder pistons down. At this time, PA is also available at valve P.

The large-area side of the hatch-cover cylinders port to tank through valve Y, the hatch-cover counterbalance valve; valve X, the hatch-cover directional valve; and valve D of L.

When the hatch cover nears its fully opened position, cams close valve M and open valve P, the elevator directional valve. The PA at valve P now ports through a check valve in valve R, the elevator counterbalance valve; through valve S, the elevator slowdown valve; and through valve T to the large-area side of the elevator cylinder. This action raises the cylinder piston.

The small-area side of the elevator cylinder ports to tank through valve U; valve Z, the elevator slowdown valve; and valve D.

When the elevator nears the end of the raise cycle, a cam shifts valve S, the elevator slowdown valve, to close off PA and decelerate the elevator.

2. Lower cycle:

This explanation starts with the conditions ending the raise cycle.

Hydraulic pressure from the accumulator is present at valve J of solenoid A and through valve G (open path) to valve L of solenoid A.

When the operators depress A1 and B1 to DOWN, latch solenoid B of valve L energizes to position the valve-L plunger to the left. In this position, valve L ports PA to the left end of valve C. This action shifts valve C to the right.

The PA at valve C of L goes through valve Z, the elevator slowdown valve, and valve U to the small-area side of the elevator cylinder. At this time, PA is also at valve X.

The large-area side of the elevator cylinder ports to tank through valve T; valve S, the elevator slowdown valve; valve R, the elevator counterbalance valve; valve P, the elevator directional valve; and valve C.

When the elevator nears the end of the lower cycle, a cam closes valve Z, the elevator slowdown valve, to close off PA and decelerate the elevator. And when the elevator is fully down, a cam opens valve X, the hatch-cover directional valve, to port PA through check-valve Y and on to the large-area side of the four hatch-cover cylinders. This action moves the cylinder pistons up, closing the hatch cover.

The small-area side of the four hatch-cover cylinders ports to tank through valve M, the hatch-cover directional and slowdown valve, and valve C.

When the hatch cover is in its fully closed position, limit-switch N trips, closing the circuit to solenoid B of valve J. The valve J plunger shifts to the left and ports PA to the small-area side of the latch cylinder. This action closes and locks the hatch dogs.

## Chapter 6—HYDRAULICS AND PNEUMATICS

## MAGAZINES

Each magazine accumulator power system in the Mk 10 GMLS has four accumulators and supplies hydraulic power for operating the following:

1. Ready service ring (RSR) motor.
2. Tray-shift mechanism.
3. Ready service ring (RSR) hoist.
4. Magazine doors.

The RSR may be rotated in either direction by the hydraulic motor. The tray-shift mechanism, used to transfer a missile either from the RSR to the hoist, or the reverse, is moved by extending a hydraulically operated piston rod. The magazine doors are hydraulically opened upward, and may be locked open or closed by means of latches which are operated hydraulically. The hoist has a hydraulic drive unit with upper drive transmission, lower drive transmission, and drive shafts. The accumulator power supply system consists of an electric motor, piston pump, supply tank, header tank, control-valve block, and four accumulators.

The location of each of the power units varies somewhat with the mark and mod of the system, and the ships on which they are installed. The Mk 10 Mod 7 system is installed forward on the main deck of CG 26-class ships and it has three RSRs instead of the usual two. Mods 3 and 4 are installed aft and athwartship on CV 63-class ships, which imposes a different placement of the RSRs (and their power units).

It is not so much the location of the different units that you need to know, as it is the action and interaction of the different components in the system.

The Mk 10 GMLSs have two magazine accumulator systems, one for each side, to supply power to the magazine components. Figure 6-3 shows the location of components in the Mk 10 Mod 7 system. The accumulator flasks are not shown in this illustration; the high pressure accumulator pump is inside the fluid-filled supply tank, and is therefore not visible. This system has one header tank common to both A- and B-sides.

The supply tank holds the hydraulic fluid, the motor-driven pump produces the hydraulic pressure, the valves in the valve block regulate system pressures, and the accumulator flasks store energy, absorb hydraulic shocks, and prevent excessive pressure fluctuations. A solenoid assembly, attached to the valve block, actuates the dump valve to dump pressure fluid to the tank when the system is deactivated.

Systems that require a large volume of pressure fluid usually make use of a parallel piston pump instead of a gear pump. In addition, all power drives have an auxiliary pump, which may be power-driven or manually operated. This is used to provide a limited supply of hydraulic fluid for emergency operation. During maintenance or installation, the hand pump is used.

Several types of valves and a parallel piston pump were illustrated and described in *Gunner's Mate M 3 & 2*, NAVEDTRA 10199-C. Not all power drives have all the associated valves in one valve block but may have several separate assemblies. The valves control the volume of fluid, the direction of flow, and the pressures at which the system operates.

### Power Drive

Figure 6-4 is a schematic of the magazine accumulator system used in the Mk 10 Mod 0 Terrier system. The accumulator flasks, one of which is shown with the internal bladder and the poppet valve, are charged with nitrogen through nitrogen valves. The pressure required varies with the temperature; refer to the data chart on the charging valve block when checking the pressure. To accomplish this daily maintenance procedure, consult the MRCs.

Checking the fluid level in the header tank is another daily maintenance requirement. Before you do this, however, shut down the system and consult the MRCs for step-by-step instructions.

The electric motor which operates the pump (inside the supply tank) is coupled to the pump through a mounting flange. The pump draws the hydraulic fluid through the intake screen and discharges the fluid under pressure to the valve block. From the valve block, the pressurized

GUNNER'S MATE M 1 & C

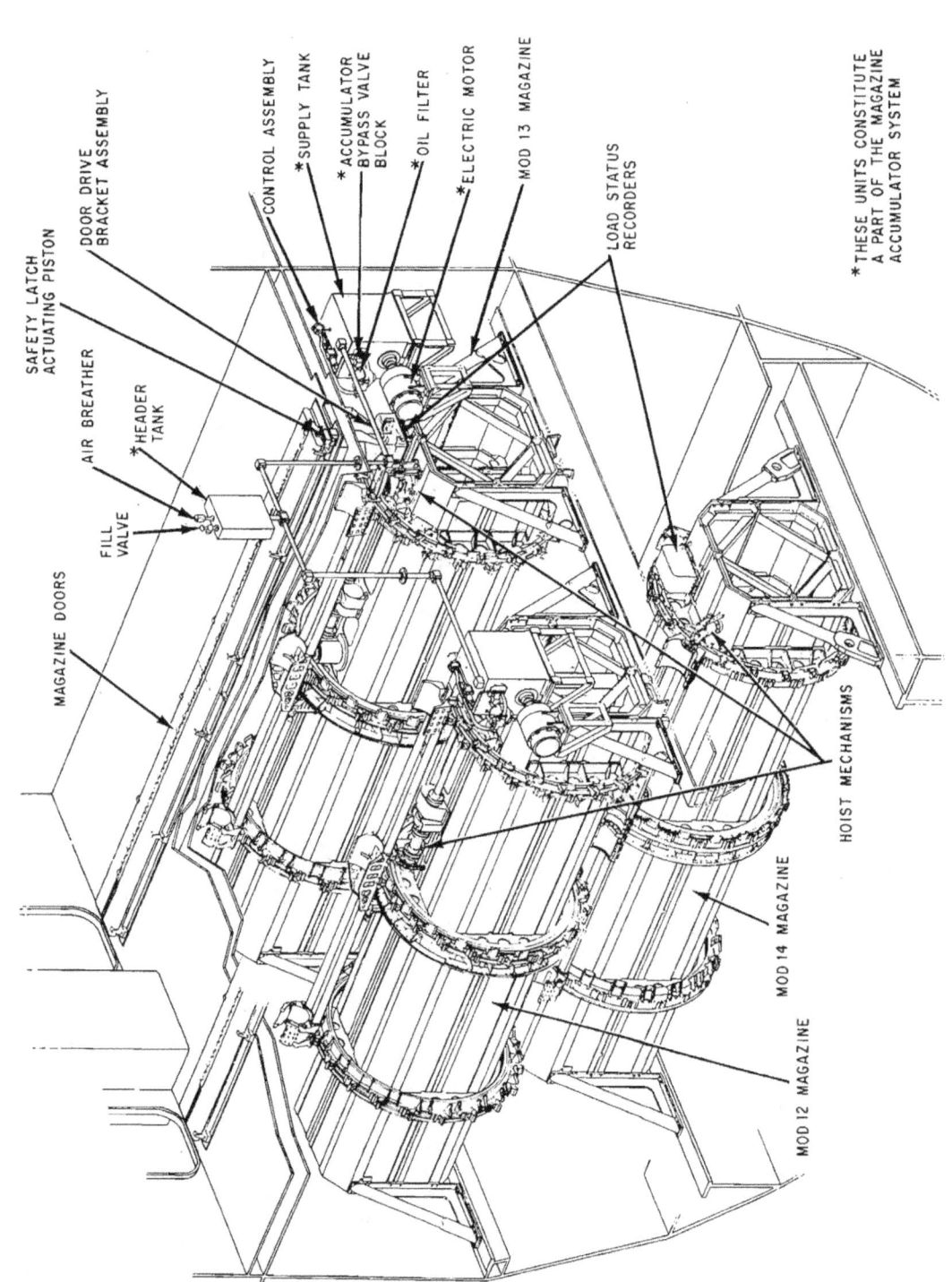

Figure 6-3.—Mk 10 Mod 7 guided missile magazine launching system: location of magazine accumulator components.

## Chapter 6—HYDRAULICS AND PNEUMATICS

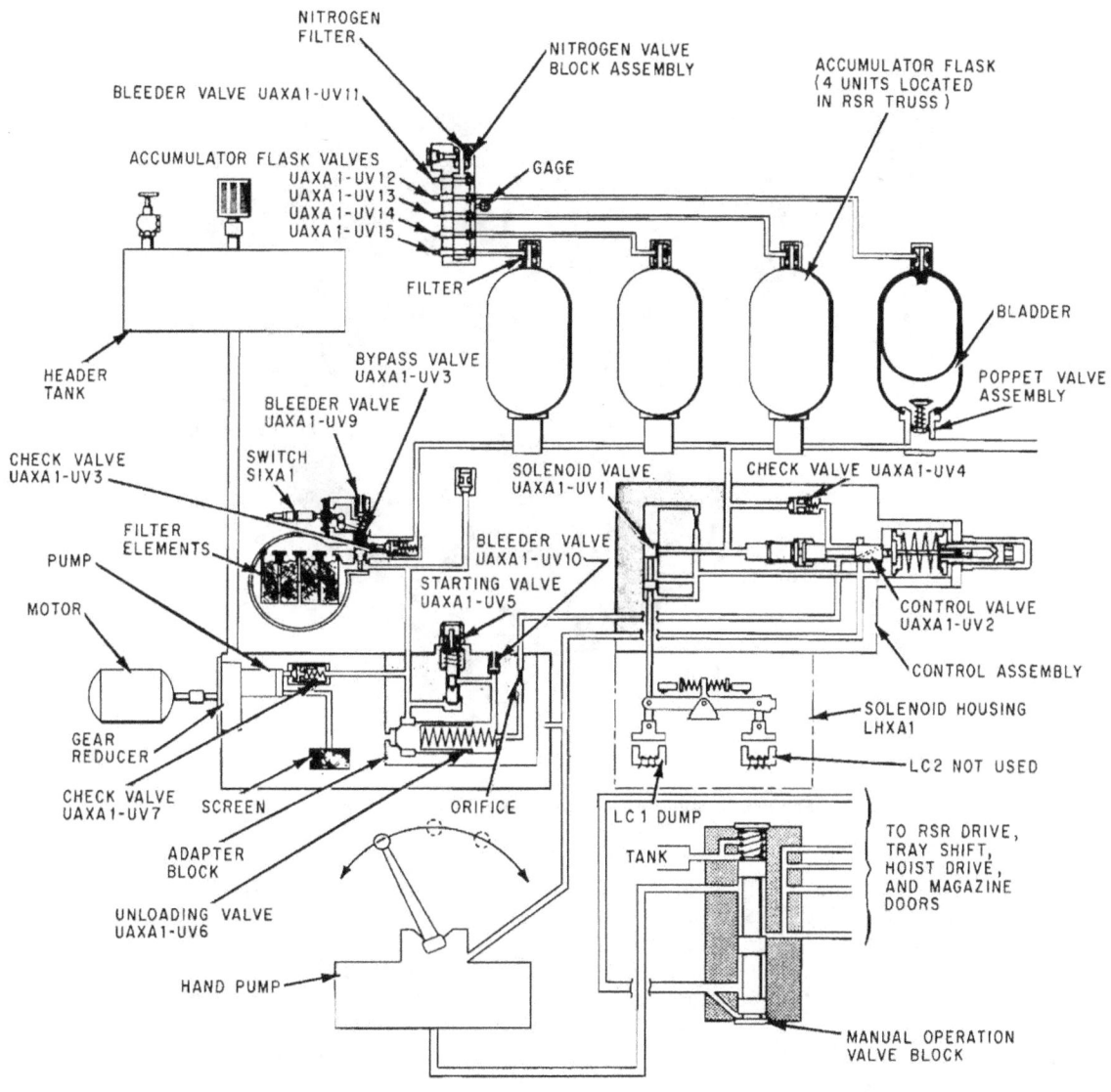

Figure 6-4.—Hydraulic schematic of Mk 64 Mod 0 power drive.

94.133

fluid is passed through the filter elements and the filter bypass valve to the unloading and starting valves. If the filters are clogged and the fluid must pass through the filter bypass valve, a light on the EP2 panel comes on, showing that the filters must be replaced. In an emergency, it is possible to continue operating the power drive when the filters are clogged, but in practice you should stop the operation and replace the filters. In servicing filters (figure 6-5), remove the 12 filter elements from the multi-element filter assembly, inspect the elements, replace them,

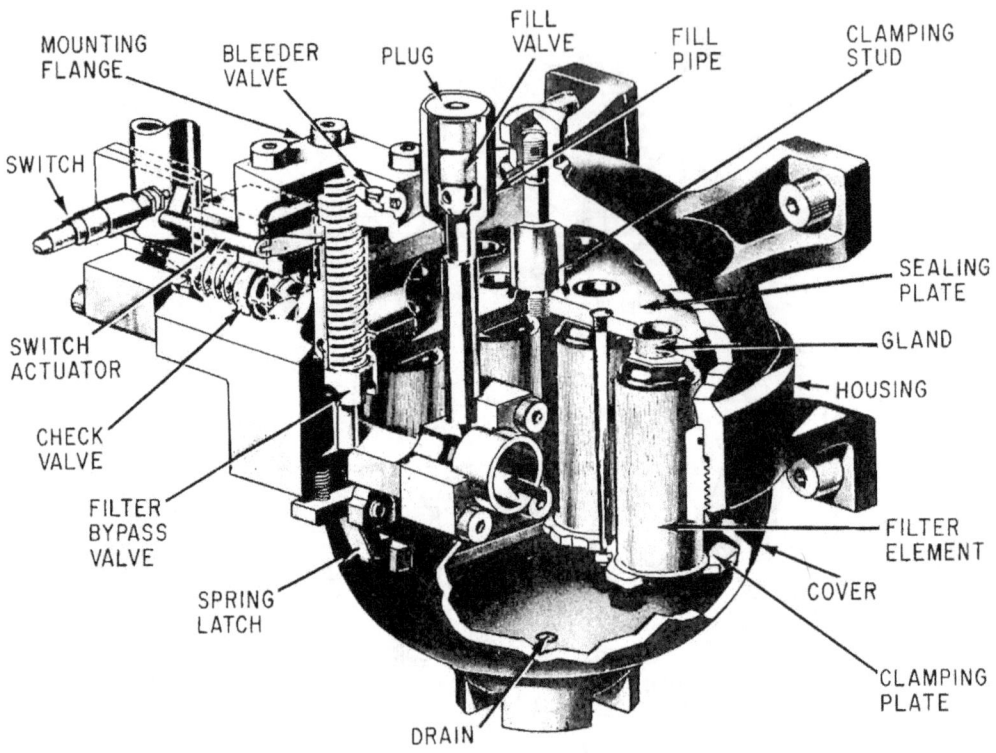

Figure 6-5.—Multiple element hydraulic filter assembly.

and reassemble the unit. Follow the MRC instructions. This assembly filters the pressure fluid passing from the adapter block to the accumulator flask assembly. The filter assembly is mounted on the side of the supply tank.

The pressure fluid passes around the unloading valve, which is operated by the control valve, and goes on to the starting valve. The check valve allows the fluid to go in only one direction. The control valve maintains the pressure between the set limits (for example, 1300 to 1500 psi), by control of the unloading valve, opening or closing it as needed to keep the pressure within limits. Excess pressure fluid is ported to the tank. Pressurized fluid passes through the check valve into the accumulator flasks, where it is stored to maintain a steady pressure in the system.

The connection of the hand pump to the system is shown in figure 6-4.

### Ready Service Ring Drive

The RSR power drive (figure 6-6) consists of five major subassemblies: the gear reduction, the B-end, the control unit, the power-off brake, and the control valve block. The drive, which rotates the RSR, acts as a gear reducer between the B-end and the output shaft (figure 6-7).

The gear reduction (figure 6-7) contains a gear train enclosed in a housing and a splined shaft. The shaft protrudes from both sides of the gear reduction housing and connects to the coupling drive shafts. A gear pump provides lubricant to the gears and bearings not submerged in oil.

## Chapter 6—HYDRAULICS AND PNEUMATICS

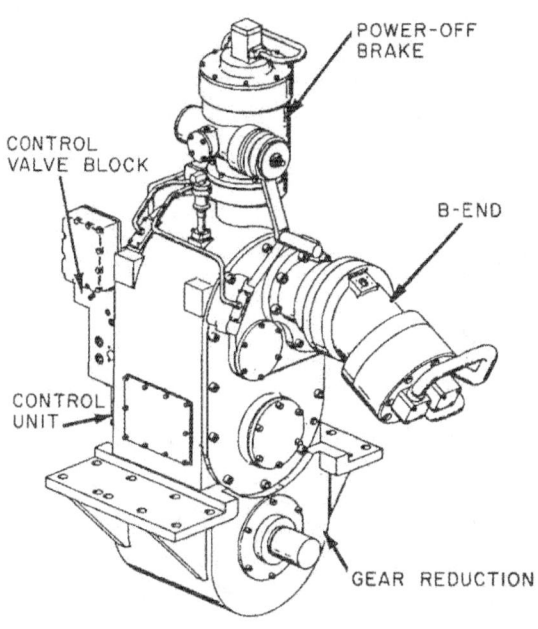

94.216
Figure 6-6.—RSR drive.

The B-end is a conventional hydraulic motor with a spline shaft extending into the gear reduction assembly. The B-end is flange-mounted to the gear reduction housing.

The control unit, fastened to the side of the gear reduction, regulates the speed of RSR rotation during an indexing cycle by limiting the discharge from the B-end. It consists of a housing which contains a camshaft, a cam, a gear, a bevel gear, and a pinion. One end of the 20X camshaft (figure 6-7) mounts the cam that operates flow control valve UVCA(B)6 in the valve block; the other end is splined to a gear that engages a bevel gear mounted on the 30X pinion inside the gear housing.

The power-off brake is used for manual operation and has two functions. It transmits rotary motion from the handcrank to the input shaft of the gear reducer. It also restrains the RSR while handcranking so ship motion or unbalanced loads do not rotate the unlatched RSR. PA to the RSR drive automatically unlocks the brake. The power-off brake's handcrank is coupled to the worm through a slip clutch. The slip clutch yields and prevents handcranking the RSR when the force required to rotate the handcrank exceeds 22 to 30 ft-lb torque. This prevents damaging the RSR or drive train in case of an obstruction.

### Tray Shift Mechanism

The tray shift mechanism (positioner) is located at the top of the RSR truss at station 1 (transfer station), and functions to shift the tray between the RSR and the hoist positions. Its fluid power is provided by the magazine accumulator system.

The following step-by-step description of the hydraulic-mechanical operation of the positioner can be followed in figure 6-8. When all conditions in the positioner's electrical circuit are satisfied, solenoid LHDA(B)1-LC1 energizes and shifts valve UVDA(B)2.

NOTE: Fluid PA (1400 to 1500 psi) is not available at UVDA(B)2 (figure 6-8) until the hoist is down and latched in the lowered position and the RSR positioning latches are engaged. When the hoist is down and latched, hoist-lowered position latch piston UCHA(B)3 ports PA to UVDA(B)1. The pressure shifts UVDA(B)1 against spring pressure and permits PA from RSR latch extended interlock valves UVCA(B)9 and UVCA(B)10 to pass through UVDA(B)1 to UVDA(B)2. Valves UVCA(B)9 and UVCA(B)10 ensure that the RSR latches are extended before the tray is positioned to the hoist position.

When UVDA(B)2 shifts, PA from UVDA(B)1 ports through UVDA(B)2 to the spring end of UVDA(B)3. This action shifts UVDA(B)3 and allows PA to pass through UVDA(B)3 to UVDA(B)4. At the same time, UVDA(B)3 ports fluid from the bottom of UVDA(B)5 to tank.

The PA shifts the plunger of UVDA(B)4 against spring pressure. The plunger of UVDA(B)4 is linked to the tray-at-RSR latch that secures the tray slide at the RSR position.

173

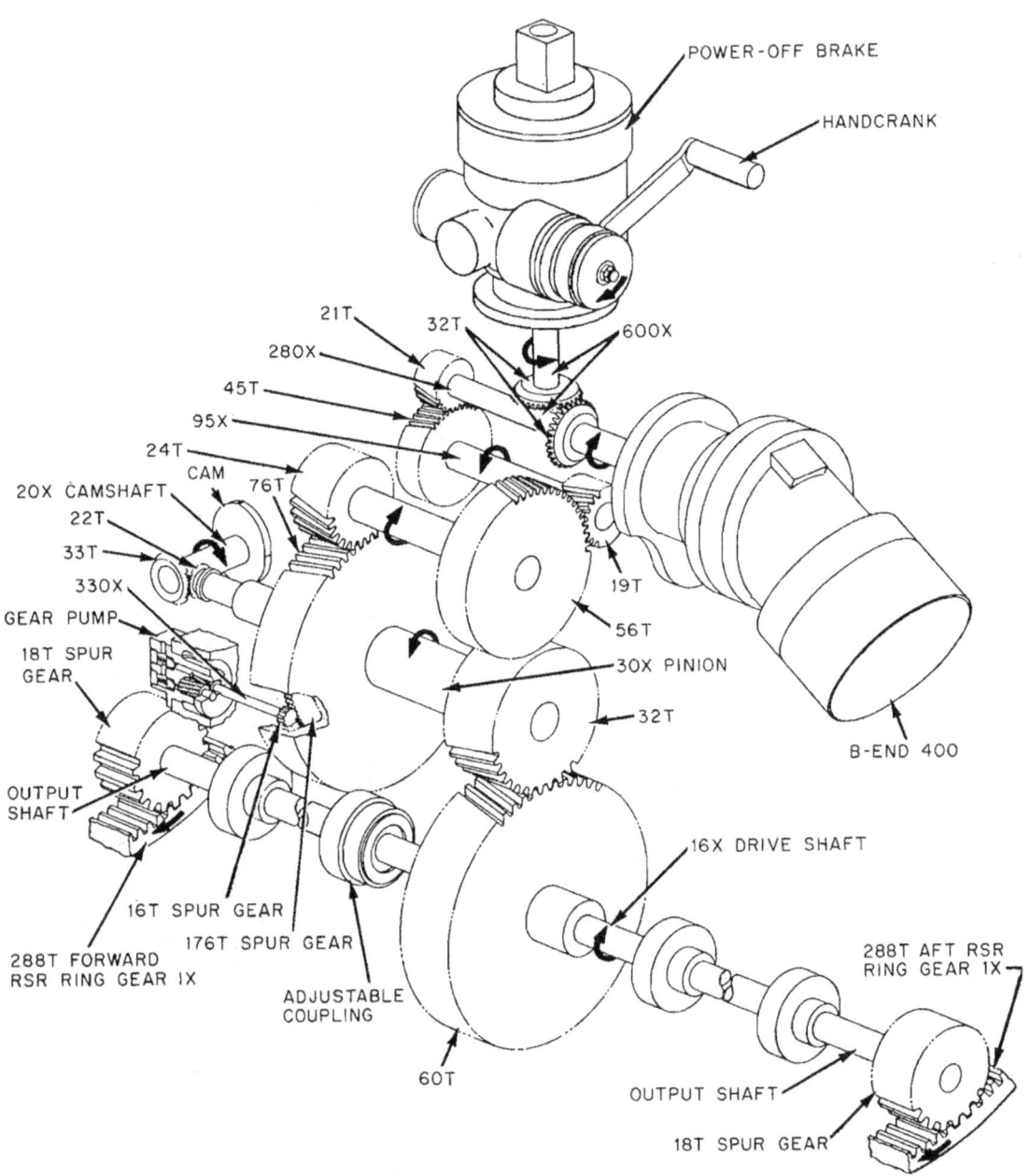

Figure 6-7.—RSR drive mechanical schematic.

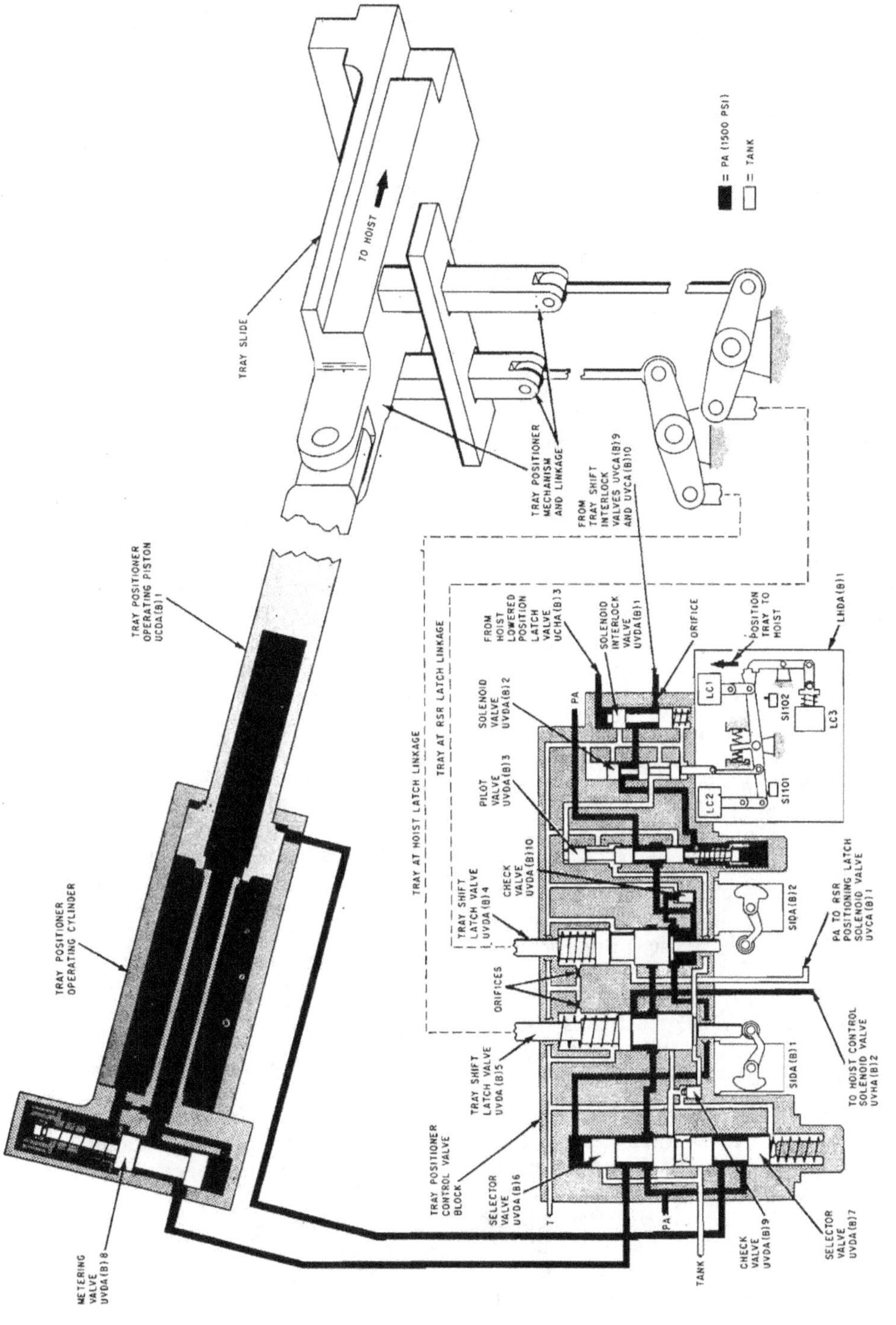

## Chapter 6—HYDRAULICS AND PNEUMATICS

Figure 6-8.—Station 1 tray shift control, tray to hoist, in the Mk 10 launching system.

The movement of UVDA(B)4 retracts the latch, freeing the tray slide. When UVDA(B)4 retracts the latch, it opens a port allowing PA to flow to UVDA(B)6. The PA shifts the plungers of UVDA(B)6 and UVDA(B)7 against spring pressure. Valve UVDA(B)6 ports PA through metering valve UVDA(B)8 to the large-area side of UCDA(B)1; UVDA(B)7 ports PA to the small-area side of UCDA(B)1.

When the plunger of UVDA(B)4 shifts and retracts the tray-at-RSR latch, the lower end of the plunger deactuates tray positioned to RSR switch SIDA(B)2A. Moreover, with the plunger of UVDA(B)4 seated against spring pressure and the tray-to-RSR latch retracted, UVDA(B)4 blocks PA to RSR positioning latch solenoid valve UVCA(B)1 and ports fluid from UVCA(B)1 to tank. This hydraulic interlock ensures that PA is not available at UVCA(B)1 to initiate an RSR positioning latch retract cycle if the tray is not latched at the RSR position.

Both UVDA(B)6 and UVDA(B)7 port PA to UCDA(B)1. The pressure on the large-area side of UCDA(B)1 extends the piston. Valve UVDA(B)8 and the grooved orifice rod of the cylinder regulate the acceleration, constant speed, and deceleration of UCDA(B)1.

When UCDA(B)1 positions the tray slide at the hoist, the tray-at-hoist latch engages the detent in the slide and secures the slide at the hoist position. As the slide is moving to the hoist position, the tray-at-hoist latch rides on the bottom of the slide. When the slide reaches the hoist position, the latch is extended into the detent cutout on the slide by spring tension on the plunger of UVDA(B)5. The spring forces the plunger of UVDA(B)5 down, which extends the tray-at-hoist latch into the detent. Shifting the plunger of UVDA(B)5 actuates switch SIDA(B)1A and opens a PA line to hoist control solenoid valve UVHA(B)2.

When SIDA(B)1A is actuated, its contacts close to energize KCDA(B)1A. Energizing KCDA(B)1A opens its contacts, thus opening the circuit to deenergize LHDA(B)1-LC1. Because LHDA(B)1-LC3 energize after the latch engages its detent in the tray slide, the solenoid rocker arm spring-returns UVDA(B)2 to neutral. Valve UVDA(B)2 ports fluid from UVDA(B)3 to tank and moves to its spring-held neutral position. When UVDA(B)3 returns to neutral, UVDA(B)10 opens a PA line from UVDA(B)6 to tank. Then both UVDA(B)6 and UVDA(B)7 return to neutral, opening tank lines from both ends of UCDA(B)1; this completes the position-tray-to-hoist cycle.

### Hoist Mechanism

The location of the hoist mechanisms was pointed out in figure 6-3. These, too, operate on hydraulic power from the magazine accumulator power system. The function of the hoist mechanism is to raise missiles from the RSR to the loader, or to return them during unloading. Figure 6-9 shows the mechanisms of the A-side hoist. The telescoping columns that raise or lower the missile are driven by chains and gearing. The hydraulic motor (B-end) drives the gearing. The upper drive is a gearbox driven by the intermediate drive shaft from the lower drive. The hoist heads contact the shoes (forward or aft) that support the missile round in the tray. The probe on the aft hoist head aligns the head with the loader trunk in the hoist-raised position. The head of the aft shoe hoist is more elaborate than the forward one.

The hydraulic control valve block is mounted on the wall of the lower transmission housing, and two detented solenoids are mounted on top of the valve block (figure 6-9). It contains a metering valve to regulate the flow of oil, an orifice valve to control speed of the B-end by size of the orifice, a pressure-off valve to shunt pressure to the power-off brake, a selector valve to control direction of the B-end movement, a sequence selector valve to pull the appropriate latch in the hoist drive assembly, an interlock valve to prevent movement before all conditions are ready, and two solenoid valves to control the flow of fluid to the sequence selector valve. The power-off brake sets to stop the hoist during power failure and is driven through for manual operation.

### Magazine Doors

The location of the magazine doors in relation to the RSRs was shown in figure 6-3.

## Chapter 6—HYDRAULICS AND PNEUMATICS

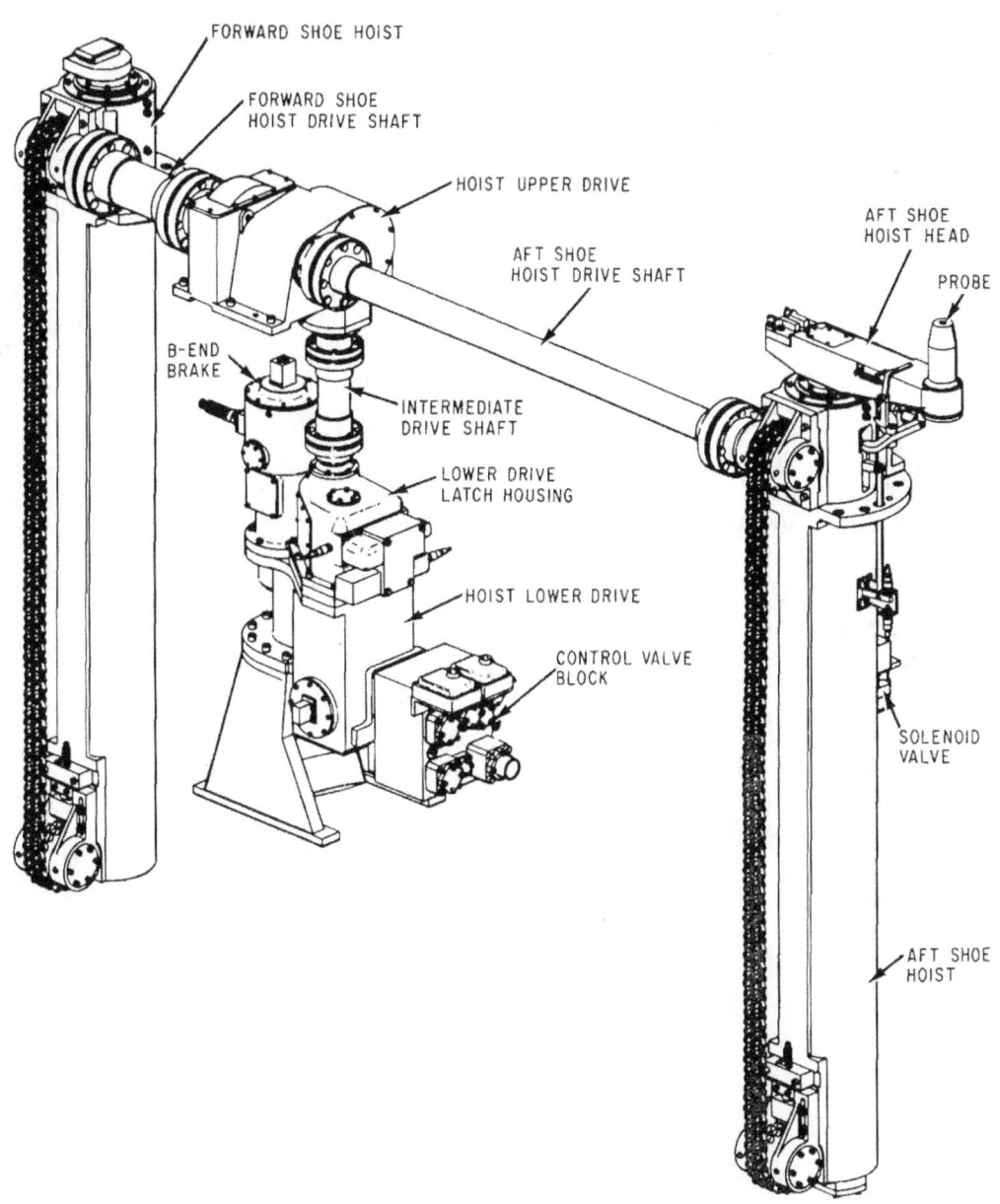

Figure 6-9.—Hoist mechanism for A-side, Mk 10 launching system.

# GUNNER'S MATE M 1 & C

The safety latch assembly, which resembles a giant zipper, may be seen in figure 6-10. At the right-hand end the safety latch actuating piston is shown, and below, the door drive bracket assembly. The door drive bracket assembly includes all the components for operating the magazine doors: the switch, the door operating piston assembly, the latch control valve, the door-open and the door-closed latch, and the solenoid valve. There is a door-closed latch at each end of the doors. Electrical and hydraulic interlocks assure that the blast doors are closed and the hoist is down and latched before the doors can be opened, and that the hoist is down and latched before the doors can be closed. The hydraulic operation is through the piston assembly. It contains a directional valve and a metering valve in addition to the piston. The solenoid valve block assembly is attached to the piston block assembly. The latch control valve block is fastened behind the piston guide. The door-open latch valve block is fastened to one of the webs of the door drive bracket. The hydraulic fluid to operate these valves and pistons comes from the magazine accumulator power system.

The opening and closing of doors may seem like a minor concern, but it is very important. The magazine doors are flametight and watertight. They must never be open when the blast doors are open. All the parts, valves, switches, pistons, etc., must act in sequence. Failure of any part can disrupt the whole series of actions. If that happens, you need to get out the hydraulic schematics and the electrical drawings and trace the cause of failure. It might be a clogged valve or a broken switch, or the pressure in the magazine accumulator might be too low. Shut down the magazine accumulator system while you are locating the trouble.

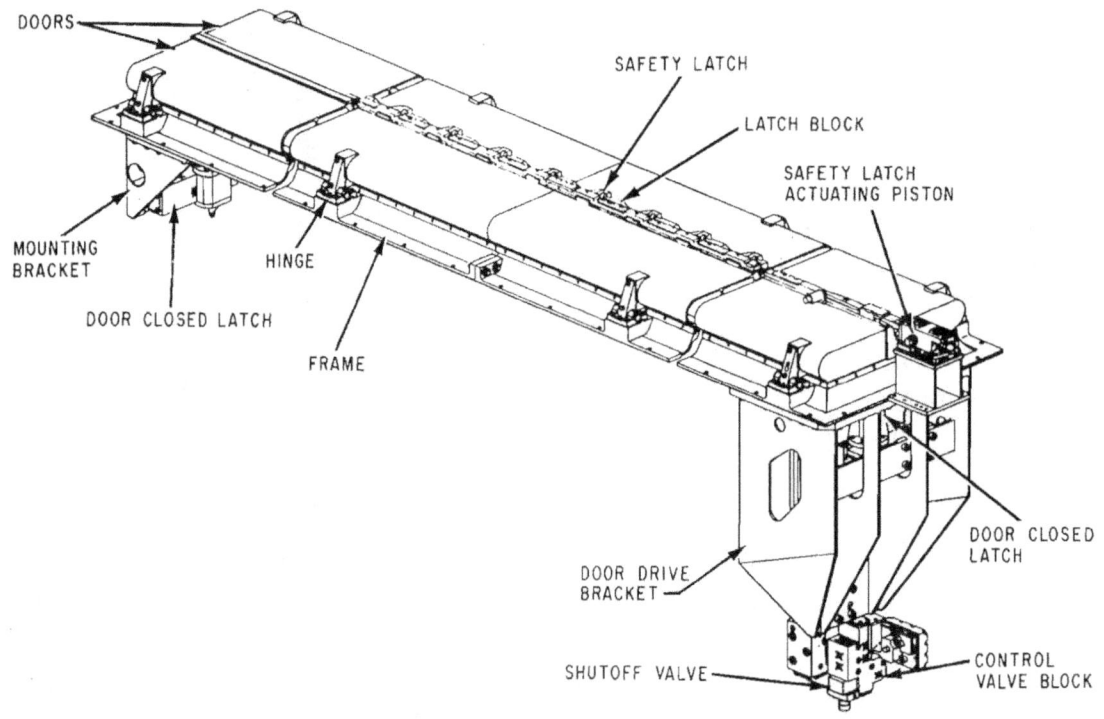

Figure 6-10.—Magazine door assembly.

94.172

## Chapter 6—HYDRAULICS AND PNEUMATICS

## LOADER COMPONENTS

The loader, figure 6-11, consists of similar A- and B-side assemblies that engage, support, and move missiles between the assembly area and the launcher, or between the assembly area and the strikedown/checkout area. Major components of the loader are the loader trunk, the tilting rail (in the Mods 5 and 7 systems), the blast doors and spanning rail, the loader chain and pawl, and the loader power drive. The loader horn automatically sounds as the loader starts to retract from the launcher (with a weapon) and stays on until the loader reaches the assembly area. The horn also sounds when SMZ9 is positioned to loading horn or horn and bell.

### Loader Power Drives

There are two accumulator-type power drives; one for the A- and one for the B-side. Each power drive includes a tank to hold the hydraulic fluid, a motor-driven pump to develop hydraulic pressure, a series of valves to regulate system pressure, and two accumulator flasks to store energy, absorb hydraulic shocks, and prevent excessive pressure fluctuations. These components operate in the same manner as the magazine accumulator power system.

The accumulator power drive furnishes hydraulic power to operate the spanning rail, the blast door latches, the retractable rails, the floating track piston assemblies, the loader pawl positioner, the interlock valve block, the tilting rail, and the strikedown gear.

A second power system, a combined A- and B-end (CAB) power unit, is located near the aft end of the loader trunk. It develops hydraulic pressure and transforms it into rotary mechanical motion to drive the loader chain.

The CAB power drive also has a power-off brake assembly. This is used to halt moving equipment, to secure driven equipment against roll and pitch when the equipment is inactive, to halt and secure equipment if there is a power failure, and to provide a means of manual operation (by handcrank) for maintenance procedures, installation, or emergency operation. A small auxiliary gear pump, driven by the same electric motor that drives the A-end, supplies the necessary hydraulic pressure for the control mechanism that controls the A-end tilt plate and, therefore, the speed and direction of rotation of the B-end. The auxiliary gear pump produces 400 to 500 psi servopressure to operate the control components of the CAB units, and delivers 100 psi (supercharge pressure) from another set of gears to replenish fluid losses from slippage and leakage.

### Loader Trunk

The loader trunk is made up of sections that are mounted to the overhead. The number of sections varies with the mod of the system. The Mk 10 Mod 0 Terrier launching system has eight sections; Mods 5 and 7 have six loader trunk sections and a tilting rail; Mod 8 has 12 sections. Figure 6-12 shows a trunk section made for the Mk 8 Mod 11 loader.

A number of the hydraulically operated components of the loader are mounted in the loader trunk sections. Only the control panels, the power drives, and the tilting rail control are not mounted in the loader trunk assembly. The cross-section view in figure 6-12 shows the channel or tracks in which the forward and after shoes and the chain can slide.

### Tilting Rail

The tilting rail (figure 6-13), mounted by trunnion supports to the overhead structure above the magazine doors, extends almost the full length of the assembly area. It raises to the elevated position or lowers to the horizontal position. In the elevated position, the forward end of the tilting rail latches to the end of truck section VI, which forms a loader track to the launcher. In the horizontal position, the forward end of the tilting rail latches to the end of truck section I and supports missiles for wing and fin assembly operations or for transfer to or from the strikedown-checkout area. Tilting rail components include the chain stowage assembly, the warmup cable reel assembly, the takeup arm assembly, and the floating tracks and extended indicator. The operating piston (figure 6-14) is the unit that elevates or lowers the tilting rail. Hydraulic power is obtained from the loader

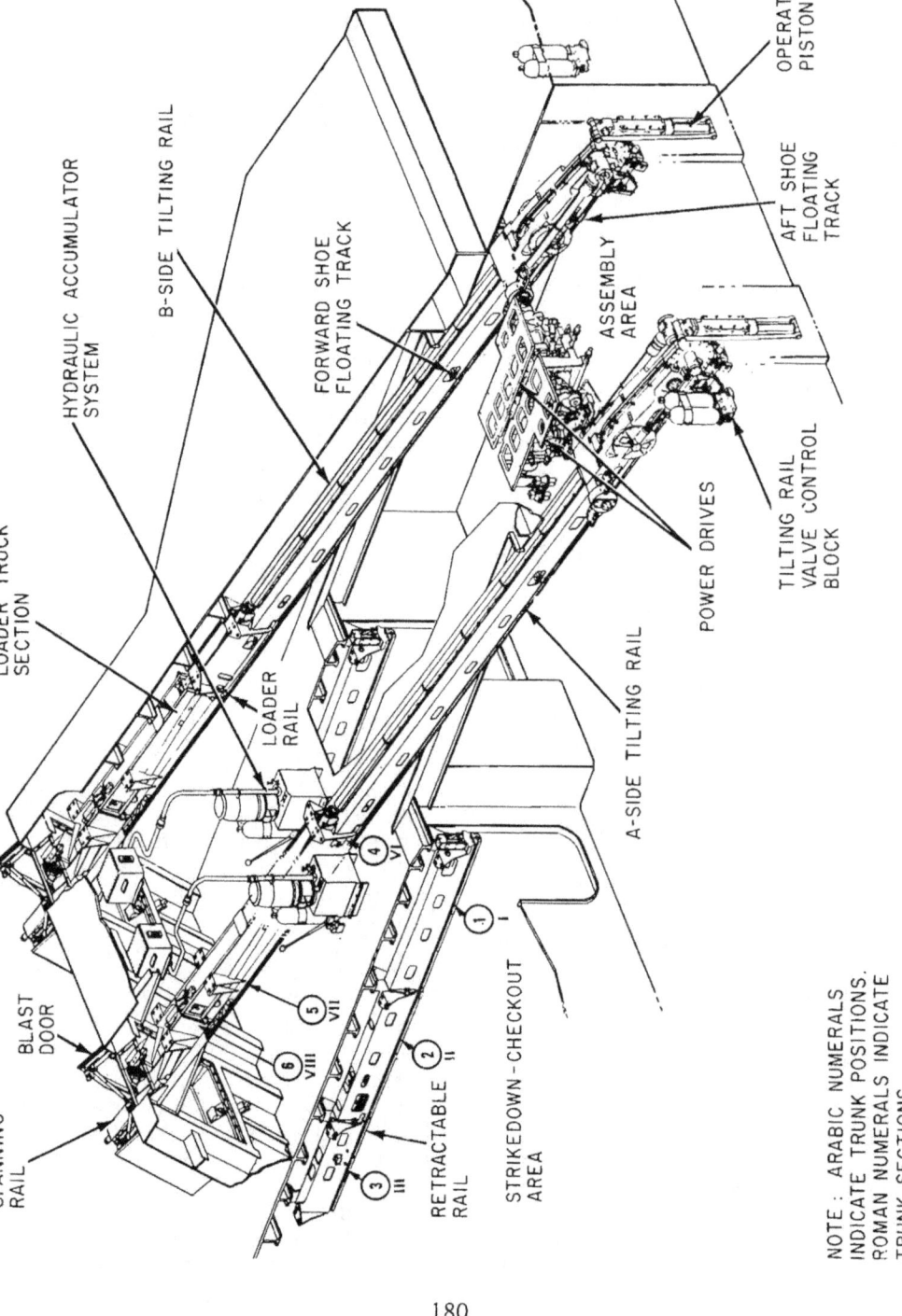

Figure 6-11.—General arrangement of Mk 5 loader system.

## Chapter 6—HYDRAULICS AND PNEUMATICS

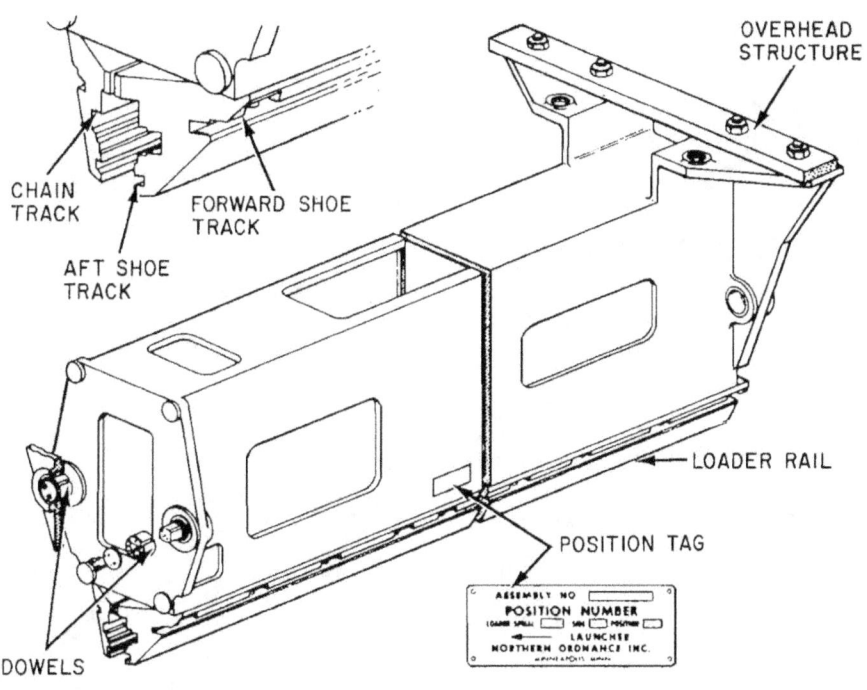

Figure 6-12.—Loader trunk, section II.

accumulator. A hydraulic transfer pin (inside the trunnion) distributes the hydraulic fluid to the floating track piston assemblies, the positioner piston and interlock valve block, and the rail-loaded indicator assemblies, through the adapter block mounted on the trunnion (figure 6-13). Mods 1 and 13 of the Mk 8 loader do not have a chain stacker. The chain links, chain sprocket gear reduction, and the chain track housing vary in the different mods. You need the OP for your launching system to study the operational details.

### Floating Tracks

The location of the floating tracks may be seen in figures 6-11 and 6-13. In mods that do not have tilting rails, the floating tracks are attached to the loader trunk in the same positions. The aft floating track assembly catches the aft booster shoe in its slide track, and the forward floating track assembly does the same for the forward booster shoe. They are piston operated by hydraulic fluid from the loader accumulator. The loader positioner moves the loading pawl forward about 3 inches, enough to slide the booster shoes out of the hoist head and into the floating track rails. The floating tracks hydraulically align the booster shoes with the track grooves of the fixed loader rail.

### Retractable Rails

The retractable rails are movable sections of the loader rail in the strikedown-checkout area which open to permit the transfer of boosters on or off the loader rail using the transfer car. With the loader at the No. 2 stop, the open-retractable-rails cycle begins when the open-retractable-rails solenoid energizes. This initiates the hydraulic-mechanical operation to

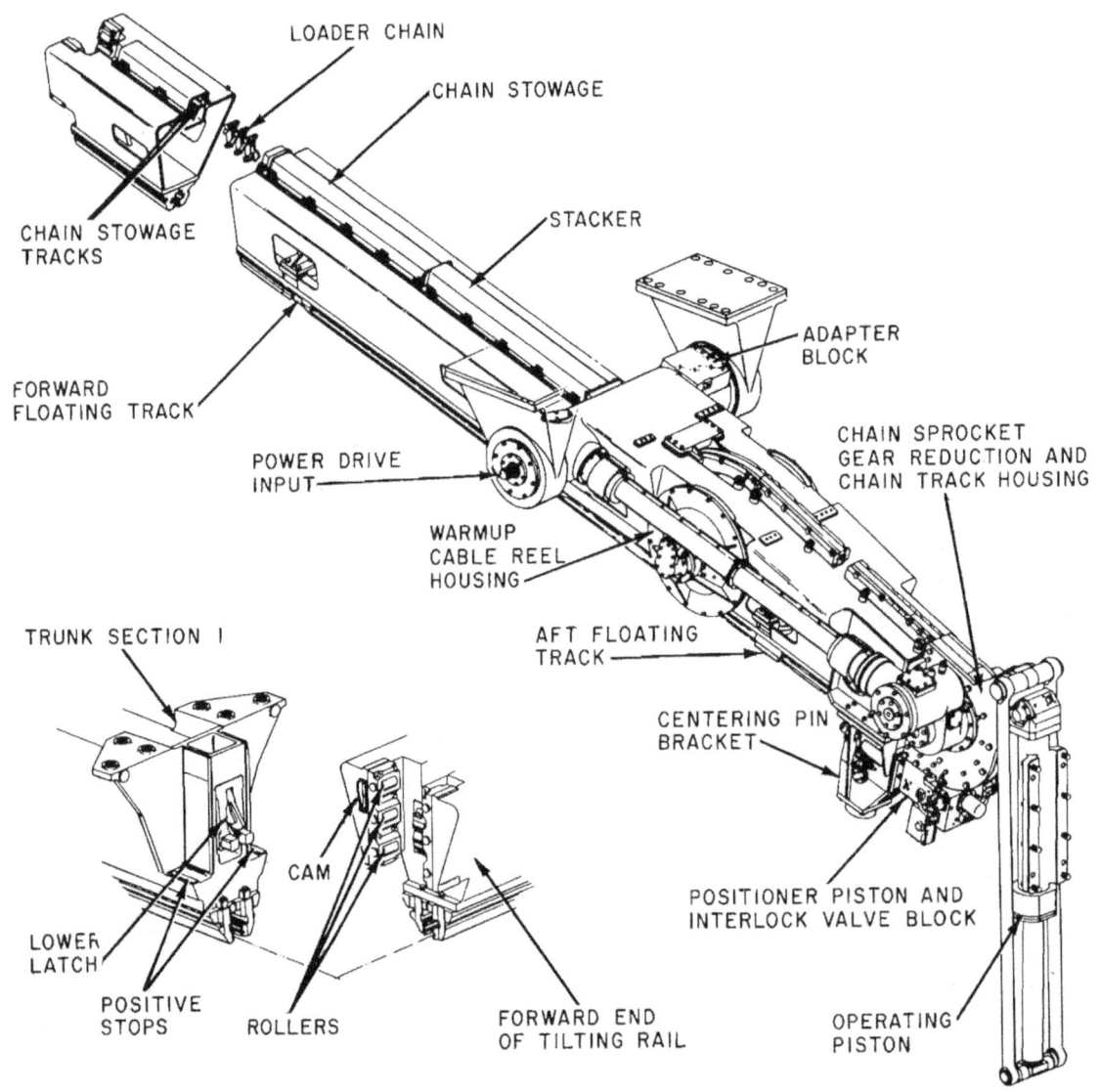

Figure 6-13.—Tilting rail and operating piston.

move the forward retractable rails inward (to allow passage of U-shaped booster forward shoes) and the aft retractable rails outward (to allow passage of T-shaped booster aft shoes). Retractable rails are not used when onloading or offloading ASROC weapons.

The hydraulic controls for the retractable rails are mounted in the loader trunk. An interlock switch prevents operation of the loader when the rails are open. The valves are typical solenoid, pilot, directional, interlock, latch (open/close), and check valves.

## Chapter 6—HYDRAULICS AND PNEUMATICS

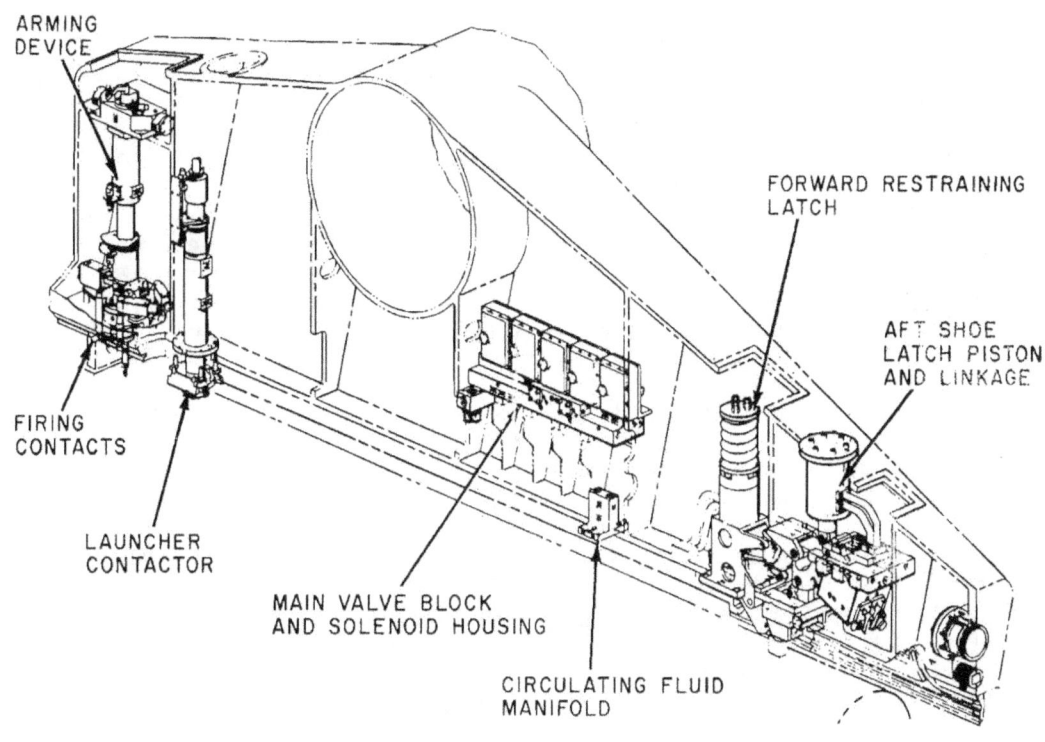

Figure 6-14.—Guide arm components, Mk 10 launching system; general arrangement.

94.56

### Spanning Rails and Blast Doors

The blast doors are a pair of doors mounted on hinges to the bulkhead of the missilehouse. The blast doors, flameproof and watertight structures, protect the strikedown-checkout area and the assembly area from weather and missile exhaust.

The spanning rail is a movable loader trunk section that bridges the gap between the loader rail (in the missilehouse) and the associated launcher guide arm. In the Mod 7 system, the spanning rail is attached to truck section VIII; in the Mod 8 system it is attached to truck sections XI and XII. When the spanning rail extends, a positioning spud on the forward end of the rail enters a recess in the aft end of the launcher guide arm to ensure proper alignment between the rail and the arm.

Operating links mechanically couple the spanning rail to the blast doors. When the spanning rail extends, it opens the blast doors. When the spanning rail retracts, it closes the blast doors.

The blast doors and spanning rail are piston operated by hydraulic fluid from the loader accumulator.

Tartar launching systems do not have a loader system like the Terrier launching system. Missiles are carried from the magazine to the launcher guide arm by a rammer type roller chain hoist. In the Mk 11 GMLS, the chain is stowed in the launcher guide arm.

### HYDRAULICALLY OPERATED LAUNCHER COMPONENTS

Although launchers contain parts that are not operated hydraulically, the interconnection with hydraulic power makes it impossible to consider them apart. All missile launching systems have a fixed stand—a steel weldment on

# GUNNER'S MATE M 1 & C

the deck, which mounts the carriage. The carriage is rotatable horizontally, to position the launcher in train. Most of the launcher components are mounted in or on the carriage. It supports the trunnion tube which holds the guides (or guide).

**Guide and Guide Arms**

After a Terrier/Standard ER weapon has been brought from the magazine by the hoist, and the missile control surfaces (tails) are unfolded and the booster fins installed, it is placed on the launcher guide arm by the loader. Terrier launchers have two guide arms, as does the Tartar launcher in the Mk 11 GMLS, but the Mk 13 and Mk 22 GMLSs have only one guide arm on the launcher. The guide arm supports the missile during the last stage of weapon handling, arming the weapon and holding it until it is launched. It contains arming devices, aft shoe latch, launcher contactor, forward restraining latch, and firing contacts. The arming device arms the missile booster by extending and winding the arming tool by means of hydraulically operated pistons.

The aft shoe latch mechanism has a piston-operated latch and associated linkage. When the missile is positioned on the guide arm (by the loader), the latch is hydraulically extended against the aft booster shoe to keep the missile from moving to the rear. The aft shoe latch (figure 6-14) may be called a positioner, a positioner spade, an aft lug latch, aft motion latch, or reverse motion latch. In all launchers the aft shoe latch is locked by a detent that is hydraulically interlocked to prevent accidental retraction of the latch due to ship's motion and guide arm movements.

The forward restraining latch prevents forward movement of the missile which might otherwise be caused by launcher depression or by ship's motion. It also holds the fired missile on the launcher until the booster has developed enough thrust to overcome the force of gravity plus the force of the adjustable spring that is part of the restraining mechanism.

The launcher contactor is hydraulically extended to apply warmup power to the missile before it is fired. On the Tartar systems, the contactor extends from the rear of the launcher into the stern of the missile.

Two booster firing contacts and two ground contacts are located in the forward section of the guide arm, one of each on each side of the arming device. They provide a double firing circuit for the booster.

The hydraulic power to move the components in the guide arm is provided by the guide arm accumulator power drive, located in the carriage (figure 6-15). It supplies hydraulic fluid for both the A and B guide arm components.

**Carriage-Mounted Hydraulic Parts**

The parts described in the paragraphs under this heading apply specifically to the Terrier carriage mounting, but other systems use similar ones. The location and some details may differ.

HYDRAULIC BRAKES.–The elevation brake is located on the reduction-gear housing, which is mounted to the upper center of the carriage, below the trunnion tube. The train brake is mounted to the bottom of the carriage. Train and elevation brakes are hydraulically operated, spring-loaded, friction-disc type. During power-off conditions they remain set, preventing movement of the launcher.

TRAIN AND ELEVATION LATCHES.–The elevation latch is a hydraulically operated steel fork, located below the elevation arc. When the latch is extended, it secures the launcher in the load position. The train latch is a hydraulically operated steel pin mounted to the bottom of the carriage. It functions in the same way as the elevation latch.

The elevation latch-control valve block houses the solenoids and valves which control the elevation latch. It is located below the reduction gear housing. A hand pump, mounted to the left side of the carriage, provides a means of operating the guide arm components and the train and elevation latches during maintenance operations and during power failure.

REDUCTION GEAR ASSEMBLY.–The train reduction-gear assembly is located within a housing mounted to the bottom of the carriage.

## Chapter 6—HYDRAULICS AND PNEUMATICS

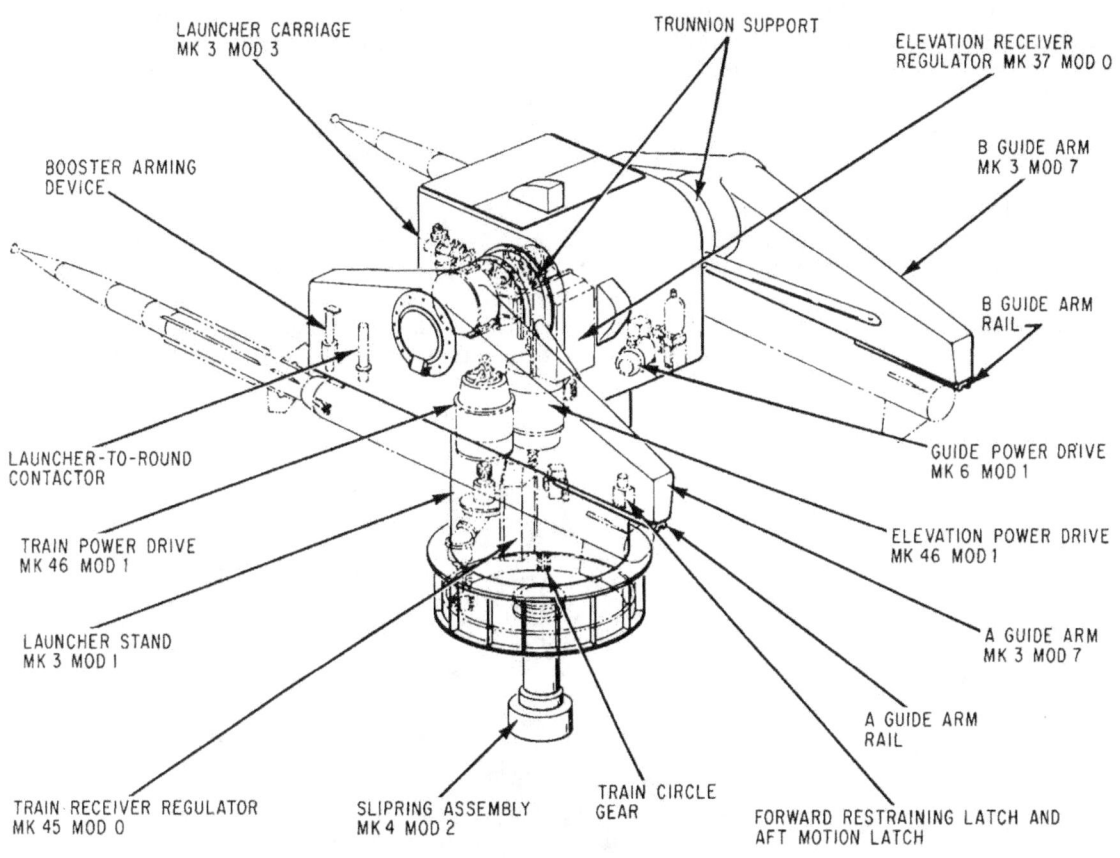

Figure 6-15.—Mk 5 Mod 8 guided missile launcher.

The gears transmit the output of the hydraulic motor, at the required speed, to the pinion gear. The elevation reduction-gear assembly, enclosed in a housing, is mounted to the upper center of the carriage, below the trunnion tube. The elevation pinion gear is meshed with the elevation arc gear and is driven by the reduction-gear assembly. The train drive pinion meshes with the training circle and causes the launcher carriage to rotate in train. The elevation drive pinion causes the elevation arc to rotate the trunnion tube which causes the guide arms to elevate or depress.

POSITIONING VALVES.—The elevation positioning valve is located below the elevation arc, and the train positioning valve is mounted to the bottom of the carriage. These valves are spring loaded and mechanically operated to ensure that the launcher is in the proper position before porting fluid to extend the securing latches (train or elevation). They actuate interlock switches when the launcher is in the LOAD position.

BUFFERS.—A buffer is anything that serves to deaden a shock or bear the brunt of a collision. Buffers are also used to slow down movement to avoid a violent shock or stop. Ordnance equipment uses hydraulic and pneumatic buffers, as well as spring buffers. The elevation and depression buffers, mounted on each side of the trunnion tube, buff the movements of the trunnion tube and thus prevent excessive stress on the missile when the guide arms reach the elevation or depression limits. An accumulator furnishes a supply of hydraulic fluid for buffer operation.

ACCUMULATORS.—Accumulators in hydraulic systems permit the use of smaller pumps than would be required if no accumulator were present. The fluid stored under pressure in the accumulator can assist fluids in motion to accomplish work when the demands of the hydraulic system require more fluid than the pump can supply. Accumulators may be used in hydraulic systems to supply fluid to compensate for leakage in closed or pressure-regulated circuits, as an emergency source of power for short periods, to operate secondary hydraulic systems, and as an auxiliary source of energey in intermittent duty systems.

Two basic types of accumulators—bag and piston—are used in launching systems. The piston type is used with missile strikedown equipment and is shown in figure 6-16. The bag type was shown in figure 6-4 in the hydraulic schematic for the Mk 64 Mod 0 power drive, and a cutaway view is shown in figure 6-17, view A.

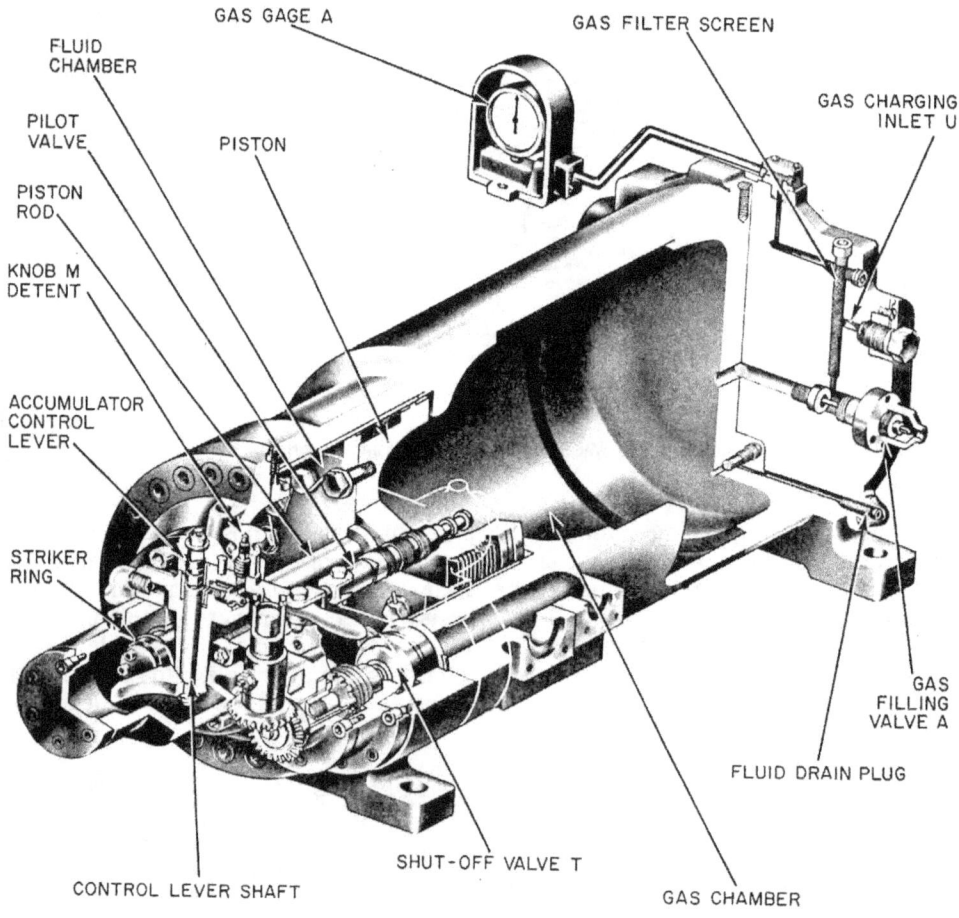

Figure 6-16.—Piston type hydraulic accumulator, cutaway view.

## Chapter 6—HYDRAULICS AND PNEUMATICS

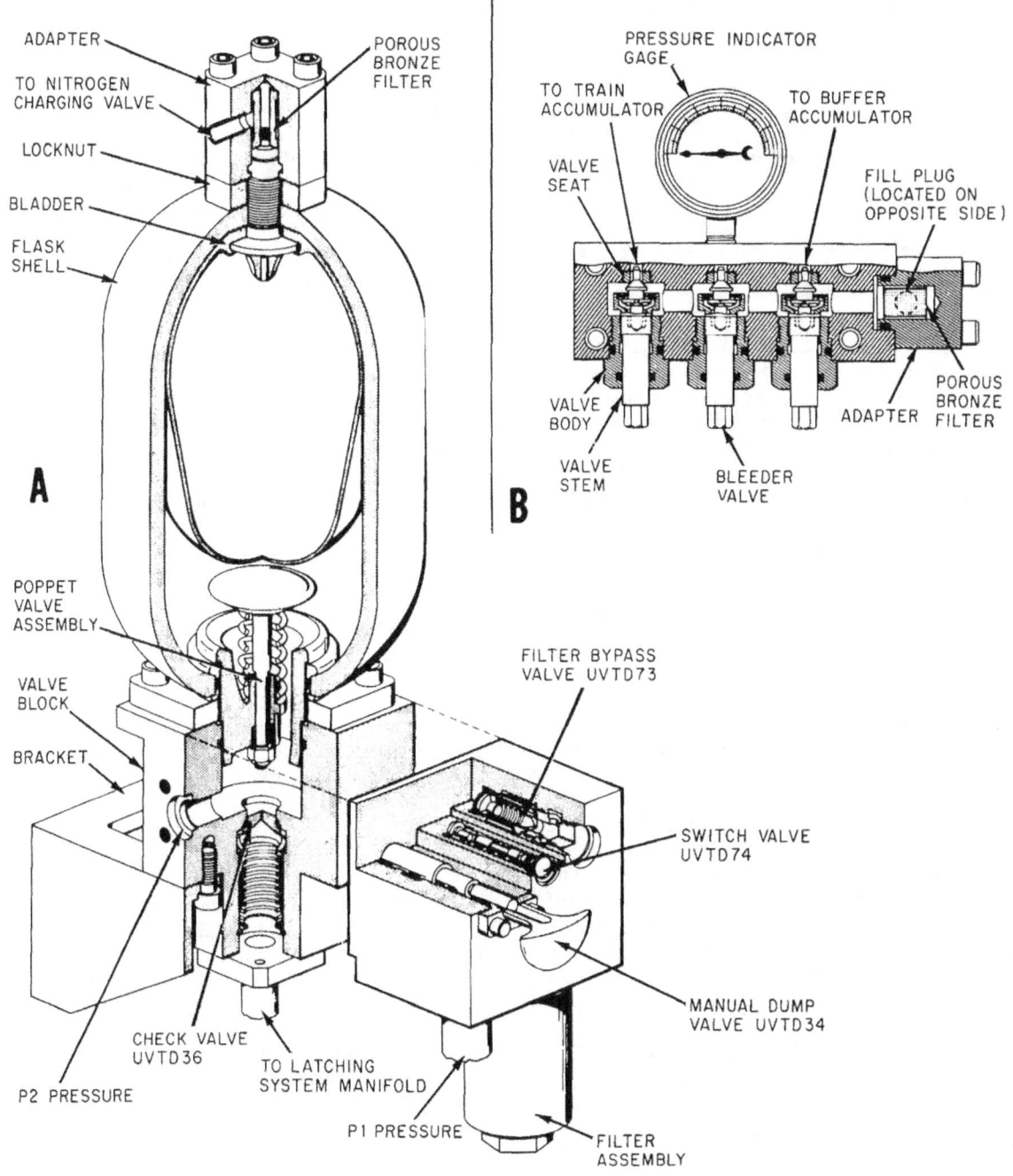

Figure 6-17.—Hydraulic accumulators: A. Bag type accumulator, cutaway view; B. Nitrogen valve block assembly.

Nitrogen is used to pressurize both types. The outside of the bag type is a metal shell; the bag, made of neoprene, is inside and contains the nitrogen. The bladder will fill approximately three-fourths of the inside area of the cylinder when the hydraulic pump forces oil into the flask. A spring-loaded poppet valve at the bottom of the flask prevents the bladder from expanding down into the manifold if there is no hydraulic fluid in the flask.

The flask is mounted on a manifold (figure 6-17, view A) and the valve block and gage (figure 6-17, view B) are mounted nearby. The location of the gage and the type of nitrogen valve assembly will differ on accumulators of special ordnance systems. The nitrogen valve assembly, (figure 6-17, view A) controls the compressed nitrogen. The nitrogen charging valve block consists of a nitrogen fill plug, a porous bronze filter, and nitrogen fill valves to the train and buffer accumulators, as shown in figure 6-17, view B.

Piston accumulators perform the same functions as the bag type although they are constructed differently. The steel cylinder is divided into two chambers by a movable piston (figure 6-16); one side for the hydraulic fluid, and the other for the nitrogen.

The cylinder head of the hydraulic fluid chamber has a manifold through which the flow of fluid is automatically controlled by a pilot valve, whether charging or discharging the accumulator. The pilot valve (figure 6-16) is actuated by the piston rod through a linkage operated by cams in a control housing. The housing includes a manually operated shutoff valve by which the accumulator pressure can be cut out of the power drive system when the drive is not in operation, or the accumulator fails.

The facilities for admitting nitrogen under pressure are at the other end of the accumulator cylinder. These include a gas filling valve, gas charging inlet, and gas pressure gage. The arrangement may vary from that shown in figure 6-16.

ACCUMULATOR OPERATION.—An accumulator valve block, containing a control valve and check valves, maintains the desired operating pressure in the accumulator by controlling the output of the hydraulic pump. The control valve is adjusted to limit the maximum pressure in the accumulator, and is designed to control the minimum pressure. (Typical accumulator pressure is 1300-1500 psi.) When the accumulator is being charged, fluid from the pump flows through check valve 1 (figure 6-18), around the lower land of the control valve, and into the accumulator. When the accumulator becomes fully charged, the pump output is ported through check valve 2 to the tank.

The control valve has a spring and seat, a two-land plunger, and a piston. The piston is larger in diameter than the lower land of the plunger against which it bears. A cross-port in the lower land of the plunger ensures equal pressure on each end of the lower land chamber. Accumulator pressure is always available at the bottom of the piston. As the charging cycle begins, the spring holds down the plunger and piston, and the hydraulic fluid passes into the accumulator. As the accumulator fills, the piston and the plunger are pushed upward, compressing the spring. When the spring is compressed enough, flow to the accumulator is blocked off and check valve 2 lifts and vents the pump output to the tank. When the accumulator pressure drops to the minimum set for it, the charging cycle is repeated.

### Power Drives

As noted in the preceding paragraphs, each launching system has several power drives. To distinguish them, the location or use of the power drive is included as part of the name; i.e., hoist power drive. The train and elevation power drives on the launcher are the ones most often referred to simply as the launcher power drives. They are two separate electrohydraulic systems which control the movements of the launcher in train and elevation.

The function of the power drives is to make the launcher position correspond to the ordered positions (orders from fire control, under normal automatic operation) with the least

Chapter 6—HYDRAULICS AND PNEUMATICS

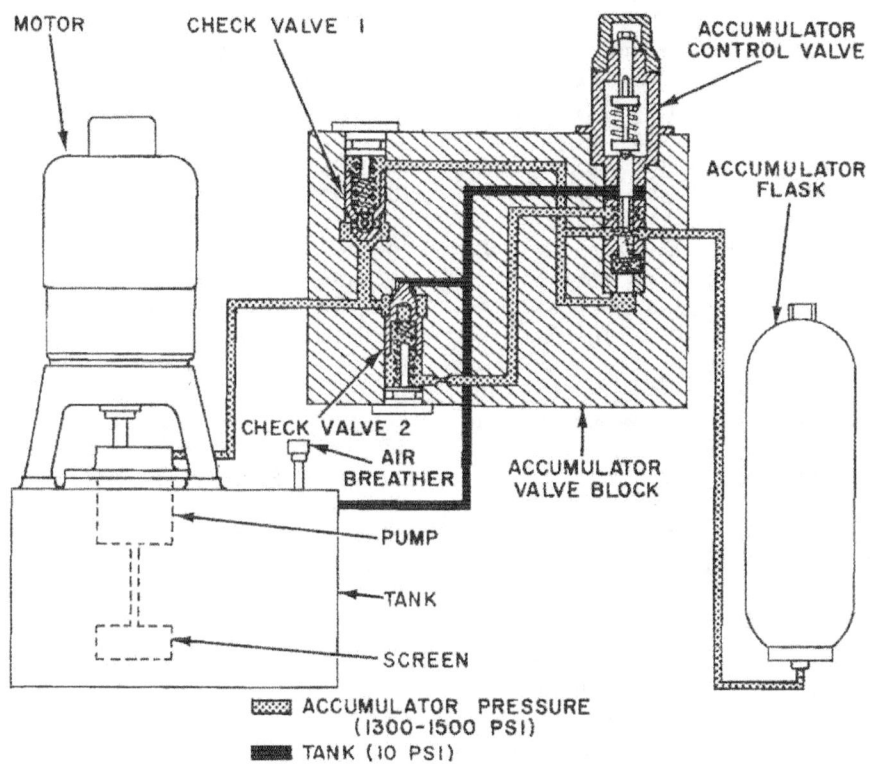

Figure 6-18.—Schematic, accumulator unit, charging cycle.

possible error at all times. The design of the train and elevation power drives is very similar, but they are not interchangeable.

The electric power for the power drive is supplied from the ship's power supply through the power panel.

The power unit consists of an electric motor, an A-end hydraulic pump, an enclosed gear train, a stroke control assembly or control cylinder, a transfer valve, and the B-end motor. The receiver-regulator is located with these components, and is functionally a part of the power drive. How they operate together is described later in this chapter.

The train and elevation power drive controls of the Mk 13 GMLS are located off-mount but the power drives are mounted inside the stand in the inner structure. The magazine power drives are also located in the inner structure of the stand. The RSR is in the outer structure of the stand. The location of components of the train and elevation power drives of the Mk 13 launching system is shown in *Gunner's Mate M 3 & 2*, NAVEDTRA 10199-C. That text also illustrates a number of devices used in hydraulic systems, such as various types of valves, buffer, dashpot, filter, strainer, and gear pump. Differences in the number of power units used by the one-armed Mk 13 and the larger Terrier launching system were also pointed out.

**Train and Elevation Power Drives of Mk 22 Tartar System**

The overriding difference between the Mk 22 GMLS and other missile launching

189

systems is its small size. It handles Tartar and Standard (MR) missiles, but fewer of them than the Mk 11 and the Mk 13 GMLSs. Compact arrangement of components necessitated some changes in placement of parts and some combination of functions. The Mk 67 Mod 0 train power drive also drives the hoist. It has a shift and clutch mechanism that enables it to drive the launcher or the hoist. In the Mk 22 GMLS, it is the launcher that moves to a position above the missile to be loaded; the RSR does not rotate. The elevation power drive elevates and depresses the launcher guide. The major components of the train/hoist and the elevation power drives are the same: an electric drive motor, a hydraulic system, a CAB unit, and a drive train. The hydraulic systems have the same type of auxiliary pump, auxiliary relief valve assembly, and accumulator assembly; they differ only in capacity. The train/hoist power drive has a speed reducer and a lubrication pump that are not duplicated in the elevation power drive. Both systems use a common supply tank and common header tank.

The main supply tank, which holds about 110 gallons of fluid, is integral with the skirt adapter assembly (figure 6-19). This is not to be confused with the main tank of the launcher guide power unit, which is fastened on the underside of the base ring and protrudes above it. That tank holds only about 20 gallons of hydraulic fluid. The location of the launcher guide power unit is also shown in figure 6-19, as are other launcher components. The train/hoist and elevation header tank is mounted in the base ring adjacent to the train/hoist power-off brake. The header tank for the launcher guide power unit is in the front end of the guide arm. A header tank provides a head of fluid to prevent entrance of air into hydraulic lines, which would cause erratic behavior of the hydraulic components. It also serves as an expansion and heat dissipation chamber for returning fluids. A strainer in the return-flow pipe strains out solid particles to keep them from getting into the servo and supercharge systems.

The major components of the train and elevation power drives are: (1) electric drive motors, (2) hydraulic systems, (3) CAB units, and (4) drive trains.

ELECTRIC DRIVE MOTORS.—Mechanical inputs to the CAB units, auxiliary pumps, and the lubrication pump are provided by the electric drive motors. They are mounted on the center column of the carriage (figure 6-20). The motors are activated by switches on the EP1 and EP2 panels. The train lubrication pump is driven directly by the electric motor. It furnishes lubricant to the speed reducer. If the pump fails, the pressure in the discharge line drops. The lowered pressure, deactuates the switch, stops the motor and prevents motor burnout. An excess of pressure in the CAB unit also will stop the electric motor by actuating a pressure-cutout switch. When either the train/hoist motor or the elevation motor stops because of pressure cutout, the ALL MOTORS STOP light on the EP2 panel starts blinking. This indicates why the motors have stopped and warns the operator to look for the cause of the pressure buildup and correct it before restarting the motors.

HYDRAULIC SYSTEMS.—The main components are an auxiliary pump, an auxiliary relief-valve assembly, and an accumulator assembly. The train/hoist and the elevation hydraulic systems are identical except in capacity. The train/hoist system is the larger one; it supplies power for operation of the train and the hoist systems. The pumps are of the type described in *Fluid Power,* NAVPERS 16193-B as gear pumps; they may have helical or spur gears or a combination. The pumps operate the power-off brake, the receiver-regulator, the CAB unit, and the main relief valve of the CAB unit. The train/hoist system also operates the hoist selector valve block assembly, and the hoist control assembly. The relief valve block contains filters, valves, switch-actuating pistons, switches, and a solenoid. The accumulators are of the flask type, containing a bladder charged with nitrogen, a manifold, and an accumulator charging valve assembly.

CAB UNITS.—The train/hoist and the elevation CAB units are of the same type but differ in size and output. They are mounted on the center column of the rotating structure (figure 6-20). The A-end is an axial parallel-piston pump, driven by the electric motor. The B-end is an axial-piston motor; it

Chapter 6—HYDRAULICS AND PNEUMATICS

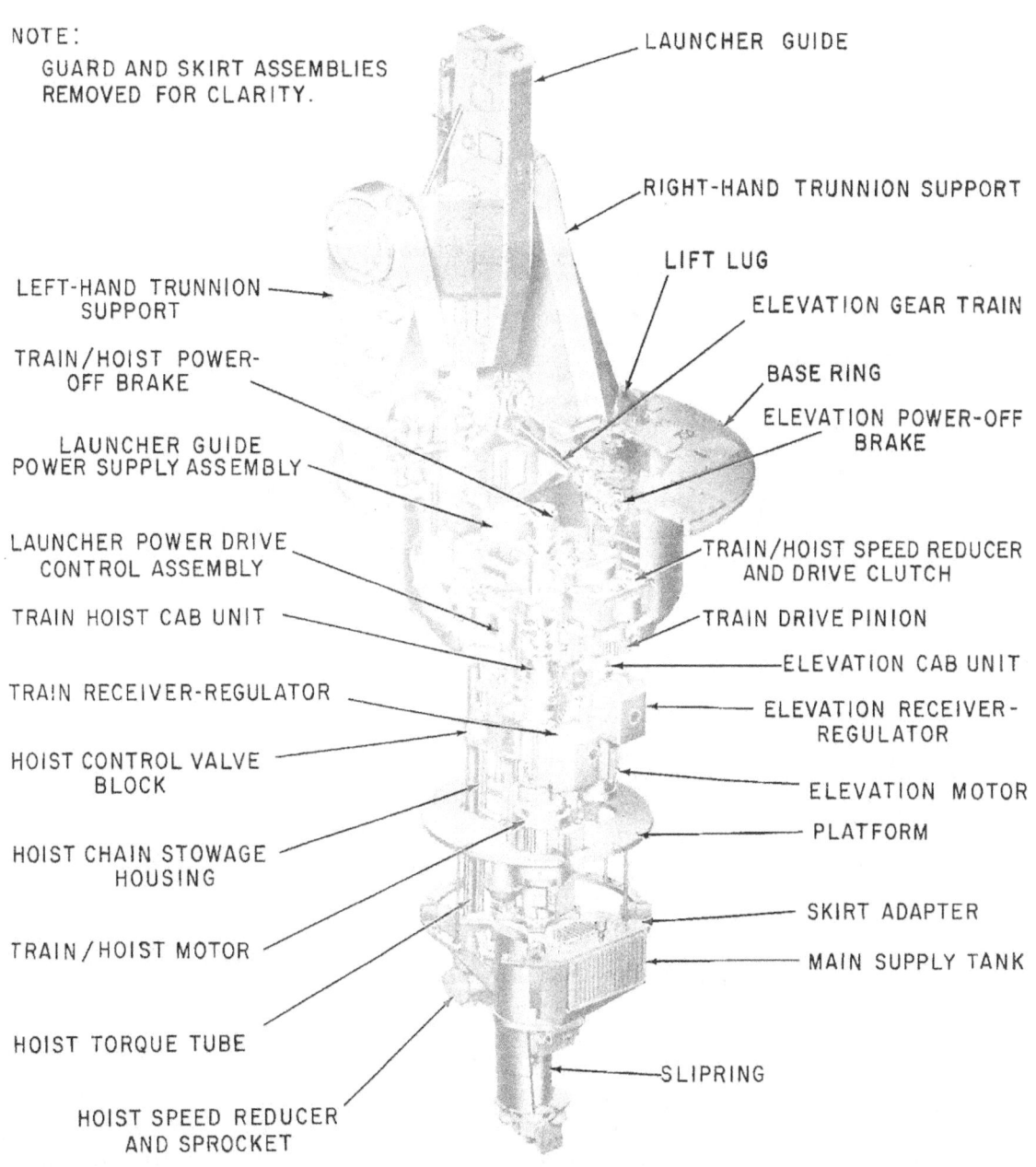

Figure 6-19.—Mk 123 Mod 0 guided missile launcher (Mk 22 GMLS).

# GUNNER'S MATE M 1 & C

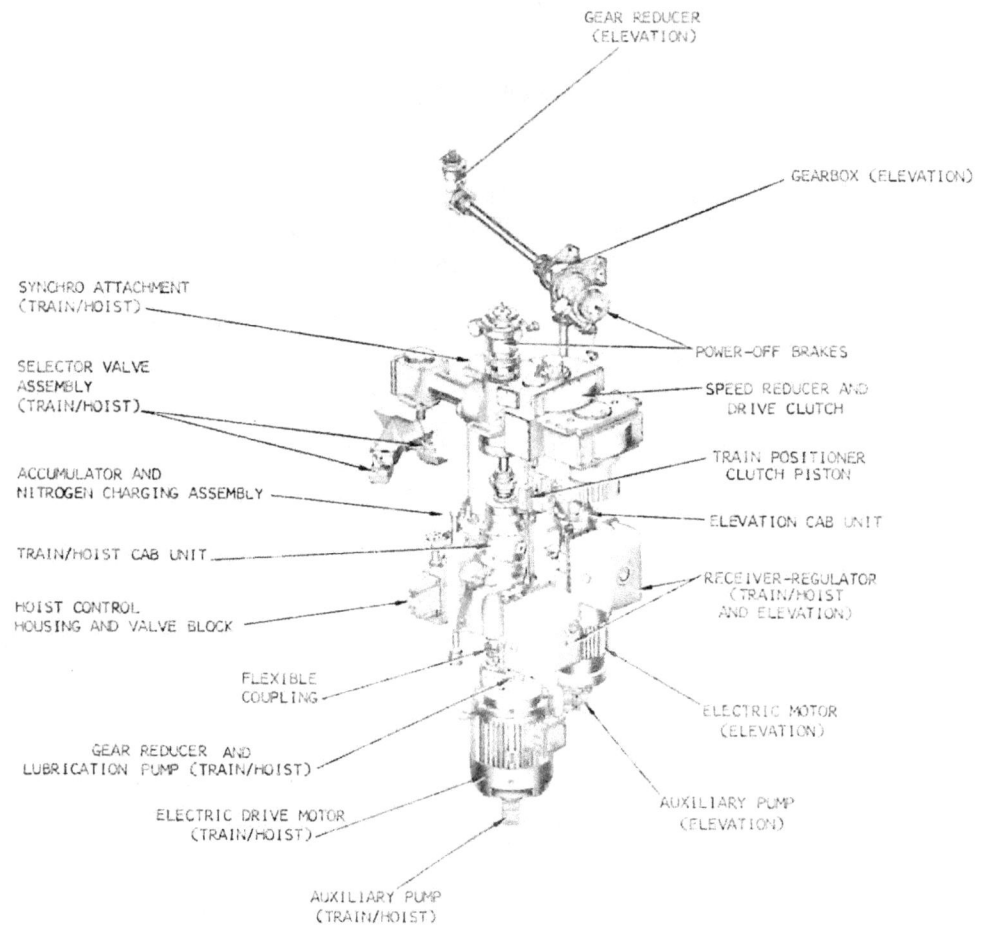

Figure 6-20.—Train/hoist and elevation power drives.

converts the power from the hydraulic fluid to mechanical motion, transmitted through the drive shaft to train/hoist (or elevation) drive train to move the launcher in train or elevation or to raise or lower the hoist chain and associated components. The operation of CAB units is described in *Fluid Power,* NAVPERS 16193-B.

DRIVE TRAINS.—The train/hoist drive train transmits mechanical movement of its B-end output shaft to either the drive pinion of the launcher carriage or the hoist drive shaft. The elevation drive train transmits the mechanical movement of its B-end output shaft to the elevation arc and, in turn, the launcher guide. For the most part, the components of both drive trains are contained within the base ring of the rotating structure. Some parts of the elevation drive train are in the launcher guide. The two drive trains are of different design but contain some similar components.

## Chapter 6—HYDRAULICS AND PNEUMATICS

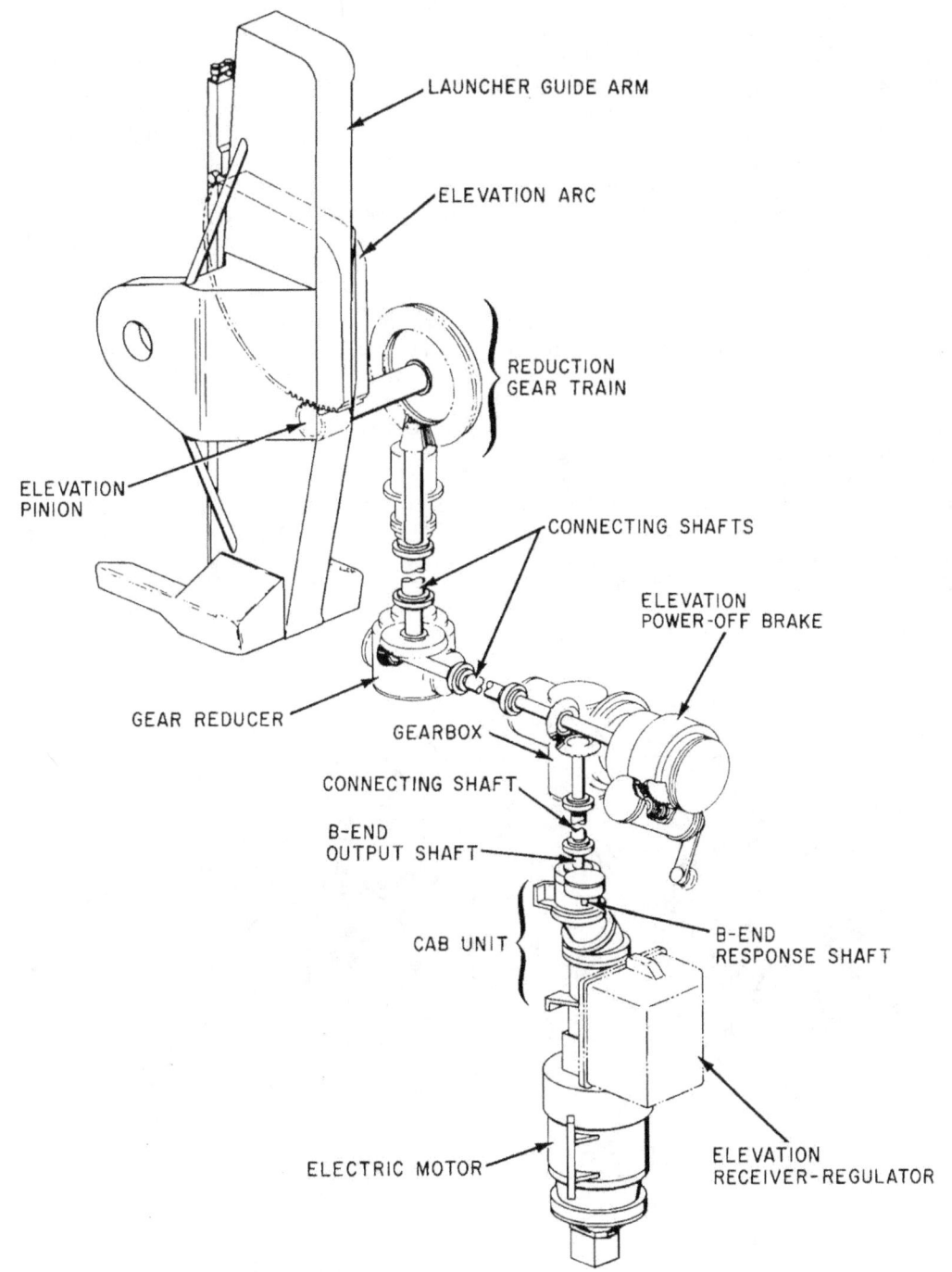

Figure 6-21.—Elevation drive: Mechanical schematic.

The train/hoist drive train uses a common gearbox and clutch to drive the rotating structure and the hoist chain. The main components are: (1) a B-end coupling, (2) a speed reducer and drive clutch assembly and associated retractable-rail assembly, (3) a power-off brake, (4) a synchro attachment assembly, (5) a selector valve assembly, and (6) a train-positioner assembly. Several of these are pointed out in figure 6-20.

The elevation drive train main components are a vertical shaft and couplings, a gearbox, a power-off brake, a horizontal shaft and couplings, and a gear reducer (figure 6-21). They transmit the output of the elevation CAB unit to move the elevation arc and the launcher guide. Unlike launchers with two guide arms, where the elevation arc is mounted on the carriage between the two guide arms, the elevation arc in the Mk 22 is inside the right-hand support trunnion. When the elevation arc is driven (by the elevation drive train), the launcher guide moves accordingly. Clockwise and counterclockwise movements of the elevation gear train elevate or depress the launcher guide. The elevation and depression buffer is a hydraulic safety device that prevents the launcher from moving beyond its design limits (figure 6-22). The buffer decelerates and stops the launcher guide as it moves beyond the 90° elevation or -10° depression. The elevation buffer piston is the shock absorbing component contacted when the launcher guide elevates beyond 90°. The depression buffer piston is contacted when the launcher depresses beyond -10°. They are of the same type, with compression springs seated in the piston recess and the other end in the sleeve.

## HYDRAULIC SCHEMATICS

Chapter 4 shows the electrical circuits used in step operation of fin openers and contactors in the Mk 13 system. The hydraulic operation of the fin opener cranks and the contactor to the missile are actuated by the electrical system. When you are performing trouble analysis of a component, you also need the hydraulic

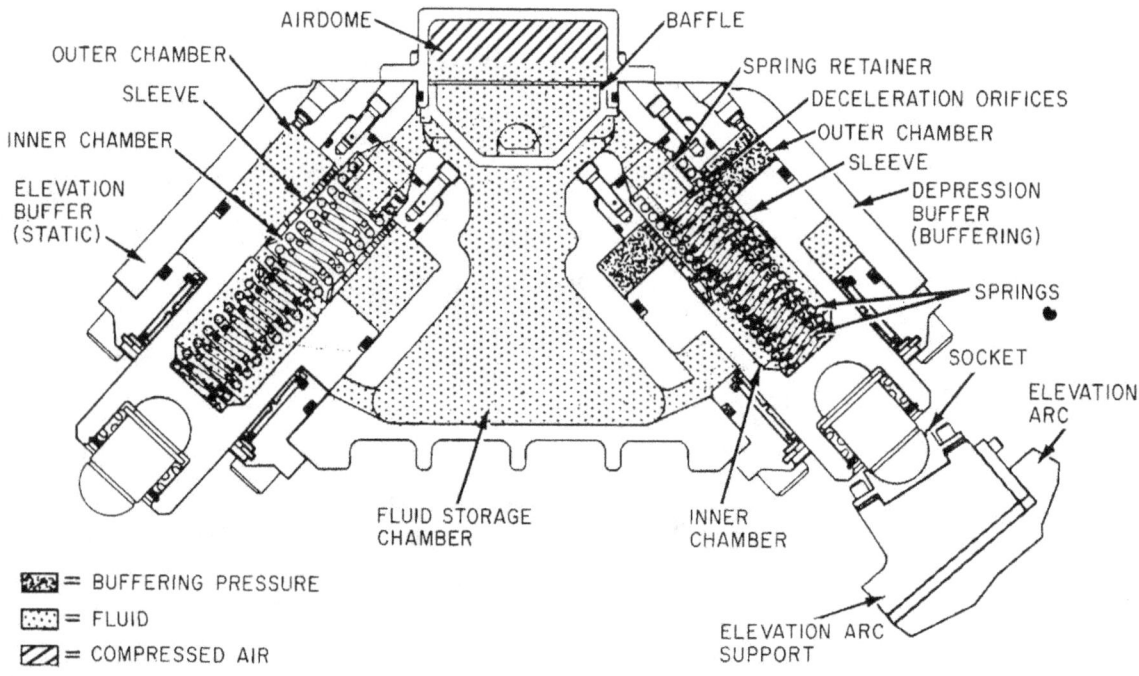

Figure 6-22.—Elevation and depression buffers: hydraulic schematic.

## Chapter 6—HYDRAULICS AND PNEUMATICS

schematic to trace the actions of the hydraulic parts. Figure 6-23 shows a hydraulic schematic of the fin opener assembly in the Mk 22 and Mk 13 systems. Primary hydraulic control of the fin opener assembly originates with two solenoid operated valves UVU6 and UVU7. These two pilot valves initiate each of the four operations of the fin openers: (1) engage fin openers and contactor; (2) extend fin opener cranks; (3) retract fin opener cranks; and (4)

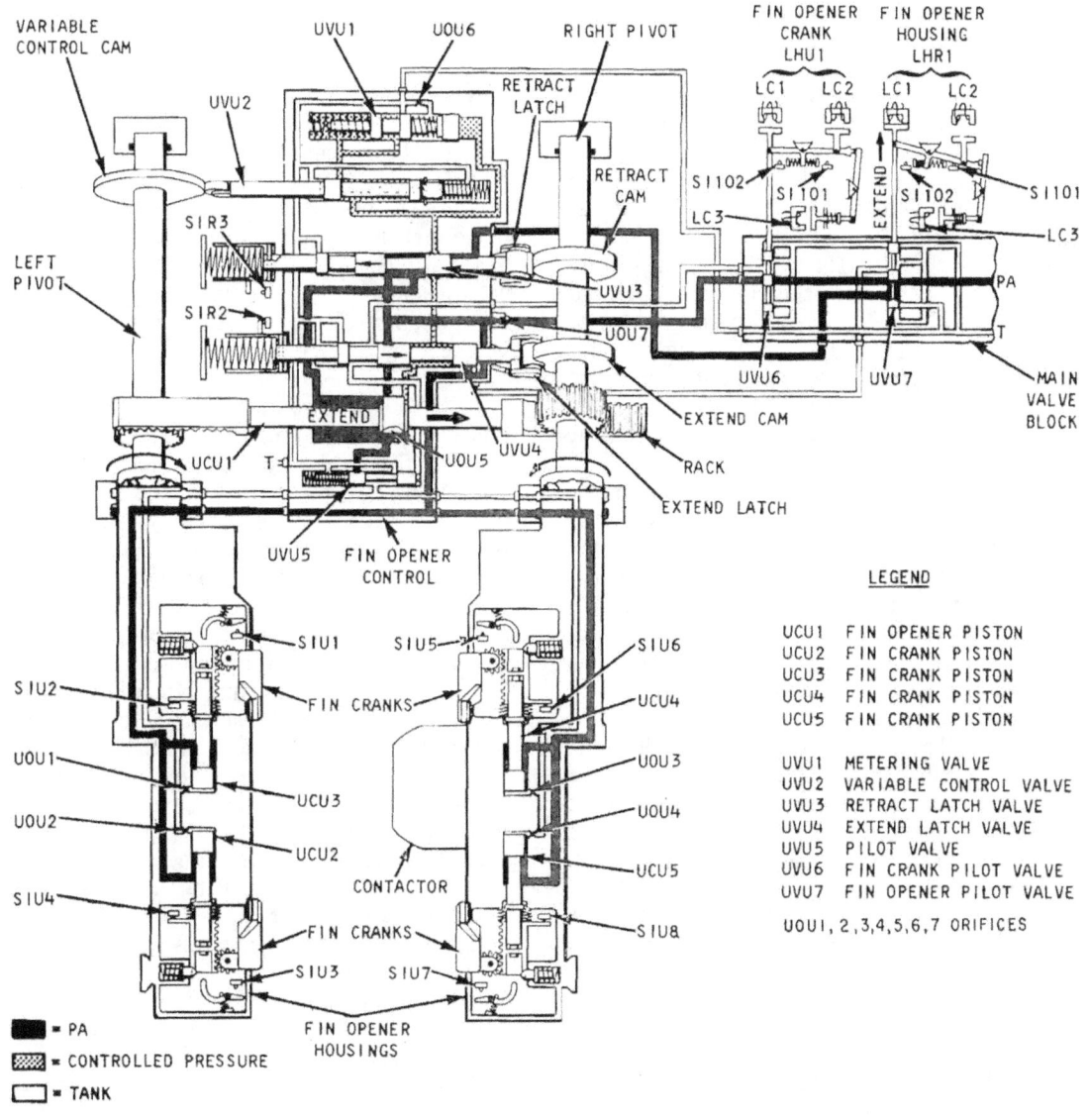

Figure 6-23.—Hydraulic schematic; fin openers and contactor extended (engaged); Mk 13 and Mk 22 GMLSs.

disengage fin openers and contactors. Five valves and a piston in the control valve block operate and extend and retract latches, control the speed of the pivots, open and close various ports, and operate some of the interlock switches. The block is located on the launcher guide, between the fin-opener shields.

The fin openers swing behind (engage) the missile as soon as it is hoisted onto the guide rail. This permits missile warmup to be resumed and missile-to-launcher interconnections to be completed almost immediately for the weapons control system. While the launcher is synchronizing to the remote order, the fin opener cranks reach their fully extended positions and actuate the interlock switches SIU1, -3, -5, and -7, which complete the circuit to energize relay KCU1, which energizes relay KCU3A. This completes the circuit to solenoid LHU1-LC3, which then releases the detent of solenoids LHU1-LC-1 and -2, and allows pilot valve UVU6 to spring-return to neutral and port pressure fluid from the extend side of the fin opener cranks to the tank. This causes the cranks to retract.

The contactor extends at the same time the fin opener housings rotate with the pivots. As the contactor mates with the receptacle on the missile, the force of engagement causes the sharp electrical pins of the contactor to puncture the seal that protects the missile contacts. This requires a pressure of 450 to 500 pounds. The contactor remains in position until the order to fire is received.

This small sample of a hydraulic schematic shows how electric and hydraulic schematics must be considered together. The electric components start the action; the hydraulic components carry out the electrical orders.

Study the OP and follow through on hydraulic schematics so you will understand the flow of hydraulic power through the system and its translation into mechanical movement. This knowledge will be invaluable to you in troubleshooting the system.

## ADJUSTMENT AND REPAIR OF HYDRAULIC SYSTEMS

To tell in detail how to adjust and repair hydraulic systems used in missile launching systems would require several large volumes. To work on any system, you need the drawings for that system installed on your ship, and the applicable OPs and ODs. The types of valves used in the power drive for the RSR may be the same as those in the power drive for the train or elevation system on the launcher, but their numbers or other designations and the locations differ. In a general course like this, we cannot name the particular valve to adjust.

### Shipboard Maintenance

Initial adjustments were made at the time of installation. Later adjustments aboard ship should be undertaken only after competent personnel have determined that adjustment is necessary. Servo valves, replenishing pump relief valves, and the control pressure valves are some components of a drive system that are adjusted whenever necessary.

FILTERS.—The indicating lights on the control panels pinpoint some troubles, such as clogged filters. Let us use as an example the Mk 64 Mod 0 power drive in the Mk 10 Mod 0 launching system shown in figure 6-4. This is the power drive used to operate the RSR, the tray shift mechanism, the magazine doors, and the magazine hoist mechanism. The oil filter assembly is mounted on the supply tank, above the electric drive motor. The assembly contains 12 filter elements that filter out foreign particles of 10-micron size or larger before the oil is pumped to the magazine components. When the filters become clogged and the filter bypass valve opens, a clogged-filter switch mechanism illuminates a light on the EP2 panel, indicating that the filter needs changing. After the filters have been changed, the filter assembly is filled with oil through the filler plug on the filter, which forces out air that was in the filters and keeps air out of the system.

Some power drives use disposable, cartridge-type filters. These require no maintenance except replacement of the cartridge, as in the system above. If the filter element has to be cleaned, follow the instructions for the type of filter and observe the safety precautions for the cleaning method and materials used.

## Chapter 6—HYDRAULICS AND PNEUMATICS

Full-flow types filter all the oil that passes through the pumps. Such filters may have a relief valve to allow bypassing of the oil if the filter element is clogged. A bypass filter is one which filters only a portion of the oil passing through the pump. Figure 6-24 is an example of a bypass filter. More correctly, it is called a proportional flow filter. It consists of a cylinder containing a filter element made up of a number of packs of perforated paper discs. Spring action maintains a uniform pressure on each of the packs. Oil passes from the outside of the pack, where the foreign matter is deposited, to the center passage, and through the outlet at the head of the filter.

To clean this type of filter, lift the head, with the filter packs attached, from the body of the filter. Connect a low pressure air supply to the oil outlet of the system. This allows the air to blow back through the element. When a white foam appears along the entire length of the pack, it is clean.

If you observe the rules for keeping contaminants out of hydraulic systems, the filters will seldom need cleaning. Keep the containers of hydraulic fluid tightly closed except when actually transferring the fluid. Strain the fluid into the hydraulic system, even though you are pouring from a freshly opened can. Keep all openings on the hydraulic system closed so water, dust, dirt, or any other contaminant cannot get in. Even with the greatest care, however, it is not possible to keep out every bit of foreign matter. Filters are installed at selected places within the system to catch bits of metal that might wear off while the pumps, gears, valves, etc., are operating. The MRCs and the OPs on maintenance tell you how often each filter is to be routinely checked. By regularly checking and replacing filters, you can greatly reduce the down time of hydraulic systems. Texts based on samples of hydraulic fluid taken from the accumulator supply tanks detect deterioration of the fluid. Use the MRC instructions for obtaining the samples.

A typical micron-type filter unit is shown in figure 6-25. It consists of a single element filter

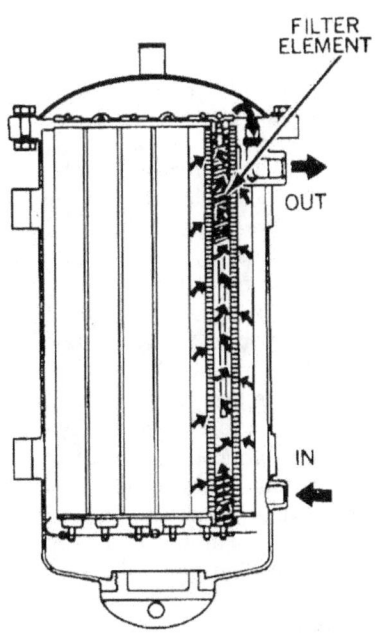

Figure 6-24.—Auxiliary bypass filter.

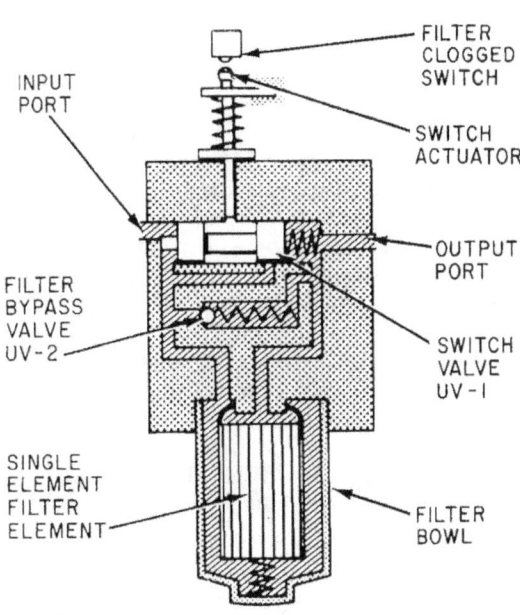

Figure 6-25.—Filter unit.

197

assembly, a filter bypass valve, a switch valve, a filter clogged switch, and switch actuator. If the filter element starts to clog, it retards the flow of hydraulic fluid and causes a pressure differential on the opposite ends of switch valve UV-1 (figure 6-25). When the pressure differential reaches a preset setting, the higher pressure at the input end of switch valve UV-1 causes the valve to shift to the right against the spring end of the valve which opens a pressure port to the switch actuator. If the ALL MOTORS—CHECK FILTERS lamp blinks, it warns the EP2 operator that a filter has clogged in one of the hydraulic systems. At this indication, the EP2 operator watches the motor STOP switches while pressing the ALL MOTORS—CHECK FILTERS pushbutton. When pressed, the ALL MOTORS—CHECK FILTERS pushbutton causes the motor STOP indicating lamp of the motor associated with the clogged filter to go on. After determining the location of the clogged filter and when operations allow, the EP2 operator shuts down the appropriate motor and has the clogged filter elements replaced. As long as the filter remains clogged and the pressure differential reaches a preset setting above that regulated by the UV-2 filter bypass valve's tensioned spring (figure 6-25), the bypass valve will open and port hydraulic fluid around the filter element and flow directly into the system through one output port. The replaceable filter element is a treated cellulose paper formed in vertical convolutions which catch and hold dirt and other solid particles. (Some filter elements are made of other fibrous material or of metal discs and rods which can be cleaned and reused if a new filter element is not available, or it may be set aside to be cleaned later and placed in storage for reuse. A filter element must be thoroughly dry before reuse.)

Before you change a filter element, turn off the hydraulic system and release the system pressure.

WARNING: Be sure that system pressure is relieved before disassembling or removing hydraulic components. High system pressure can cause serious injury to personnel.

Arrange to catch hydraulic fluid that will spill from the bottom of the filter unit when the bowl is removed. Unscrew and remove the filter bowl, then pull the filter element off the head. Install a new, or properly cleaned, filter element and an O-ring, seating them firmly on the filter head. Examine the O-ring in the filter bowl; replace it if it is faulty. Coat the external threads of the filter head with petrolatum and reassemble the filter bowl to the head. Install the safety wire. If the filter continues to clog, the hydraulic system may need to be drained, flushed, and refilled.

VALVES.—Daily and weekly operational checkouts may reveal a need to adjust one or more valves. The types of valves used in hydraulic systems are described and illustrated in *Fluid Power,* NAVPERS 16193-B, along with some general instructions for installation and maintenance. See also the illustrations and descriptions in chapter 8 of this text and in *Gunner's Mate M 3 & 2,* NAVEDTRA 10199-C.

Numerous valves, simple and compound, are used in the hydraulic components of missile launching systems and in the missiles. Note the number and variety of valves in that small segment of a launching system shown in figure 6-23. The same principles of operation apply to all, regardless of complexity, but the components and their methods of assembly may differ. When a valve has to be disassembled, be sure to get the drawing showing all the parts and the order of assembly. The automatic valves, such as regulators, relief valves, and safety valves, should not be disturbed except at overhaul unless found faulty. Foreign matter in the valve seat, scoring and grooving of parts, or plugging of openings may cause the valve to stick or fail to close completely. The usual remedy for such conditions, as for practically all serious valve troubles, is to dismantle the valve, thoroughly clean all parts, replace those that are damaged, and reassemble. You may have done this with the simpler valves, or under supervision. Maintaining compound relief valves is fundamentally the same as maintaining simpler types. You can generally tell that a compound relief valve is not functioning properly because it will overheat, operate sluggishly or erratically, or will function at the

## Chapter 6—HYDRAULICS AND PNEUMATICS

wrong pressure. The valve will usually clean itself if you start the pump and back off the adjustment screw on the pilot valve a little by turning it counterclockwise so that the pressure control spring responds to a lower pressure. The adjustment screw should never be removed completely while the system is under pressure. (Some instructions require complete release of system pressure before adjusting: check the instructions for your launching system.) After a flow of liquid has cleaned the valve, carefully reset the adjustment screw using a pressure gage. The relief valve should be set to open at about 25 percent more than the maximum normal operating pressure.

Check valves require little attention over long periods of time. Leakage may be caused by a tiny particle of foreign matter between the checking device (ball, cone, or poppet) and its seat. It will be necessary to remove the valve and disassemble it completely for cleaning. Remember the warning about high pressure systems—shut off power and release the pressure before removing any part from the hydraulic system.

If no scratches are found on the valve seat or the checking device, wash all parts in clean hydraulic fluid of the same type as used in the system. Inspect the housing and checking device for evidence of corrosion. A slightly rough surface can be smoothed by buffing. Replace the valve if there is corrosion or excessive roughness. A cone type check valve may have a tendency to lean to one side, in which case the movable part may dig into the soft aluminum body of the housing and stick there.

Remember that the arrow on the housing must point in the direction of the flow of liquid through the valve. Before removing a check valve from a line, mark the adjacent structure, indicating the direction in which the arrow points. When installing the check valve, grip the wrench flats of the check valve at the end to which the connecting tubing is being installed. Do not grip the opposite end. This will prevent the possibility of distorting the valve body, which would cause the valve to leak.

When a valve has to be disassembled for cleaning and replacement of broken or worn parts, such as a broken spring, deteriorated O-ring, or scored valve plunger, it is important that the correct parts be used, and that they are assembled in the proper order. The MRC gives all the steps in order. The OP with the illustrated parts breakdown (IPB) identifies every part by name and stock number. A neat, orderly workbench is essential so parts can be laid out in order. A dust-free area helps in keeping dirt out of the valve when reassembling.

Through constant use, working parts may become worn, springs may be weakened or cracked, and O-ring and backup rings may become deteriorated. Vibrations can cause metal parts to crystallize and crack. However, keeping dirt, moisture, and air out of the hydraulic system is the best preventive measure. Daily inspections will detect leaks that can be corrected by simple tightening; daily checking of pressures and fluid levels can detect other troubles before they become major ones.

Valves are not disassembled as a routine maintenance procedure; they are disassembled only if they are not functioning properly.

PUMPS.—As a rule, the pumps in hydraulic power drive systems require little maintenance other than proper lubrication and a clean hydraulic system in which to operate. Overheating, unusual noise, or failure to deliver the designed output are signs of trouble. A frequent cause of noise is failure of oil to reach the pump; the oil level in the reservoir may be low, or there may be clogged lines or filters. Since the pump depends on the hydraulic fluid for lubrication, failure of the supply will soon cause the pump to overheat and will probably cause its parts to bind.

Another cause of abnormal noise is poor alignment between a pump and its driving mechanism. This condition will cause worn parts and possible leakage, reducing the pump's efficiency. Correcting the misalignment can be a major repair job, but without repair, the problem will become worse.

Pounding or rattling noises in axial piston pumps may be unavoidable because of a partial vacuum produced in the active system during high speed operation or under heavy loads. The noise should stop when the load is reduced. If it does not, bleed air from the system at the vents.

Hydraulic systems which perform satisfactorily and show no evidence of sludge, corrosion, etc., should not be opened. Cover plates should be kept tightly secured, and should not be opened without good reason.

Use special care when you reassemble a rotary gear type pump. The rotors operate in a pump casing or body. End plates enclose the rotors on each side. When tightening the screws that hold the sections together, use only moderate force. Make them just tight enough to allow free movement of the rotor with no leakage.

The routine inspection, lubrication, and checking of fluid level and pressure are tasks for the GMM3 and GMM2. They use the MRCs for instructions, and check off each job on the work schedule after completion. Your job is to supervise and check the work, and make sure it is done at the intervals scheduled.

MOTORS.—Hydraulic motors are activated by receiving fluid flow from the power pump. This fluid, under pressure, forces the motor pistons away from the flow source, thus resulting in a rotation of the motor drive shaft. The pressure buildup in the high pressure line between the pump and motor will be in direct proportion to the mechanical output or work required of the motor. The motor speed will vary directly with the amount of fluid pumped to the motor. The direction of rotation can be instantly reversed without harming the motor. The direction of fluid flow is normally controlled by a servovalve.

A fixed displacement hydraulic motor may be used with a variable displacement hydraulic pump. A radial piston motor is usually used with a radial piston pump, and an axial piston motor with an axial piston pump. See *Fluid Power*, NAVPERS 16193-B for descriptions.

Hydraulic motors are self-lubricating; daily inspection for leakage is usually all the maintenance needed. If the motor must be removed for overhaul or corrective maintenance, be sure to plug all openings of connecting pipes so no dirt will get into the system. Use a lifting device to transfer the motor to the workbench. Figure 6-26 illustrates a typical hydraulic piston-type motor. Disassemble the parts carefully to avoid marring or scratching smooth surfaces. This is especially necessary where the fit must be exact to prevent oil leakage, as at oil seals. Follow the disassembly and reassembly instructions in the OP for the power system. The correct order makes the work easier, and is less likely to result in damage to parts by excessive use of force.

When removing the roller bearings, take care that the rollers do not fall out. As you remove each part, carefully place it on a cloth or paper-covered space in the order of removal. There are several small parts that can easily be lost. Do not throw away a part until you have a replacement part for it. You may need it for comparison, even though the stock number for the new part is the same as the old. A flaw or defect in a part may not be visible until after the part is cleaned. Inspect each part after you have cleaned it. Do not leave bare parts exposed any longer than necessary without the protection of a coat of oil. Rust can develop quickly and mar the polished surface of a precision-fitted part.

On some later model B-end motors, a hydraulic equalizing valve is mounted on each side. These valves were set when the power drive was tested. If the motor is removed or replaced, adjustment can be made so exactly equal pressures can be developed in each B-end motor high pressure line. Gages can be mounted on the B-end motor when pressure tests are to be made. The snubber and fittings that accommodate the gages are located beside the equalizing valves.

**Troubleshooting**

When scheduled maintenance or system testing (DSOT) reveals a fault, system troubleshooting procedures begin. The use of troubleshooting charts and procedures contained in PMS/SMS manuals (OPs) will help isolate the fault. After finding and isolating the fault, system manuals direct maintenance personnel to

## Chapter 6—HYDRAULICS AND PNEUMATICS

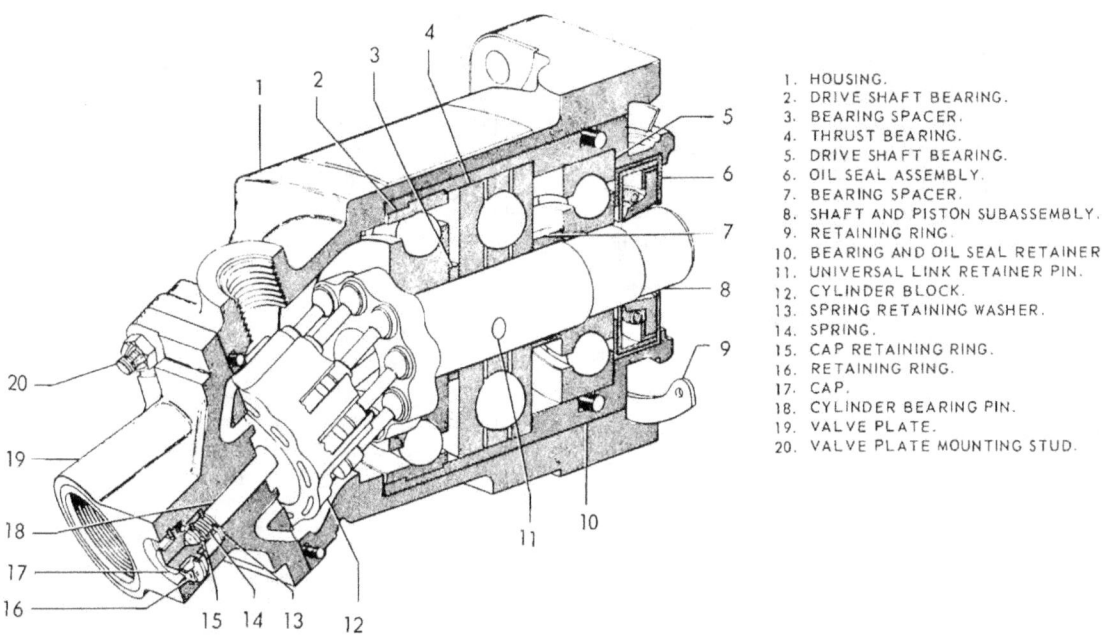

1. HOUSING.
2. DRIVE SHAFT BEARING.
3. BEARING SPACER.
4. THRUST BEARING.
5. DRIVE SHAFT BEARING.
6. OIL SEAL ASSEMBLY.
7. BEARING SPACER.
8. SHAFT AND PISTON SUBASSEMBLY.
9. RETAINING RING.
10. BEARING AND OIL SEAL RETAINER.
11. UNIVERSAL LINK RETAINER PIN.
12. CYLINDER BLOCK.
13. SPRING RETAINING WASHER.
14. SPRING.
15. CAP RETAINING RING.
16. RETAINING RING.
17. CAP.
18. CYLINDER BEARING PIN.
19. VALVE PLATE.
20. VALVE PLATE MOUNTING STUD.

94.63
Figure 6-26.—Typical hydraulic motor, piston type.

the appropriate instructions for correcting the malfunctions. Corrective maintenance cards (CMCs) supply corrective maintenance instructions. These CMCs provide instructions for the alignment, adjustment, repair, and replacement of parts and components. These cards also cover dismantling, repair, replacement, and alignment of major assemblies and subassemblies.

### Adjustment and Repair at a Navy Yard or Repair Tender

Given careful daily maintenance and inspection, hydraulic systems can be used for long periods of time without needing major repairs. Breakdowns can occur, however, in spite of the best care you can give. Size and weight of some components make repair aboard ship difficult or impossible; alignment of such components may require the facilities of a shipyard or repair tender. Attempts by unskilled personnel to overhaul or repair components can result in serious damage to costly equipment, and possible personnel injury. Before disassembling any part, be sure you can put it together again correctly, and understand how it should operate. The illustrated parts drawings are essential for less experienced personnel, and help even the most experienced to check themselves on reassembly.

Alignment of large components may require yard or tender facilities. Critical adjustments may need to be deferred yard or tender work. Train and elevation power drive units normally are not removed or installed by ship personnel. While it is possible for shipboard personnel to remove the train or elevation power unit, it is recommended that such removal be accomplished during major overhaul. If the power unit has to be removed before yard or tender facilities are available, it is better to remove it by disassembling than to try to remove it as a unit when adequate handling facilities are not available.

Realigning a launcher rail is a task of considerable proportions. Arrange for tender or yard assistance, if possible. Do not readjust the launcher rail unless it is absolutely necessary.

Removal of excessive backlash in train and elevation drives is best reserved for overhaul. Excessive backlash causes misalignment between the launcher and the weapons system; insufficient backlash causes galling and binding.

Repair and overhaul of train and elevation gearboxes are tasks for tender and shipyard personnel. Although ship personnel can remove, disassemble, and reassemble the gearboxes, getting the proper alignment and backlash within the gearbox, and properly aligning the gearbox to the sector gear or training circle would not be possible with the equipment normall available on board.

Do not let these paragraphs give the impression that few adjustment and repair jobs are done aboard ship; look over the list of maintenance procedures for the system now assigned to you. The maintenance index pages (MIPs) for your launching system's CRCs will list the tests, repairs, alignments, adjustments, and servicing of a launching system that can be accomplished aboard ship. While this list includes such minor items as changing a light bulb, there are many complicated adjustments, such as adjusting firing cutout switches, position plus lead switches, and the entire alignment of the receiver-regulators.

## RECEIVER-REGULATORS

Receiver-regulators are located on the launcher with the power drive units (figure 6-15 and 6-21). The two control systems—train and elevation—comprise similar electric, hydraulic, and mechanical equipment. Each system receives its own order signals; the train system receives train order signals through its receiver-regulator, and the elevation system receives elevation order signals through the elevation receiver-regulator. During normal (automatic) operation, the train and elevation systems convert electrical signals received from a remotely located computer into hydraulic movements. These hydraulic movements control the velocity, acceleration, and position of the launcher carriage and the guides or guide arms.

Basically, the power drive consists of a CAB unit and a receiver-regulator. The CAB unit is composed of a B-end (hydraulic motor) and an A-end (hydraulic pump). The B-end converts fluid flow into mechanical motion. The output shaft of the B-end drives the launcher through reduction gears. Therefore the speed and direction of launcher movement is determined by the speed and direction of the B-end output.

The A-end is the hydraulic pump that supplies pressure fluid to the B-end. The fluid flow supplied by the A-end determines the speed of the B-end, while the direction of fluid flow from the A-end to the B-end governs the direction of the B-end rotation.

The A-end output is determined by two stroking pistons controlled by the receiver-regulator. These pistons "stroke" the A-end tilt plate, and thus regulate both the quantity and the direction of fluid flow from the A-end. The A-end is driven at a nearly constant speed by a unidirectional electric motor. The receiver-regulator controls the hydraulic fluid ported to the stroking pistons, and thereby regulates the CAB unit operation. The regulator components position the A-end tilt plate so the B-end output is in accordance with the electrical signal input order to the receiver-regulator. The position signals are sent from the computer to both the train and elevation systems. Synchro transmitters (CX) at the computer initiates the signal voltages; synchro receivers (CT) in the receiver-regulators receive the signals.

Figure 6-27 shows a functional diagram of the synchro control system for the Mk 13 and Mk 22 GMLSs. The launcher position orders are transmitted by CX to CT, indicating the desired launcher position. The CT acts as a differential, combining the actual position of the launcher—indicated by B-end response—with the ordered position. In these GMLSs, the velocity order is transmitted from a tachometer generator in the computer to the launcher's amplifier and is applied electrically to the error signal. The velocity system improves the synchronizing ability of the launcher. One of the

## Chapter 6—HYDRAULICS AND PNEUMATICS

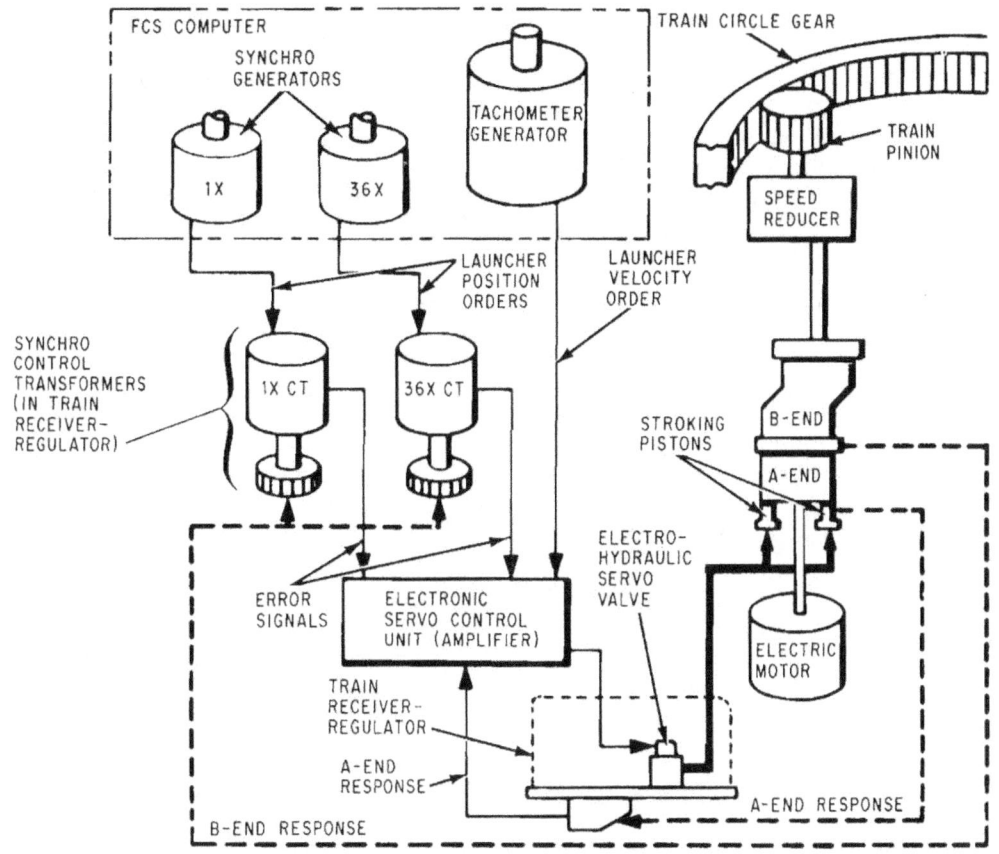

Figure 6-27.—Launcher synchro control system and amplifier.

major differences between the launcher power drive receiver-regulators of these systems and those of the Mk 10 GMLS is that these systems use only one electrohydraulic servovalve (primary). The receiver-regulators in the launchers for the Mk 10 GMLS use two servovalves (primary and velocity). The launcher receiver-regulators for the Mk 10 GMLS also use a hydraulic integration system that functions to improve the accuracy of the launcher.

The CT output is a signal voltage proportional to the B-end error. The B-end error is the difference between the ordered position of the driven equipment (launcher) and its actual position. The CT output is transmitted to the amplifier where it is amplified and sent back to the receiver-regulator. There it drives an electrohydaulic servovalve which transforms the amplified electrical input into a proportional hydraulic movement, and moves the launcher to the position ordered. If the target is moving, the position of the launcher has to be corrected with each new signal from the computer.

To enable the launcher to be positioned more accurately, the train and elevation synchro control systems have CXs and CTs in pairs. Each pair consists of a coarse synchro called 1X (one-speed) and a fine synchro called 36X

(36-speed). The CXs are located at the computer and the CTs are at the receiver-regulator. The rotor movement of the 1X coincides with the launcher movement, while the rotor of the 36X CT will move 36 degrees for every degree of launcher movement. As long as the system is on automatic operation, the launcher is moved in train and elevation on signal from the computer. In local operation, the launcher is moved by moving the train and elevation dials. Figure 6-28 shows the train and elevation local control dials, 1-speed and 36-speed, on the EP3 panel for the Mk 5 launcher.

The EP3 panel is a combination test unit and local control director. The upper part of the panel, called the local control station, contains synchro transmitters and tachometer generators, control knobs and dials, and illuminating lamps. After the EP2 operator activates the EP3 panel and retracts the launcher train and elevation latches, the EP3 operator can train and elevate the launcher using the local control knobs. The test unit at the lower part of the EP3 panel (not shown) contains switches, indicating lamps, relays, test receptacles, terminal boards, and phone jacks for conducting tests, checks, and adjustments on the launcher train and elevation systems. During test procedures a dummy director, frequency generator, oscillograph, and other test equipment can be plugged in as required.

Synchro systems are discussed in the preceding chapter and in the preceding course, *Gunner's Mate M 3 & 2,* NAVEDTRA 10199-C and in the *Navy Electricity and Electronics Training Series (NEETS).*

**Mk 13 and Mk 22 GMLS Receiver-Regulators**

The following description of Tartar receiver-regulators is for those used with the

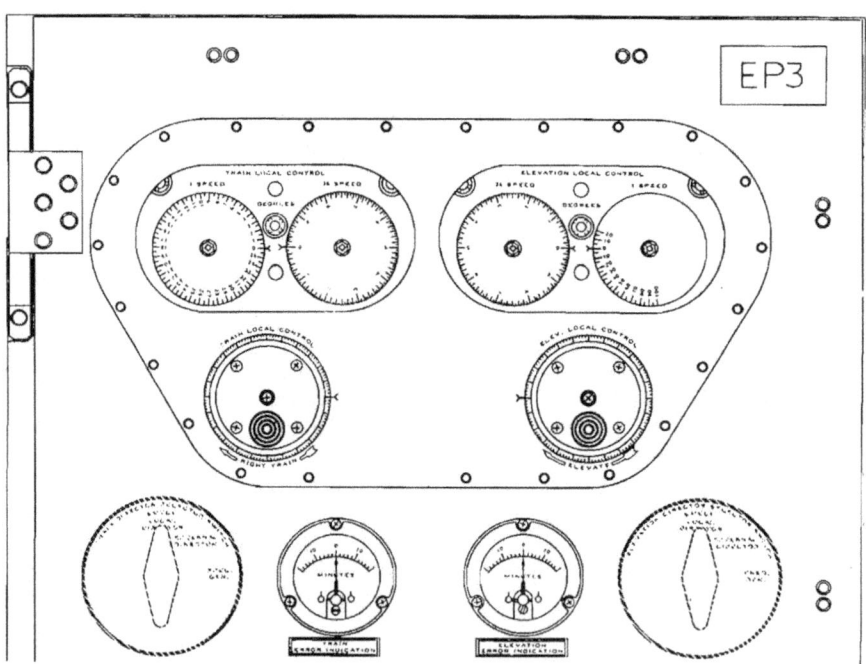

Figure 6-28.—Train and elevation dials, local control station, EP3 panel, Mk 5 launcher.

## Chapter 6—HYDRAULICS AND PNEUMATICS

Mk 13 and Mk 22 GMLSs. The receiver-regulator is mounted less than 2 inches from the main supply tank. The heating and cooling of the hydraulic fluid tends to create a vacuum in the receiver-regulator cases, which causes air from the main supply tank to be sucked into the receiver-regulator cases. The air breather on the main supply tank allows salt or humid air to enter the tank. This air would cause damage if it reached the receiver-regulator parts. To prevent this, an expansion chamber with a quantity of inert gas is connected to the top of the main supply tank. The receiver-regulator cases port through this chamber in such a way that only the inert gas can ever enter the receiver-regulator cases when a vacuum is created.

Remote train and elevation order signals originate in the missile fire control system. Order signals may also originate in the control transmitters within the launching system (EP3 panel). These signals are used in the local control mode of operation which is used primarily for daily workout or routine maintenance.

The receiver-regulators on the Mk 116 launcher of the Mk 13 Mod 0 GMLS are different from preceding models in several ways. The Mk 116 launcher uses an amplifier (electronic servocontrol unit) which electrically performs many of the functions that were previously performed hydraulically by other types of receiver-regulators. A modified synchro system is used, with the B-end response positioning the synchro rotors. There is no rotary piston response. Only one modified electrohydraulic servovalve is used. Both chambers of the electrohydraulic servovalve plunger are used, and each chamber is directly connected to the A-end stroking piston. The two stroking pistons have equal areas for hydraulic pressure to act upon.

The A-end response is transmitted electrically by a potentiometer to the amplifier and mechanically to the limit-stop and automatic tracking cutout systems. A modified limit-stop system is used to mechanically return the electrohydraulic servovalve, and thus the A-end, to neutral.

The automatic tracking cutout system uses the limit-stop system to stop the power drive.

There are no hydraulic velocity and integration systems in the regulator. However, a velocity signal is electrically applied to the amplifier.

The Mk 13 Mods 1, 2, and 3 guided missile launching systems use the Mk 116 Mod 1 launcher, and a number of changes have been made in the associated equipment. The principal change in the train and elevation systems is the redesign of the electronic servocontrol units. Minor modifications have also been made in the train and elevation drive motors, the servo and supercharge hydraulic systems, and the receiver-regulators.

The Mk 48 Mod 1 train receiver-regulator is the same as the Mk 48 Mod 0 receiver-regulator except for the synchro gear, the stroke response assembly, and the automatic-tracking-cutout valve block assemblies. The Mk 49 Mod 1 elevation receiver-regulator differs from the Mk 49 Mod 0 regulator in the same way.

The modified stroke response assembly includes an electrical connector to facilitate replacement of the stroke response potentiometer. A resistor, which is wired to the tachometer generator, has been added to the synchro gear assemblies. The automatic tracking cutout valve block assembly has been mounted with rollers above and below the limit-stop actuating cam to prevent binding.

Some of the check valves that are on the auxiliary relief valve assembly in the Mk 13 Mod 0 system have been relocated on the header tank cover in the Mods 1, 2, and 3 systems.

The train and elevation auxiliary pumps in the Mk 13 Mods 1, 2, and 3 furnish supercharge pressure at approximately 150 psi instead of 100 psi as in the Mod 0. The servopressure remains the same, approximately 525 psi at 3 gallons per minute (gpm).

The only changes in the CAB units (hydraulic transmission) involve the safety relief valves. They are compound valves mounted on the valve plate between the A-end pump and the B-end motor. The assembly consists of the valve block, six valves, and two orifices. The valves serve to limit maximum pressure buildup in the high pressure output line of the A-end pump and prevent cavitation by porting hydraulic fluid to

# GUNNER'S MATE M 1 & C

the low-pressure (suction) line of the A-end pump. (This compensates for fluid lost through slippage and leakage.)

## Mk 10 GMLS Receiver-Regulators

The receiver-regulators for the Mk 10 GMLS are mounted to the center of the launcher carriage, directly above the CAB units. Servopressure at 400 psi is supplied to the receiver-regulator valves and pistons. There are two major servosystems in the receiver-regulators—the power drive and the velocity drive systems.

The signal for moving the launcher to the correct train and elevation position for dud jettisoning comes from fixed synchros in the EP2 panel. Air motors provide a means of training or elevating the launcher for maintenance.

Figure 6-29 shows a receiver-regulator with the cover removed, and some of the main

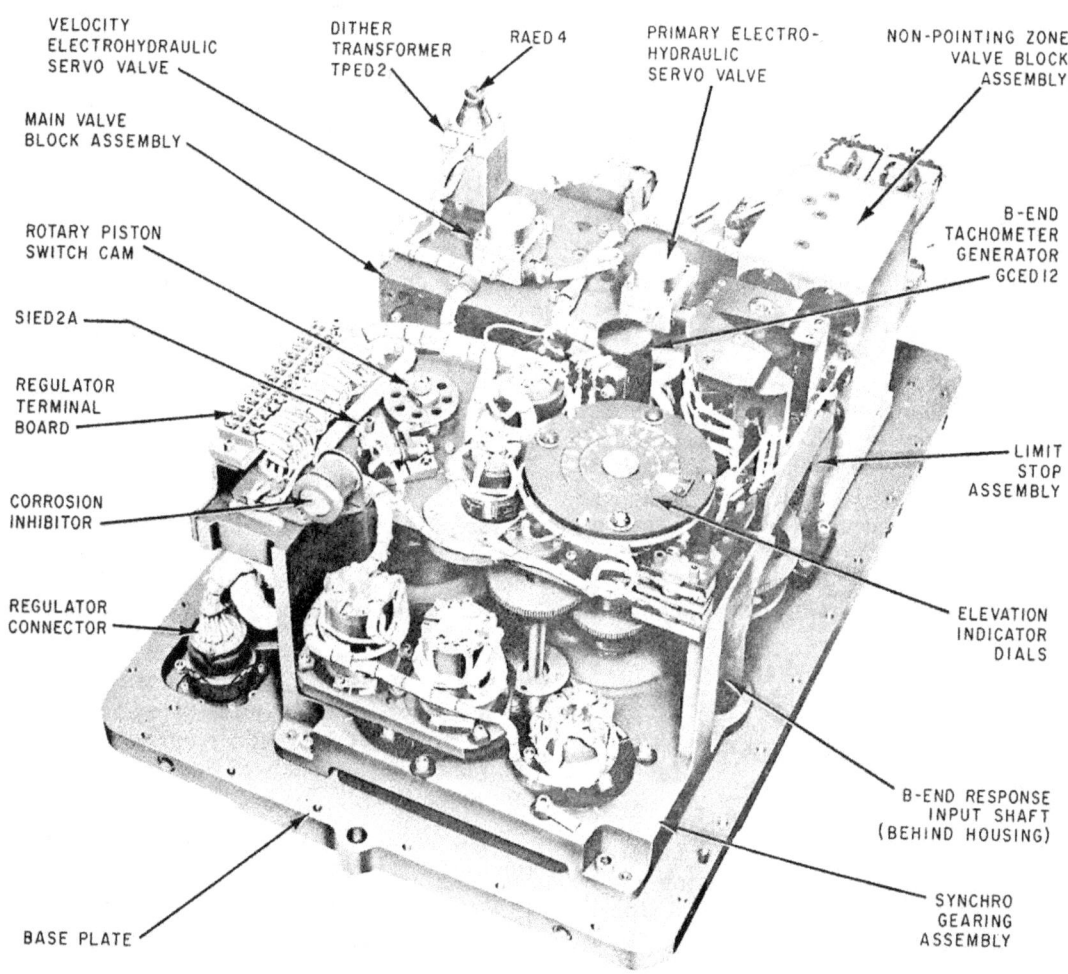

Figure 6-29.—Elevation receiver-regulator, Mk 5 launcher; cover removed.

## Chapter 6—HYDRAULICS AND PNEUMATICS

components are named. It contains a multitude of components, here grouped into seven logical sections to simplify location and identification:

1. The main valve block (including the electrohydraulic servovalves).
2. The nonpointing zone valve block.
3. The limit-stop assembly.
4. The B-end response.
5. The synchros and their accompanying gearing.
6. The rotary piston assembly.
7. The A-end response assembly.

Between the limit-stop assembly and the elevation indicator dials are two B-end response gears. One of the B-end response gears drives the synchro gearing assembly and the other gear drives the limit-stop assembly. The B-end response shaft that drives the gears leads through an opening in the regulator base plate. The B-end tachometer is driven by the limit-stop gearing.

The synchro gearing assembly includes indicator dials and five visible synchros with the gearing immediately below them. The dials are visible through the window in the cover of the receiver-regulator.

The A-end response shaft leads into the receiver-regulator through an opening in the base plate, and is not visible in figure 6-29. The rotary piston assembly is attached to the inboard side of the main valve block and lies below the rotary switch cam

Except for minor differences, the train receiver-regulator is identical to the elevation regulator. The train rotary piston cam has a slightly different contour than the elevation rotary piston cam, but it operates similarly. The contour is different because of the different acceleration and velocity specifications in train.

The nonpointing zone components of the train and elevation receiver-regulators prevent the launcher guide arms from training or elevating into any part of the ship's structure. The train limit stop does not have a gear and rack as does the elevation limit-stop assembly, but contains a nonpointing zone cam. The cam may be halted by pistons in the nonpointing zone valve block. By means of the gear and rack, the power drive can elevate the launcher arm above the nonpointing zone.

The train B-end response assembly differs from the elevation B-end response assembly in design because of mounting position. The train B-end response is coupled to the train gear reduction.

The function of the electrohydraulic servovalve is to convert an electrical signal from the train (or elevation) primary amplifier into a proportional hydraulic order. It does this with a minimum of friction and negligible time delay. Figure 6-30 is a cutaway view of an electrohydraulic servovalve. There are four ports in the base of the valve. One port supplies hydraulic fluid at 400 psi to the servovalve, one port leads to the tank, and a third port supplies control pressure from the servovalve to the rotary piston neutralizing valve. The fourth port is blocked off and is not used.

The hydraulic pressure ordered by the electrical input is applied to the end of the servovalve plunger and positions it. The plunger position results in the output of control pressure proportional to the electrical input, which is sent through the control port.

The force motor (figure 6-30) consists of two permanent magnets, two pole pieces, two coils, and a reed. One end of the reed is centered in the air gap between the two pole pieces and the other end is centered between two nozzles in the mixing chamber. The reed is the armature of the magnetic circuit and is polarity conscious. The force motor transforms the electrical input, a differential current, into a proportional force on the motor reed. The hydraulic amplifier converts the reed movements into corresponding differential pressures. The differential pressures cause the plunger to shift. A decreasing order signal causes a shift to the left; an increasing order signal causes a shift to the right.

Adjustment can be made on the adjustment screw (figure 6-30). The filters may need to be cleaned, or the whole valve may need to be disassembled, cleaned, and reassembled. Servovalves are used in several of the hydraulic systems, and you need to be able to maintain and repair them.

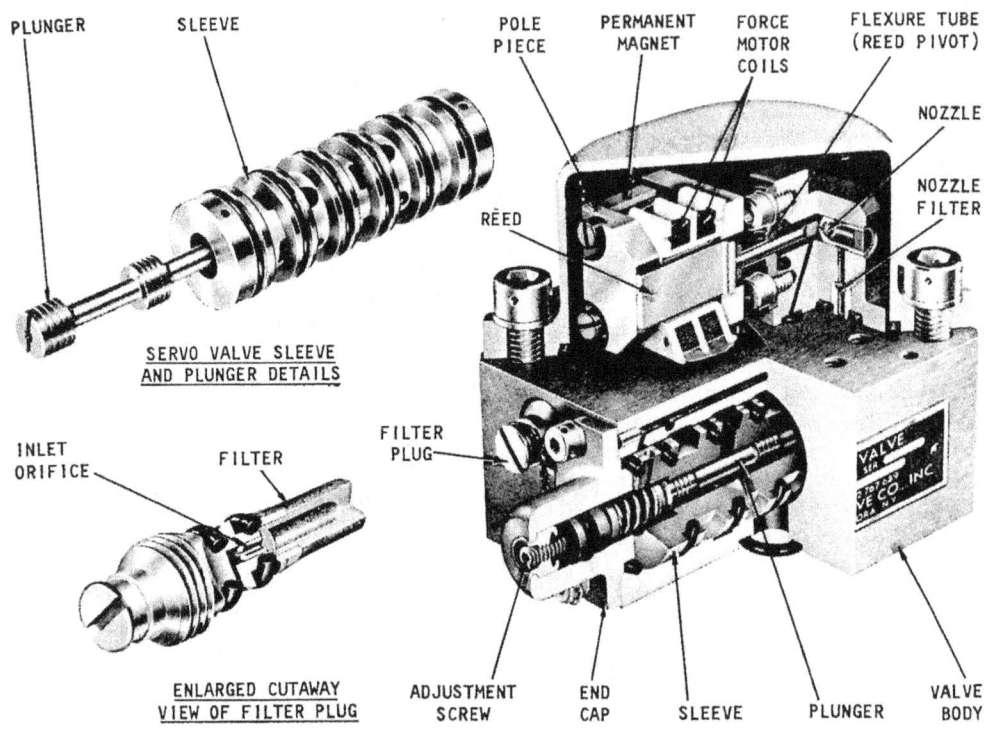

Figure 6-30.—Electrohydraulic servovalve, cutaway view.

## HYDRAULICS IN MISSILES

Since the missiles presently used aboard ship are under a NO-TEST program, you will be interested mainly in whether the hydraulic sump for the missile is leaking or if the sump is low on fluid. The Terrier HTR, BT, the Tartar ITR, and the Sparrow missiles have hydraulic sumps. The Standard ER and MR missiles do not use hydraulics. Instead, an electric tail actuator is used, as explained in chapter 5 of this manual.

Guidance and control are sometimes spoken of as though they were one and the same. They are two parts of the problem of getting the missile to the selected target after it is fired. The main reason for controlling a missile in flight is to gain increased accuracy for missiles with extended ranges.

A missile guidance system keeps the missile on the proper flight path from launcher to target in accordance with signals received from a control point. The missile control system keeps the missile in the proper flight attitude. Together, the guidance and control components of any guided missile determine the proper flight path to hit the target, and control the missile so that it follows this determined path. Missiles accomplish this "path control" by the processes of tracking, computing, directing, and steering. The first three processes of path control are performed by the guidance system; steering is done by the control system.

*GMM 3 & 2*, NET 10199-C, described the external control surfaces of guided missiles such as wings and fins and explained the effects of natural forces acting upon them and how the missile compensates for them. The remainder of this chapter explains the characteristics of some of the numerous mechanical, hydraulic, and pneumatic systems used to control the steering

## Chapter 6—HYDRAULICS AND PNEUMATICS

components to maintain a stable missile flight. We will deal with the general principles rather than the actual design of any specific missile.

A missile control subsystem is a servomechanism. A servomechanism takes an order and carries it out. In carrying out the order, it determines the type and amount of difference between what should be done and what is being done. Having determined this difference, the servomechanism then goes ahead to change what is being done to what should be done. To perform these functions, a servomechanism must be able to:

1. Accept an order which defines the result desired.

2. Evaluate the existing conditions.

3. Compare the desired result with the existing conditions, obtaining a difference between the two.

4. Issue an order based on the difference so as to change the existing conditions to the desired result.

5. Carry out the order.

For a servomechanism to meet the requirements just stated, it must be made up of two systems—an error detecting system and a controlling system. The load, which is actually the output of the servo, can be considered part of the controller.

By means of servosystems, some property of a load is made to conform to a desired condition. The property under control is usually the position, the rate of rotation, or the acceleration of the load. The system may be composed of electrical, mechanical, hydraulic, pneumatic, or thermal units, or of various combinations of these units. The load device may be any one of an unlimited variety; a missile control surface, the output shaft of an electric motor, and a radar tracking antenna are a few typical examples.

### OVERALL OPERATION

Before studying the individual components of the missile control system, let us take a brief look at the operation of the system as a whole. Figure 6-31 shows the basic missile control system in block diagram form. Free gyroscopes provide physical (spatial) references from which missile attitude can be determined. For any particular missile attitude, free gyro signals are sent from the gyroscopes to the computer network of the missile.

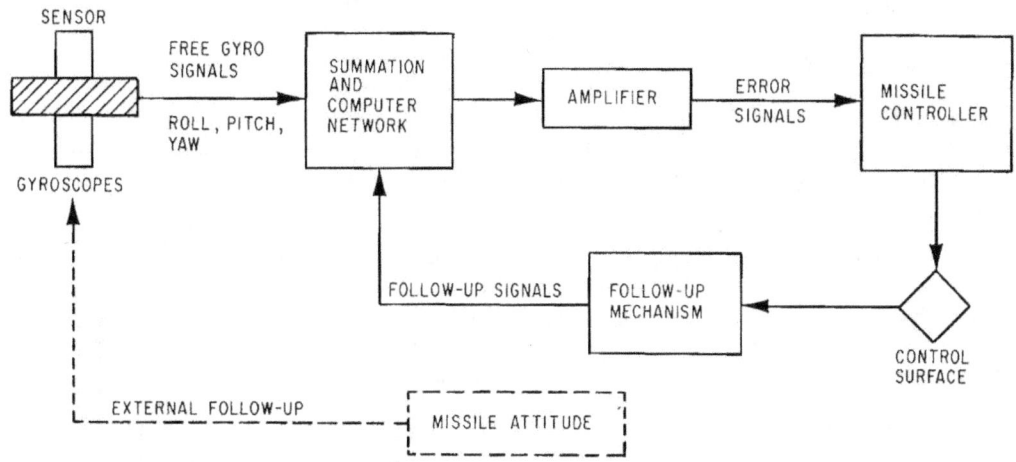

Figure 6-31.—Basic missile control system.

33.62

209

These signals are proportional to the amount of roll, pitch, and yaw at any given instant. After these signals have been compared with other information (for example, guidance signals), correction signals result. The correction signals are orders to the controller to position the control surfaces. The purposes of the amplifier is to build the weak correction signals up to sufficient strength to cause actuation of the controller. As in any closed-loop servosystem, followup information plays an important role. A followup mechanism continuously measures the positions of the control surfaces and relays signals back to the computer network.

In addition to the internal followup which is actually measured by a mechanism, we can think of the missile's movement detecting devices as providing an external followup feature. The fact that the gyroscopes continuously detect changing missile attitude introduces the idea of external followup. This is represented by the dotted line in figure 6-31.

## COMPONENTS OF MISSILE CONTROL SYSTEMS

Figure 6-31 names parts of a missile control system and some of the components which have been discussed. The components may be grouped according to their functions. They cannot be strictly compartmentalized as they must work together; there is some overlapping. Devices for detecting missile movement may be called error-sensing devices. The amount and direction of error must be measured by a fixed standard; reference devices provide the signal for comparison. Correction-computing devices compute the amount and direction of correction needed and correction devices carry out the orders to correct any deviation. Power output devices amplify the error signal, but the prime purpose is to build up a small computer output signal to a value great enough to operate the controls. The use of feedback loops provides for smooth operation of the controls.

Do not confuse the missile control system with the weapons control system. The weapons direction system and the fire control systems and their related components comprise the weapons control system. This shipboard equipment controls all weapons aboard including guns, missiles, and torpedoes. The missile control systems are in the missile and may receive direction from shipboard equipment.

## CONTROLLER UNITS

A controller unit in a missile control system responds to an error signal from a sensor. There are several types of controller units, and each type has some feature that makes it better suited for use in a particular missile system.

### Solenoids

A solenoid consists of a coil of wire wound around a nonmagnetic hollow tube; a movable soft-iron core is placed in the tube. When a magnetic field is created around the coil by current flow through the winding, the core will center itself in the coil. This makes the solenoid useful in remote control applications since the core can be mechanically connected to valve mechanisms, switch arms, and other regulating devices. Two solenoids can be arranged to give double action in certain applications.

### Transfer Valves

Figure 6-32 shows an application in which two solenoids are used to operate a hydraulic transfer valve. The object is to move the actuator which is mechanically linked to a control surface or comparable device.

The pressurized hydraulic fluid, after it leaves the accumulator, is applied to the transfer valve shown in figure 6-32, view B. The valve is automatically operated by the response of the solenoids to electrical signals generated by the missile computer network.

If solenoid 1 in the figure is energized, it will cause the valve spool to move to the left. This will permit pressurized fluid to be ported to the right-hand side of the actuator and cause its movement to the left. If solenoid 2 is energized,

## Chapter 6—HYDRAULICS AND PNEUMATICS

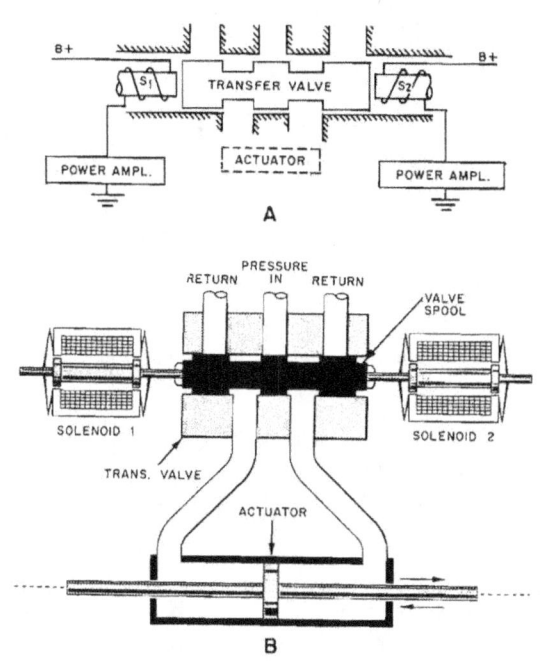

33.184
Figure 6-32.—Transfer valve: A. Closed position (schematic); B. Hydraulic transfer valve and actuator.

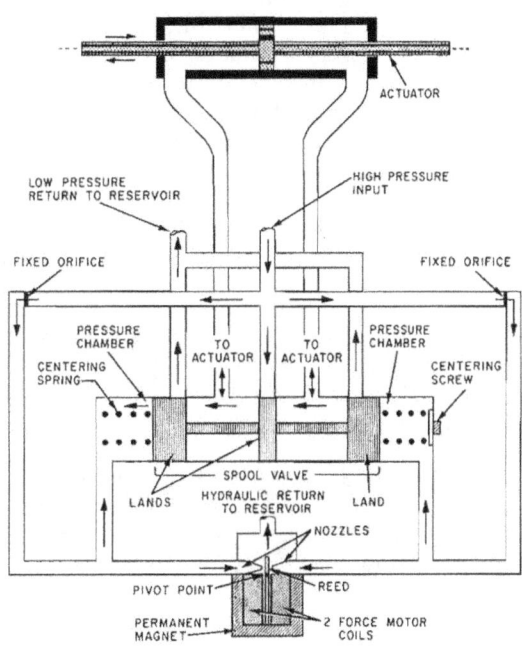

33.185
Figure 6-33.—Servovalve.

the valve spool will move to the right, causing actuator movement to the right in the same manner. When neither coil is energized, the valve is closed (figure 6-32, view A).

The transfer valve just described has one disadvantage in that it operates in an on-off manner. This means that it provides positive movement of the control surfaces, either full up or full down, full right or full left. A finer control is usually more desirable in missile systems. The servovalve (figure 6-33) provides this control. With neither of the windings energized (or a balanced current flowing through both), the magnetic reed is centered as shown (figure 6-33). In this condition, high pressure hydraulic fluid from the input line cannot pass to the actuator since the center land of the spool valve blocks the inlet port. The pressurized fluid flows through the alternate routes, through the two restrictors (fixed orifice), passes through the two nozzles, and returns to the sump without causing any movement of the actuator. If the left-hand force motor coil is energized, the magnetic reed will move to the right, blocking off the flow of high pressure fluid through the right-hand nozzle. Pressure will build up in the right pressure chamber. This will move the spool valve to the left. In moving left, the center land will open the high pressure inlet and permit fluid flow directly to the right-hand side of the actuator. At the same time, the left-hand land of the spool will open the low pressure return line and permit flow to the sump from the left-hand side of the actuator. This process will cause actuator movement to the left. By energizing the right-hand force motor coil, the reed will move to the left, and the entire process will be reversed, the actuator then moving to the right. The actuator can be used to physically position a control surface.

211

## ACTUATOR UNITS

Actuating units use one or more hydraulic or electrical energy transfer methods. Each of these methods has certain advantages, as well as certain design problems mentioned earlier in this chapter. Control devices make use of more than one method of energy transfer but are classified according to the major one used. Combinations are hydraulic-electric and electric. Mechanical linkages or gearing are used to some extent by all of them.

Pascal's law states that whenever a pressure is applied to a confined liquid, that pressure is transferred undiminished in all directions throughout the liquid, regardless of the shape of the confining system.

This principle has been used for years in such familiar applications as hydraulic door stops, hydraulic lifts at automobile service stations, hydraulic brakes, and automatic transmissions.

Generally, hydraulic actuator units are quite simple in design and construction. One advantage of a hydraulic system is that it eliminates complex gear, lever, and pulley arrangements. Also, the reaction time of a hydraulic system is relatively short because there is little slack or lost motion. A hydraulic system does, however, have a slight efficiency loss due to friction.

The hydraulic-electric method of actuating movable control surfaces has been used more than any other type of system. As previously mentioned, the more important advantages of this type of system are the high speed of response and the large forces available when using hydraulic actuators.

You have studied several of the components shown in the simplified block diagram of a hydraulic-electric controller (figure 6-34). This system comprises (1) a reservoir which contains the supply of hydraulic fluid, (2) a motor and a pump to move the fluid through the system, (3) a relief valve to prevent excessive pressures in the system, (4) an accumulator which acts as an

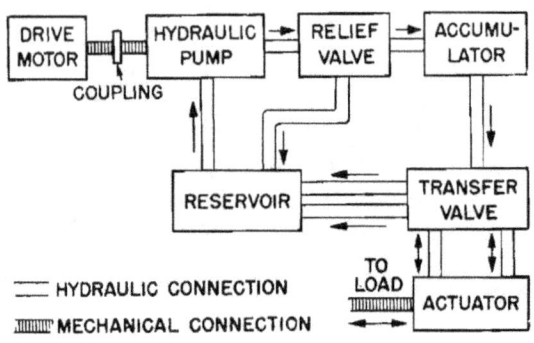

33.182
Figure 6-34.—Basic hydraulic controller.

auxiliary storage space for fluid under pressure and as a damping mechanism which smooths out pressure surges within the system, and (5) a transfer valve which controls the flow of fluid to the actuator.

Most of these components of the system have been covered in the preceding pages. Pumps used in a missile's control system are normally either of the gear or piston type, and are driven by an electric motor. The theory of operation for these pumps is explained in *Fluid Power*, NAVPERS 16193.

## PNEUMATICS IN A GMLS

Air, pressurized and unpressurized, is used in many ways in connection with guided missile launching systems, and in tools and equipment. The effects of air on missile flight were discussed as part of the fire control problem in the preceding manual. This information will not be discussed further here: refresh your memory when necessary by reviewing the preceding manual in this series, *Gunner's Mate M 3 & 2*, NAVEDTRA 10199-C.

The use of pressurized air in the launching system and handling equipment will be covered in this chapter. The dud-jettisoning equipment described in chapter 4 is one example of a pneumatic component you have learned to operate, test functionally, disassemble, inspect,

## Chapter 6—HYDRAULICS AND PNEUMATICS

clean, and lubricate. Your knowledge of this component must now be expanded to include overhaul, repair and adjustment of the equipment, and planning and supervising the maintenance and repair program for the equipment.

The pneumatic parts of the equipment are intricately connected to electrical and hydraulic components, so it is difficult to discuss the pneumatic features separately.

Compressed air is supplied to various systems by high pressure, medium pressure, or low pressure air compressors in the ship's engineering department. Compressed air outlets are located in the spaces where needed, such as checkout and repair spaces. Low pressure is 150 psi or less; medium pressure is 150 to 1000 psi; pressures above 1000 psi are classed as high pressure. Reducing valves reduce higher pressure to a lower pressure for a specific system. Compressed air has many uses aboard a modern Navy ship such as for operating pneumatic tools and handling equipment, charging and firing torpedoes, and for operating the dud-jettisoning unit and other parts of the missile launching system. On most ships the air is dried. If you require dry compressed air, as for blowing out or drying out electrical components, check to be sure that the air at the outlet is dried. Use only rubber or insulating hose in portable air lines for blowing out electrical equipment. Also, pressure must be low, not over 30 psi on motors and generators up to 50 horsepower or 50 kilowatts.

## AIR DRIVEN MISSILE HANDLING EQUIPMENT

In this section we will discuss the chain drive fixture used on the Mk 13 and Mk 22 GMLSs and the adapter loading fixture assembly (for ASROC) used in the Mk 10 Mods 7 and 8 GMLSs. It is important that dirt, metal chips, filings, and other extraneous material be kept out of an air system. Moisture in an air system is hard to control and will ultimately cause premature failure of air motors. Once an air motor seizes up, you will have to disassemble it. Ensure that before dismantling any component, the air supply is turned off, bled off, and tagged out. Do not let grease or oil accumulate on the flexible lines.

### Chain Drive Fixture

An air-driven chain drive fixture with a manual air control valve, and a strikedown hand control box, is used for strikedown and for offloading Tartar missiles. The chain drive fixture is used with the Mk 13 and Mk 22 GMLSs. Figure 6-35 shows the fixture attached to a launcher. The launcher captain, using local control operation, positions the launcher to a convenient position to attach the missile handling equipment.

The chain fixture is attached manually to the front of the launcher when preparing for strikedown or offloading. When the latch lever (figure 6-36) is pushed, the quick-release pins can be inserted to attach the fixture to the launcher guide. The latch engages a block on top of the retractable rail.

Inside the fixture housing, or attached to it, are a chain, an air motor, a chain drive sprocket and gears, a pressure regulator, and an air throttle valve (figure 6-36). The strikedown chain pulls the missile onto the launcher guide during strikedown and controls the missile during offloading. There are four cams in the chain which actuate linkages to the throttle valve and interlock switch SIN2. The stop cam stops the air motor (through linkage to the air throttle valve) when the chain is fully retracted. The air motor shaft drives a simple gear train which drives the chain drive sprocket.

The pressure regulator reduces air pressure in the extend cycle of the chain. It is mounted in parallel with a check valve in the air line between the throttle valve and the air motor (figure 6-36). The regulator is factory adjusted to a static pressure of 20-22 psi which must not be changed.

The air throttle valve regulates the speed of the air motor and determines its direction of rotation. Two inlets are connected to the manual control valve (figure 6-35) and one to the ship's air supply. Two outlets connect to the air motor and two others port exhaust air to the atmosphere. Cams on the chain shift the valve through linkages to open or close air inlets or

213

# GUNNER'S MATE M 1 & C

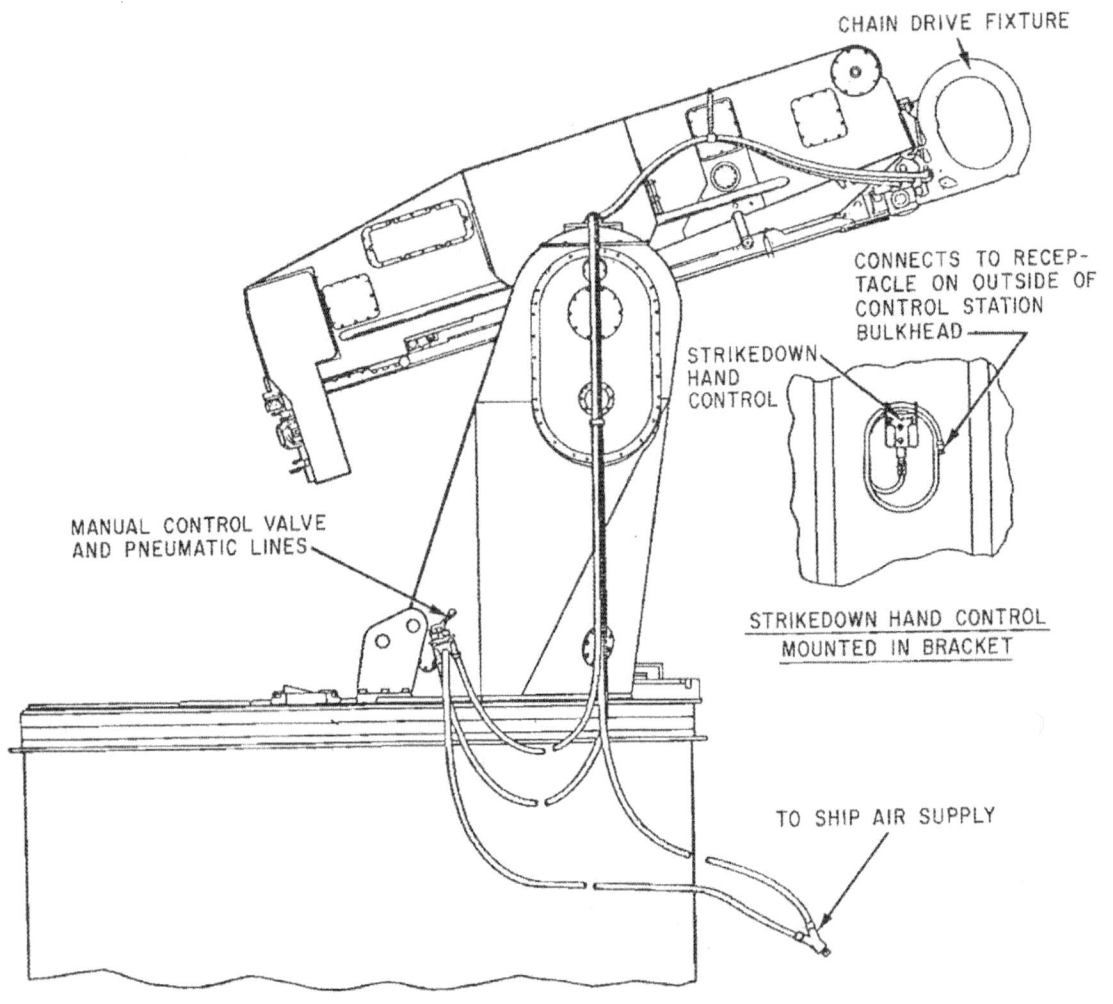

Figure 6-35.—Strikedown gear.

outlets and thus control the speed of the air motor and chain.

The manual control valve (figure 6-35) ports air pressure to the air throttle valve to shift it to retract or extend the strikedown chain. The position of the control handle on the manual control valve for "retract" or "extend" has to be determined by trial (for each ship installation) and then marked. When not in operation, the plunger is centered to "neutral" by a double-acting spring.

The hand control box (figure 6-35) positions the launcher in train and elevation for strikedown, checkout, or missile component removal. The launcher captain operates the hand control box on deck where the launcher and the operations are in full view. To position the launcher for mounting the strikedown gear and chain drive fixture, the EP2 panel operator and the launcher captain follow the procedure as for checkout operation. The launcher is trained to a convenient position by local control, and the

Chapter 6—HYDRAULICS AND PNEUMATICS

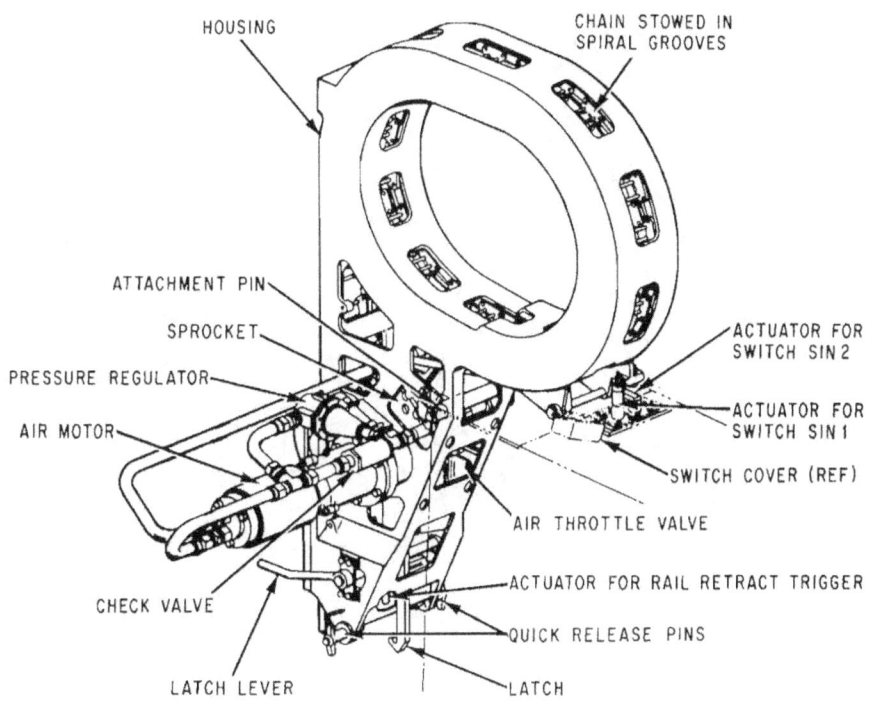

Figure 6-36.—Chain drive fixture, strikedown gear, Mk 13 and Mk 22 guided missile launching systems.

guide is lowered to zero degrees elevation. The firing safety switch handle must be removed from the EP2 panel before anyone is permitted to begin mounting the strikedown gear to the launcher. This is to make certain that the launcher cannot be started while someone is working on it. When the fixture is attached and air line hoses are connected (two hoses between the throttle valve and the manual control valve, one between the throttle valve and the ship's supply Y, and another between the manual control valve and the ship's supply Y), the firing safety switch handle can be returned to the EP2 panel and the system reactivated.

Consult the publication for the Mk 13 GMLS (OP 2665) for a complete description of the steps in strikedown, offloading, checkout, and deactivation. OP 3115 is the publication to consult regarding the Mk 22 GMLS.

### ASROC Loading Fixture

Before an ASROC weapon can be stowed in the magazine, it must be placed into an adapter rail. The loading fixtures (figure 6-37) are mounted to each loader truck assembly in the checkout area and are used to position the ASROC weapon into the adapter rail. The loading fixture has three major components: the stowing mechanism, the drive assembly, and the chain assembly (figure 6-37). The stowing mechanism is mounted to the loader truck in the strikedown/checkout area. It consists of two mounting brackets, a worm, a gear quadrant, four supporting arms, and an extended-retracted latch mechanism. A special handcrank is needed to crank the fixture down from stowed position. The latch mechanism serves to lock the latch pin on the drive in either the drive-retracted or the drive-extended

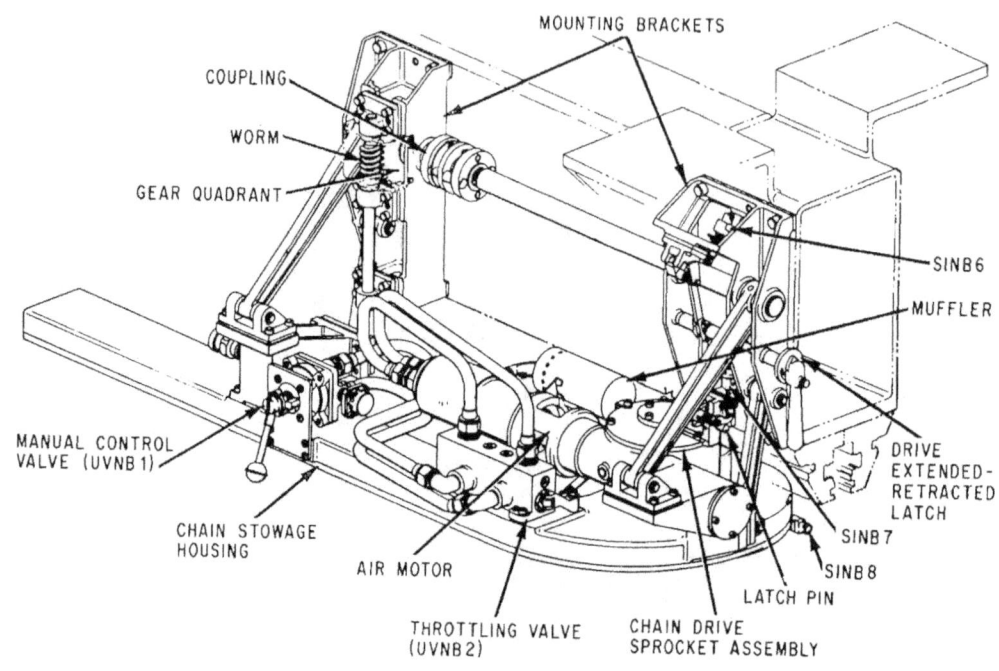

Figure 6-37.—Adapter loading fixture assembly (for ASROC).

position, and to actuate the interlock switch SINB6 or SINB7 to indicate in the loader electrical circuits either the retracted or the extended position of the drive assembly. The latch handle has two positions, LATCHED AND UNLATCHED. It is positioned manually.

The loading fixture drive assembly is mounted on the chain stowage housing (figure 6-37). It consists of a manual control valve, an interlock valve (not shown in figure 6-37), a throttling valve, an air motor, a speed reducer and chain drive sprocket assembly, a chain stowage housing, and chain-retracted interlock switch SINB8.

The manual control valve (figure 6-37) is a three-position valve spring-loaded to OFF, EXTEND, and RETRACT. The throttling valve is a three-land, three-position valve. It is initially shifted by air pressure. It is returned to neutral and held there by spring-loaded linkages. The reversible air motor drives the chain drive sprocket through a worm and worm gear speed reducer.

The J-shaped chain stowage housing (figure 6-37) serves as a mounting base for the drive assembly. The retracting chain is drawn into the housing by the drive sprocket, and as it passes around the sprocket section, it enters the stacking section where the chain is folded link-on-link to stow it. The chain assembly is a rammer-type roller chain and pawl.

The throttling valve is between the manual air control valve and the air motor in the pneumatic circuit (figure 6-38). The chain-extended cam or the chain-retracted cam returns it to neutral after it is activated by the air motor. The loading fixture pawl is pivoted to the end of the chain. The interlock valve is in the pneumatic circuit between the ship's air supply and control valve. It is actuated by a cam that is shifted when the ASROC adapter contacts it.

Chapter 6—HYDRAULICS AND PNEUMATICS

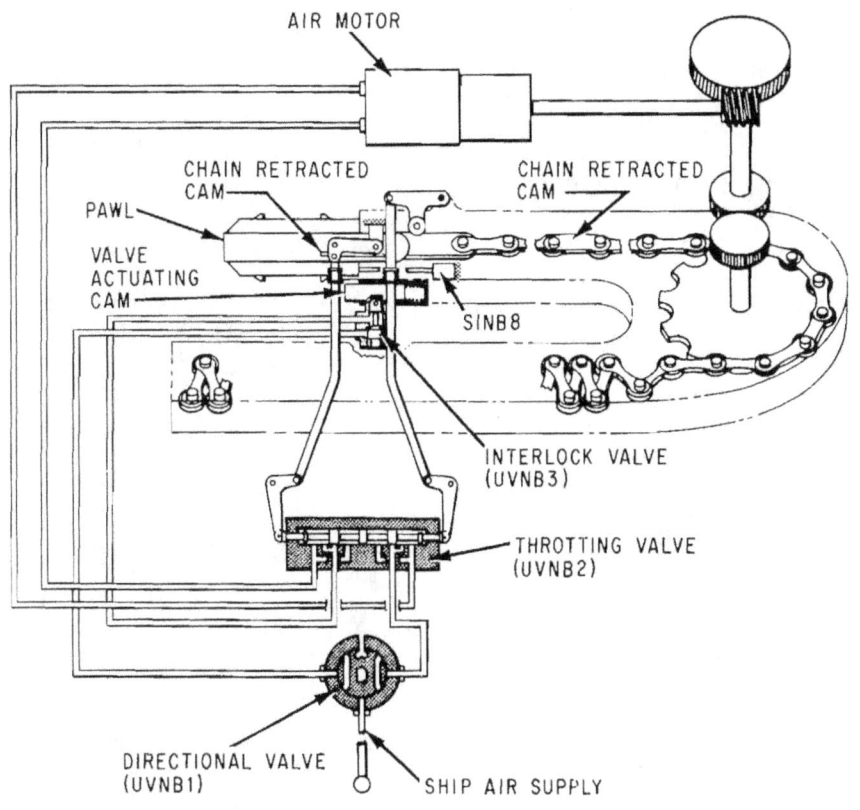

Figure 6-38.—Loading fixture (for ASROC); pneumatic schematic.

To load an ASROC weapon into the adapter rail, the appropriate adapter rail (depth charge or torpedo) is brought up from the RSR and positioned at the #1 stop, and the loading fixture is cranked down manually to a position that aligns with the adapter rail. Before proceeding, ensure that the snubbers on the adapter rail are retracted, and the Mk 10 cable connector is clear for receipt of the weapon. Using the transfer car, position the ASROC weapon under the adapter rail and raise the weapon until its shoes are aligned with the adapter rail. Next, move the carriage of the transfer car approximately 4 inches toward the rear of the adapter rail, ensuring that the weapon's shoes engage with the adapter rail and the loading fixture pawl. The transfer car can now be released from the weapon, lowered, and positioned clear of the work area. Before proceeding, remove the thrust neutralizer from the weapon. Now, position the adapter fixture's loader lever to the EXTEND position. This action causes the air motor on the loading fixture to move the ASROC weapon to its stowed position on the adapter rail. When the weapon's after lug contacts the stop block on the adapter rail, release the loader lever and ensure that the weapon is locked into the adapter rail by resetting and locking the restraining latch. Next, using the manual pump handle, close the snubbers.

NOTE: When closing the snubbers, be sure to follow all applicable safety precautions listed in the OP and OD for the system.

# GUNNER'S MATE M 1 & C

The only loading procedures that remain are to connect the Mk 10 cable assembly to the weapon receptacle and position and lock the cable retractor in the cantilever beam. Then retract the adapter rail, with the ASROC weapon, to the assembly area and stow in the RSR. The final step in this procedure is to secure the loading fixture.

## LAUNCHER AIR DRIVE SYSTEMS

Air motors have been mentioned in connection with the missile component handling crane, monorail overhead air hoist, receiving stands, deck fixtures, and chain drive fixture. They are also used to train and elevate missile launchers in manual control. The air drive motors used on the Mk 10 GMLS are described below. In case of a power loss, the air motors may be used in conjunction with hand pumps and handcranks to perform essential operations with the launching system. For example, the loader chain can be retracted by use of a handcrank. A hand pump can be used to furnish hydraulic fluid directly to a component deactivated by a power loss. The blast doors, for example, can be closed by this means in an emergency.

### Power Drives

Manually controlled air motors are attached to the power-off brakes of the train and elevation systems. If manual operation is to be used, the power for the side (A- or B-side) is turned off at the EP1 panel. Figure 6-39 shows the location of the air drive motor in relation to the power-off brake. The location is similar for train and elevation systems. The air motor drive is used during power failure or during installation and maintenance procedures.

When the air motors are to be used, the power system must be turned off. The air motors drive the associated gear reduction. The air pressure to operate the air motors is supplied from the ship's air lines, using 100 psi. No electrical control is used. An air control valve assembly (figure 6-40) controls the flow of air to

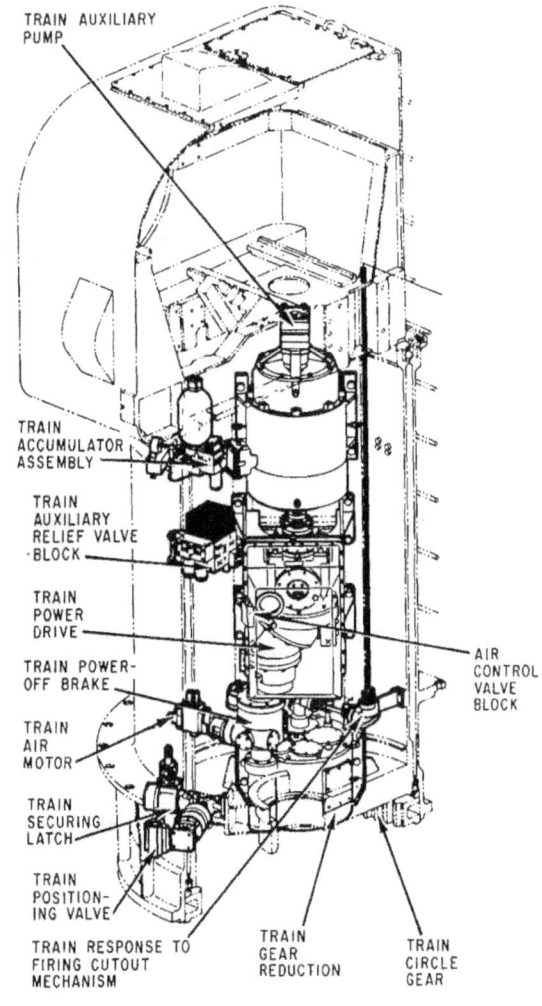

94.146
Figure 6-39.—Train system, general arrangement of components.

the train and elevation air motors. The assembly is fastened to the left side of the base ring, above the train power-off brake (figure 6-39). The valve assembly has two identical sections (figure 6-41); one section controls the elevation air motor, and the other the train air motor.

CAUTION: When operating the launcher with the air motors, normal safety interlocks are

Chapter 6—HYDRAULICS AND PNEUMATICS

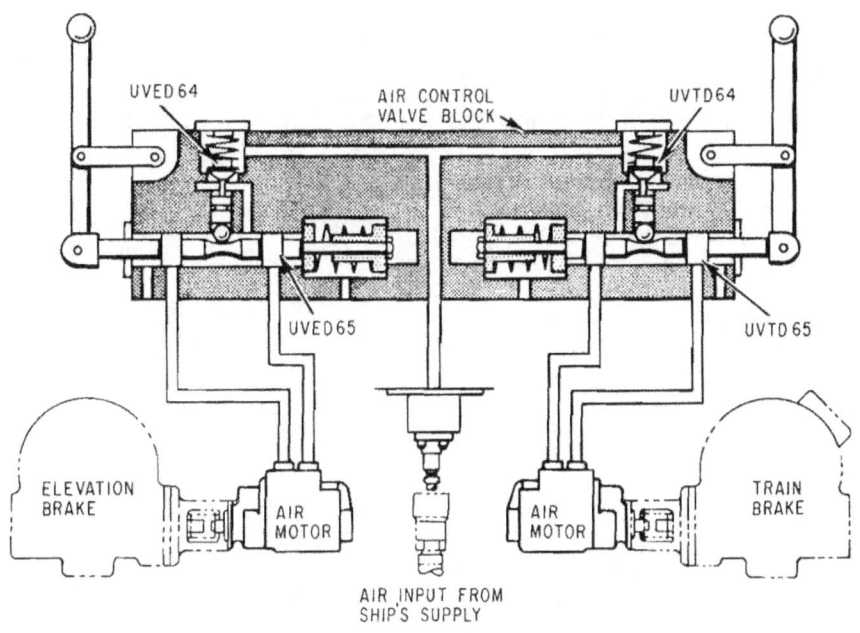

Figure 6-40.—Air control valve assembly.

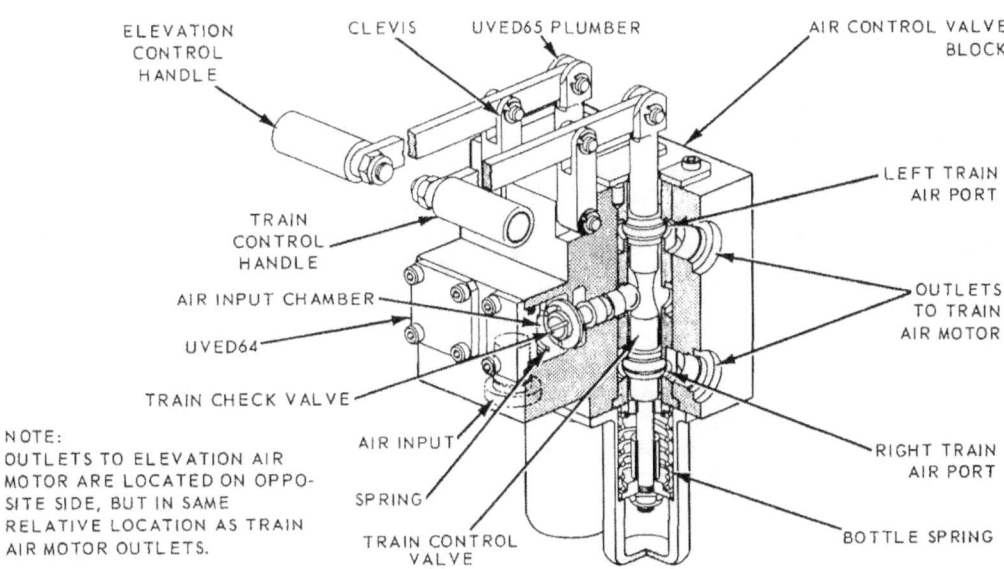

Figure 6-41.—Schematic of air motor drive.

219

bypassed. Use extreme caution; specifically, never move the launcher if the blast doors are open, and never open or close the blast doors by use of the hand pump if the launcher has been moved off the stow position with the air motors.

The train section of the train control valve assembly (figure 6-40) consists principally of a control handle, air control valve, and check valve. The control handle, employed as a first class lever, operates the air control valve. If the handle is moved up and down, air is ported through one of the two outlet ports to the train air motor. The outlet port through which the air is supplied determines the direction of rotation of the air motor.

The air control valve is linked to the control handle at the upper end. The lower end of the valve is attached to a bottle spring that holds the valve at neutral until displaced by the control handle. With the valve at neutral, the two output ports of the train air motor are closed. The check valve prevents passage of air from the supply source to the center chamber of the control valve unless the check valve plunger is unseated.

When the control handle is moved, it forces the train control valve plunger off neutral which unseats the train check valve plunger and allows air to flow through the central chamber of one of the two outlet ports of the air motor.

Before using either the train or elevation air motors, be sure train and elevation latches are retracted. In manual operation, they are retracted by use of the hand pump, and power drives must be off.

If lubrication is scheduled, follow instructions on the MRCs and review safety precautions for manual operation. When you are making use of the air motors, you will also use the hand pumps for hydraulic actuation of components, and handcranks for mechanical actuation. Be sure electrical power is off in each case.

Although we have illustrated and discussed the use of air motors only in the Terrier Mk 10 GMLS, similar air motors are used in the Mk 11 GMLS. For manual operation of train and elevation power drives in the Mk 13 GMLS, a handcrank is attached to the splined end of the worm shaft of the power-off brake assembly.

When repair or maintenance is required on the train or elevation system of a launcher, the air motors are used for moving the launcher. They are also used if the automatic system for operating the launchers is disabled. Air motors, therefore, must be kept in operating condition by proper maintenance, repair, and overhaul. Solenoid-operated control valves, which have identical components in hydraulic or pneumatic control, must actuate on signal. All O-rings and gaskets should be replaced with new ones whenever the valve is overhauled. The caution to remove gaskets carefully is intended chiefly as a caution against scratching or gouging the seat for the gasket. Take the valve to a dirt-free area to disassemble, clean, replace parts, and reassemble. Do not use waste for cleaning; use clean, lint-free rags. Be sure to secure all power to the launcher before removing any parts.

WARNING: Position the main power circuit breaker and the train and elevation air drive motor manual control valves at OFF. Place warning tags on these controls.

To remove the air drive motor, disconnect two air lines. Plug lines to prevent entrance of dirt or foreign particles.

Air motors used in Tartar and Terrier launching systems are of similar construction. Obtain and study the maintenance instructions for those on your equipment before attempting repair work on them.

Assembly is the reverse of disassembly. Do not disassemble any more than necessary. Careful alignment and snug fit are important; frequent disassembly tends to destroy these. If replacement of the head gaskets or the rear gasket is necessary, scribe the cylinder and cylinder-to-rear housing for alignment or reassembly. Make alignment scribe marks on the head, distributor, and drive shaft before disassembling these parts from the head. These scribe marks are important because rotation of

## Chapter 6—HYDRAULICS AND PNEUMATICS

the distributor by 180° changes motor rotation to the opposite direction.

Use new gaskets and new cover plate screws when reassembling, and replace any pitted or worn balls, and worn oil seal.

### Air Lubricators

Air lubricators (figure 6-42) are of the micro fog type which convert the oil to a vapor which is carried along with the air to give internal lubrication to the air drive motors. One lubricator supplies oil vapor to the train and elevation air drive motors; the other supplies the guide pneumatic cylinders. Do not disturb the factory adjustment of the lubricators unless you are positive a malfunction is caused by lubricator maladjustment. The oil level may be seen through a sight glass on the lubricators; if the level falls to the lower third, replenish the oil (check NAVSEA drawing for correct type).

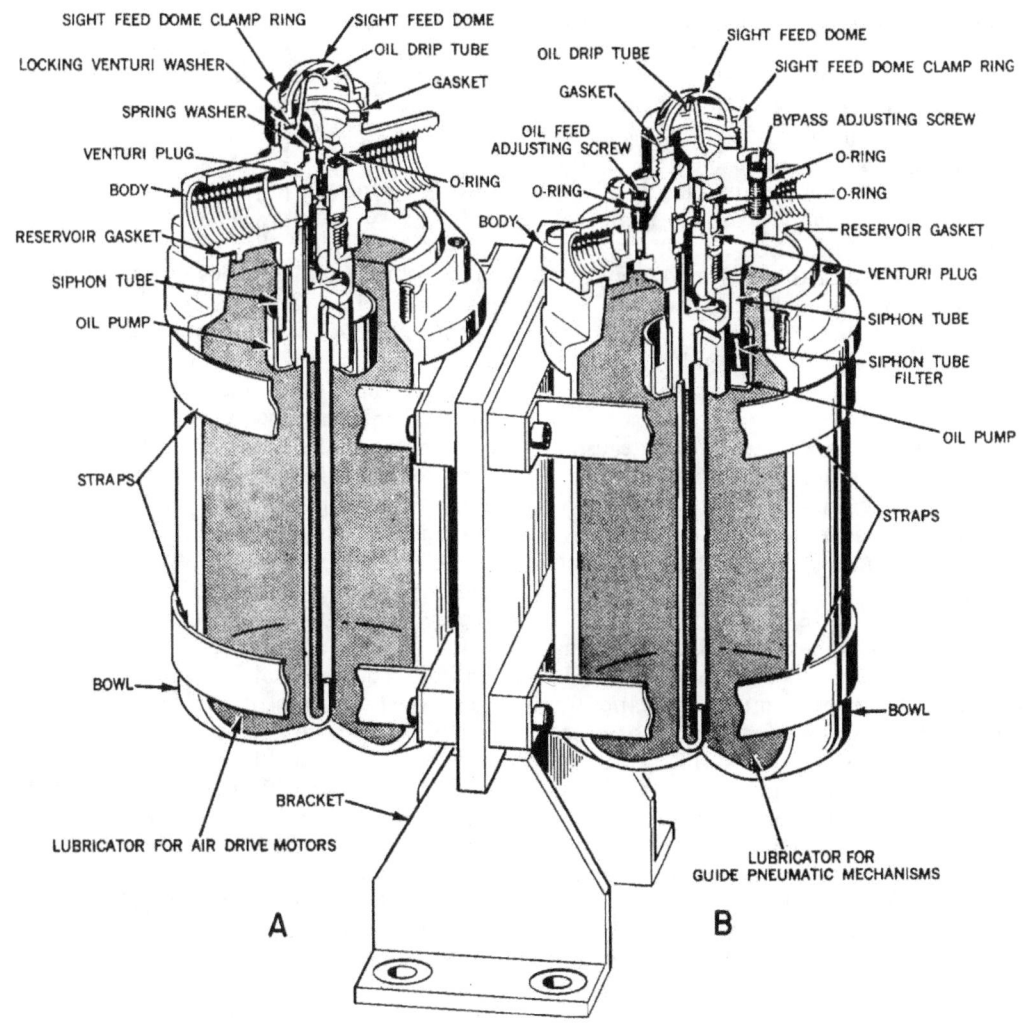

Figure 6-42.—Air lubricators, cutaway view; types A and B.

Remember to secure all power to the launcher before doing this.

The air-driven chain drive fixture used with the Mk 13 GMLS for strikedown, described earlier in this chapter, also has an air lubricator in the air supply line to the air motor, air throttle valve, and manual air control valve components. Although not always mentioned, other air motors also have air lubricators to supply air that lubricates.

The air lubricators, which provide a fine mist of oil to the air motors, sometimes need cleaning or adjustment. Check the fluid level monthly and replenish with the proper grade of oil if the level falls to the lower third of the sight glass on the lubricator. Check the applicable drawing for your equipment. For example, drawing 1600594 (for the Mk 5 Mod 3 guided missile launcher) directs that train and elevation air motors need to have their reservoirs filled after 4 hours of operation. Symptoms of maladjustment are:

1. Oil accumulation in the guide (or near the lubricators).
2. Noisy air drive.
3. Sticking pneumatic cylinders causing jerky piston operation, binding, or slamming.
4. Loss of power in air drive motors.
5. Lubricator has to be filled too often.
6. Lubricator needs no oil after extended launcher operation.

Each air lubricator has an oil feed adjusting screw (figure 6-42) by which you can adjust the drip to 8 to 10 drops per minute (this varies for lubricators in different systems). The position of the plate under the sight feed dome should be checked; remove the clamp ring and the sight feed dome. The bottom horizontal plate in the sight cavity should have the cast arrow pointing to position B for full air flow to the air motor and the lubricators. Adjust the plate with a screwdriver if it is not in position B. Type B lubricators have a bypass screw for adjusting the air flow (figure 6-42, view B).

Operate the equipment to check the success of the adjustment. Several adjustments may be necessary before you achieve good results.

Although air lubricators are of rather simple construction, adjustments must be made with care. It is NAVSEA equipment if it is installed on the ship's air lines.

Air control valves may require adjustment or overhaul. Any time a control valve is disassembled, all O-ring and backup seals should be replaced.

## OTHER USES OF PNEUMATICS

The origin of compressed air used in various equipment is the ship's compressed air supply. Inline reducers are used to lower the ship's low pressure (LP) air to a usable valve for the air bladders.

### Dud-Jettisoning Equipment

Chapter 4 described and illustrated methods of jettisoning dud missiles in Tartar and Terrier systems. In the Mk 10 Mods 7 and 8 Terrier/ASROC systems, the same jettisoning equipment is used for all types of missiles loaded. The three main components of the jettison unit are the A-side ejector, the B-side ejector, and the control panel. A manually operated shutoff valve on the panel shuts off the 4500-psi air supply to the jettison controls. The positioner air supply valve admits (or shuts off) air at 100 psi from the ship's air supply to the A and B positioner control valves. The 4500-psi air supply is used to charge the air chamber (figure 6-43) and to operate the firing valve. The air pressure gage is usually set at 3500 psi, and the panel operator cuts off the air supply when this pressure is reached. This is done by moving the operating lever of the charge and fire control valve to the READY position. (Although there are some differences in control panels for different systems, the one shown in figure 4-3 is typical, and can be referred to for location of parts.) The B ejector controls are on the right-hand side of the panel, and duplicate controls are on the left-hand side for the A ejector. Each side had four indicating lights, a positioner control valve, a charge and fire control valve, and a pressure gage. The shutoff valve and the positioner air supply valve control both sides.

## Chapter 6—HYDRAULICS AND PNEUMATICS

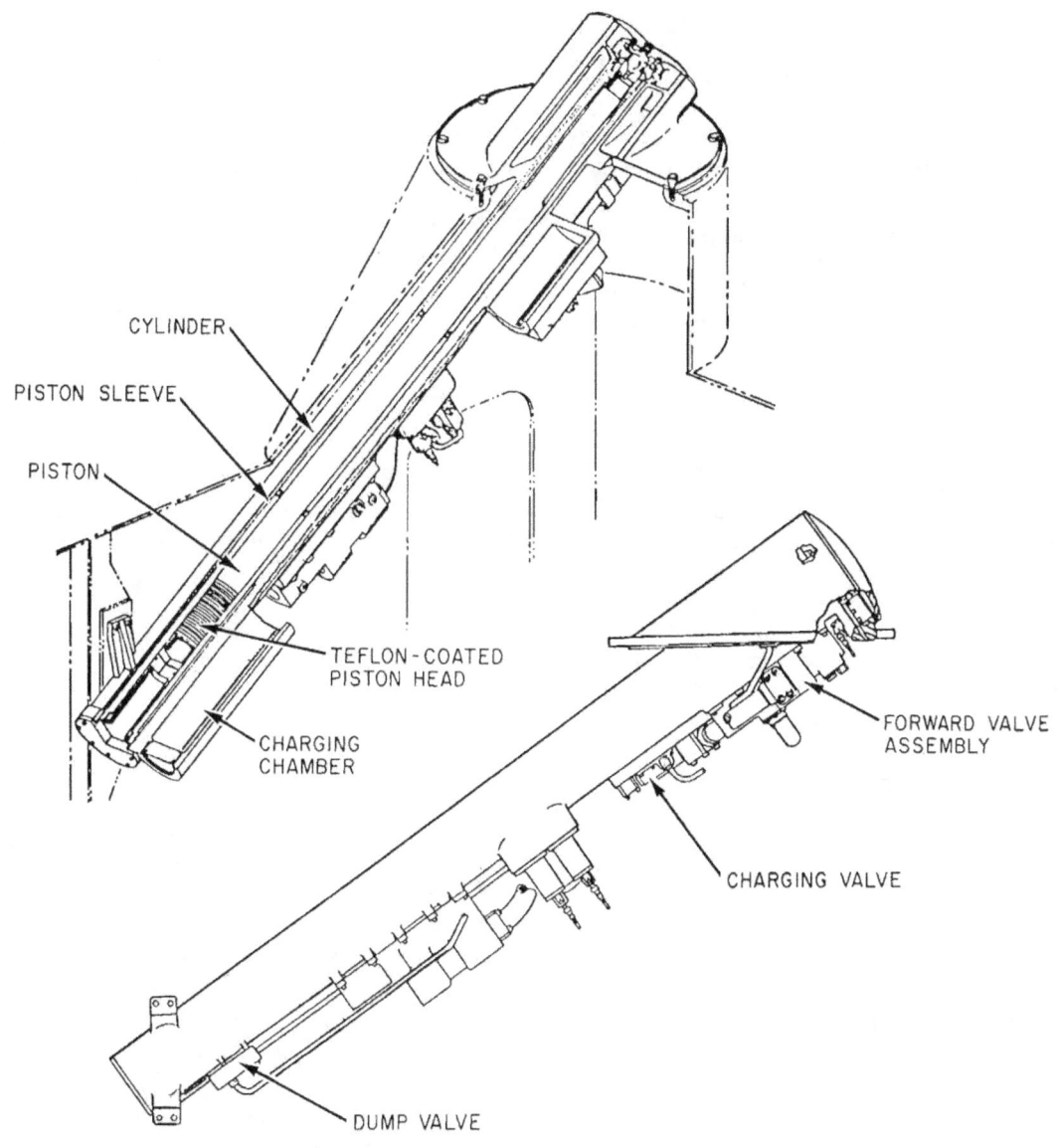

Figure 6-43.—Dud-jettison ejector.

Figure 6-42 shows the location of some of the valves and other components on the cylinder weldment that are actuated when you move levers on the control panel. The dud jettison order synchros, of which there are four for each unit, are mounted in a housing in the EP2 panel. The rotors of the dud jettison order synchros are adjusted and fixed at predetermined positions so they train and elevate the launchers to preset dud-jettison positions when the launcher captain positions SMY1 train and elevation control switch to DUD JETTISON and SMX3 A or B selector switch to the desired rail. The dud jettison normal relay (KCB1) and associated electrical circuitry must be energized before the ejector will move to POSITION I (POSITION II is no longer utilized on the Mk 10 launching systems).

When the operator of the dud-jettison panel is signalled that everything is ready at the EP2 panel, the steps listed on the dud-jettison panel's instruction plate are executed. The operator first turns the positioner air supply valve to OPEN which directs 100 psi air to the positioner control valves. Since only POSITION I is to be used, the positioner control valve (in the upper right-hand corner of the panel) is turned to POSITION I. The dump valves (figure 6-42) then port 100 psi air pressure to the rear of the ejector sleeve which causes the sleeve to move forward about 24 inches. Meanwhile, 4500 psi air is going to the charge and fire control valve which you set at CHARGE. Air is ported to the firing safety valve (not shown), the charging valve (figure 6-42), and both ends of the firing valve which is in the forward assembly.

When the indicating needle on the pressure gage reaches the stationary needle (which was preset at 3500 psi), move the lever of the charge and fire control valve to READY. This shuts off the 4500-psi air supply so the pressure will not go higher.

Check all the indicating lights on the panel and the pressure gage to be sure everything is in readiness, and observe the inclinometer which is adjacent to the control panel. If the roll of the ship is more than 20°, the weapon must be ejected only on the downroll in the direction the weapon is pointing.

Now move the charge and fire control lever to JETTISON AND OFF. The lines to the firing safety valve, the charging valve, and the spring-loaded end of the firing valve are vented to atmosphere. Charging pressure is vented from the firing valve to the check valve and the shuttle valve in the forward valve assembly. This supplies air to both forward and rear ends of the ejector piston and to the plungers and the check valve in the rear valve assembly. As the piston moves forward, the pan at the front contacts the booster. The resistance causes an immediate pressure buildup behind the piston which causes the control valve in the rear valve assembly to open and send air behind the piston to force the missile overboard. The rate of piston movement is controlled by orifices and the position of the control valve. In practice sessions, when you are not using a missile, the control valve assures that the piston will move at about the same rate as if loaded.

### Anti-icing Systems

Anti-icing systems keep vital areas of the launching system ice-free during freezing conditions. The anti-icing fluid is circulated by a motor independent of the rest of the launching system. Air bladders in the heat exchanger tank (figure 6-44) maintain a constant head of pressure on the anti-icing fluid, compensating for expansion and contraction of the fluid under varying temperature conditions.

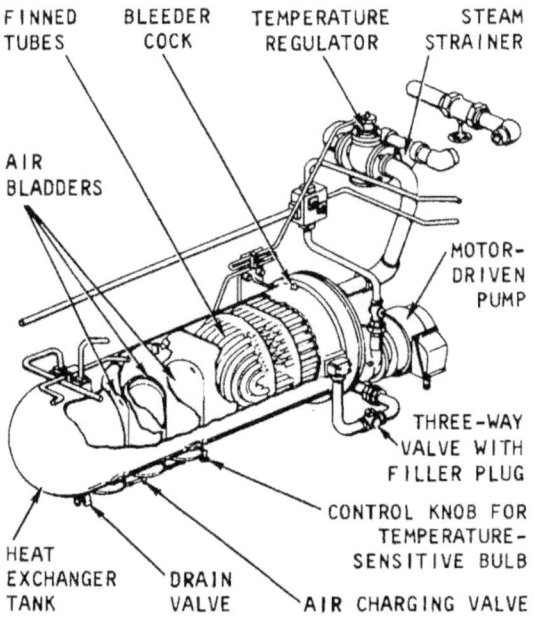

94.152
Figure 6-44.—Major components of anti-icing system.

## Chapter 6—HYDRAULICS AND PNEUMATICS

The heat exchanger tank and components shown in figure 6-44 are the type used in the Mk 10 launching system. The anti-icing heater tank shown in figure 6-44 is used in the Mk 13 launching system. The principles of operation are the same for both systems. The same special tool, air-charger-and-gaging assembly, is used to check the air pressure in each bladder and to charge it to the correct amount. The frequency of checking and charging the air bladders will depend a great deal on the temperatures in which the ship is operating.

The anti-icing system shown in figure 6-45 consists of an electric motor, a centrifugal pump, a fluid heater, a transfer manifold, 16 flow control valves, and nine air bleed valves.

Circulating system motor BPX3 is a constant speed 1-1/2 horsepower unit. It drives a centrifugal pump that circulates heated anti-icing fluid from the heater tank to 16 anti-icing circuits.

The fluid heater uses ship's steam to maintain the returning anti-icing fluid temperature at approximately 40°F. The amount of steam that the fluid temperature regulator allows to enter the finned tubes of the heater determines the temperature of the fluid. A bulb (partially filled with a volatile liquid) expands or contracts in the regulator to control the flow of steam. The three bladders in the heater are charged with air by means of the air charging valve. The bladders maintain pressure and allow fluid expansion within the anti-icing system.

The flow control valves, one for each of the anti-icing circuits, regulate fluid flow from the circuits into the return lines. Each valve is set for a different flow rate (depending on the area to be heated) so that the temperature for all the areas is approximately the same.

The air bleed valves (eight on the topmost part of the anti-icing circuits, one on the heater

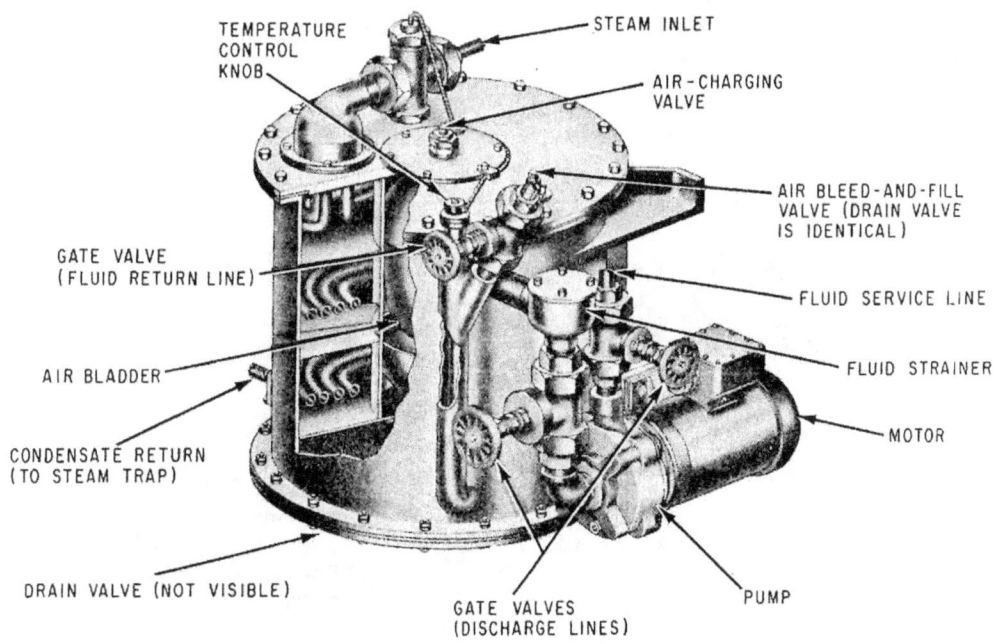

94.153

Figure 6-45.—Anti-icing heater tank, Mk 13 launching system (Tartar).

tank) allow trapped air to be bled out of the anti-icing system. Air in the system causes air locks that hamper the flow of fluid.

Figure 6-46 shows the charging and gaging assembly used for air-charging and gaging the anti-icing bladders in the Mk 10 GMLS. The bladder pressure for the anit-icing system on the Mk 10 GMLS is 10 psi. Check the requirement for the system you have aboard.

Compressed air is also used in the compression tank for the water injection system in the Mk 11 and Mk 13 GMLSs described in the next chapter.

Accumulators in the hydraulic system of the missiles and the launchers are pressurized with nitrogen.

## Thermopneumatic and $CO_2$ Sprinkler Systems

A thermopneumatic control system is designed to actuate a magazine's sprinkler system in response to either a rapid rate of rise in temperature or a slow rise to a fixed temperature in a protected space. The automatic thermopneumatic system is installed as an adjunct to a hydraulic control wet or dry type of magazine sprinkling system or an independent carbon dioxide ($CO_2$) system. Some missile magazines also use a water injection system containing a compression tank which supplies freshwater under air pressure as part of a sprinkler system. The basic procedures for testing, operating, and maintaining magazine sprinkler systems are explained in *GMM 3 & 2*, NET 10199-C, and chapter 7 of this text. An essential reference for the maintenance and repair of a magazine sprinkler system is the Instruction Book for Magazine Sprinkling Systems, NAVSHIPS 0348-078-1000.

The pneumatically operated components of a $CO_2$ system are heat sensing devices and pneumatic control heads; for a sprinkling system, they are heat sensing devices and pneumatic release pilot (PRP) valve.

The heat sensing devices detect temperature increases and transmit pneumatic pressure changes to a PRP valve or $CO_2$ pneumatic control heads.

A pneumatic control head reacts to pneumatic pressure from a heat sensing device by opening a discharge head and releasing liquid $CO_2$ from a supply cylinder. A control head consists of an air chamber and a diaphragm.

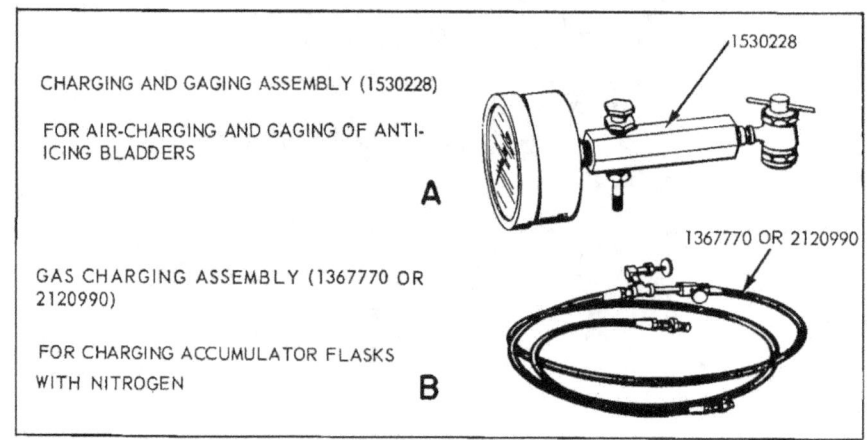

Figure 6-46.—Charging assemblies: A. Charging and gaging assembly for anti-icing bladders; B. Nitrogen charging assembly.

## Chapter 6—HYDRAULICS AND PNEUMATICS

When pressure in the control head chamber increases, the diaphragm expands and trips a lever that releases a trigger mechanism which activates the $CO_2$ fire suppression system.

The PRP valve connects pneumatically to a heat sensing device. In addition to a diaphragm, the PRP valve contains a lever, a spring mechanism, and a compensating vent. The diaphragm expands in response to sudden pressure changes and moves a lever to release a spring mechanism which opens the PRP valve. Saltwater from the ship's firemain flows through the PRP valve and opens the sprinkling system main control valve which admits firemain supply water to sprinkle the magazine. The compensating vent functions to leak off normal temperature fluctuations within the pneumatic piping system and heat sensing device to prevent inadvertent tripping to the PRP valve. The compensating vent is calibrated and adjusted at the factory. No adjustments should be undertaken by ship's force.

Most of the systems installed on board ships consist of separate rate of rise and fixed temperature circuits. The rate of rise circuit uses heat actuated devices (HADs) as sensing devices, and the fixed temperature circuit uses fixed temperature units (FTUs) as sensing devices. A recent modification to some of the pneumatic control systems has been the replacement of HADs and FTUs by heat sensing devices (HSDs). The HSD combines the functions of the HAD and the FTU.

The rate of rise circuit is the primary circuit in most control systems. The operation of the rate of rise circuit is based on the following principles:

1. Air expands when heated.
2. Pressure is created when air expands in a closed system.
3. Pressure can be converted to mechanical energy.

A differential pressure of at least 8 ounces per square inch across the release diaphragm is necessary to trip the PRP valve. A heat sensing device creates pneumatic pressure in two ways.

First, a rapid temperature increase in a missile magazine heats and expands air in a bellows to increase air pressure. Second, a fusible slug melts and releases a spring which collapses the bellows producing a sudden increase in pressure in the pneumatic lines leading to the PRP valve or the $CO_2$ control heads. The pressure is converted to mechanical action by the expansion of a diaphragm. When the diaphragm expands, it

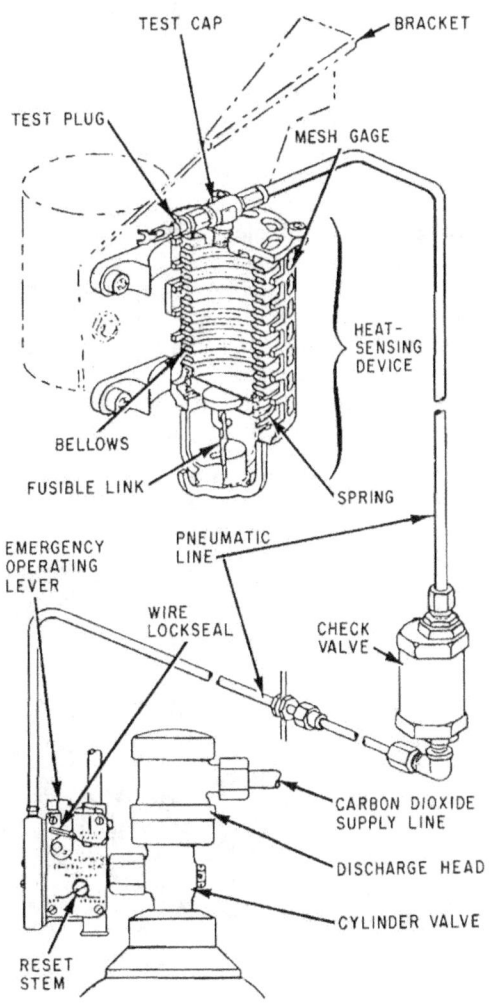

94.183
Figure 6-47.—Carbon dioxide system: pneumatic components.

releases a spring mechanism which opens a valve in the sprinkling system or shifts a plunger in the $CO_2$ system, thus activating a magazine's fire suppression system. Figure 6-47 shows the pneumatic control components of a $CO_2$ system.

## PNEUMATICS IN MISSILES

The Terrier HTR, BT and the Tartar ITR missiles all use air taken in through the missile nose to operate parts of the missile internal system. Figure 6-48 shows a cutaway view of the nose section of a Terrier BT-3 missile. The probe shield is a protective covering to prevent entrance of dust and moisture. The shield is blown away by air pressure against its face when the missile is launched. The nose orifice admits air to the transducer. The gas pressure transducer is a variable-reluctance device that senses total air pressure (static pressure plus pressure caused by missile velocity) and converts it to a voltage which regulates the servo gain in the roll and steering systems to compensate for changes in the control surface effectiveness caused by changes in missile velocity and altitude. The potentiometers associated with the gas pressure transducer are located in the signal control package. Changes in the ram and static air pressure are signalled to the signal control package to effect changes in the missile attitude.

The Terrier HT-3 missile has similar parts with slightly different names. The nose section is called the radome section, and it has a ram pressure probe that supplies the pressure transducer with ram air pressure. The transducer, in turn, supplies an electrical output that drives the servometer. As the servometer turns, it positions the ganged potentiometers as a function of missile ram pressure. The potentiometers act as fractional multipliers for various signals in the guidance computer so the steering system gain is correctly controlled for variations in air density and missile velocity. The same technique is used in Tartar missiles.

NOTE: In the Standard MR and ER missiles, the roll channel of the autopilot contains instruments to sense deviations from correct missile attitude and to establish a continual dither of the tails, resulting in a limit cycle oscillation of the missile. The magnitude of the oscillation, which is directly proportional

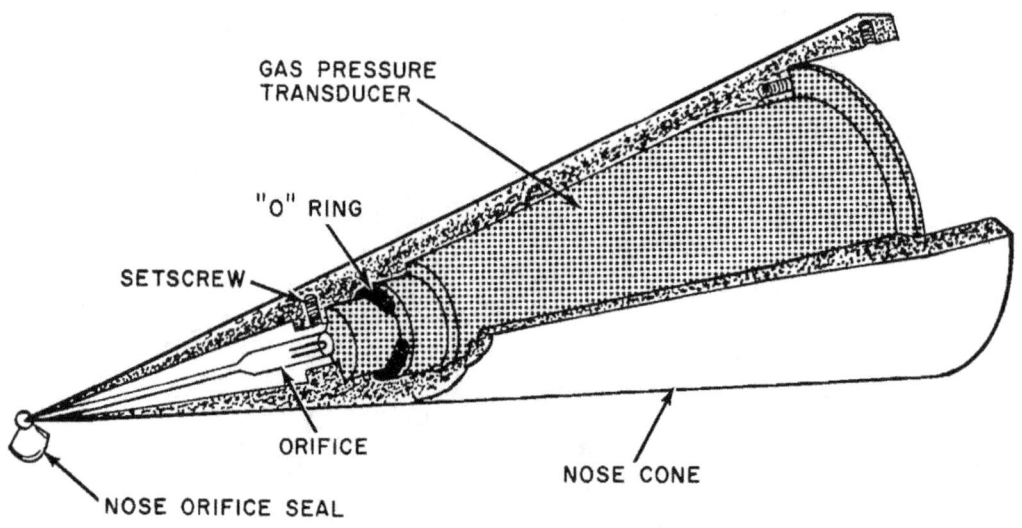

Figure 6-48.—Nose section of Terrier BT-3 missile; cutaway view.

## Chapter 6—HYDRAULICS AND PNEUMATICS

to the aerodynamic gain, controls the gain of the steering channels.

The purpose of this "gain adaptivity" is to cause the steering channels to command greater tail deflections as required for a called-for acceleration from the guidance computer as the missile ascends into thinner air.

## REFERENCES

1. NAVTRA 10200-B, *Gunner's Mate (M) 1 & C.*
2. NAVPERS 16193-B, *Fluid Power.*
3. NAVTRA 10784, *Principles of Guided Missiles and Nuclear Weapons.*
4. NAVEDTRA 10077, *Blueprint Reading and Sketching.*
5. OD 3000, Lubrication of Ordnance Equipment.
6. OP 3114, Guided Missile Launching System, Mk 10 Mods 7 and 8, volume 1.
7. OP 2593, Guided Missile Launching System, Mk 11 Mod 0, volume 1.
8. OP 2665, Guided Missile Launching System, Mk 13 Mods 0, 1, and 2.
9. OP 3115, Guided Missile Launching System, Mk 22 Mod 0, volume 1.
10. OP 4470, Guided Missile Launching System, Mk 13 Mod 4.

# CHAPTER 7

# WEAPON SURVEILLANCE AND STOWAGE

Basic information on explosives, with some definitions of terms used in relation to explosives, is contained in the preceding manual, *Gunner's Mate M 3 & 2*, NAVEDTRA 10199-C.

This chapter will amplify your previously gained safety knowledge regarding missile surveillance and stowage requirements, with emphasis on handling and stowage hazards of ammunition and explosives.

## SAFETY CONSIDERATIONS

Supervisory personnel must ensure that all safety precautions set forth in OP 4, Ammunition Afloat, and related ordnance reference publications are strictly observed during shipboard handling and stowage operations. Ashore handling and stowage regulations are set forth in OP 5, Ammunition and Explosives Ashore.

Explosives are intended to be destructive. While some are more dangerous than others, all explosives must be treated with respect. Since familiarity with any work is apt to lead to carelessness, all supervisory personnel who are concerned with the inspection and use of explosives shall—

1. Exercise the utmost care that all regulations and instructions are observed.
2. Carefully instruct and frequently warn those under them of the necessity for using the utmost care in the performance of their work. A lack of vigilance is not permissible.
3. Explain to their subordinates the characteristics of the ammunition, explosives, and other dangerous materials; the equipment; the precautions to be observed; and the hazards of fire, explosion, and related catastrophies safety precautions are intended to prevent.

Supervisors are required to maintain high standards of good housekeeping in ordnance spaces. All ammunition and explosives shall be protected from excessive temperature, humidity, and hazards of radio frequency (rf) energy. Moisture and/or heat extremes may increase the hazards of ammunition and explosives or degrade their operational performance.

In each weapon space where missiles are stowed or handled, or where missile equipment is operated, such safety orders as apply should be posted in conspicuous places. Conditions not covered by these safety instructions may arise which, in the opinion of the supervisor, may render missile stowage or missile handling unsafe. The supervisor may at any time use such additional safety instructions as may be deemed necessary.

## RADIO FREQUENCY (RF) HAZARDS

Modern radio and radar transmitters produce high intensity radio frequency (rf) fields. Such fields can cause premature actuation or degradation of sensitive electroexplosive devices (EEDs) contained in weapon systems. Rf energy can also ignite flammable fuel-air mixtures and cause biological damage to exposed personnel. Because of these dangers, NAVSEASYSCOM has established a safety program aimed at

## Chapter 7—WEAPON SURVEILLANCE AND STOWAGE

controlling rf hazards. OP 3565/NAVAIR 16-1-529, Radio Frequency Hazards to Ordnance, Personnel, and Fuel is the primary Navy reference publication source. This manual provides information about the Navy's hazards of electromagnetic radiation to ordnance (HERO) program. The HERO program sets forth rf problem areas, establishes EED susceptibility classifications and prescribes control measures for ordnance in rf energy environments. Once safety and reliability requirements are established, all EED-contained ordnance is assigned a HERO SAFE, HERO UNSAFE, or HERO SUSCEPTIBLE classification based on design features, environmental exposure, and handling and testing configuration. These classifications are readily available as tables in OP 3565 and should be consulted prior to any ordnance loading, handling, or testing evolution. In this regard, both ships and stations are required by OP 3565 to develop and maintain a HERO emission control (EMCON) bill which must be activated when exposed ordnance is present. The HERO EMCON bill therefore provides a plan to control rf radiation depending on location and HERO classification of ordnance. The HERO EMCON bill sets forth radiation hazard (RADHAZ) safety device requirements plus HERO conditions which allow the continuation of shipboard operational commitments. Operating personnel shall be fully cognizant of all the HERO requirements set forth in OP 3565 and HERO EMCON bill.

Rf transmitting systems, with high power output and high gain antennas located throughout a ship, have increased the possibility of biological injury to personnel working in the vicinity of radiation patterns. Effects of overexposure is thermal in nature due to body or organ absorption of rf waves. Nonthermal effects have not been fully determined. Radiation safety precautions for personnel are included in OP 3565.

OP 3565 also sets forth safety precautions for fueling operations since volatile fuel-air mixture may be present. The hazards of rf energy arcing, therefore, need to be considered in an rf safety program.

## WEAPON INSPECTIONS

Receipt (acceptance) inspections of all explosive components are necessary.

If you are stationed ashore, you may be concerned with individual components such as igniters, fuzes, and auxiliary power supplies. A checklist is essential to properly conduct a receipt inspection regardless of the component. Memorizing the operational steps on a checklist accomplishes very little. Becoming too familiar with any task involving explosive ordnance is a dangerous proposition. Before you handle any component, inspect the safety device to be sure it is in the safe position. If you are assigned to one of the Mk 10 GMLSs on board ship, refer to NAVSEASYSCOM OP 4093 which provides documentation support for all shipboard Terrier/Standard (ER) missiles (conventional and nuclear) and training rounds.

The purpose of OP 4093 is to (1) update and consolidate Terrier nuclear weapon handling procedural guides; (2) list a number of reference documents that are used simultaneously in shipboard missile handling operations; (3) reduce redundancy; (4) eliminate conflicting data; and (5) add detailed handling and maintenance procedures and checksheets. This OP takes precedence over all documents that cover the handling of Terrier/Standard (ER) missiles, except when procedures contained in OP 4093 for handling BTN missiles conflict with those in the special weapons ordnance publications (SWOPs); then the SWOP takes precedence.

OP 4093 consists of five volumes and the arrangement is as follows:

Volume 1 presents information applicable to all Terrier ship classes. The volume introduces the reader to shipboard operations involving the Terrier/Standard (ER) missiles and training rounds, and provides continuity to the safety descriptions and detailed missile handling and maintenance procedures presented in volumes 2 through 5. Chapter 1 generally describes commonality to all ships; typical shipboard systems and their roles in the handling and use of missiles; safety requirements applicable to conventional and nuclear missiles and to missile handling operations; and the various missile types, their capabilities and limitations. Chapter

1 also provides tabular listings of reference data. Chapter 2 describes physical features and highlights functional purposes of the missiles and training rounds, including their propulsion system, and the containers and handling equipment used to strikedown/offload the complete round. Chapter 3 briefly describes maintenance requirements for the missiles and training rounds on a combatant vessel, supplementing the detailed maintenance procedures in volumes 2 through 5.

Volume 2 describes the launching and handling equipment and contains the detailed missile handling and maintenance procedures and checksheets for use aboard CG, CGN-25/35, and DDG classes of ships. The volume also contains, in addition to a safety summary and specific safety precautions, a safety briefing which is required prior to performing missile handling procedures.

Volume 3 describes the launching and handling equipment and contains the detailed missile handling and maintenance procedures and checksheets for use aboard CGN-9. In addition to a safety summary and specific safety precautions, the volume also contains a safety briefing which is required prior to performing missile handling procedures.

Volume 4 describes the launching and handling equipment and contains the detailed missile handling and maintenance procedures and checksheets for use aboard CV-66. The volume also contains, in addition to a safety summary and specific safety precautions, a safety briefing which is required prior to performing missile handling procedures.

Volume 5 describes the launching and handling equipment and contains the detailed missile handling and maintenance procedures and checksheets for use aboard CV-64. In addition to a safety summary and specific safety precautions, the volume also contains a safety briefing which is required prior to performing missile handling procedures.

## EXPLOSIVE HAZARD CLASSIFICATION

All ammunition and explosives are assigned to a hazard class based on the United Nations Organization (UNO) hazardous materials system. The Department of Defense (DOD) further divides these classes into more specific hazard categories, i.e., mass-detonating, fragment, mass fire, moderate fire, chemical, etc. The combined classes, divisions, and groups are then applied to storage, compatibility, and quantity-distance safety considerations ashore as described in OP 5, volumes 1 and 2.

Ammunition and explosives, which include missiles and their components, are further divided into hazard classes based on Department of Transportation (DOT) evaluation of potential destruction characteristics and compatibility factors. For example, DOT classifications for highway and railroad transportation safety are:

Class A—Explosive materials which detonate or are otherwise maximum hazard. Class A items explode violently when contacted by spark or flame or when subjected to excessive heat or shock. All class A explosive components therefore must be handled with extreme care and protected from fire sources and shock. Examples of class A items are missile warheads and fuze boosters filled with high explosives.

Class B—Explosive components which, in general, function by rapid combustion rather than detonation, are generally considered a flammable hazard. The personal hazards created by class B explosives are fire, heat, and noxious gases. Missile components such as flash signals, rocket engine grains, igniters, etc., are examples of class B items.

Class C—Explosive components are placed in the class C category either because they contain restricted amounts of class A or B explosives, or because their damage potential is insufficient to be classified higher. Typical examples of class C items are safety and arming (S&A) devices, and guidance and control (G&C) groups.

DOT has further separated ammunition and explosives into shipboard hazard classes which take into consideration afloat, evironmental, and capability factors. Shipboard ammunition and explosives stowage requirements are listed by Coast Guard classes in CG-108, Rules and Regulations for Military Explosives and

## Chapter 7—WEAPON SURVEILLANCE AND STOWAGE

Hazardous Munitions, which is a DOT manual. For example, the Mk 90 Mod 0 warhead for the Standard MR missile would be shipped by truck or railcar as class A and be stowed on board ship as CG class VII. A class B rocket engine might be classed as CG class X-E, and a class C S&A device might be classed a CG class VI.

A listing of UNO/DOD, DOT, and CG classes may be found in OP 5, volume 2.

## MISCELLANEOUS EXPLOSIVE HAZARDS

Black powder has been called the most dangerous of all explosives. It must be protected against heat, moisture, sparks, rf radiation, and friction. Only very small quantities are used in modern naval ordnance—in fuzes, igniters, and primers. The largest quantities are contained in impulse charges.

The cast propellants used in rocket motors and sustainers must be protected against heat, moisture, and physical damage from dropping, abrading, etc. A crack in the cast propellant can cause failure of the missile because it prevents continuity of the burning rate. Powdered or crumbled propellant is more dangerous than the undamaged material. Dragging boxes over propellant powder grains on concrete decks or docks has caused fires. Powder grains that have fallen into cracks and crevices are believed to have been the cause of many fires. The explosive ordnance disposal (EOD) team should be called immediately if powder is spilled. Work must be suspended until the spilled or broken explosive has been collected and placed in plastic containers. Report all accidents or incidents in accordance with OPNAVINST 5102.1 (latest revision).

Some high explosives used in warheads look very much like harmless chunks of clay or pieces of rock. Scraping, striking, or dropping them can cause them to explode. When handled with bare hands, some high explosives can cause dermatitis. Some high explosives give off poisonous gas when they burn; one type leaves a white powdery residue; another type leaves a residue that is explosive if moved even a little. As short a drop as 5 inches can cause PETN to explode; tetryl has a drop sensitivity of 12 inches. The EOD team is trained in procedures to follow in emergencies involving explosives; untrained personnel should not move damaged explosives.

A rocket motor that has been dropped must not be fired. It must be returned to the depot or disposed of according to instructions in the OP or from the commanding officer.

Never use power tools on the rocket motor. Never apply heat to the motor, or to any of its associated components.

In case of a rocket motor misfire, wait at least 30 minutes to make sure the firing circuits are open before you approach the rocket.

Missiles not expended in live runs must be safed at the first opportunity in accordance with the instructions for the missile.

The self-destruct charge contains composition B and tetryl, both high explosives, connected by two explosive leads. The explosive leads are detonated by an electric primer. The primer leads must be shorted at all times until just before firing. Handle all electroexplosive devices (EEDs) in accordance with HERO requirements. Always check the visual indicators for SAFE condition of the unit prior to installation.

## NUCLEAR WEAPONS HAZARDS

Though nuclear weapons are so designed as to prevent a nuclear yield in the event of accidental detonation, there is still a probable hazard. The two components of a nuclear weapon that constitute the most probable hazard in the case of an accident are (1) high explosives and (2) plutonium. Other components may produce hazards, but they are of such a nature that precautions taken against explosives and plutonium are more than sufficient for their control. Keep in mind that accidents involving nuclear weapons or components usually will involve other materials in more widespread use, such as gasoline or other volatile and explosive fuels. If fire occurs, acrid, suffocating, and toxic fumes and smoke will probably be generated by the combustion of surrounding materials. In that event, normal procedures and precautions applicable to the specific type of fire should be taken.

### High Explosives

Most atomic weapons contain conventional high explosives in varying amounts up to many hundreds of pounds. These high explosives comprise the major hazard associated with accidents involving atomic weapons. Because high explosives are usually present in any atomic weapons' shipment, accidents or fires involving such shipments must be treated the same as accidents or fires involving conventional high explosives. The following information should be applied to atomic weapons where appropriate.

DETONATION.—In any accident involving high explosive there is some possibility of a detonation. The detonation may range from a very small one to one of considerable magnitude or to a series of small explosions. The breakup of the weapon due to impact or a small explosion will probably result in the local scattering of small pieces of high explosive. Rough handling, as well as accidents, may produce powered explosives. Most explosives are more unstable in these conditions and are most apt to detonate due to changes in temperature and/or shock. Exposure to sunlight likewise increases the sensitivity of the explosive and changes its coloring, usually making small pieces and powder difficult to distinguish from their surroundings. Thus, it is unwise for anyone other than trained demolition personnel to attempt clearing an area of broken high explosives.

FIRE.—If a nuclear weapon is involved in a gasoline fire, the high explosive may ignite, burn, and in most cases, detonate. These detonations may also range from one large to several small ones. It is extremely difficult to extinguish large quantities of burning high explosives. Whenever burning high explosives are confined, as in an intact weapon, detonation may occur at any time. When high explosives burn, torching (jets of white flame coming out of the weapon) may be observed; however, torching is not always evident before detonation. High explosives may melt at comparatively low temperatures, flow out of the weapon, and resolidify. In this state, they are extremely sensitive to shock. If unconfined, high explosives may burn with the production of toxic gases and leave a poisonous residue. Ignition or detonation of the high explosives in a nuclear weapon involved in a fire can be prevented if the temperature of the explosives is kept below 300°

NUCLEAR YIELD.—While it is not feasible to predict the exact effect of an accident involving high explosives, the possibility of the accidental explosion of a nuclear weapon is so remote as to be negligible.

### Plutonium

Plutonium may become dispersed as small particles as the result of impact or detonation of the high explosives, or as fumes if a fire occurs.

EFFECTS ON THE BODY.—When small particles of plutonium are suspended in the air, it is possible to inhale them, thus depositing plutonium into the lungs. Plutonium also may be swallowed, but it is in a highly insoluble form and only a small percentage will be retained in the body. However, cuts in the skin do provide an entry into the bloodstream for plutonium.

If it remains outside the body, plutonium is no hazard because it is an alpha emitter. The alpha particles have a very short range and lack the ability to penetrate the skin. This characteristic of plutonium radiation makes its behavior markedly different from that of fallout from an atomic explosion in that it does not emit the more penetrating beta and gamma radiation. Conventional survey meters are of little use in detecting alpha radiation, and only special teams trained for handling nuclear accidents are capable of evaluating the radiological situation at the scene of an accident.

AMOUNT AVAILABLE.—Field experiments indicate that the principal potential source of intake of plutonium into the body is by inhalation. Safety restrictions have been placed on the number of atomic weapons per shipment so that the plutonium inhaled from an accident would not result in serious injury. Once the fine particles have been deposited on the ground, the hazard is markedly reduced. Whereas it is always

## Chapter 7—WEAPON SURVEILLANCE AND STOWAGE

desirable to reduce to a minimum the intake of plutonium, where necessary, one may enter or remain in a highly contaminated open area for short periods of time (up to several hours) after passage of the cloud.

HOW TO AVOID.—Conventional mechanical breathing apparatus or dust filter masks, goggles, and protective clothing will reduce contamination for those who must enter the smoke. Members of the special radiological team wear full protective clothing since they usually remain in the area for considerable lengths of time; however, the nonavailability of any or all of these items should not hold up rescue operations. Potential hazards in buildings can be reduced by shutting doors and windows and turning off ventilation equipment. The potential danger from plutonium to those engaged in rescue work is no more serious than the danger from the other products of combustion.

### INSPECTION AND DISPOSAL OF EXPLOSIVE COMPONENTS

The general surveillance programs conducted by units afloat consist primarily of visual inspections of the exterior of explosive components and the observation of the environmental conditions under which these components are stowed and used. Unsafe or damaged weapons must be safed, if possible, or disposed of in accordance with current directives. A semiannual inspection of Terrier weapons is required by the Planned Maintenance Systems in accordance with OP 4093. ASROC depth charges with a nuclear warhead are inspected in accordance with Navy SWOP 44.34.1, which bears a classification higher than that of this publication.

### RECEIPT INSPECTION

Inspections which must be made on receipt of the missiles and missile components at replenishment were discussed in chapter 2.

Missiles are delivered to the firing ship in the assembled condition. The components were inspected at the assembly facility. If the missile is delivered in a container, inspect it only for evidence of damage from rough handling or water. After unpackaging the missile for stowage, inspect the exterior more closely for evidence of rough handling, water damage, mildew or other fungus growth, and for broken or missing parts. Parts, such as wing and fin assemblies, should be inspected when they are unpackaged for stowage. Check the position of safety switches to be certain the missile is not armed. Check the humidity indicator if there is one; if the humidity is too high (the OP for the component lists the humidity and heat limits), unpackage the component and inspect it for damage. Heat damage is seldom visible. The missile's record provides the evidence of overexposure to heat and cold.

On the Terrier BTN missile, fungus growth, corrosion, and superficial scratches or abrasions on a warhead are not cause for rejection; these conditions can be remedied. However, dents and deformations on a warhead are cause for rejection.

If a Terrier BTN missile (Mk 22 warhead) is received aboard with the Mk 7 safety switch actuator in the ARMED position, use the arming tool H3114 (which your commanding officer has) to rotate it back to the SAFE position, in accordance with SWOP W45.21-9, section 6.

### DISPOSAL OF EXPLOSIVES

Do not jettison a missile unless it is absolutely necessary to do so for the safety of the ship and crew. If the missile is a dud, return it to the magazine unit until it can be returned to a weapons station for refurbishing. In case of a misfire in which the APSs are expended, the aft section of the missile must be replaced. This is done at a weapons station. After waiting the minimum required time after the misfire (30 minutes for Terrier), apply external power for at least 1 minute in the HT/HTR and BT/BTN missiles to ensure caging of the gyro. Return the missile to the magazine as a dud. Enter all information and findings on the guided missile service record (GMSR), propulsion unit history sheet (PUHS), and the nuclear ordnance record card (NORC), if a BTN weapon is involved.

### Boosters and Sustainers

Procedures for disposal of missiles and boosters in peacetime because of a dud or misfire have been covered in previous chapters. If a condition exists which is considered hazardous to the ship, even in peacetime, dud-jettison the weapon. A dud weapon should not be jettisoned in peacetime unless, through a circuitry malfunction, the weapon is in fact a misfire and is potentially hazardous to the ship and its personnel. Smoke emission from the booster is an excellent indication of a hazardous condition. A dud Standard ER missile with its battery expended must be handled as a misfire.

Black smoke exhausting from the battery vent on a misfire or dud Standard ER missile indicates a battery failure and the existence of an extremely hazardous condition; OP 4093 states emergency firing or dud jettisoning is the correct course of action. The established waiting period for a misfired Standard ER missile in which the battery has activated normally (no black smoke) is 4 hours.

When a 30-minute minimum waiting period after the misfire has been observed, and the batteries have activated normally, personnel wearing protective clothing, goggles, and safety gloves may approach the weapon with care.

NOTE: Should personnel come in contact with the battery electrolyte potassium hydroxide (KOH), instruct them to wash the affected area with large quantities of water and seek medical aide. Personnel using a garden hose can wash down the missile and launcher using freshwater to clean the area of electrolyte.

Instructions for disposition of the misfired Standard ER should be requested from NAVSEASYSCOM. If you find an S&A device in an ARMED condition, your command will again request disposition of the weapon from NAVSEASYSCOM. While awaiting these instructions, dud the tray or cell on which the weapon is located.

### MISSILE COMPONENT IDENTIFICATION

Navy guided missiles, as other ammunition, are classified as service (tactical) missiles and nonservice missiles. Tactical missiles, or rounds, are fully functional and fully explosive loaded rounds. Nonservice missiles may be further segregated into practice (exercise) rounds, training (or inert operational) rounds, and dummy (or shape) rounds. Each type of nonservice missile carries an identifying ammunition color code.

The external surfaces of all Navy guided missiles (service rounds), except radomes and antenna items, are painted white. White has no identification color coding significance when used on guided missiles. The significant color-coding colors—yellow, brown, and blue—are used on guided missiles and their components. The three colors are applied to the external surface of guided missiles to indicate explosive hazards and uses.

### COLOR CODE INTERPRETATION

Yellow identifies high explosives and indicates the presence of an explosive which is either sufficient to cause the ammunition to function as a high explosive, or particularly hazardous to the user.

Brown identifies rocket motors and indicates the presence of an explosive which is either sufficient to cause the ammunition to function as a low explosive, or is particularly hazardous to the user.

Light blue identifies ammunition used for training or firing practice. Blue painted items may also have a yellow or brown band painted on them to indicate explosive hazards or they may be an overall blue color without bands, indicating it is a training item that is nonexplosive loaded. Any missile with external surfaces painted all blue is a fully inert training item.

Light green identifies smoke or marker ammunition.

### MISSILE AND COMPONENT MARKINGS

Guided missiles containing compressed gas components fitted with an explosive squib are indicated by a brown band on the component

# Chapter 7—WEAPON SURVEILLANCE AND STOWAGE

and external surface of the missile section in which the gas flask is contained. The brown band alerts the user to a potential hazard. Figure 7-1 illustrates color coding for a typical missile configuration.

Guided missile warheads and their associated fuze mechanisms may be loaded and configured for service (tactical) or nonservice use. Some large surface-to-air missiles have more than one explosive type of warhead, while practice warheads for all missiles may be either inert loaded with an in-flight destructor charge installed, or completely nonexplosive loaded. Service tactical warheads for all missiles are painted overall white. A high explosive warhead painted overall white has a solid yellow band no greater than 3 inches wide painted around the warhead.

Warheads fitted with pyrotechnic components to indicate fuze activation are painted with a 1-inch light green band adjacent to a 1-inch brown or yellow band (figure 7-1). Training warheads with an explosive destructor charge installed are marked with the symbol COMP B in yellow letters, as illustrated in figure 7-1.

## MISCELLANEOUS EXPLOSIVE DEVICES

Miscellaneous missile explosive devices encompass all independent explosive or pyrotechnic devices that are not components of the missile fuze and warhead or the propellant units and igniters. Items specifically included under this grouping are in-flight destructor charges, safe arming devices, auxiliary power units, and arming and firing devices. These devices follow the ammunition color-coding requirements. Explosive components containing high explosive or having sufficient explosive to function as a high explosive are painted yellow overall, or with a yellow band. Components

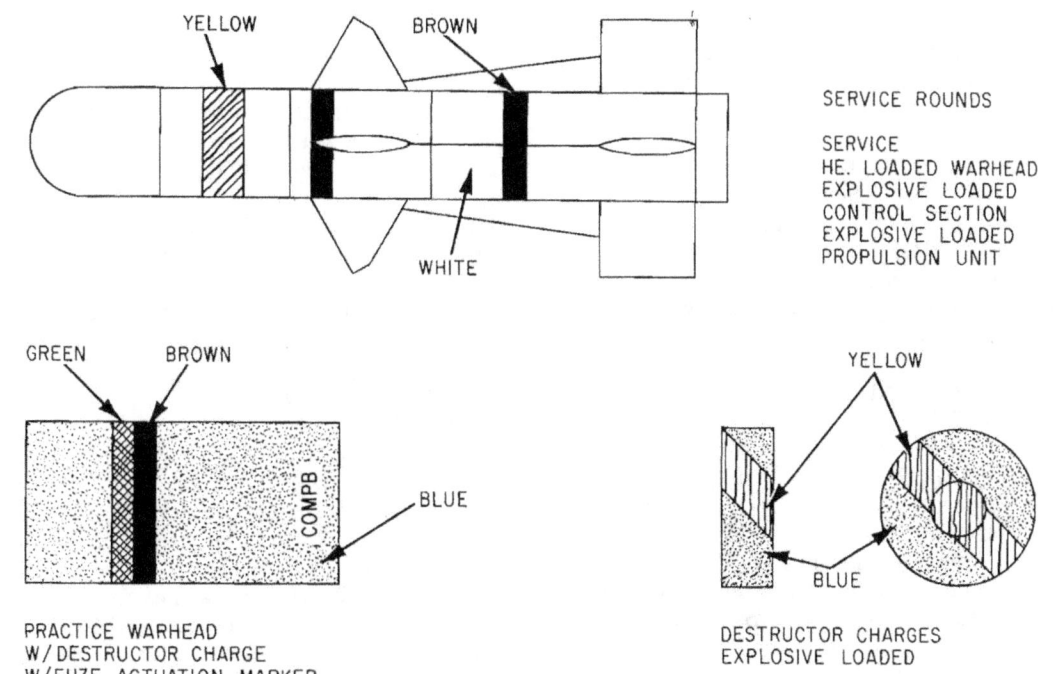

94.184

Figure 7-1.—Typical guided missile component painting.

containing explosives in amounts sufficient to function as a low explosive or to be deemed particularly hazardous to the user are painted brown or with a brown band.

Training items nonexplosive loaded are painted blue overall or with a blue band. Practice items may be explosively loaded and have a yellow or brown band painted over the blue overall background color (figure 7-1). Antisubmarine rockets (ASROCs) used with some Terrier weapon systems are painted gray overall and carry the same ammunition color code specified for guided missiles. Blue is the color used for totally inert training and handling ASROC weapons. Guided missile and rocket designations and ammunition color coding for missile and rocket components are explained and illustrated in Identification of Ammunition, OP 2238.

## MISSILE MAGAZINES

Surface-to-air guided missiles Terrier, Tartar, Sparrow III, Harpoon, and Standard are ready service complete rounds of ammunition. The complete missile represents a mixture of mechanical, electrical, and electronic equipment hazards, plus hazards due to several different explosive components. Because of the nature of guided missiles, requirements for their stowage aboard ship differ from the conventional ammunition magazine requirements. Surface launched missile magazines are usually located above the ship's waterline.

### CONSTRUCTION AND ARRANGEMENT

Missile magazines are constructed so each missile is segregated in cells or trays for easy handling and maximum protection against fire and shock. Missile magazines contain the necessary electric, hydraulic, and pneumatic power-operated equipment to stow, select, and deliver a missile from the magazine to the launcher rail for firing. The location and general arrangement of the various types of missile magazines differ with the type of missile and the type of ship in which the missile system is installed. In some missile magazines, restraining gear is provided to prevent movement of an inadvertently ignited missile motor while the missile is stowed in a cell. Special care is taken with the magazine vent systems to ensure that magazine pressures do not build up to a dangerous level if a missile rocket motor is accidentally ignited. A plenum chamber and vent are provided in Tartar magazines which vents the exhaust gases from an accidentally ignited missile to the atmosphere.

In some missile magazines, flame barriers are installed between each cell to make them a separate, enclosed compartment open only at the top through which a missile passes during loading and unloading. Figure 7-2 shows this type of arrangement used in a Tartar magazine of a Mk 11 GMLS. Missile magazines also contain firefighting equipment which consists of built-in sprinkler systems, water injection systems, carbon dioxide ($CO_2$) systems, foam systems, or a combination of these systems.

Missile magazine access doors, flametight blast doors, and compartment doors should be kept closed at all times except when they must be opened to permit passage of missiles, missile components, or personnel. Special emphasis is placed on this requirement during periods of weapon assembly, disassembly, system testing, system firing, or other operations involving missile movements. The same precautions observed in magazine areas must also be observed in all areas of a missile system where weapons are handled or tested.

### MAGAZINE SAFETY REQUIREMENTS

Missile magazines contain automatically controlled missile handling equipment which can be hazardous to personnel if safety precautions are not observed. Hazards from moving equipment within the magazine areas can be eliminated by removing or positioning safety switches from a controlling station which stops all equipments within a magazine area. Other hazards, such as a suffocation hazard from a $CO_2$ firefighting system, can also be safed by securing valves which feed the system.

Chapter 7—WEAPON SURVEILLANCE AND STOWAGE

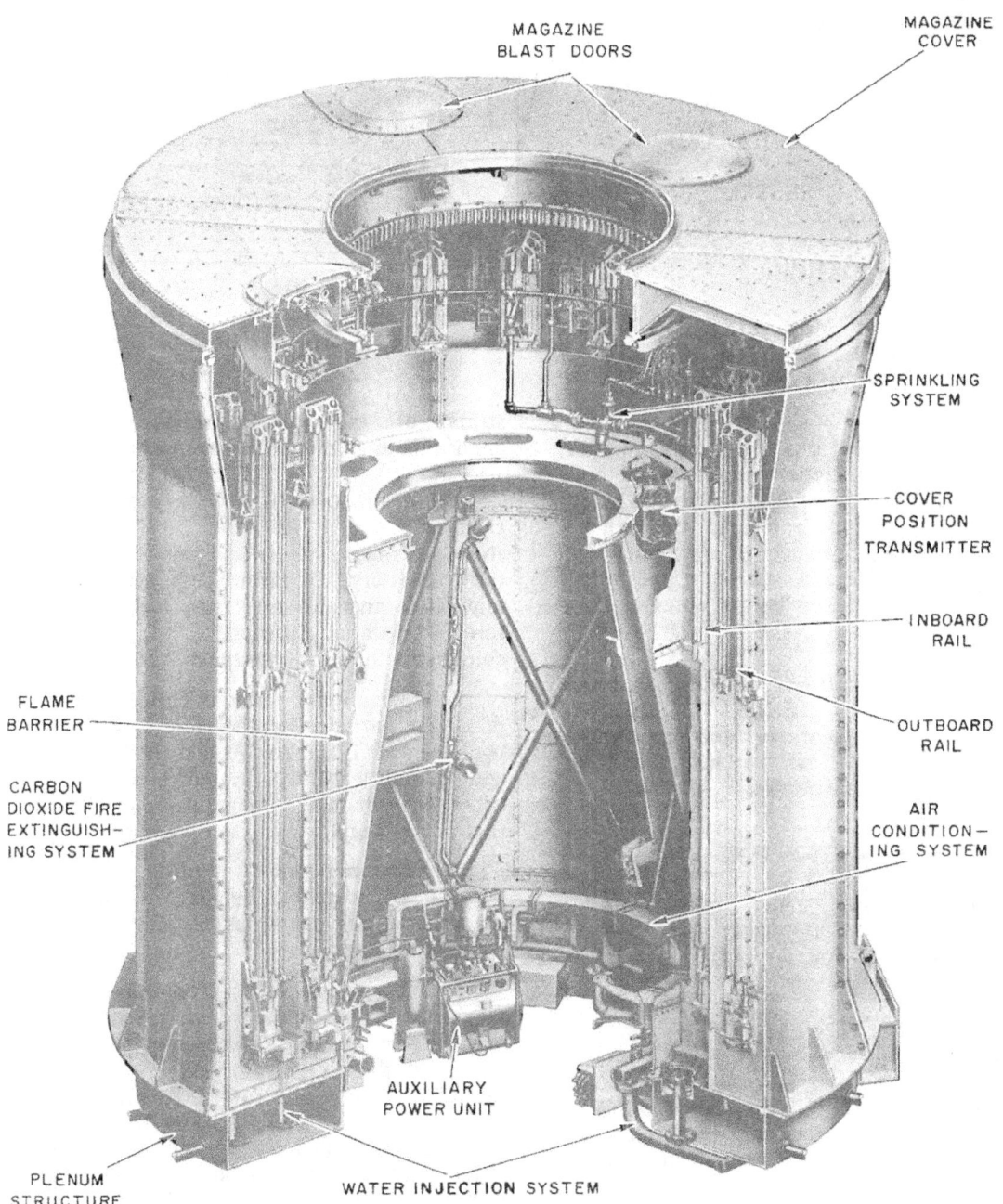

Figure 7-2.—Mk 11 GMLS missile magazine.

Safety instructions posted near the entrance of magazines are very effective if they are easily understood and can be complied with easily Some standard safety warnings are:

SUFFOCATION HAZARD

SECURE $CO_2$ SYSTEM BEFORE ENTERING MAGAZINE AREAS; and,

DANGER

TO PREVENT MAGAZINE MOTOR ACTIVATION, REMOVE SAFETY SWITCHES FROM CONTROL PANELS

These warnings point out the potential danger, but do not give instructions about the methods of eliminating the dangers. Where safety methods are not fully explained, the launcher supervisor should instruct all personnel who have access to magazine spaces of the proper procedures to be taken before entering these spaces.

Additional instructions may be posted near the warning signs (figure 7-3) to give information on the location of and action which must be taken to safe a magazine area. Instructions can read as follows:

SUFFOCATION HAZARD

BEFORE ENTERING MAGAZINE AREA, CLOSE THE TWO SHUTOFF VALVES THAT SERVE THE CARBON DIOXIDE ($CO_2$) SYSTEM

These valves are located outside of the launching system structure. They are normally secured (but not locked) in the open position; accordingly, disengage and close both valves and secure in closed position before entering the magazine area.

Specific safety precautions relating to shipboard stowage of guided missiles are presented in launcher systems OPs and in chapter 3 of OP 4, volume 2. Listed below are

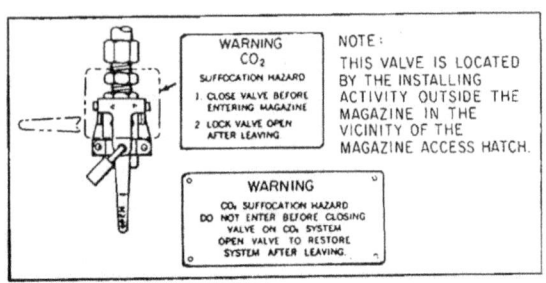

Figure 7-3.—Additional safety instructions.

some of the general safety precautions applicable to all missile magazine areas.

1. All magazines shall be kept scrupulously clean and dry at all times. Nothing shall be stowed in magazines except missile rounds and the necessary magazine equipment. It is imperative that no oily rags, waste, or other foreign material susceptible to spontaneous combustion be stowed in the magazines.

2. To minimize environmental hazards, the missile magazine is temperature and humidity controlled. It is imperative that the temperature and humidity control systems operate at all times. Inspect missile magazines daily to verify that proper humidity and temperature exist.

3. Personnel must remove from their persons all matches, lighters, or any other firemaking or sparkmaking devices before entering a magazine space.

4. Blowout discs and hatches are provided as safety measures to relieve pressure in the magazine in case of rocket motor ignition. The discs and hatches should be inspected periodically to make sure they are operable. They should be clearly marked to show their locations. Personnel should stand clear of the hatch area and the area directly beneath the hoods where the discs are ejected.

5. In the event a booster or sustainer accidentally ignites in the magazine, stand clear of the magazine until exhaust gases have been completely vented. The gases are toxic;

## Chapter 7—WEAPON SURVEILLANCE AND STOWAGE

consequently, lethal if inhaled in sufficient amounts. If you are not wearing special equipment, wait a minimum of 10 minutes after burnout before you approach the area.

6. Precautions should be taken to ensure that heat detectors, as well as the sprinkler and $CO_2$ heads, are not covered, damaged, or subjected to any environment that might falsely activate them or impair their utility. Because of the suffocation hazard represented by $CO_2$ in a closed area, personnel should disable the $CO_2$ system before entering a magazine area.

7. If a magazine has been flooded with carbon dioxide, allow 15 minutes for all burning substances to cool down below their ignition temperatures, then thoroughly ventilate the area for an additional 15 minutes to make certain that all portions of the magazine area contain only fresh air. If it is necessary to enter the installation before it is thoroughly ventilated, use a fresh air mask or other type of self-contained breathing apparatus.

## MAGAZINE FIRE SUPPRESSION EQUIPMENT

*GMM 3 & 2,* NAVEDTRA 10199-C, describes the types of missile magazine firefighting equipment presently installed on board naval ships. Since fire and explosions are the chief dangers in a magazine where missiles and their explosive components are stowed, prevention of conditions that can cause fire and explosions and the means of fighting fire if it occurs are included in every missile magazine. During the daily inspection of missile magazines, examine them carefully for cleanliness, ventilation, temperature, and the general condition of the missiles stowed in the magazine. Also visually inspect the sprinkler system for any leaks in the control lines. On wet-type magazine sprinkler systems, the pressure tank should be checked. The required pressure of 50 psi should be maintained. The temperature and the moisture content of the magazine's atmosphere must be constantly watched. Temperatures are read daily and the maximum and minimum readings are recorded in a magazine temperature record book. A magazine sprinking system has to be inspected and tested monthly. Magazine flooding control systems, quenching systems, and installed missile handling equipment must also be inspected for security, safety, and operation periodically.

## Test and Maintenance Hazards

In missile magazines that have both $CO_2$ and sprinkling systems, the activating control units could be the same type. An example, illustrated and explained in *GMM 3 & 2,* NAVEDTRA 10199-C, is the magazine firefighting system used with the Mk 13 GMLS. In this system two control circuits, one for $CO_2$ and the other for sprinkling systems, are activated by heat sensing devices.

A common hazard of a heat sensing device is its method of operation. It is activated by a fusible slug which melts at a predetermined temperature. This action causes a mechanical action to take place which activates either the $CO_2$ system or the sprinkling system. If a heat sensing device is located too near an operating electric motor or hydraulic unit, the fusible slug could melt from the excessive heat emitted from the units, and accidentally activate one of the systems. Because of this hazard, the slugs are checked periodically to ensure their condition. Fusible slugs come in many types which melt at different temperatures. In the Mk 13 GMLS, two types are used—one which melts at 174°F for the sprinkling circuit and one which melts at 158°F for the $CO_2$ circuit. Since all heat sensing devices are identical (except for their fusible slugs), extreme caution must be observed when conducting maintenance on these units. If an inspection reveals that a slug must be replaced, a maintenance requirement card (MRC) will explain all the steps necessary to perform this task and list the safety precautions related to the task. A launcher supervisor should research the MRC to ensure that the required actions listed include all additional safety requirements for entering a magazine area. The supervisor should also ensure that all safety instructions are understood by the personnel performing the task. Most MRCs include a statement to observe all standard safety precautions. A standard safety precaution is one that pertains to all types of magazines and is not listed as a specific instruction on the MRC for the maintenance action being performed. A standard magazine safety precaution is to ensure that no matches, or other flame producing apparatus, are taken into the magazine while it contains explosives. In cases where similarity of systems may cause confusion, the launcher supervisor must take all the necessary additional precautions, even though they are not listed on an MRC, to ensure personnel safety.

## Fusible Slug Installation

Before installing a fusible slug in the heat sensing device of a sprinkling or carbon dioxide system, both systems should be secured regardless of which system is being maintained. When a damaged fusible slug is removed from a heat sensing device, it releases a compressed spring that forces a bellows to collapse. This action causes a sudden pressure change in the heat sensing device. The pressure change causes a mechanical action to take place which actuates either a control head of a $CO_2$ system or a PRP valve for the sprinkling system (both systems are explained and illustrated in *GMM 3 & 2*, NAVEDTRA 10199-C). To prevent accidental activation of either system, both must be secured prior to removing a fusible slug from any heat sensing device.

To secure the carbon dioxide system, disconnect the control heads from the supply cylinder and close off all valves that serve the carbon dioxide system. Also secure all firemain water pressure valves that serve the sprinkling system and install a sprinkling system test casting into the sprinkling system saltwater control valve. When either a PRP valve or a $CO_2$ control head is activated, its position is shown by an indicator on either unit (figure 7-4). An activated condition of a PRP valve is shown as the trip position and, for the $CO_2$ control head, a released position. The position of the control mechanisms is an important factor when you are performing a maintenance action. In normal operation, the position indicator on a $CO_2$ control head will move when the bellows of a

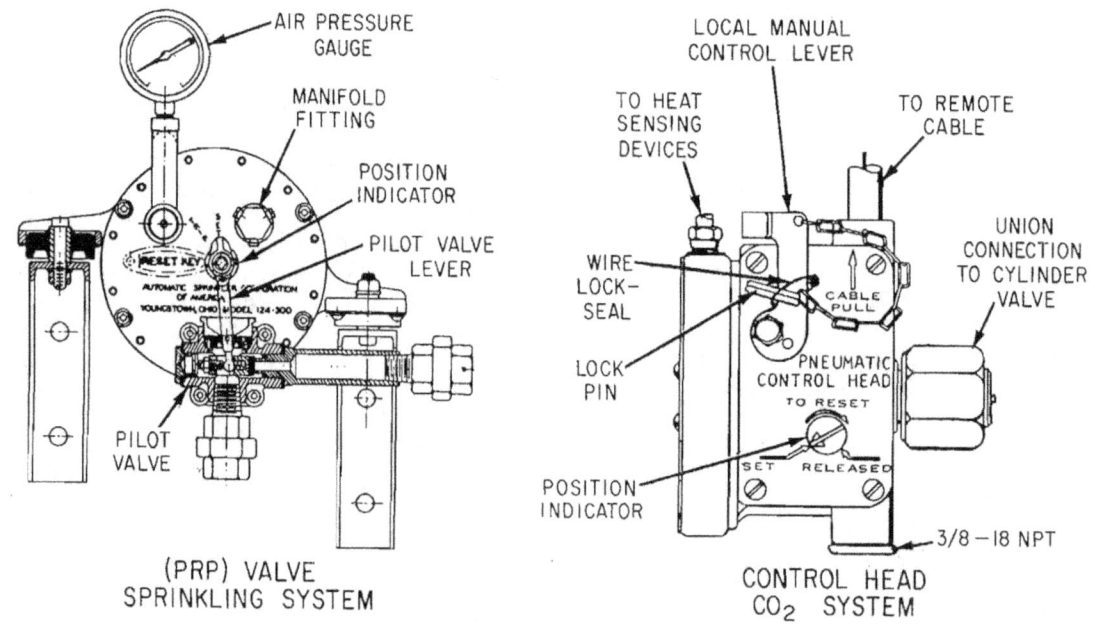

Figure 7-4.—Automatic control devices.

## Chapter 7—WEAPON SURVEILLANCE AND STOWAGE

heat sensing device collapses to produce a sudden pressure increase in the pneumatic lines leading to the $CO_2$ control heads. The pressure differential causes a diaphragm mechanism to trip an actuating lever which releases a compressed spring. The spring shifts a plunger in the control head mechanism and opens a pilot seat in the cylinder valve (figure 7-5). Liquid carbon dioxide flows through the pilot seat to the upper chamber of the discharge heads, forcing the piston down and opening the control cylinder valve. Opening the cylinder valve releases liquid carbon dioxide from the supply cylinders through shutoff valves and into the magazine area where a gaseous snow is produced which quickly reduces temperature and extinguishes fires. During maintenance, closing the shutoff valves prevents carbon dioxide from entering the magazine.

Though all precautions have been taken, an accidental activation of either system could occur. When a new fusible slug is installed in a sensing device, as shown in figure 7-6, the bellows must be expanded by a special tool. This tool, called a pull rod, is attached to a section called the collet. When the pull rod is pulled out, the bellows attached to the collet is reset in a position to collapse when a fusible slug melts. The fusible slug holds the extended collet in place, and the collet holds the reset bellows. Resetting the bellows does not automatically reset the tripping mechanism of either the $CO_2$ control head or the PRP valve; they must be reset manually.

Before reactivating the $CO_2$ system, check to see if the visual indicator on the control heads is in the SET position (figure 7-4). To reset the tripping mechanism on the control head, turn the stem on the visual indicator clockwise with a screwdriver from the released position to the SET position. Slight resistance will be met just before the stem locks.

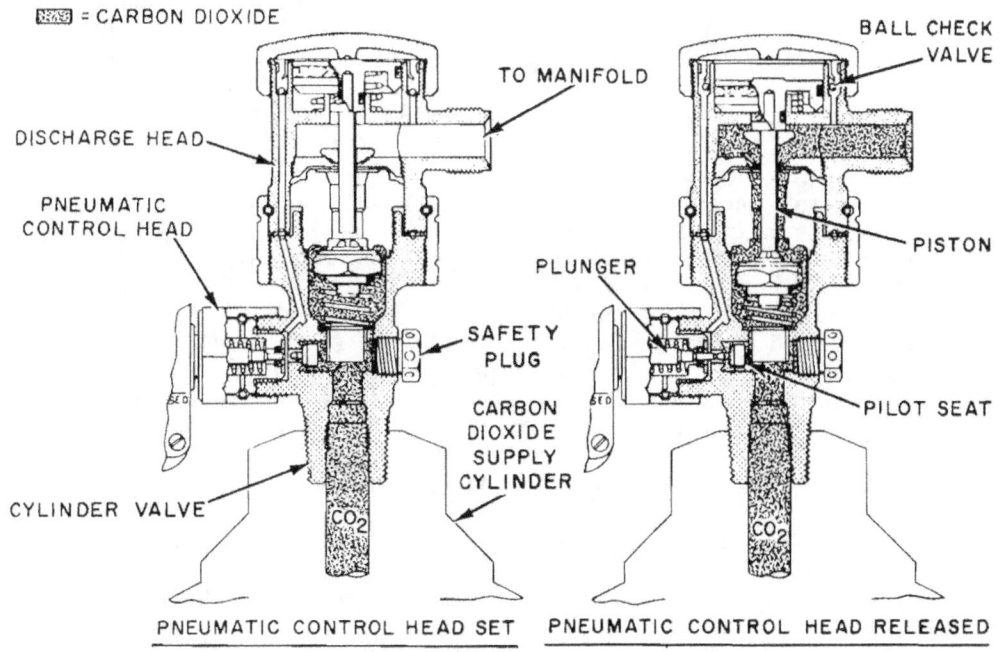

94.186

Figure 7-5.—Cylinder valve, pneumatic control head, and discharge head.

243

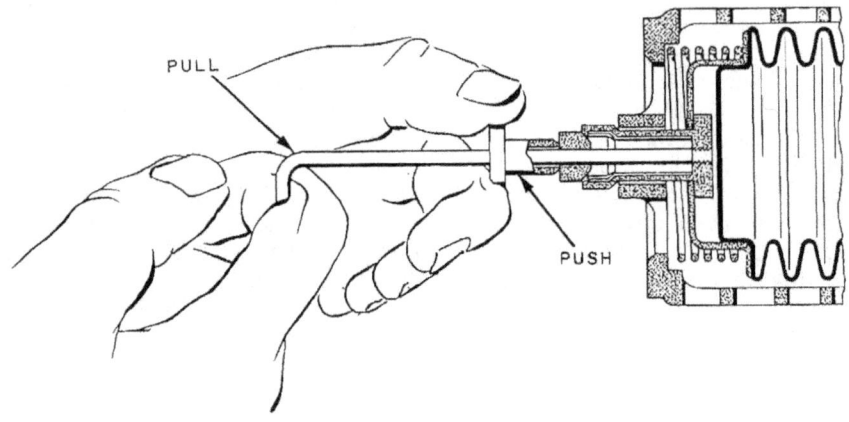

Figure 7-6.—Installing fusible slug in heat sensing device.

## OTHER MAGAZINE PROTECTIVE DEVICES

All means should be taken to maintain weapons in their best condition to help to prevent accidents. The air-conditioning and ventilation systems in the magazine may be considered protective devices in this sense. The heavy construction of the magazines gives protection against blast and fire from accidental ignition of a weapon. The blast doors on the magazine (Tartar) provide protection against blast and exhaust when a missile is fired from the launcher; interlock switches prevent firing of a missile until the blast doors are closed, the launcher is synchronized, etc. Flametight magazine doors are installed between the Terrier magazine and assembly area so the blast and flame from an accidental ignition in the magazine cannot spread to the assembly area.

### Alarm Systems

There are several alarm systems installed in missile magazines which warn personnel of danger to themselves and to alert them to take preventive or corrective action to protect the missiles stowed in magazines. A high temperature alarm, for example, alerts personnel that a possibility of a fire exists. When alarm systems are activated, personnel must investigate and correct the problem before any damage is done to the missiles. There are other types of alarm systems used in a missile magazine which indicate either $CO_2$ or sprinkling system activation, or a security violation; other alarm systems warn personnel when handling equipment is activated or when missiles are being moved. Missile magazine alarm systems are explained in *GMM 3 & 2*, NAVEDTRA 10199-C, chapter 10.

### Water-Injection Systems

Water-injection systems are used in the Terrier and Tartar magazines. The water-injection detector nozzle is discussed in chapter 10 of the *GMM 3 & 2* rate training manual. The major difference in the water-injection systems is that Tartar uses a pressurized freshwater supply tank.

### Plenum Chambers

All Tartar missile magazines have a plenum chamber which carries off gases and exhaust fumes from an accidentally ignited missile in the magazine. The plenum chamber is in the base of the magazine, beneath the missile. Each cell of a Tartar magazine has a blow-in plate assembly which gives way under pressure when a rocket motor is accidentally ignited and permits high

Chapter 7—WEAPON SURVEILLANCE AND STOWAGE

pressure exhaust gases to escape to the atmosphere through the plenum ducts. Figure 7-7 shows a magazine base structure of a Tartar missile magazine used with the Mk 22 GMLS. All missile magazines have some type of blowout plate which gives way and vents high pressure exhaust that escapes upward in a magazine.

### Anti-icing System

A major difference between a standard shipboard magazine and a missile magazine is the location aboard ship. Most standard magazines are located belowdeck and are not subject to outside weather conditions. Missile magazines, because of their function, are located adjacent to or below their launchers. In some GMLSs, missiles, such as Tartar, are loaded directly from their magazine onto launcher guide arms for firing. Others, such as Terrier, must pass through an assembly area prior to being loaded onto the launcher guide arm. Both systems have a type of blast door between the launcher and the missile stowage area which must be opened for loading and remain closed during firing. These doors and other exposed areas of a launcher and magazine must have an anti-icing system to prevent ice from accumulating on exterior surfaces during cold weather operations. Heated anti-icing fluid pumped through a closed piping system warms designated areas to keep them free of ice which could interfere with missile loading operations or prevent the functioning of a safety device. Some of the areas of a Tartar system serviced by an anti-icing system are shown in figure 7-8. These areas—the blast door, blowout plates, magazine cover, and magazine base ring—are considered necessary for both safety and system operation. The Terrier magazine areas are located within the ship's structure. The magazines have no openings requiring passage of missiles onto a weather deck area. However, the blast doors, launcher guide arms, launcher

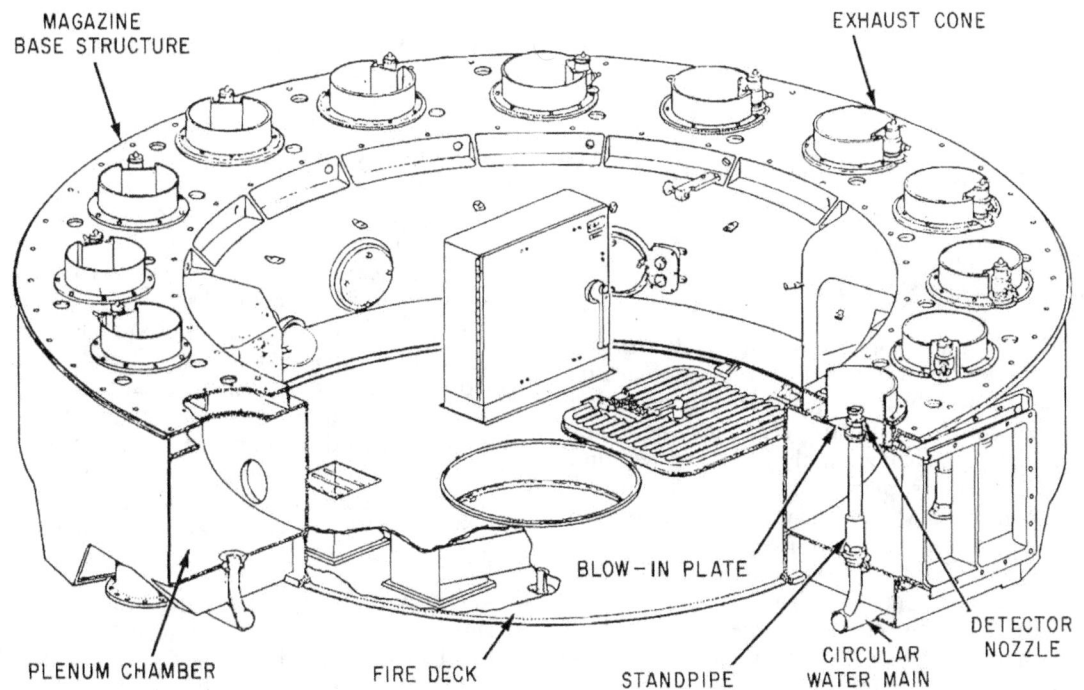

94.87(94B)

Figure 7-7.—Magazine base.

245

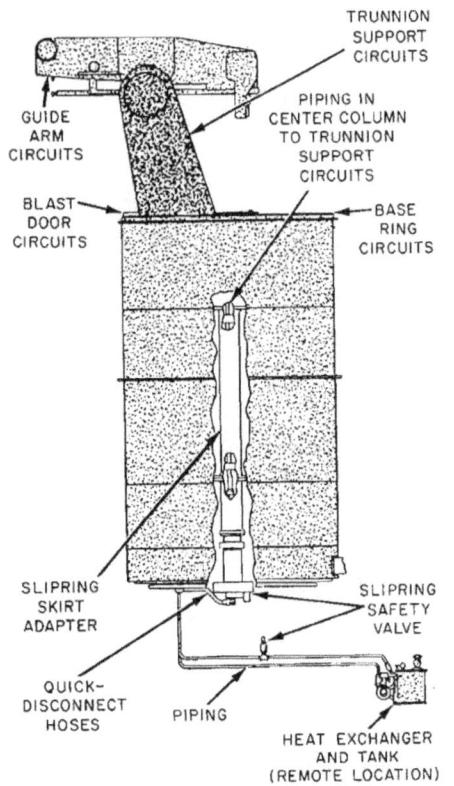

Figure 7-8.—Anti-icing system arrangement.

trunnion, and dud-jettison units are provided with anti-icing system piping.

The location of the components for an anti-icing system varies with the type of missile system in which it is installed. All anti-icing systems contain a heat exchanger, a motor driven pump, steam supply lines, valves, and manifolds and supply lines to distribute the heated fluid. The heat exchanger is a tank which heats and stores the anti-icing fluid which is made up of a 1 to 1 mixture of distilled water and ethylene glycol. The motor driven centrifugal pump circulates the heated fluid through the system. Fluid in the system is heated by ship's steam in the heat exchanger and recirculated throughout the system by a network of piping and flexible hoses. A temperature control unit (thermostat) is installed on the heater tank and is set at a temperature which is adequate to prevent icing.

## MAGAZINE AND WEAPON INSPECTIONS, RECORDS, AND REPORTS

The missile magazines that are part of the missile launching system are in use every day for some part of the training, maintenance, and repair operations. The magazines can be inspected during the course of daily work. Before anyone enters the launcher magazine, be sure to deactivate the power equipment and remove the switch handle from the control panel so its machinery cannot be activated while someone is within.

On the missiles, look for leakage of hydraulic fluid, lubricants, or missile fuel (for Harpoon and Tomahawk), and report leakages. ASROC torpedoes use a liquid fuel in the propulsion system.

### VISUAL MAGAZINE INSPECTIONS AND RECORDKEEPING

Daily inspections should be scheduled to coincide with the taking of magazine temperature reports. Teach your personnel what to look for when they inspect the magazine. Inspect the general condition of the magazine daily for cleanliness, ventilation, dryness (note any signs of dampness or "sweating"), lighting, presence of unauthorized tools and gear, temperature, presence of any odors indicating decomposition of explosives, escape of gases, or other indication of anything amiss. Find the source of any odor and remedy the trouble. You may need the assistance of the EOD team. With modern air-conditioned magazines, explosives usually remain in good condition for a long time.

Detonators, of course, should not be in the same magazine with rocket boosters, propellants, etc., unless they are in an assembled warhead in a missile. Little pools of exudate may form in detonator wells, and must be removed with the greatest care. This exudate is extremely sensitive, and removal should be

## Chapter 7—WEAPON SURVEILLANCE AND STOWAGE

prompt. Forward a sample of the exudate to the Commanding Officer (Code TC), Naval Ordnance Station, Indian Head, Maryland 20604, with a report of the circumstances describing the munitions from which the exudate was obtained, the mark, mod, manufacturer, lot number, loading activity, date of loading, and all other pertinent information. A copy of the report also should be sent to NAVSEASYSCOM.

Notice the condition of the containers. There must be no open or damaged containers, no spilled powder or broken propellant grains, and no dust, dirt, trash, or combustible materials about. Containers should be neatly stacked and fastened down so there is no shifting, slipping, or rolling about. Remove damaged containers to another compartment for repair.

If there are windows or ventilation openings, be sure that sun and rain are kept out and that the screening is intact. In an air-conditioned magazine, the ventilation openings are secured during general quarters except for blowout purposes or for emergency use. Be sure the lights are working and that there are no shorts to cause sparks. The door must be flametight; if it is even slightly sprung, have it repaired at once. The fire extinguishers, firehoses, and water buckets should be in ready condition.

A checkoff sheet listing all the items to be checked in the magazine area is a practical necessity. The sheet may contain spaces for up to a month of daily checking, depending on how often the weapons officer wants the report turned in.

Magazine temperature records and magazine logs are maintained by each command for every magazine and ready service locker aboard ship and selected magazines ashore. On shore stations a representative number of magazines in each group, containing representative quantities of each type of explosive material, are inspected each day. Different magazines within the respective groups are inspected on successive workdays until all magazines have been inspected. The frequency of inspections ashore may be varied on the basis of prevailing outside temperatures. The date and hour of each inspection shall be noted and recorded in a magazine inspection log over the signature of the person who makes the inspection. Substandard or abnormal conditions shall be reported promptly to the supervisor in charge for correction and the conditions observed noted in the log. When conditions are satisfactory and normal, this also shall be noted in the log by the entry "normal". Magazine inspection logs ashore may be destroyed when they are 1 year old. Magazine inspections ashore shall be made during daylight hours when there is sufficient light to assure that any existing substandard conditions can be seen and reported. Magazines are inspected to determine if repairs are needed; to make sure that the safety regulations, particularly those with regard to cleanliness and elimination of fire hazards, are being observed; to ascertain that materials are not deteriorating into an unsafe condition; and that they are stored in an orderly, approved manner as specified in OP 5, volumes 1 and 2.

Aboard ships, daily entries of temperatures are recorded on magazine temperature cards, department record magazine logs, and the ship's official deck log and serve to document the magazine inspection records. Identifying the magazines which have experienced temperature changes since the previous inspection, and noting any abnormal physical conditions observed, completes the report. If no abnormal conditions are observed, the notation "conditions normal" should be made. When abnormal conditions in the magazines are discovered and recorded, the facts shall be reported in person to the commanding officer or the command duty officer (CDO).

## INVENTORY RECORD OF SMALL ARMS AND PYROTECHNICS

Small arms will be issued to you and your personnel for guard duty, and must be strictly accounted for on an individual basis. However, before small arms can be issued, you and your personnel must be qualified in accordance with OPNAVINST 5510.45 series. SECNAVINST 5500.4 series gives the instructions for reporting lost, stolen, and recovered small arms. A letter report is required in each instance.

*Gunner's Mate M 3 & 2*, NAVEDTRA 10199-C, contains material on pyrotechnics with illustrations and diagrams. Ammunition Afloat, OP 4, volume 2, gives the official rules for care and maintenance, surveillance, stowage, and disposal of various types of pyrotechnics. Be sure to get the latest revision of this volume; rules have been made more precise and stringent because of disasters caused by mishandling of pyrotechnics.

## AMMUNITION RECORDS AND REPORTS

Because ammunition is esential to naval operations, and because of its high cost and unique logistic characteristics, ammunition status is under careful and continuous study at the highest echelons of the defense establishment, as well as by operational and logistics commanders. It is vital that an accurate and prompt method of reporting ammunition stock status be available to commanders of naval forces. For this reason, commanding officers are responsible for submitting reports regarding all receipts, transfers, expenditures, and quantities of all ammunition components within their command.

A quarterly ammunition report is made to the Navy Ships Parts Control Center (SPCC), Mechanicsburg, Pennsylvania, so an accurate inventory of assets and expenditures of expendable ordnance items throughout the naval service can be maintained. A quarterly ammunition report includes all conventional expendable ordnance material (including gun type), bombs, rockets, ASW weapons, guided missiles, military chemicals, mines, torpedoes, and demolition and pyrotechnic materials assigned a four digit Navy ammunition logistics Code (NALC) in accordance with NAVSEA OD 16135. The report excludes nuclear ordnance. The frequency with which ammunition assets and expenditures must be reported to fleet commanders and other command authority is outlined in COMNAVSURFLANTINST 8015.1 series and COMNAVSURFPACINST 8015.5 series as appropriate. Most ships report monthly to Commander Surfaces Forces Atlantic or Pacific who, in turn, reports to the SPCC. These reports, when processed by the SPCC, provide Naval Sea Systems Command with information concerning expenditure rates, ammunition availability, and facts from which fleet requirements can be determined.

There are numerous reports which must be made periodically concerning ammunition afloat and ashore. These reports contain complete ammunition identification data including lot number, mark and modification numbers, and the NALCs of all ammunition and components. Some of the information required in these reports is—

1. Available stowage space of the activity.
2. Types and numbers of missiles used for training.
3. Performance of ordnance equipment.
4. Performance of pyrotechnics and other ammunition components used.

By NAVSEAINST 8510.5 series (torpedo firing reporting system), a reporting system for fleet firing of torpedoes has been established to determine operational readiness and material reliability. Reports on unsatisfactory or defective torpedoes or equipment (RUDTORPE) has been superseded by the conventional ammunition integrated management systems (CAIMS) in accordance with NAVSEA-INST 8510.3.

OPNAV Instruction 8110.16 details the notification procedures for nuclear accidents, significant incidents, and nuclear incidents. Procedures for reporting accidents, malfunctions, and incidents involving nonnuclear explosive ordnance and material are promulgated in OPNAVINST 5102.1.

## NUCLEAR WARHEAD REPORTS

If you work with missiles or depth charges with nuclear warheads, the stowage conditions must meet the requirements for the nuclear component. For specific information on the temperature and humidity limitations for certain nuclear weapons, general information on stowage requirements for nuclear components,

## Chapter 7—WEAPON SURVEILLANCE AND STOWAGE

and the stowage limitations of the missiles and depth charges you have aboard, consult the Navy SWOPs for those weapons. The classified publications custodian in your division has charge of those publications. SWOP 5-8 contains instructions that pertain to NAVORD Form 8110/4, Special Weapons Unsatisfactory Report. This report is used to report material discrepancies and dangerous conditions.

Missiles containing war reserve nuclear warheads are never loaded on the launcher except when actual firing is anticipated.

Training weapons are of several categories, depending on the completeness of the weapon. If you are going to use it only for practice in putting a missile into the magazine and bringing it up to the launcher, the size, weight, and conformation of a real missile are all that are needed.

### REFERENCES

1. NAVTRA 10200-B, *Gunner's Mate M 1 & C*

2. OP 4093, Terrier BT and HT, Standard ER Missiles, and Training Rounds, Handling and Maintenance Procedures, volume 1

3. OP 4, Ammunition Afloat

4. OP 5, Ammunition and Explosives Ashore

5. OP 1014, Ordnance Safety Precautions; Their Origin and Necessity

6. OP 3051, Guided Missile Complete Round RIM-2D-3-4-5-6 Terrier BT-3B General and Functional Description

7. OP 3726, Terrier/Tartar/Standard Guided Missile Warheads and Fuse Components Disc/Oper/Maint

8. OP 3347, U.S. Navy Ordnance Safety Precautions

9. OP 2238, Identification of Ammunitions

10. OP 2793, Toxic Hazards Associated with Pyrotechnic Items

11. OP 3565/NAVAIR 16-1-529, Radio Frequency Hazards to Ordnance, Personnel, and Fuel

12. CG 108, Rules and Regulations for Military Explosives and Hazardous Munitions

# CHAPTER 8

# FIRE CONTROL AND ALIGNMENT

The effective use of any weapon requires that a destructive device be delivered to a target, usually a moving target. To deliver the weapon accurately, we must know the location of the target, its direction of travel, and its speed. Many targets now travel faster than sound, and must therefore be engaged at great distances.

## WEAPON SYSTEM CONFIGURATION AND DATA FLOW

A weapons system is the combination of a weapon (or multiple weapons) and the equipment used to bring their destructive power against an enemy.

A weapons system includes:

1. Surveillance units that provide initial detection, location, and identification of the target.

2. Command and control units that provide target evaluation and issue command and control signals to other units of the weapons system.

3. Fire control units that direct and aim the delivery and destruction units.

4. Delivery units that deliver or initiate delivery of the destruction unit.

5. Destruction units that destroy the target when in contact with it or near it.

Figure 8-1 (foldout at the end of this chapter) shows how components of a representative weapons system might be arranged to fit into these categories.

Now, let us look at each type of unit individually; first, with respect to its general functions in the weapons system, then in terms of representative Terrier components that perform these functions.

## SURVEILLANCE

The first steps in using a weapons system to solve the fire control problem are to detect, locate, and identify the target. Initial detection of a surface or air target may be made either visually or by radar. It is difficult to detect a target visually at long range, or even at short range when visibility is poor. For that reason, the first contact with an airborne target usually is made by air search radar. These radars are designed to keep a large aerial volume under continuous surveillance. To cover the necessary area, search radars use a wide beam. In addition, most search radar antennas rotate as they search. Targets show up on the radar's target display indicators as alternately fading and brightening spots.

After the search radar has detected a target and determined its approximate location, the next step in the development of the fire control problem is to identify the target. The problem of recognizing and identifying a friend or foe is as old as warfare. Passwords, flaghoist signals, and even uniforms are identification devices that have been developed through the years.

In modern warfare the identification problem is urgent. Radar systems present targets in the form of spots or spikes (called echoes or blips) on a radar screen; but friendly and enemy targets look alike on the screen. Furthermore,

## Chapter 8—FIRE CONTROL AND ALIGNMENT

high-speed planes and guided missiles give us very little time to solve this problem. When friendly fighter aircraft pursue enemy planes to within weapon range of our ships, the identification problem is acute.

Therefore, once we have detected and located a target, we must identify it. Positive identification is established by IFF (identification, friend or foe) equipment. The IFF equipment can challenge an unidentified target, and determine from the answer whether the target is friendly. The equipment consists of two major units: the challenging unit which asks the question (friend or foe?); and the transponder in the aircraft, which answers the question. The IFF equipment is used in conjunction with the air search radars. Briefly, this is how it works. To challenge a target, you press a switch that controls the IFF transmitter. When the ship's IFF transmitter is turned on, each target is challenged by a pulse of low power radio energy as the IFF antenna is turned toward it. If the target is friendly, it will carry a transponder which consists of a receiver and a transmitter. When the receiver picks up a challenge, it causes the transmitter to send out an answering pulse or pulses. The answer is usually a coded message. It is picked up by the challenging unit's receiver and sent to the indicator of the search radar.

## COMMAND AND CONTROL

Once the air search radar detects and locates the target, and the IFF equipment has determined whether it is a friend or foe, the target information from these sources is sent to the command and control units. These units include computers, weapon direction equipment, stable elements, and many other mechanical, electrical, and electronic instruments.

As a target approaches, the command and control units evaluate the tactical situation and perform certain preliminary processes that enable the fire control units to establish a line of fire. These processes are: (1) evaluation of the target, and (2) designation of a selected target to the appropriate fire control unit.

### Evaluation

Target evaluation is concerned with these questions:

1. What does the target intend to do? Is it going to pass close to the ship for observation purposes or is it going to launch an attack?
2. How threatening to the ship's safety is the target? If its obvious intent is to attack, how much time does the ship have to launch a counterattack?
3. What kind of attack is the target capable of launching and what weapons should the ship use to repulse the attack? For example, if the target carries missiles, what kind of weapon must the ship use to destroy the target before it can launch its missiles?

There are other factors involved in evaluating a tactical situation, but these sample questions should give you some idea of what the term "evaluate" means.

The command and control equipment presents a complete visual picture of the tactical situation. It displays all the targets that have been detected by the search radars. Each target must be evaluated with respect to the overall defense picture. Decisions are made to bring the ship's weapons to bear on the most threatening targets. These selected targets must be designated to the appropriate fire control systems.

### Designation

Designation is the step taken to assign the tracking element (director's radar or optical equipment) of a fire control system to a particular target. On the basis of target evaluation and the availability of fire control systems (some of which may be disabled, or busy with other targets), a decision is made to assign a fire control system to the target. This is usually done by pressing a button to activate circuitry that transmits target position information from the command and control equipment to the antenna positioning circuits of a radar set, or the power drives of a director. These units automatically move the radar antenna to the designated position. If the designation is inaccurate, the director must search for the target.

This searching process may last for a second or longer, depending on the accuracy of the designation information and other factors. Once the director has found the target and starts to track, it can be said that it has acquired the target. When the fire control radar has acquired the target, a more accurate determination of target position is provided.

## FIRE CONTROL SYSTEM (FCS)

The primary functions of the fire control units in a missile weapons system are: to acquire and track targets; to develop launcher and missile orders; to guide missiles to the target; and in some instances, to detonate the missile's warhead. Secondary functions of the system are to provide target information such as target speed, target course, range to the target, and system and weapon status information to the display units of the command and control system. This information is used to evaluate the tactical situation and to aid the weapons assignment.

An FCS provides or establishes three basic elements of the fire control problem—a line of sight, prediction quantities, and a line of fire. The line of sight is established by the radar set. The fire control computer calculates the prediction quantities needed to determine the amount of offset to the line of sight to establish the line of fire and to produce weapon orders. The orders are transmitted to the missile launcher to position it to the line of fire.

The fire control radar used in present day missile FCSs is usually an automatic target tracking and missile guidance radar. It normally receives target designation signals from the command and control system, through the fire control computer. The designation signals position the radar set to the designated range, bearing, and elevation. If the radar set does not acquire the target immediately, the fire control computer originates a search program that enables the radar set to seek out the target. Designations between missile fire control systems are possible by designating a nontracking system to a system that is tracking a target.

When the radar set acquires the target in range, bearing, and elevation, the radar set locks onto the target and starts to track it. Tracking circuits within the radar set automatically keep its tracking beam on the target. Target position is continuously transmitted to the computer. The computer and the radar set working together solve for the target's rate of movement about the ship by calculations based on the line of sight movements.

A guided missile fire control computer is designed to operate automatically. It is located in the missile plotting room and is used with the fire control radar set. The computer can be described in terms of three basic methods (called modes) of operation—air ready, designation, and track.

### Air Ready Mode

In this mode the computer is energized, but is receiving no information. It generates orders only to put the radar and launcher in predetermined air ready positions, so the director can respond rapidly to a target designation and the launcher can respond rapidly to a launcher assignment. For example, the air ready position of the radar director set may be at zero degrees elevation and either zero degrees or 180° train, depending on the location of the director on the ship (fore or aft).

### Designation Mode

The computer goes into this mode of operation when it receives a "director assigned" signal from the command and control unit(s) in the weapons control area of the naval tactical data system (NTDS). The computer directs the radar to the designated target position so that the radar line of sight will point at the target. It also sends a search program to the radar. The search program causes the radar beam or beams to move in a preset pattern about the designated target position. The radar searches for the target, and when the target is gated, the computer automatically goes into the track mode of operation.

## Chapter 8—FIRE CONTROL AND ALIGNMENT

### Track Mode

When the radar acquires the target in range, bearing, and elevation, the track mode starts. The radar then transmits an on-target signal to the computer. The computer sends signals to the radar that cause it to drive at a rate that will keep it locked on the target. The computer determines the proper lead angles for the launcher, and transmits these quantities in the form of electrical signals. These signals drive the launcher to the proper aiming position when the launcher is assigned to a missile fire control system.

Before the missiles are launched, the computer determines and transmits signals to the missiles that program the operation of the missile gyros and also provide selected missile orders pertinent to the homing missiles. The computer also transmits tactical data such as present target position, future target position, and missile time to target intercept (time of flight) to the various display consoles of the weapons direction system (WDS).

### DELIVERY

The delivery units in a weapons system are either a gun mount or the missile launching system. They deliver or initiate delivery of the destruction unit. The purpose of the missile launching system is to stow, load, aim, and launch missiles.

### DESTRUCTION

The end purpose of the surveillance, command and control, fire control, and deliver units is to cause the destruction unit (commonly referred to as the weapon) to intercept or pass near the target. It is then the function of the destruction unit to destroy or inflict maximum damage on the target.

### TERRIER WEAPON SYSTEM

The Terrier guided missile weapons system provides air defense for ships against medium-range air targets. It is also designed for conducting combat operations against low altitude air targets, surface targets, and for bombarding shore installations. Aboard Terrier ships, the weapons system will be either an analog or a digital system.

Areas common to both systems are the search radars, the NTDS/WDS, and launching system. The areas of greatest difference are the fire control systems. We highlight digital difference later in this chapter.

### SEARCH RADARS

Three search radars are used as a source of target data for the guided missile weapons system. The search radar data is supplied to the missile weapons system by means of the NTDS/WDS. Radar set AN/SPS-10 is a short-range, surface-search, two-coordinate radar used primarily for surface-search display data. The AN/SPS-29, AN/SPS-37, AN/SPS-40, and AN/SPS-43 radar sets are long-range, air-search, two-coordinate radars used as a secondary source of target data. (Installation of radar type will differ from ship to ship.) The AN/SPS-48 radar set is a long-range, three-coordinate, hemispherical-scan search radar used as the primary source of own ship air target data.

### Navigational Radar

The AN/SPS-10 radar set is a short-range, two-coordinate, unstabilized fanbeam radar capable of good discrimination for surface surveillance. Installed aboard all Terrier ships, the AN/SPS-10 is used for both surface-search operations and harbor navigation. The AN/SPS-10 operates at C-band frequencies and is highly effective as far as the radar horizon.

### Two-Coordinate Air Search Radar

The two-coordinate air search radar provides secondary air target data for the Terrier weapons system and NTDS. The radar is a long-range, broad-beam, early warning search radar capable of detecting fast-moving air targets of small cross-sectional areas. A pulse compression feature on the AN/SPS-37, 40, and 43 radars permits radar set operation at a lower transmitted peak power but with the same average power as conventional systems without sacrificing range detection capability or range resolution. An IFF antenna is mounted on and rotates with the antenna to provide an integrated display of target echo and IFF response when desired. Physical appearance of the AN/SPS-29, 37, and 40 radars is similar to the AN/SPS-43 radar.

### Three-Coordinate Radar

The primary air search radar aboard Terrier ships is the AN/SPS-48. The AN/SPS-48 is a digital, high data rate, computer-controlled pulsed radar that provides range, azimuth, and elevation data. The radar set operates in the upper E-band through lower F-band frequencies, and uses multiple pencil-beam frequency scanning to provide long-range and high-target resolution.

This frequency scanning technique produces a search pattern that provides three-coordinate target position data consisting of rectangular position coordinates (from which range and bearing quantities are developed) in the slant plane and a target elevation angle. An electronic 45° elevation scan with beam stabilization to a maximum motion of 20° of deck plane is accomplished by computer control of the transmitted frequency.

As with the two-coordinate radars, IFF interrogation information is synchronized and displayed concurrently with the air search data. The AN/SPS-48 is also fully integrated with NTDS data processing and display equipment.

Figure 8-2 shows the antennas of the three aforementioned radars as they appear in shipboard installations. Notice the small elongated IFF antennas located at the top of the -43 and -48 radar antennas.

You can compare the operational characteristics of these and various other two- and three-coordinate radar sets used aboard Terrier ships of older design in the rate training manual *Fire Control Tehcnician M 3 & 2 (Terrier)*, NAVEDTRA 10209A-S1.

### NTDS/WDS

Command and control functions are performed by the Mk 11 naval tactical data system/weapon direction system (NTDS/WDS-11). It consists primarily of data processing equipment, display equipment, and data transmission and communication equipment.

The NTDS unit computers are the central store for the NTDS/WDS-11 system, receiving target data from display consoles and making this data available to the data links as required by the task force communication system. The computers also receive track data from other NTDS-equipped ships and special inputs from various stations in the combat information center (CIC).

The PPI display consoles accept, process, and display radar data from a shipboard or aircraft early warning (AEW) radar; sonar data from own ship sonar; and computer track data from the unit computer. Thirty-two modes of operation are available to the computer; i.e., air tracker, ship weapon coordinator, engagement controller, etc. The PPI display consoles use input data that presents to the operator a continuous display of the tactical situation. Controls on the PPI display consoles permit the operator to enter tactical data and decisions into the unit computer, thus making the information available to other display consoles in the data display group. In addition, electronic countermeasure (ECM) and electronic counter-countermeasure (ECCM) operations are monitored and coordinated to minimize the effects of jamming and other forms of interference.

Radio transmission equipment provides automatic, high-speed digital communication links between ships and aircraft.

### MK 76 MOD 5 GUIDED MISSILE FIRE CONTROL SYSTEM

The Mk 76 Mod 5 GMFCS consists of the AN/SPG-55B fire control radar and the Mk 119 Mod 5 computer. The system acquires and

Chapter 8—FIRE CONTROL AND ALIGNMENT

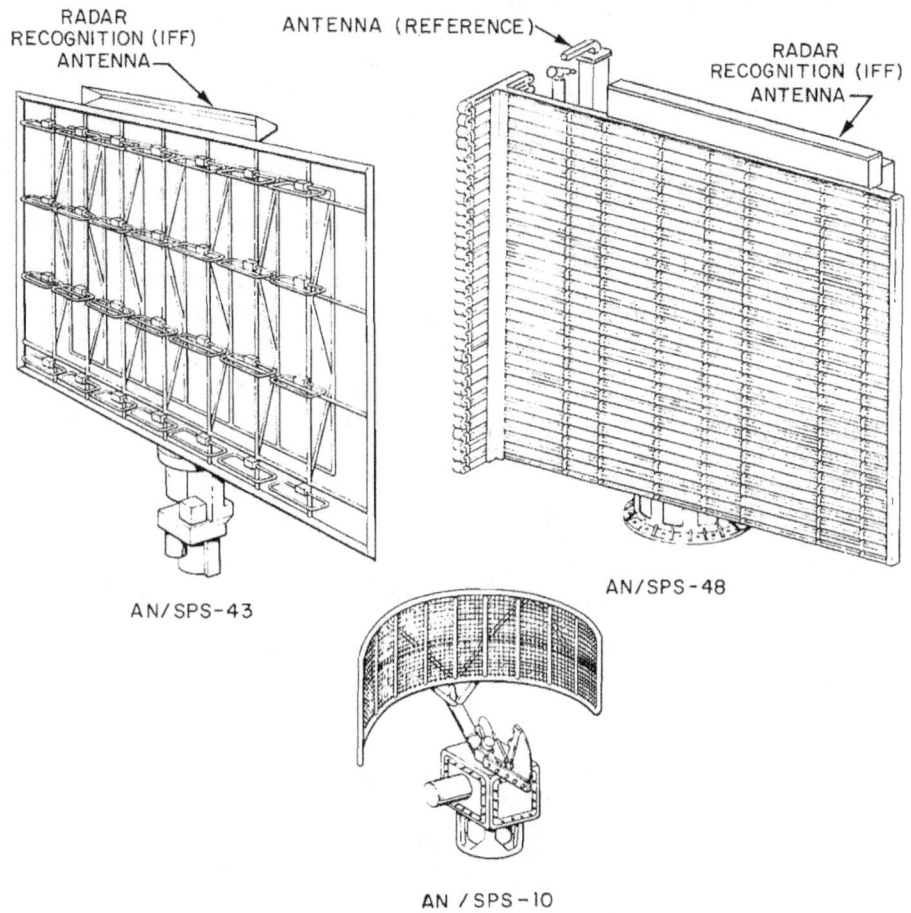

Figure 8-2.—Search radar and IFF antennas.

tracks designated targets, computes launcher and missile control orders, transmits missile guidance signals, and develops information for tactical evaluation by NTDS/WDS-11.

### AN/SPG-55B Radar Set

The AN/SPG-55B radar set contains a G-band tracking transmitter, a capture beam transmitter, a guidance transmitter, and a continuous wave (CW), J-band illumination transmitter. The capture and guidance transmitters are used with beam-riding missiles, and the CW illuminating transmitter is used with homing missiles.

The radar set can be operated in either an automatic or a semiautomatic mode. In the automatic mode, target designations are received from the WDS by way of the fire control computer. Target acquisition, tracking, and missile guidance are performed automatically. In semiautomatic operation, target designations are received by interphone or other means and

255

inserted manually, but the target acquisition, target tracking, and missile guidance are automatic.

For beam-riding missiles, guidance functions are performed by focusing RF energy into a conically scanned capture beam to initially acquire and control the missile after launch, and a guidance beam to guide the missile to the target. By riding along the nutation axis of the guidance beam, which is coincident with the center of the track beam locked onto the target, the missile is able to seek out the target regardless of its maneuvers during missile flight.

The CW illumination transmitter provides two beams of RF energy (main and missile reference) for use by the homing missile. The axes of the main (target illuminator) beam and the missile reference beam are both coincident with the electrical axis of the track beam. The target illuminator beam illuminates the target, thereby providing reflected signals that guide the homing missile to the target. The missile rear reference beam provides identification of the proper illuminating radar to the missile.

The continuous wave acquisition and track (CWAT) modification updates the Mk 76 guided missile fire control system to the Mod 6 configuration, and is used only with the digital fire control systems. This modification provides the radar with a second (Doppler) tracking system for use when the pusle system encounters clutter, jamming, chaff, or a pulse system failure occurs.

**Mk 119 Computer**

This is a transistorized analog computer which is used to automatically solve Terrier fire control problems. In addition, it generates launcher position and missile orders, supplies the AN/SPG-55B radar with the proper angular rates to maintain target tracking, and supplies NTDS/WDS with repeat-back data and status information for FCS operations.

The Mk 119 computer operates in five modes: air ready, designation, track, shore bombardment, and surface-to-surface.

In the air ready mode of operation, the computer receives no target information. The launcher and director remain at predetermined air ready positions.

The computer switches into the designation mode of operation when a designate alert signal is received from the weapon direction system. The computer accepts the target data from a designating source and drives the director to the designated position. On a signal from the radar, in certain designation submodes, the computer generates and transmits a search pattern to the director to assist the radar in target acquisition.

The computer automatically switches into the track mode of operation when the radar locks on the target. Target present position quantities are transmitted to the computer from the radar. The computer drives the director at the tracking rates required to hold the radar on target. From information received from the radar, stabilization quantities from the gyrocompass, and anticipated ballistic corrections preset into the computer, the correct aiming point for the launcher is determined and transmitted to the assigned launcher. The computer also transmits tactical data for display at the weapon direction system. The launcher transmits signals to the computer that indicate the type of missiles on the launcher rails, and the computer generates and transmits the type of orders required for the missiles indicated.

The shore bombardment mode is used only with the BTN missile. During shore mode, the target is a desired burst point and therefore, is not tracked by the radar. Target position relative to own ship position is plotted in the surface operations area of CIC. Target range, bearing, and burst height data from NTDS is used as the designation input to the fire control computer. The missile follows the guidance beam which remains at a constant elevation angle until missile-to-target range (R-Rm) reaches a programmed value, at which time the computer programs the beam downward to the burst point.

The surface-to-surface mode can be used with either a BTN or a homing missile. In the

## Chapter 8—FIRE CONTROL AND ALIGNMENT

BTN surface mode, the director tracks the surface target in range and bearing and the computer programs the director in elevation. The director is held at a fixed elevation angle until R-Rm decreases to a predetermined value. At this time, the computer programs the antenna downward so that its elevation angle is zero when the missile reaches the target. In the surface homing mode, the director tracks in range and bearing but the elevation is fixed at a low angle (Low E). The missile sustainer and proximity fuze are disabled (the SM-1 missile sustainer is not disabled). The missile then homes to the target and warhead detonation occurs when the missile contacts the target.

MISSILE LAUNCHER ORDERS.—Launcher orders for the HTR and SM-1 missiles are generated in the computer and are based on the assumption that the target maintains a straight line constant velocity course during time of flight of the missile. For a given set of present position values and relative motion rates, there is only one set of values for future target position and time of flight. The linear rates of target motion are multiplied by time of flight to determine target movement during time to flight. Adding the target movement coordinates to their corresponding present position coordinates determines the target position at the end of time of flight (called future target position). Ballistic corrections are combined with the future target position coordinates to obtain aiming position coordinates that define the point in space at which the launcher must be aimed. These coordinates are referenced to the horizontal plane. A trunnion tilt correction is performed to obtain corresponding launcher orders referenced to the deck plane.

Launcher orders for the BTN missile are also generated in the computer, and are based on the assumption that the target maintains a straight line constant velocity course during time of flight to capture beam intercept (T6). The linear rates of target motion are multiplied by the time to beam intercept to determine target movement from time of launch to time of capture beam intercept. Adding the target movement coordinates to their corresponding present position coordinates determines target position at time of beam intercept. Ballistic corrections are combined with the intercept position of coordinates to obtain aiming position coordinates that define the point in space at which the launcher must be aimed so that the missile can intercept the capture beam. These coordinates are referenced to the horizontal plane. A trunnion tilt correction is performed to obtain corresponding launcher orders referenced to the deck plane.

HTR/SM-1 Launcher Orders.—Launcher elevation order Edgl' and launcher train order Bdgl' are synchro voltages transmitted to the launcher that aim the launcher at a point in space so that the missile can intercept the target at the future target position. The HTR and SM-1 missiles are neither guided nor controlled during their boost phase of flight. Consequently, the missile deviates from a straight-line trajectory through the line of launch because of the effects of external forces of gravity, wind, and launcher rotational velocity as the missile leaves the restraint of the launcher. Therefore, it is necessary that launcher orders include corrections for anticipated ballistic deviations. In addition, a superelevation adjustment is added to future target elevation so that the missile will fly an up-and-over trajectory. With this type of trajectory, the missile is launched at a higher elevation than that required for a collision course. Thus as the missile approaches intercept, its internal servos cause it to descend upon the target, resulting in greater maneuverability and probability of intercept.

BTN Launcher Orders.—Launcher elevation order Edgl' and launcher train order Bdgl' are synchro voltages transmitted to the launcher that aim the launcher at a point in space so that the BTN missile can intercept the radar capture beam at intercept T6. However, since the BTN missile is neither guided nor controlled during its boost phase of flight, the missile will deviate from a straight-line trajectory in a similar manner as discussed for the HTR/SM-1 missiles. Consequently, launcher orders for the BTN missile include corrections for the anticipated ballistic deviations. It is also necessary to compensate for power modulation of the capture beam and beam rate.

MISSILE ORDERS.—The computer also generates and transmits missile orders that are stored by the missile for use during flight. These are not the same for homing and beam-riding missiles, as shown in table 8-1. The launcher transmits signals to the computer to indicate the type of missiles loaded, and the computer then switches into the appropriate missile submode.

HTR/SM-1 Missile Orders.—While the missile is on the launcher, the following orders are transmitted to, and stored by, the homing missile for use after launch:

1. Missile roll order.
2. Seeker head orders.
3. Variable navigation ratio (N order).
4. Gravity bias.
5. Clutter reject band and sweep select.
6. Doppler predict.

The missile roll order is stored in the missile to provide a vertical reference. This order compensates for the roll of the missile while it is

Table 8-1.—Missile Orders

| HTR AND SM-1 | FUNCTION |
|---|---|
| MISSILE ROLL ORDER | CORRECTS FOR SHIPS ROLL AND PITCH EXISTING AT TIME OF LAUNCH. |
| SEEKER HEAD ORDER | POSITIONS SEEKER HEAD IN THE DIRECTION OF THE TARGET. |
| VARIABLE NAVIGATION RATIO | REGULATES GAIN OF THE MISSILE STEERING SYSTEM SERVOS. |
| GRAVITY BIAS | COMPENSATES FOR EFFECTS OF GRAVITY ON THE MISSILE. |

| HTR ONLY | FUNCTION |
|---|---|
| CLUTTER REJECT BAND | PREVENTS DOPPLER SEARCH IN THE CLUTTER FREQUENCIES. |
| SWEEP SELECT | DETERMINES POSITION OF THE START OF THE SPEEDGATE SWEEP. |

| SM-1 ONLY | FUNCTION |
|---|---|
| DOPPLER PREDICT | DETERMINES POSITION OF THE START OF THE SPEEDGATE SWEEP. |

| BT(N) | FUNCTION |
|---|---|
| MISSILE ROLL ORDER | SAME AS THE HTR AND SM-1 |

## Chapter 8—FIRE CONTROL AND ALIGNMENT

on the launcher (due to the roll and pitch of the ship). The missile roll gyro provides a reference system for guidance and roll stabilization. Before launch and during the boost phase of flight, the roll gyro is caged (locked). Therefore, the missile cannot tell which way is up. The fire control computer provides this vertical reference in the form of missile roll order. At the end of the boost period (about 4 seconds), the roll gyro is uncaged by a servomechanism device, the missile roll stabilizes, and its vertical reference is the reference that was supplied prior to launching. The net effect of missile roll order is to align the vertical axis of the missile guidance reference system with the guidance beam reference system.

The HTR missiles look for the target in much the same way that you would look for a program on the radio if you did not know on what station it was. You would probably start sweeping the tuning knob from one end of the dial to the other while you listened for some identifying sound from the program. The HTR missile seeker uses similar search technique. The seeker circuits (called a speed gate) sweep a narrow band of Doppler frequencies that represent a narrow range of target speeds. To shorten the search time, the missile fire control computer determines where in the speed range the seeker should look. This sweep select information is sent from the fire control computer to the missile before it is launched. Essentially, the sweep select signal tells the missile receiver circuits to look for the target Doppler signal on the low end of the dial or the high end, depending on target speed. The Doppler frequency of a particular target, once acquired, should change very little unless the target executes violent evasive maneuvers which would change the missile-target range rate. The Doppler predict order for the SM-1 missile performs basically the same function as the sweep select order performs for the HTR missile. The Tartar missile also uses this method of seeking the target.

The HTR and SM-1 missiles are launched toward a point in space so they can intercept the target. The fire control computer predicts where the target will be with respect to the missile (seeker head order) at booster burnout. This information is transmitted to and stored in the missile while it is on the launcher (figure 8-3). The target position information is used later to position the seeker on the target. From launch until booster separation, the missile follows a ballistic trajectory. Except for maintaining the control surfaces streamlined, the missile steering system is inactive during the boost phase. Before launch and until a short time after booster separation, the seeker head is aligned with the fore and aft axis of the missile. At booster separation, the missile is roll stabilized, for the missile must know which way is up so it can determine the position of the target. Before launch, vertical reference information is put into the missile gyro system. Before the missile is launched, the fire control computer tells the missile where to look at booster dropoff. This prelaunch order is called seeker head order.

For maximum maneuverability, the missile is aimed and launched in such an attitude that it ascends to a high altitude and then plunges downward to intercept the target. This is sometimes called a hyperbolic trajectory. (No guided missile in current use follows a hyperbolic trajectory.) At target acquisition, however, the missile normally will perform a sharp turning maneuver and proceed in a straight line to intercept the target (figure 8-4). Such a maneuver would cause a large loss in the velocity because of extreme control surface aerodynamic drag, and further loss of velocity because a large portion of the missile flight would occur in the denser air (with increased drag) of lower altitudes. So, to prevent straight-line intercept at target acquisition, the missile is made to turn gradually (figure 8-4) toward the direction of target interception, thereby greatly conserving the boost velocity for the terminal phase of the missile flight and thus increasing the kill probability. The amount of turn required of the missile (N order) is calculated by the computer on shipboard and transmitted to the missile's guidance and control system, where it is stored until the homing phase. This type of missile maneuver is also called proportional navigation, a type of homing guidance. The rate at which the missile turns toward the intercept point for a given error signal from the seeker head is

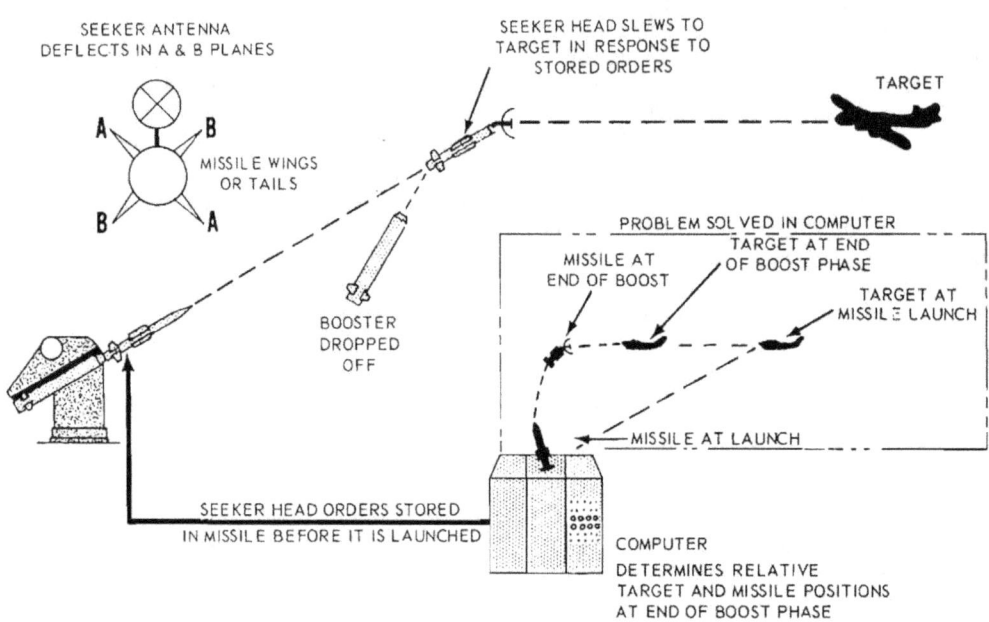

Figure 8-3.—Seeker-head order to missile.

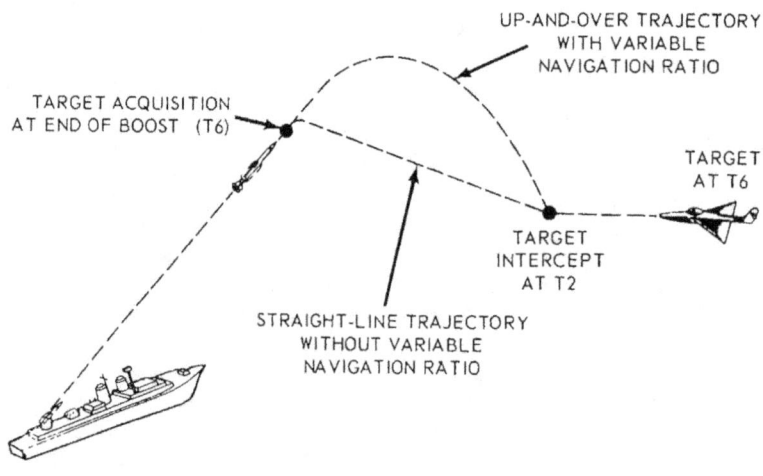

Figure 8-4.—N order to missile.

## Chapter 8—FIRE CONTROL AND ALIGNMENT

determined by a function called the variable navigation ratio.

BT Missile Orders.—Figure 8-5 shows the general path of BT missile orders. They originate in the missile fire control computer and flow through the missile fire control switchboard to the launching system. Within the launching system, they flow through the launching system control circuits to the launcher contactor (figure 8-5) into the booster, and finally end up in the missile itself.

Fuze delay time determines when the warhead will detonate. This preflight order sets the proximity fuze (in the warhead) so it will detonate at a distance from the target calculated to get maximum destructive effect. The best distance to get maximum destructive effect from the warhead depends principally on the target size. Large targets, because the proximity fuze will detect them earlier (the large target has more reflecting area for electromagnetic waves), require relatively greater delay than do small targets.

## MK 76 MODS 6, 7, AND 8 GUIDED MISSILE FIRE CONTROL SYSTEMS

Digital conversion of the Mk 76 FCS is accomplished by removing the Mk 119 analog computer and installing the Mk 152 digital computer with associated peripheral equipment. In the Mod 6 system, a continuous wave acquisition and track (CWAT) module is installed in the AN/SPG-55B radar. In the Mod 7 system, a track module is installed in the AN/SPG-55B radar. A combination of both CWAT and track modules is installed in the AN/SPG-55B radar of the Mod 8 system. The digital fire control system (DFCS) consists of the AN/SPG-55B radar set, Mk 152 digital computer (C152), the Mk 75 signal data converter (SDC), the Mk 19 digital data recorder (DDR), and the Mk 77 input/output console (IOC). A basic DFCS is shown in figure 8-6.

The Mk 152 digital computer is a general-purpose, stored program, real-time digital data processing device used to solve fire control problems and to control operations of

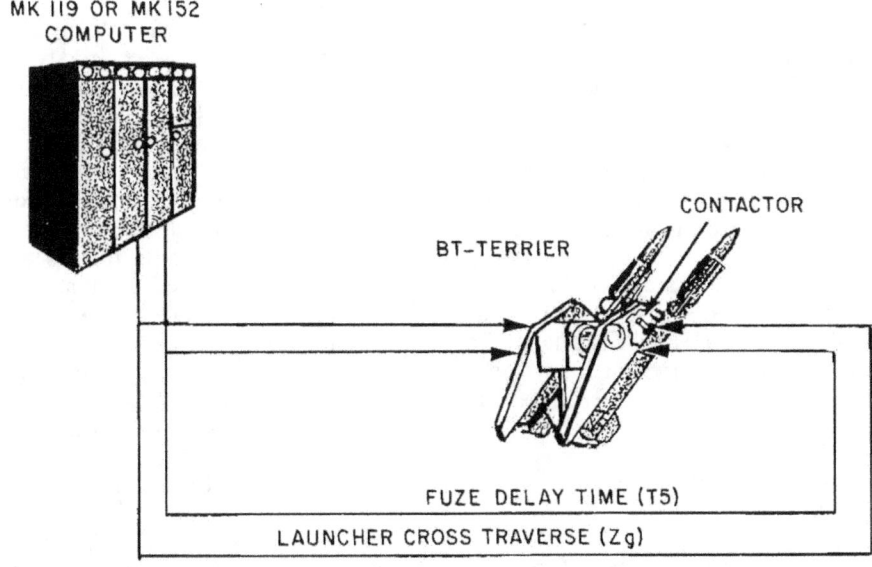

Figure 8-5.—BT missile orders from fire control computer.

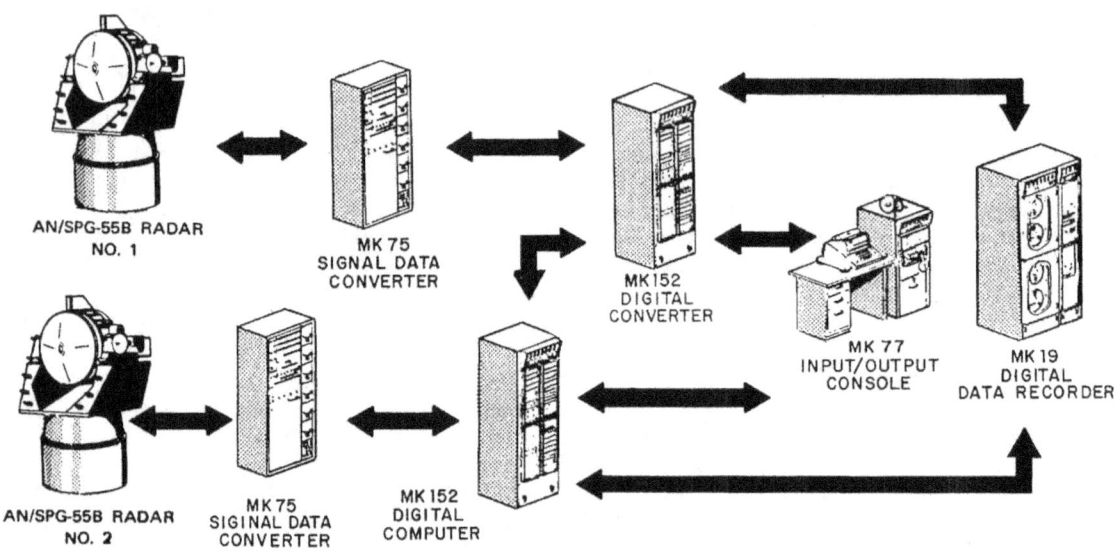

Figure 8-6.—Basic digital fire control system.

the weapons system. The SDC serves as the interface unit between the C152 and the other analog and digital components of the Terrier weapon system. The DDR is used to load tactical and test programs stored on magnetic tape into the C152 memory. It also provides storage for data which is transferred from the C152 and is used for evaluation purposes. The IOC is used to load programs and the other data into the C152, monitor program execution, and to provide a printed record of various events.

All ships with Terrier weapon systems, except aircraft carriers, are scheduled for DFCS installation. Also the AN/SPS-48 search radars will be updated to the AN/SPS-48A. The Mk 76 Mods 3, 4, and 5 are analog fire control systems, and the Mk 76 Mods 6, 7, and 8 are digital fire control systems.

## MK 11 NTDS/WDS FUNCTIONS IN AAW OPERATIONS

We will now summarize the major functions of the Mk 11 NTDS/WDS in antiaircraft warfare (AAW) operations using Terrier missiles. The sequence begins with target detection under conventional battle conditions and ends with target destruction. To simplify the description, it is assumed that the ship is operating independent of NTDS data links and that the attacker is a single hostile aircraft approaching at high altitude and high speed with no IFF. It is further assumed that all equipment is operable and that tracking is by conventional means. Enemy electronic countermeasures are to be considered negligible during the sequence.

So that the discussion may be kept within reasonable bounds, only the essential alerts and responses are described. To make the description applicable to the CG-26, DDG-37, and CG-16 class ships with or without DFCS, and because the NTDS consoles are multipurpose, the text and the accompanying AAW sequence diagram (figure 8-7) refer to operating positions rather than to specific equipment. Also, operating positions aboard the CG-16 and the CG-26 class ships include separate data readout consoles, whereas aboard the DDG-37 class and CG-17 thru -N25 ships the data readouts are incorporated into the display consoles. Hence, data readouts are referred to without identifying the console. Finally, certain positions include

## Chapter 8—FIRE CONTROL AND ALIGNMENT

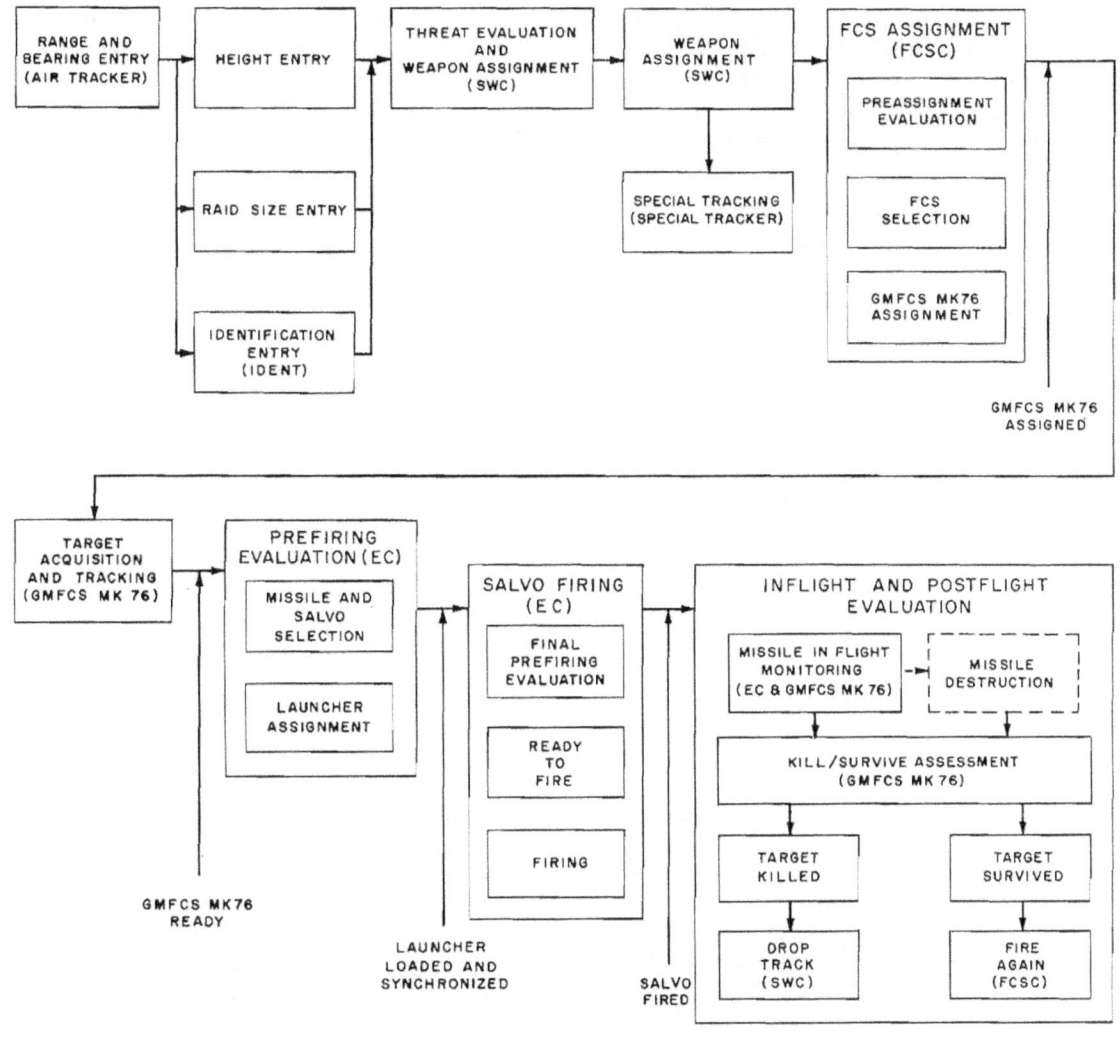

Figure 8-7.—AAW operational sequence.

NTDS digital data introducers (keysets) for manual entry of special data into the system, but their use is not described here.

### Target Detection and Tracking

The approaching aircraft is first detected by the air tracker in the detection and tracking area of CIC assigned AN/SPS-48A master. The air tracker observes the target as video on the PPI display with the three coordinate AN/SPS-48A radar set selected by the radar select switch. The track supervisor (TRK SUP) position takes note of the target and monitors the tracking actions of other tracker positions in CIC.

The air tracker enters the coordinates of target range and bearing into the NTDS computer memory by using the manual track ball to position the ball tab symbol over the video and then presses the NEW TRACK pushbutton to make the data entry. Then a tentative track symbol encircled by a close control symbol replaces the ball tab symbol. As the aircraft approaches, the video moves away from the symbol. The operator restores video and symbol coincidence and enters the corrected coordinates into computer memory by means of the position correction (POSIT CORR) pushbutton. Consequently, the tentative track symbol begins moving with the video.

After a minimum of three position corrections is made, the computer program establishes a firm track. An air-unknown symbol and track-heading/speed line (velocity leader) replace the tentative track symbol at all display consoles concerned and the track is assigned to identification (ID) and height/size consoles. Video and symbol should then remain coincident until the aircraft changes heading or speed, at which time a position correction must be made as before.

### Target Height and Raid Size

The altitude and number of aircraft in the raid are initially detemined at the height/size console. Since the track is firm and under close control of the computer program, target video appears on the raid size analysis indicator (SAI) and on one of the two raid height indicators (RHIs) provided on the height/size video display console (or AN/SPS-48A RSC). The operator, noting that the SAI shows only one target, uses a pushbutton to enter the number ONE into computer memory. Other possible size analyses are FEW (two to four), MANY (five or more), and UNKNOWN.

Target height is not measured directly, but is derived (by the computer) from target range/elevation angle of the video supplied by the radar set console.

### Target Identification

Aircraft identity (hostile, friendly, or unknown) is established at the identification (ID) position. Information available for this purpose includes data readouts of track altitude and velocity data provided by the computer, IFF/SIF interrogation displays, intelligence reports, ordered-maneuver responses, current operational orders, and orders from the officer-in-tactical-command (OTC). Identification begins when an "unknown" identification alert message is received at the ID position. The operator uses a track-sequencing pushbutton to place the track in close control and reviews the available information. The track is judged to be a "confirmed hostile," and the operator enters that information into computer memory. The computer program causes the air-unknown symbol to be replaced by a hostile-air-unengaged symbol at all display consoles concerned. Other possible identifications include "assumed hostile," "assumed friend," "confirmed friend," and "unknown."

### Target Threat and Interceptability Evaluation

After target identification and preliminary evaluation, the next action on the track is performed at the ship weapons coordinator (SWC) position. The SWC is responsible for overall coordination of the ship's combat system including reviewing hostile tracks, pairing each track with the most suitable weapon, ordering engagement with own ship weapons or by interceptor aircraft, ordering termination of individual or general engagements, and handling emergency alerts. Additionally, the SWC is responsible for effecting the orders of the OTC relayed through the force weapons coordinator (FWC) position.

By this time, the hostile aircraft has come closer. This distance is still beyond the maximum tracking range capability of the Terrier Mk 76 guided missile fire control system (GMFCS), but the track is headed toward own ship. The computer program evaluates the threat level as intermediate and transmits an alert-pending message to the SWC position. The SWC places the track in close control by means of the ALERT REVIEW pushbutton, whereupon the computer program causes data

## Chapter 8—FIRE CONTROL AND ALIGNMENT

readouts to indicate both basic track data (speed, course, altitude, range, track number, threat level number, and time-to-go-weapon-release line) and the program recommendation to assign special tracker, and intercept geometry is displayed with respect to the track in close control on the PPI. If concurring, the SWC presses the ASSIGN pushbutton, and the computer program, placing the track in a special tracker list, causes action at the special tracker console.

With no air intercept assigned as the target closes, the computer program recognizes that Mk 76 GMFCS tracking is now possible, and transmits a second alert to the SWC position. The SWC presses the ALERT REVIEW pushbutton and the data readouts present the program recommendation to engage missiles along with current track data. If concurring with the recommendation, the SWC, with the target in close control, presses the ENGAGE MISSILE pushbutton to assign the track to the fire control system coordinator (FCSC) console. At all display consoles concerned, the track symbol is modified to indicate assignment to missiles.

### Precision Tracking

Special (precision) tracking applies to the process of refining track data for subseequent GMFCS or intercept control assignment. The computer program does not make mandatory the use of special tracking, but precise track data reduces GMFCS acquisition time.

Special tracking is performed in range and bearing at the special track (SPL TRK) position and in elevation at the RSC. Special tracking procedures are identical to those previously described for the air tracker position except that the SPL TRK position is operated at low-range scale (32 miles or less) and the computer program automatically offsets the display to the coordinates of the track in close control, which places the track symbol in the center of the PPI. The program drops the track from the SPL TRK list when the SWC terminates engagement. With the SPL TRK on line throughout the entire engagement, the SWC is able to provide accurate target position data to the tracking GMFCS if that system should lose lock on.

### FCS Assignment

Assignment of an air track to an FCS for tracking is accomplished at the FCSC position. The FCSC is responsible for ensuring the most efficient employment of the missile fire control systems, including evaluation of track threat priority and engageability, and assignment of the FCS best suited for the track. Then to maintain fire control radar tracking, the FCSC operator initiates interdirector designation or recommends a change to own ship's course as appropriate. Additionally, the FCSC is responsible for coordination of optical tracking stations and reporting FCS operability status to the NTDS/WDS operational program.

With missile engagement ordered at the SWC console position, an alert-pending message appears at the FCSC position, to which the FCSC responds by using the sequence pushbutton. The track is placed in close control with additional track data on the digital readout (DRO) and the MISSILE ALERT, ENGAGE, and ORDER indicators are lit. The FCSC then presses the appropriate DESIG FCS pushbutton to assign a Mk 76 GMFCS to the track.

Designation data is transmitted to the assigned FCS computer in the proper format by the NTDS converter group. The GMFCS computer converts the designation data to missile director positioning orders, and the director begins slewing to the designated position. Director position is repeated back via the NTDS data system converter group to the Mk 11, NTDS/WDS. Then at all display consoles concerned, the FCS number symbol begins flashing and starts moving from the air ready position toward the track symbol.

### FCS Acquisition and Tracking

Upon arriving at the designated position, the director begins searching in a pattern programmed by the GMFCS computer. When sufficient target echoes are detected, the director radar acquires the target, and the system switches from designate mode to an automatic track mode, sending an FCS on-target signal to the NTDS computer program. The FCS number symbol becomes steady and should be coincident with the track symbol. The GMFCS computer is now computing the solution to the fire control problem.

**Missile Selection, Launcher Loading and Assignment, and Missile Firing**

The engagement controller (EC) position is responsible for final threat and engageability evaluation of targets designated for missile engagement, selection of missile type to be loaded on the launcher for the engagement, assignment of the loaded launcher to the tracking FCS, and salvo firing. The EC is also responsible for reporting launching system status and missile inventory to the NTDS/WDS operation program. The EC position includes a display console and the launching system module (LSM), or a display console and half of the weapons control panel (WCP) on CG-26 class ships.

After the EC position has received a missile alert-engage order message and the EC has placed the track in close control (as described for the FCSC position), the geometry of target engageability is displayed on the console. The display for this illustrative engagement is computed for one of the two types of homing missiles by the tracking Mk 76 GMFCS. The display is similar to that described for the FCSC position but, in addition, shows future height (in thousands of feet) and launcher clearance lines. The EC may choose to call up for display the engageability geometry of beam-riding missiles by using a pushbutton on the display console. By this time the intercept point, symbolized by the ball tab symbol and height numerals moving along the engageability line, is entering the homing missile outer range circle and an FCS READY (or TGT ENGAGE for DFCS) indication appears on the LSM and fire control system module (FCSM) (or WCP). Also, the NTDS symbol is modified to show this condition. The NTDS computer program transmits a launch system-assign FCS (number) message, and the MISSILE and SALVO SELECT indicators on the LSM (or WCP) begin flashing to indicate program recommendations. The EC reviews the engageability display, the track data readouts, and the program recommendation to determine the most suitable missile type. The EC then selects the missile type, orders launcher loading (if not previously accomplished), and selects the salvo, using controls on the LSM (or WCP). Indicators on the LSM (or WCP) permit the EC to monitor the launcher loading sequence. By pressing a pushbutton on the LSM (or WCP), the EC assigns the launcher to the GMFCS and the launcher slews to the computed line of fire. When the launcher is synchronized with the launcher orders from the assigned GMFCS computer as indicated on the LSM (or WCP), and if all conditions are favorable, the computer program transmits a recommend-fire alert to the EC display console.

If the program recommendation is accepted, the EC fires a missile salvo by closing the firing key on the LSM (or WCP). When the salvo leaves the launcher, SALVO-IN-FLIGHT indicators are lit at the LSM and FCSM (or WCP). At all display consoles concerned, the track symbol is modified to signify that the target has been fired on.

**Target Destruction**

Missile salvo effectiveness is judged by GMFCS radar director operators and is based on scope and meter indications of missile and target motion presentation at the director control console (triconsole). When the operator determines that the target has been destroyed, the TARGET KILL pushbutton on the triconsole is used to transmit the information to the Mk 11 NTDS/WDS. A KILL indication appears on the FCSM (or WCP), and a data readout alert appears at the SWC and FCSC display consoles. When the SWC or FCSC sequences to the track, the data readout indicates target killed. Concurring with the report and based on the PPI display, the SWC presses the BREAK ENGAGE pushbutton to initiate transmission of missile-alert cease fire order to the FCSC position. The FCSC releases the FCS assignment by means of pushbutton and number entry on the display console. The director returns to the air ready position. The SWC presses the DROP TRACK pushbutton,

## Chapter 8—FIRE CONTROL AND ALIGNMENT

causing the track to be cleared from computer memory. The track symbol disappears from all display consoles.

If, based upon a fire-again alert from the FCSC, it is determined that the target has not been killed, another salvo may be launched by the EC.

## ALIGNING THE MISSILE BATTERIES

Alignment may be of two types—the alignment of all parts of a component so it functions correctly and smoothly, and the alignment of all the components of a weapons system so they function properly as a whole system. Most of the information on alignment that you find in OPs is of the first type—adjustment of and alignment of the parts of a component. If the weapons system has been manufactured and installed properly and is functioning as intended, it is best not to tamper with it. Adjustments may need to be made to correct for wear or damage.

The alignment of all the components of a guided missile weapons system is done originally by the shipbuilder. Refinements and adjustments may be necessary as the system is worn in. Some of these realignments are made on the shakedown cruise.

### SOURCES OF ALIGNMENT INFORMATION

Prior to undertaking any alignment tasks, you should become thoroughly familiar with the contents of SW225-AO-MMA-010/OP 762 ALIGNTHEORY, Theory of Combat System Alignment manual, formerly OP 762. This manual will assist you in obtaining a general understanding of the total combat system alignment methods. It defines combat system alignment, why it is needed, and what it does and does not accomplish. The principles of alignment and the general reasoning behind the procedures involved in alignment are explained. Detailed instruction for the alignment of specific installations are not covered in this manual.

The SW 225/OP 2456 series, Total Combat System Alignment manual contains specific alignment procedures for each class of ship. This manual is intended to be used as a guide when performing combat system alignment. It is to be used in conjunction with planned maintenance system (PMS) testing and maintenance. PMS tests are designed to check the proper operation of all subsystems, either as a single entity (total combat system), or as individual subsystems. Although most PMS tests are not developed solely for alignment purposes, review of the results of those tests that provide alignment verification over a period of time, will indicate trends toward out-of-tolerance alignment conditions.

When an alignment problem is suspected, reference to this manual will enable the alignment technician to identify the alignment problem, isolate the problem, and perform the corrective alignment. Prior to performing an alignment, study the step-by-step outline. Each step represents an objective that must be accomplished during the alignment. These steps must be performed in the sequence indicated unless otherwise specified. It is recommended that checkoff lists be prepared corresponding to the procedural steps. Check off each step as it is completed before performing the next step.

Alignment data must be documented upon completion to provide information for future checks and to inform responsible personnel of equipment and subsystem alignment status. A complete and accurate alignment data package is essential for effective combat system alignment.

The importance of maintaining an alignment smooth log cannot be overemphasized. (Refer to COMNAVSURFLANT/PAC Instruction C8000.1 (LANT) series and C8000.2 (PAC) series Gunnery Notes for an explanation of what this log should contain.) The baseline document for the alignment smooth log is the shipyard alignment report.

### SYSTEM ALIGNMENT

All alignments that can be accomplished in drydock also can be done at sea. The techniques

# GUNNER'S MATE M 1 & C

may differ slightly but they are sound and efficient. All final alignments are accomplished when the ship is afloat with 80 percent loadout (minimum), water, fuel, stores, and ammunition or equivalent weight.

System alignment requires orienting and adjusting several components to each other so they function properly as a whole. No alignment work should ever be undertaken without first making careful tests to make certain that adjustment is necessary. Before changing any adjustment, make a careful analysis to determine alignment errors and calculate the adjustments necessary. An incorrect or unnecessary adjustment can cause serious trouble in the system.

## Shipyard Alignment

Alignment of a weapons system is primarily concerned with the directions in which the equipment (launchers, directors, guns, etc.) are pointed. To establish directions, a definite and complete set of geometric references must be used. The necessary references are contained in the geometeric structure, called a reference frame, which consists of a reference point, a reference direction, and a reference plane (figure 8-8, view A).

Directions are expressed by giving instructions from a specific point. Any desired point may be selected as the starting point; on the CGN-38 class ships, the reference point is at the intersection of the roll axis and pitch axis and these numbers apppear in the Mk 86 GFCS digital program. Once this selection has been made, it becomes a part of any measurement. Since the measurement must refer to the starting point, it is called the reference point.

After a reference point has been selected, it is also necessary to have a reference direction (figure 8-8, view A), from which to measure angles. The angles are measured about the reference point, starting from the reference direction. In naval ordnance, a fore-and-aft line pointing in the direction of the ship's bow is the most frequently used reference direction.

Angles expressing direction cannot be described completely unless a means is available

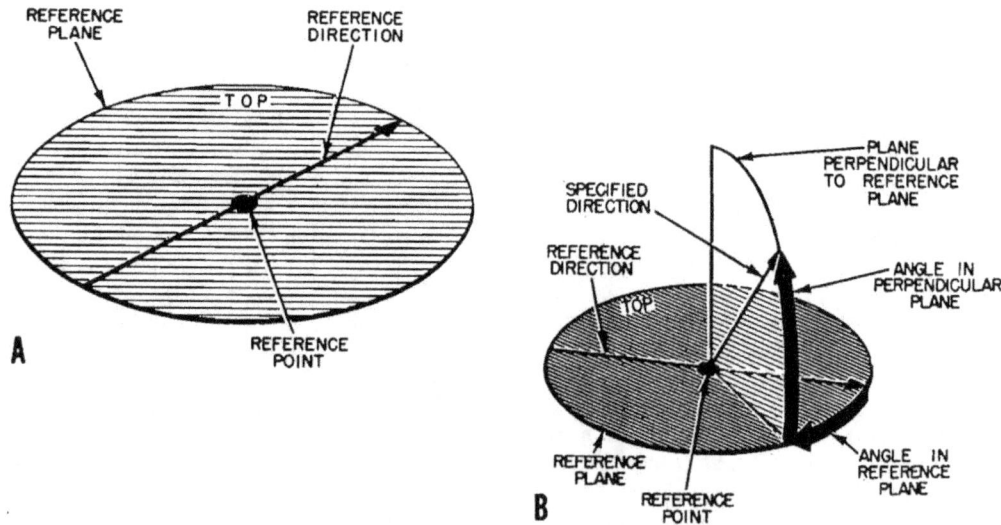

Figure 8-8.—Reference frames: A. Representation of reference frame; B. Expressing direction with respect to reference frame.

94.103

## Chapter 8—FIRE CONTROL AND ALIGNMENT

for specifying the particular planes in which the angles are to be measured. This condition is met when a reference plane is selected. The horizontal plane (also called deck plane) is one of the most commonly used reference planes. When the ship is afloat and you are comparing the horizontal plane to several other planes, two spirit levels are necessary for each comparison (inclination of one plane with respect to another).

The three references described in the preceding paragraphs must all be used when measurements are given to describe directions. In the complete reference frame, directions are specified by two angles measured about the reference point. One angle is in the reference plane and is measured from the reference direction; the other angle is a plane perpendicular to the reference plane and is measured from the reference plane (figure 8-8, view B).

Before any alignment can be accomplished on a new ship, the reference frame must be established. During the construction of a ship, one base plate is installed within the hull of the ship. This plate is referenced to a similar plate on a fixed ground installation. The plate is leveled as accurately as possible before the ship is launched and imaginary base plane is figured from the average readings taken from the base plate. The foundation, and thus the roller paths for the fire control directors, launchers, and gun mounts, are machined so that they are as nearly as possible parallel with the base plane. The fire control reference plane or weapons control reference plane (WCRP) is the horizontal plane to which all combat system elements are related. The weapons control reference plane is perpendicular to the ship centerline plane (SCP), and parallel to the ship base plane (SBP). In practice, it is defined by the roller path plane of one, and at times two, of the major elements of the combat systems. On the CG-26 class ship, the missile launcher is selected as weapons control reference plane (WCRP) for missile related elements, while the Mk 68 director is the WCRP for the gun related elements. A vertical plane perpendicular to the WCRP and lying along the ship's centerline is the zero train reference for the combat systems equipment.

After battery alignment in train comes alignment in elevation. The purpose of alignment in elevation is to set all elements so that when they are positioned in elevation with their lines of sight parallel to their own roller path plane, the elevation dials of the elements will read zero and the elevation synchros will be at electrical zero.

So that guns and launchers can be realigned to the same position, bench marks and tram readings are provided. Upon completion of initial alignment or subsequent realignment by shipyard or support activities, a shipyard alignment report must be submitted to the commanding officer of the ship. Included in this report are the alignment data, tolerances, demonstration results, and other pertinent data for all combat system equipment and subsystems aligned by shipyard personnel.

BENCH MARK.—For purposes of checking director's zero train at sea, a bench mark and bench-mark reading are established. The bench mark usually is a small brass plate with crosslines etched on it. This plate is welded to a secure part of the ship within vision of the director's sights. After zero director train has been established and the dials set, train the director and put the crosswires of the boresight or telescope on the bench mark and read the train angle-read dials. This is the bench-mark reading which should be recorded, and which will remain the same until such time as new waterborne data is obtained. The same telescope must be used for obtaining all settings and readings. The launcher is trained to position its rails parallel to the zero train reference plane and the train indicators are adjusted to the indicated zero train. Then the launcher is elevated to position its rails parallel to its roller path plane and all elevation indicators are adjusted to indicate zero elevation.

TRAM READINGS.—A reference point for each turret, mount, or launcher must be established to check the accuracy of launcher or gun train dials at sea. The original tram readings are taken after zero mount or launcher train has been established and the dials set. However, unlike bench-mark readings, which never change, tram readings will change each time response is broken and any alignment correction is made to the mount or launcher. New tram readings must be taken and recorded after making any alignment correction between mount or launcher and director.

### Shipboard Alignment Requirement

The alignment requirements for a weapons system include internal alignment of each of the components and system alignment of the different components or equipment with each other. The internal alignment of an ordnance component is established at manufacture. A high degree of machining and fitting of structural parts assures good internal alignment. If any basic alignment is necessary because of faulty manufacture, overhaul at a navy yard usually is necessary. Each director should be internally aligned with the ship's references (figure 8-8). All parts of the weapons system are aligned to the reference while the ship is being outfitted or in drydock, and the whole system is tested. When the ship is afloat, the operation of the system must be rechecked. If there are serious distortions, the ship is returned to the shipyard for adjustments.

The launchers and guns must be aligned to the directors in train and elevation.

Alignment work done while the ship is afloat consists principally of tests and adjustments required for keeping the ship's ordnance equipment in readiness to deliver fire of maximum effectiveness.

We will not describe all the procedures of battery alignment that apply to the different types of ships. However, if you understand the following procedures, which are based chiefly on procedures given for the Tartar system on DDG-2 class ships, you should not have much difficulty on any other type of ship.

On DDG-2 class ships, the Mk 68 gun fire control system is aligned first, and the Mk 74 missile fire control system is aligned to it. The Mk 68 gun fire control director is the reference director and the Mk 73 missile fire control director is aligned to it in train and elevation. The Mk 68 director is used to determine the alignment condition of all rotating equipments with the exception of the missile launcher and the gyrocompass, which are aligned to the Mk 73 director. The work of aligning the directors is not done by GMMs, but that work must be completed before the launchers can be aligned. Your alignment checks should be done soon after the directors have been aligned.

We should mention here a preliminary check which must be made before any alignment afloat work is undertaken. This is the transmission check. Synchro and dial errors corrected at this point will keep you from compounding the errors, or introducing errors to correct for errors in the ensuing alignment procedures. (Of course, these errors, even if initially undetected, would be revealed before you completed your alignment work. But by then you would be faced with the task of redoing one or more of the alignment phases.)

Do not proceed with synchro alignment unless the preliminary check shows a misalignment. If the synchro is close to zero, make only the fine adjustment.

### ALIGNMENT OF LAUNCHER

Precise alignment of the launcher requires extreme accuracy in the performance of alignment checks and adjustments. The manual train and elevation features of the Mk 11 launching system make checking very difficult when there is motion of the ship. It is suggested that the checks be made with the ship moored to a pier or at anchor in a calm sea. If the safety warnings are heeded, the checks and tests can be made without damage to the equipment or injury to personnel.

When ready to proceed with launcher and gyrocompass alignment, man the launcher and the gyrocompass, and establish telephone communications.

### Train Alignment

The train alignment check provides an accurate method of determining the degree of parallelism between the zero train lines of all

## Chapter 8—FIRE CONTROL AND ALIGNMENT

equipments of the system. When the director is trained to any point and the launcher dial pointers are matched with zero settings, the director and launcher lines of sight are parallel in train.

Since the ship is now afloat, it is impracticable to use multiple targets to obtain parallelism between the launcher and director. However, if the lines of sight of both director and launcher are aligned on a target at infinite range, for all practical purposes they will be parallel. The most accurate method of alignment is to a celestial body.

When train alignment is performed simultaneously for several equipments, the train dial readings from all stations should be transmitted to a central station (such as the missile plotting room) for systematic recording. The recorders at the individual equipments should cross-check all readings to eliminate possible errors in recording the readings. Rotation of the Earth and ship motion may cause the line of sight to drift from the target, but this drift is not detrimental as long as the line of sight is on the target when the reading is taken.

Install the Mk 75 boresight telescope (modified) in the Mk 8 launcher on the stationary guide, and insert a T-lug in the front guide of the B-rail. The T-lug prevents the front guide from tripping, and thus prevents personnel injury. Install the peepsight on the forward end of the guide arm at a point above five-eights of an inch from the rear of the front guide.

Be sure the launcher is not energized when it is being manned for alignment and test purposes. Clear all unnecessary personnel from the area. An observer is stationed inside the launcher to read the train and elevation dials.

BENCH MARK CHECK.—The bench mark should be in the boresight operator's field of vision. The launcher is manually trained until the vertical crosshair of the borescope coincides with the vertical index on the bench mark. This should be done several times, training to the right and to the left of the bench mark to check on the first alignment.

CORRECTING TRAIN ALIGNMENT ERRORS.—Before attempting to correct the equipment error, carefully analyze the results of the train alignment check. Generally, a small deviation from zero (say, 2') is acceptable. A careful analysis of the launching system is required to determine whether the misalignment is caused by the components of the launching system or those of the fire control system. Misalignment within the launching system may cause serious casualties in the equipment. To isolate the cause of any misalignment, check the operation of the launching equipment to the fixed mechanical positions of stow, dud jet A, dud jet B, load, and transfer. If the launching system operates correctly in those positions in train and elevation, the trouble is in the fire control system. Emergency adjustment can be made at the computer. Afterward, a transmission check must be made between the computer and the launcher. The launcher train dials should indicate launcher train order plus the correction applied. The results should be recorded in the battery alignment log.

NOTE: Once a ship has received a thorough optical, mechanical, and electrical battery alignment, do not change the dial settings before establishing a pattern to verify the alignment.

To correct the train dial, which is in the train receiver-regulator under the launcher shield, deenergize the launcher. Locate the adjustable (vernier) coupling on the B-end response shaft of the receiver-regulator, loosen the lockscrew of the coupling, and adjust the coupling to correct the train response dials by the amount of the train dial error. In other words, make the train dial reading the same as the director train dial reading when both are on target and in manual operation. Tighten the lockscrew on the adjustable coupling and recheck according to the previously described procedure. Continue to make adjustments until the error is within the allowable tolerance.

### Elevation Alignment

The launcher is aligned in elevation to the director. It is elevated in manual control to bring its rails into position paralle to its roller path

# GUNNER'S MATE M 1 & C

plane (at a point of known inclination) within 3' of arc. All elevation indicators are adjusted to indicate zero elevation.

The Mk 75 boresight telescope is installed on the stationary guide as for train alignment. The T-lug is inserted in the front guide of the launcher B rail. The peepsight assembly is installed on the forward end of the guide arm at a point about five-eighths of an inch from the rear of the front guide.

The launcher must not be energized for alignment or test; set the safety switch on the EP3 panel to SAFE and remove the handle. The train and elevation securing pins are released by turning the clamp screws, using handcranks. When the securing pins are fully retracted, they are locked in that position with the clamp screws. The train and elevation latches have to be retracted by hydraulic pressure. The train and elevation latches may be retracted by using the hand pump instead of lighting off the launcher power drives.

BENCH MARK CHECK.—The bench mark check for elevation is conducted in the same manner as for train, except that the horizontal index is used instead of the vertical index.

ELEVATION HORIZON CHECK.—The elevation horizon check provides an accurate method of determining elevation alignment errors between elements under normal operating conditions and a method of determining the relative inclination between the roller paths of the reference and nonreference elements while the ship is afloat. The aft missile director No. 3 is used as the reference element for the Mk 8 launcher and the Mk 19 gyrocompass.

The horizon check is conducted by comparing the dial readings at the director and at the launcher as they are aimed at a series of points on the horizon. The ship should be underway, on a steady course, with a ship's speed of between 5 and 10 knots. A clear, calm day with a well-defined horizon is necessary.

The borescope should be installed in the launcher as for train and elevation check. All stations participating in the check should be manned.

Since the various elements of the missile battery are installed at different heights about the waterline, the dip angles of the elements will differ. The dip angle is the angle by which the line of sight of an element must be depressed before the horizontal to place the line of sight on the horizon. Figure 8-9 illustrates dip angles and dip differences. The dip angle is given in minutes and the height (H) is the height in feet of the element optics above the waterline at

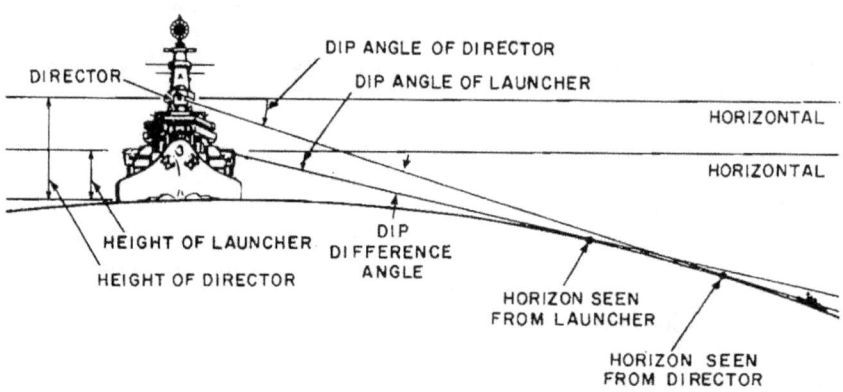

Figure 8-9.—Dip angle and dip difference.

92.48

## Chapter 8—FIRE CONTROL AND ALIGNMENT

mean ship draft. The angle is computed by the formula:

$$\text{dip angle} = .98 \sqrt{H}$$

Before the elevation horizon check is begun, record the dip angle of each element.

The director is trained to a suitable bearing and depressed to below zero degrees elevation. Each element that is being checked is trained to the same bearing and then is depressed to the horizon. Then the director is positioned so the line of sight is on the horizon. As the line of sight approaches the horizon, the operator calls "standby," to alert the operators at the other stations. When the line of sight is exactly on the horizon, the operator calls out "mark." The operators of the elements being checked correct their elevation until the line of sight is exactly on the horizon. When all are aligned, the elevation dial readings are recorded.

These readings are taken throughout the full arc of launcher train, with readings at every 15° of bearing. All data is recorded on worksheets and the information is then plotted on graph paper for analysis.

There should always be a difference between the launcher and the director elevation reading due to the difference in height of these elements aboard ship. The director, being higher, must always depress further to sight on the horizon than the launcher. This angle of depression onto the horizon is called dip angle. The difference between dip angle of the director and dip angle of the launcher is called dip difference. When the launcher and director are properly aligned in elevation, the zero axis of the horizon check curve should be above the zero axis of the graph by the amount of dip difference. If the zero axis of the horizon check curve fails at any other point on the graph, an elevation alignment error exists. This error is positive (high) when the zero axis of the curve is above the value of the dip difference, and negative (low) when below the value of dip difference.

To compute dip angle and dip difference, consult ship's plans to obtain the height of the elements above the waterline. Enter table 8-2 at this height to obtain the dip angle to horizon.

Table 8-2.—Dip of the Sea Horizon

| Height, ft | Dip ' | Angle " | Height, ft | Dip ' | Angle " |
|---|---|---|---|---|---|
| 1 | 0 | 59 | 27 | 5 | 06 |
| 2 | 1 | 23 | 28 | 5 | 11 |
| 3 | 1 | 42 | 29 | 5 | 17 |
| 4 | 1 | 58 | 30 | 5 | 22 |
| 5 | 2 | 11 | 31 | 5 | 27 |
| 6 | 2 | 24 | 32 | 5 | 33 |
| 7 | 2 | 36 | 33 | 5 | 38 |
| 8 | 2 | 46 | 34 | 5 | 43 |
| 9 | 2 | 56 | 35 | 5 | 48 |
| 10 | 3 | 06 | 36 | 5 | 53 |
| 11 | 3 | 15 | 37 | 5 | 58 |
| 12 | 3 | 24 | 38 | 6 | 02 |
| 13 | 3 | 32 | 39 | 6 | 07 |
| 14 | 3 | 40 | 40 | 6 | 12 |
| 15 | 3 | 48 | 45 | 6 | 36 |
| 16 | 3 | 55 | 50 | 6 | 56 |
| 17 | 4 | 02 | 55 | 7 | 16 |
| 18 | 4 | 09 | 60 | 7 | 35 |
| 19 | 4 | 16 | 65 | 7 | 54 |
| 20 | 4 | 23 | 70 | 8 | 12 |
| 21 | 4 | 29 | 75 | 8 | 29 |
| 22 | 4 | 36 | 80 | 8 | 46 |
| 23 | 4 | 42 | 85 | 9 | 02 |
| 24 | 4 | 48 | 90 | 9 | 18 |
| 25 | 4 | 54 | 95 | 9 | 33 |
| 26 | 5 | 00 | 100 | 9 | 48 |

Subtract the dip angle of the launcher from the dip angle of the reference director. The difference is dip difference. It should be constant at all bearings.

All elevation readings are in minutes. On those elements having response dials graduated in degrees and minutes, the dial reading must be transposed to minutes, with 2000' representing zero elevation.

Generally, an error of ±3' is acceptable. Small errors may be the result of incorrect readings. For this and other reasons, it is better policy to do no adjusting of the dials unless large errors are found, after several readings have been taken that definitely indicate that adjustment is needed.

Since most missile system elements do not have roller path tilt correctors or leveling rings, no adjustments can be made to correct for errors in roller path tilt aboard ship. When roller path tilt errors are found to be excessive, correction is accomplished at a shipyard.

## FIRING CUTOUT MECHANISMS

It is hard to overemphasize the importance of checking the firing cutout mechanisms after making the original alignment, after doing any work or repair on the launchers that would disturb the firing cutout mechanism, or in the course of routine checkups. Every casualty caused by ships firing into their own superstructures testifies to the seriousness of any misalignment of the firing stop mechanisms. In every case these casualties could have been prevented. They resulted from negligence on the part of the ship's personnel; or cams were cut improperly and in some cases misaligned; or the firing cutout mechanisms were inoperative through lack of preventive maintenance.

As you remember, firing cutout mechanisms are designed to interrupt electrical firing circuits and firing mechanism linkages whenever guns and launchers are trained or elevated to a position where firing the guns or launchers would endanger personnel or damage the ship. They should not be confused with the limit-stop cams that are used occasionally to limit the movement of some guns and launchers to a safe zone of fire, or with train or elevation limit stops. Firing cutout mechanisms do not interfere with the free movement of the gun or launcher.

The Naval Sea Systems Command has issued definite instructions for personnel responsible for plotting, cutting, installing, and checking firing cutout cams and mechanisms. These regulations must be complied with in all cases. In addition, special instructions govern particular installations.

The computations for the missile's trajectory and for the necessary safety clearances relative to the ship's structures and equipment are complicated and extensive. A high degree of precision and skill are required to make these computations, and to prepare and install the cutout cams in the launchers. The computations are now done with electronic computers at the Naval Surface Weapon Center (NSWC), and the cutout data prepared for the requesting ship. NSWC also prepares the cutout cams and assists in installing and adjusting the cams.

When a new cam is installed, it is essential that the two train reference points be reestablished. These are the train B-end stopped position and the nonpointing zone cam arrested position. The nonpointing zone switches must be set accordingly. NSWC personnel will assist you in this.

The firing cutout cams are plotted, scribed, and cut during the final stages of the initial installation or overhaul period and after all installation and alterations to the topside, superstructure, masts, and rigging are completed.

New firing interrupter switch operating cams must be scribed, cut, and installed whenever changes in the topside arrangement of the ship affect existing areas of fire.

Procedures for scribing and matching the firing cutout cams are given in the applicable OD.

Performance of the cams should be checked before each firing, whenever new cams are installed, and as prescribed by your system's PMS schedule.

The Tartar system actually has four interrelated systems to ensure safe operation of the launcher. These are the (1 and 2) limit-stop system for train and elevation; (3) the automatic-pointing-cutout system; and (4) the automatic-firing-cutout system. The nonfiring zones are identical with the nonpointing zones. The train and elevation systems are physically and mechanically separate but are electrically connected through the automatic-pointing-cutout system and the automatic-firing-cutout system. The components of these system—cams, levers, switches, brakes, etc.—are in the train and elevation receiver-regulators. The pointing cutout system prevents movement of the launcher into zones in which firing would be hazardous. The firing

## Chapter 8—FIRE CONTROL AND ALIGNMENT

cutout system opens the firing circuits so the missile cannot fire when the launcher is in a nonpointing zone or the strikedown gear is attached to the launcher.

The train and elevation limit stops restrict launcher movement under certain conditions. When activated, the limit-stop system neutralizes the associated power drive, thus limiting the movement of the launcher. The limit-stop cam controls the deceleration rate of the launcher power drive. Train and elevation require different rates of deceleration; consequently their cams differ in contour. The actuating cams are identical. When the launcher approaches a nonpointing zone, the actuating cams start the limit-stop system.

An adjustment screw is secured to the bottom of each limit-stop cam. To aid in alignment, scribe lines are scored into the cams. The cam-stacks, which indicate position-plus-lead to the automatic-pointing-cutout and automatic-firing-cutout systems, have a vernier that permits simultaneous adjustment of all the cams in the stack, and each cam can be adjusted to a vernier in its base.

Firing cutout cams, limit-stop cams, and associated shafts, switches, and components are preset by the manufacturer and checked by the installing activity. These cams do not require routine adjustment. They should be checked periodically and should be reset only if they are not within plus or minus 1° of actual launcher settings.

When the launcher is operated in the TEST mode of control, the firing cutout system is checked. You will need the MRC for your system to indicate which lights will activate on the test panel for each condition.

### RADAR ALIGNMENT

All elements of the guided missile battery are aligned in the same manner as a conventional weapons battery. There is, however, one additional step you must accomplish before beginning the battery physical alignment. You must align the radar reference beam and the optical boresight telescope of the radar antenna.

This is accomplished by using a shore tower approximately 100 feet high and at least 1300 feet from the ship, on which is located an optical target and a tunable radar transmitter.

In some missile systems, the radar beam is used as the reference for this alignment. The radar beam is trained and elevated to the tubable transmitter and electrically aligned. The boresight telescope is then adjusted to the optical target and locked in place. In other missile systems the boresight telescope is the reference. The boresight telescope is trained and elevated to the optical target on the tower and then the radar beam is aligned to the tunable transmitter. This is the most critical alignment because, in both cases, the boresight telescope, after alignment, becomes the only reference line of sight for the director.

The above explanation is for drydock alignment, performed by shipyard personnel, perhaps assisted by FTs. When the ship is afloat, the radar reference beam is again checked (by FTs). While at the pier, the shore towers are used. At sea, all guided missile ships will use bow and/or stern towers, installed in accordance with current NAVSEA instructions. Each tower will contain an optical boresight target, a capture antenna, and a track and guidance antenna.

In discussing the alignment of guided missile radar systems, we talked about the alignment of the reference beam to the boresight telescope. But there is more to guided missile radar beam alignment (collimation) than that. In some of our guided missile radar systems we have as many as four different radar beams: track, capture, guidance, and illumination. These must all be collimated to their own zero positions (beam zero indication) and to the reference beam. In some guided missile radar systems the guidance beam is used as the reference beam, while in others the track beam is used. Whatever beam is used, the problem is the same; all the other beams must be collimated to the reference beam.

### SONAR-TO-RADAR
### ALIGNMENT CHECKOUT

The sonar alignment check is performed to assure that the AN/SQS-23 and the AN/SQS-26 sonar are accurate to the degree required by the ASROC weapon system. This check is accomplished by comparing sonar range and bearing with radar range and bearing of a surface target or a snorkeling submarine.

This check should be made monthly or whenever a target ship is available. Because the complete checkout requires considerable time, the entire check may be divided into sections so that at least one section of the check can be accomplished when a target ship is available. A different section of the check should be made each time. Fewer than the recommended bearing readings may be taken for each run when scheduled operations do not allow sufficient time for the complete check.

Selection of a fire control director to be used in this alignment check varies from one class of ship to another. In general, however, the director nearest the sonar transducer should be used. If the horizontal distance between the selected director and the sonar transducer exceeds 20 yards, compensation must be made for horizontal parallax.

The sonar alignment check consists of simultaneous sonar and radar determination of range and relative bearing of a target ship while both the target ship and own ship are on a parallel course at the same speed and at a predetermined range.

It is best to perform dial alignment in calm weather and sea conditions, with good visibility, while the ship is (1) underway in company with another ship to serve as a target at selected ranges; (2) anchored in quiet water; (3) moored to a buoy; or (4) tied up to a dock.

## FINAL ALIGNMENT ADJUSTMENT PROCEDURES

The success of missile flight will depend to a great extent upon the mechanical and electrical alignment of the system. Since guided missiles are used at relatively long ranges, the accuracy with which target angles and range are measured becomes increasingly important. A pointing error of one-half a mil at 30,000 yards will result in a miss distance of 15 yards at the target. The same pointing error at a range of 90,000 yards will result in a miss distance of 45 yards at the target.

NOTE: One mil equals one yard for every 1000 yards of range.

If any error corrections were made to train or to elevation receiver-regulator dials, new alignment readings must be established. Obtain the detailed instructions for your launching system and follow them with care.

Upon completion of the train and horizon checks, the elements of the system are rechecked on their respective bench marks and new dial readings are recorded in the ship's battery alignment and smooth fire control logs.

Although both of the above tests can and should be conducted by ship's force, it is well to remember that any adjustment to either the train or elevation response requires an adjustment also to the load, stow, dud jettison, and transfer position synchros and cams. These adjustments are extremely critical and difficult to make. Before any adjusting is done to the system by ship's force, it would be wise to ask for technical assistance from a repair facility.

### Final Operational Check

Modern ordnance installations are operated almost exclusively in automatic control, except under certain special conditions or in emergencies. Therefore, it is particularly important for an installation to be aligned accurately for automatic operation. If the alignment methods described in this chapter are employed so that the dials of each element are aligned accurately with the pointing line and the synchros are aligned with the dials, a good alignment should be obtained. However, it is advisable to check the results under conditions which approximate those under which the equipment will be operated.

Perform the check with the installation in automatic control, and with the parallax equipment functioning. A boresight telescope will be necessary.

If possible, select various targets at different bearings and at ranges which will be

## Chapter 8—FIRE CONTROL AND ALIGNMENT

approximately equal to mean battle range for the equipment. For antiaircraft installations, try to use air targets which are at an elevation angle near 45°. The target should produce a slow bearing rate, so that accurate tracking is not difficult.

Train and elevate the director to track a target as accurately as possible, particularly in train. When on target, the director-trainer will call "mark" by telephone to the operators at their stations. The operator at each station observes the target through the sight telescope or the boresight, and makes a note of any train error present when the director is on the target. This is done for targets at various bearings, some moving to the right and some moving to the left. In this check, some small error is to be expected because there is always some lag and lost motion in the followup servomechanisms. However, the error observed when tracking to the left should be essentially equal to that observed when tracking to the right, and should be in the opposite direction. If the errors do not change direction when the direction of tracking is changed, or if they are considerably larger for one tracking direction than the other, a misalignment is indicated. This can be corrected by adjusting the train synchros. Before any adjustment is changed, however, a careful analysis should be made to be certain that the error is not caused by some other factor. For example, a misalignment of the the sight telescope could cause an error. This should be corrected by boresighting the telescope—not by adjusting the synchros. In this case, adjusting the synchros would bring the sight telescope on, but would result in firing errors. If after careful analysis an adjustment is made to the synchros, a check should be made to see whether or not a corresponding adjustment must be made to the dials or any other part of the equipment.

### Systems Tests

To operate and maintain launching systems effectively, you must know the relationship between the missile and the launching system. Just as important, you must know their relationship to the rest of the weapons system. This "need to know" about the relationship of each part of a weapons system to the other parts of the system is clearly demonstrated by the various systems tests. Systems tests are designed to check the overall readiness and effectiveness of the entire weapons system. These tests will reveal almost any sign of trouble, especially in the interchange of information between systems and equipment in the weapons system.

You can see that every component in a weapons system is linked either directly or indirectly to the others and so are the operators and maintainers of the equipment. You must think and act in terms of the weapons system as whole. What you do and what your equipment does affect the operation of the system as a unit.

Alignment of the Tartar system is given more coverage here than other systems. Although there are many areas of similarity, the alignment of each weapons system is specific for the ship on which it is installed. Data for your installation must be used when making any adjustments or alignments. The admonition stands: Do not tamper with it if it is working all right.

### REFERENCES

1. NAVTRA 10200-B, *Gunner's Mate 1 & C*
2. NAVEDTRA 10209-A, *Fire Control Technician M 3 & 2*
3. NAVEDTRA 10209A-S1, *Fire Control Technician M 3 & 2 (TERRIER)*
4. NAVEDTRA 10209A-S2, *Fire Control Technician M 3 & 2 (TARTAR) Supplement*
5. NAVEDTRA 10105, *Synchro, Servo, and Gyro Fundamentals*
6. SW225-AO-MMA-010/OP 762ALIGN-THEORY, *Theory of Combat System Alignment.*
7. SW225/OP 2456 series, *Total Combat System Alignment.*
8. OP 3000, *Weapons System Fundamentals; Element of Weapons Systems,* volume 1

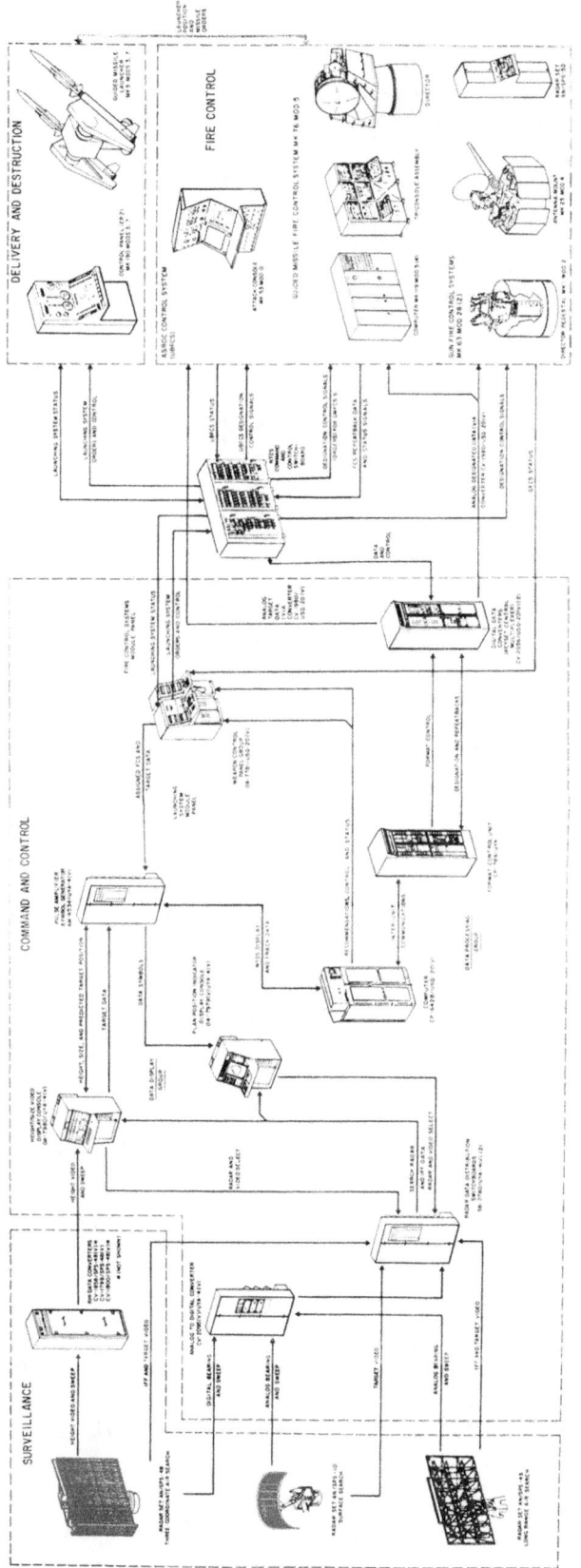

Figure 8-1.—Representative weapon system (Terrier).

# CHAPTER 9

# REPAIR AND TEST

The preceding manual, *Gunner's Mate M 3 & 2,* defined preventive and corrective maintenance and maintenance levels—organizational maintenance (on shipboard), intermediate maintenance (tender), and depot or yard maintenance. Organizational maintenance includes operational and technical maintenance.

Operational maintenance consists of on-the-job inspection, cleaning, servicing, lubrication, adjustment, and preservation of components and assemblies. It also includes the replacement of minor parts when this does not require special skills, or necessitate alignment or adjustment as a result of the replacement.

Technical maintenance is limited normally to replacing unserviceable parts, assemblies, and subassemblies, followed by aligning, testing, and adjusting the equipment.

Tender/yard or depot maintenance involves major overhaul or complete rebuilding of the principal subassemblies, assemblies, or the total equipment.

In performing any type of maintenance, GMMs must use knowledge and skills of two types: First, they must have specific information relating to the particular equipment they are to maintain or repair. Second, they must possess and use certain general skills and knowledge which apply to many kinds of equipment.

The specific information required consists of detailed step-by-step procedures approved for a specific piece of equipment. This information is almost always found in publications prepared by NAVSEASYSCOM or by a vendor of the equipment under contract to NAVSEASYSCOM.

You have acquired the general maintenance skills and information while progressing to your present rate. The procedures generally follow the personnel advancement requirements (PARs) which you must complete as part of your qualifications for advancement in rating. Procedures in soldering, use of basic handtools, performance of basic electrical measurements with devices such as voltmeters, ammeters, and ohmmeters are examples of some of the general skills in maintenance.

Now that you are preparing for GMM1 or GMMC, you must be prepared to teach the basic skills to lower rated GMMs and, at the same time, increase your knowledge and skill so you can take care of advanced work on the weapons system. You need to learn more about the overall and continuing plan of maintenance and the administration of the plan.

The technical duties include the maintenance of specialized test equipment as well as the assembly, adjustment, maintenance, and testing of missile launching components. The manuals written for each missile system and for each series of test equipments provide the equipment details and the approved procedures for repair and maintenance.

## NAVY MAINTENANCE PROGRAM

Planned maintenance for a ship's weapons system has progressed from the division level to a Navywide planned maintenance system (PMS). The PMS includes all components of a ship's weapons system and provides a scheduled maintenance program which details all necessary tests, cleaning, inspection, and lubrication of specific types of ordnance equipment for a specific type of ship. The procedures for a maintenance program for a GMLS are performed at the departmental organizational level and are part of a PMS for a surface missile system (SMS). The PMS/SMS concept involves daily system operability tests (DSOTs) and supplemental system tests and maintenance procedures. The system tests determine the overall operability of a system, whereas scheduled maintenance is performed on the individual equipment of a system to detect possible trouble areas and to maintain a high degree of readiness. General guidelines for implementing PMS are contained in OPNAVINST 4790.4, Maintenance and Material Management (3-M) Manual.

DSOTs are used to give the weapons system a quick run through each day. If any part does not function as it should, corrective maintenance is applied. Routine maintenance established for the equipment is performed on the days assigned, according to the plan. By conscientious performance of scheduled maintenance, minor difficulties can be discovered and corrected before they become serious.

The 3-M Systems do for the entire weapons system what earlier systems of maintenance did for components of weapons and for weapons. *Military Requirements for PO 1 & C*, NAVEDTRA 10057 series, contains a chapter that discusses the managerial aspects of the 3-M Systems from the standpoint of the responsibility of a petty officer first or chief.

## TECHNICAL RESPONSIBILITIES OF THE GMM

A study of the occupational standards regarding maintenance shows that the GMM1 must be able to overhaul, repair and adjust; test and adjust; ..., perform authorized maintenance; ..., record system performance, while the GMMC must be able to plan, implement, and supervise the maintenance and repair program. Two exceptions to this division of responsibility concern the indicator and receiver-regulator equipment in the power drive system. In those systems, it is the chief who must test, adjust, overhaul, and repair. Note that nearly all the basic skills are at the GMM3 and 2 level, with the exceptions of principles of receiver-regulators, and functions of fire control systems and equipment.

The GMM1 is expected to have knowledge of basic layout geometry for drawings and sketches, to be able to prepare freehand sketches, and to read and interpret diagrams and service instructions. The preceding manual explained in some detail the use of different types of ordnance drawings in your work. Drawings of electrical, electronic, mechanical, and hydraulic systems are included.

The decision to repair or replace a component often has to be made by you. This requires a knowledge of the equipment that is both detailed and broad in scope. Knowledge of the supplies or replacements available is indispensable. Before discarding any part, be sure there is a replacement aboard.

The success of any planned maintenance program depends to a very large extent on cooperation at working level. Help your personnel to understand how their day-by-day work of lubrication, cleaning, and similar routine upkeep helps to prevent costly and time-consuming breakdowns and subsequent hard repair work.

## ADMINISTRATIVE RESPONSIBILITIES OF THE GMM

The administrative duties of a GMM1 or GMMC, with regard to the 3-M Systems, are to ensure proper operation of the 3-M Systems within the work center(s). In addition to receiving 3-M Systems' training periodically, you will need to become knowledgeable with the subject matter contained in the Ships'

## Chapter 9—REPAIR AND TEST

Maintenance Material Management Manual, OPNAVINST 4790.4. Your administrative functions will not end with the 3-M Systems. Much of your work will be dictated by the type of ship, missile system, and overall shipboard organization with which you will be involved. More information on administrative functions will be presented in chapter 10 of the manual.

### STEPS IN MAINTENACE PROCEDURE

Maintenance procedures include visual inspection, tests, lubrication, equipment operation, performance tests, and cleaning parts. Preventive maintenance involves four major types of activity:

1. Periodic cleaning.
2. Periodic lubrication.
3. Periodic inspection.
4. Periodic performance checks.

Corrective maintenance is generally performed in three phases: (1) troubleshooting, (2) removal and replacement of parts, and (3) alignment and adjustment. There may be some overlapping between corrective and preventive maintenance; there is no sharp dividing line.

### VISUAL INSPECTION

All components, including explosives, receive frequent visual inspection. Although it is of limited value in detecting some types of weapons system troubles, it is the first method used in trying to find the source of trouble or potential trouble. Do not let it become a casual inspection. Normal ship vibration will cause screws and lugs to work loose; a good visual inspection will locate loosened ones. Loose terminal lugs and screws are a common source of trouble. Loose mounting bolts can be the cause of misalignment.

Cables should be inspected for looseness or damage at places where they enter equipment or at any other point in the cable run. Cables showing signs of damage or abuse should be either rerouted or protected. Particular attention should be given to the coaxial cable which is easily damaged by dents or sharp bends.

Look for signs of overheating and faulty insulation. When these signs appear, the equipment may be blackened around the area of overheating. Check the condition of plug-in and screw-in components (relays, circuit boards, indicating lights, etc.). Sometimes they become loosened by vibrations.

Inspect junction boxes and other unit covers to see that they are properly tightened. Tighten all retaining bolts and dogs evenly and firmly, alternating between diametrically opposite bolts or dogs.

Visual inspection discovers leaks in hydraulic systems, dents and similar damage in pipes, tanks, and other components. The efficient operation of any hydraulic system depends to a great extent on the effectiveness of the seals in keeping air and dirt out of the system and keeping the fluid in it. Fluid leakage can be discovered by visual inspection, though the accumulation of leaked fluid may be some distance from the leak and you have to trace it to its source. Do not sell short this simplest of troubleshooting methods; it can save much time and testing in locating troubles with your equipment.

### CLEANING OF PARTS

One of the rules of preventive maintenance that cannot be overemphasized is "keep it clean." Much of the equipment with which you work is highly machined and has close tolerances. Dirt, dust, or other foreign substances can cause the equipment to operate erratically. Grit causes excessive wear of parts; moisture causes corrosion, thus bringing about inaccuracies in operation. Excessive use of, or the wrong kind of, grease can also hamper operation.

Scheduling routine cleaning is part of your responsibility. In the PMS, as in previous systems of maintenance, intervals of cleaning are based on normal conditions. If you have a situation other than normal, such as an extreme amount of dust, more frequent cleaning may be necessary. Prepare your daily and weekly schedules in accordance with the PMS and modify them to take care of any special situations on your ship. As each job is completed, check it off.

As supervisor and instructor of the personnel doing the cleaning of equipment, be sure that all safety precautions are observed. With any kind of solvent cleaner, ventilation is necessary to carry away fumes. Heat, fire, and sparks must be kept away from solvent cleaners. The Navy has tested many types of cleaners to find the best in effectiveness and safety. Use those recommended in the OP for the equipment, and use them as sparingly as possible. Aside from cost savings, there are several reasons for this. Fumes will be less, reducing the health hazard; danger of fire is lessened; the solvent will not run into parts where it can do damage, as in electrical parts; and skin exposure is lessened, reducing the hazard of dermatitis. Because cleaners are used so frequently, the tendency to become careless with them needs to be held in check. A small amount of solvent on a clean, lintless cloth is the best way to clean small or delicate parts. Federal Specification P-D-680 Type II is the solvent most commonly used on mechanical parts to remove oil, grease, and dust, etc., embedded in them. Alcohol is used for cleaning cork and rubber parts. Always check the OP for the right type of cleaner to be used. OD 3000 describes the different types of solvents and cleaning compounds.

Arrange the layout of work so you have adequate working space, good lighting, and good ventilation. Planning the layout and the work sequence will do much to expedite the work, making it easier, and reducing mistakes and hazards.

## LUBRICATION

You know the importance of lubrication in the maintenance of all equipment. You are acquainted with lubrication charts and have used them in your maintenance work. The types of lubricants and types of lubricating tools were discussed in *Gunner's Mate M 3 & 2,* and in OD 3000.

The parts of the launching system most susceptible to corrosion are those that are entirely abovedeck and constantly exposed to sea spray and water. Maintain the paint on all painted surfaces and a protective coating of lubrication on unpainted surfaces. Flush, clean, and relubricate any bearing surfaces that have been flooded with saltwater. Be sure to use the correct lubricant for the part being lubricated and for the weather conditions. The lubrication chart on the maintenance requirement cards (MRCs) show all points requiring lubrication, give access locations, designate the required lubricants, and tell you how often to lubricate at each point.

Caution against over lubrication is especially important where electrical components are concerned. An excess of lubricant in gear housings also can be a source of trouble. When the oil heats up during operation of the unit, it expands, and it may seep out and into parts where it will cause damage. Always check the oil level during maintenance, and do not add oil above the indicated oil level.

Other cautions regarding lubrication concern cleanliness. If there is dirt, lint, or a gummy substance at the area to be lubricated, clean the area before adding fresh lubricant. When grease-lubricated bearings or bearing surfaces are disassembled, all the old grease must be removed and the bearings and housing washed with solvent before fresh grease is applied. The lubricating tool (grease gun, grease pump, oiler, etc.) also must be clean. Wipe it clean before using it and also wipe the point of application on the unit being lubricated. Before opening an enclosed unit, especially one that has gaskets to keep out dust and moisture, wipe the outside of the container. Do the work in a clean area, and place clean parts on a clean cloth or paper. Just a few grains of grit in a delicate instrument can be ruinous.

Although you are observing these cautions already, see that they are observed by the personnel who are helping you and are learning from you.

## Chapter 9—REPAIR AND TEST

Several grease guns should be available for use by your group so each one can be used for a different type of grease. If you have only one grease gun, you have to clean it thoroughly every time you have to use a different type of grease. Do not mix different types of lubricants. To reach some parts for lubrication, such as all parts of the training circle of the launcher, the train drive pinion, and elevation drive pinion, the launcher must be moved from its stowed position. Take great care to avoid injury to personnel. The air motor is used to move the launcher and the trunnion tube. Never use automatic movement to train or elevate the launcher during servicing, or to move the trunnion tube. Be sure to return the trunnion tube and launcher to stow position, using the air motor, after completing the maintenance work.

Hydraulic systems need a fluid level check at different points in the system. On some components, such as the ASROC adapter buffer used with the Terrier launching system, the fluid level may be noted through a viewing indicator. Hoist the adapter to the loader rail to inspect. If the level is too high, loosen the plug at the bottom of the reservoir and drain enough fluid to bring it down to the required level. The fluid level in the train buffer should be maintained at the height of the filler plug. Oil level plugs mark the filling level on train and elevation gearboxes and hydraulic brakes. When filling or topping off a hydraulic system in a launching system, the preferred method is to use the portable filtering equipment. If the filtering equipment is out of service, OD 3000 prescribes the use of No. 200 or preferably a No. 400 wire mesh strainer. You know the trouble just a little dirt in the hydraulic system can cause, so call upon your experience to impress on your subordinates that the need for such care is not mere fussiness.

## MAINTENANCE TESTS AND CHECKS

It is hardly an exaggeration to say that at any given moment, some part of a weapons system is undergoing test. The maintenance, test, repair, and operational checkout processes are continuous. The formal planned maintenance programs, from the Shatterwhite system to PRISM and IMP, and now the 3-M Systems, all were established to prevent forgetting some components in the maintenance and test programs. Personnel from other ratings perform some of the tests on the fire control system and the weapons direction equipment. The ship's maintenance plan for the weapons system includes all of the ratings responsible for the work. When you plan the assignments for your work week, you must coordinate the jobs with those in other units of the weapons system, and avoid interference in the performance of the work. Cooperation in planning and performance are essential for successful testing of a system. An understanding of the relationship between the parts of the system and the place of each in the whole is needed for intelligent cooperation.

The types and frequency of tests are subject to change. The analysis of results with the 3-M Systems will reveal need for greater frequency of some tests and less testing in other cases. Always check the latest MRC.

### Daily

Some tests and maintenance work must be performed every day. We will use the Tartar system as an example. Daily preventive maintenance and a daily operational checkout are required for the Tartar system. A more comprehenisve weekly operational checkout, plus monthly preventive maintenance, and pre- and postfiring checkouts are required. Each day, in addition to inspecting for leaks, etc., checking pressure, cleaning, and lubricating, certain tests must be made. Daily operation of the launching system perfects the training of the crew and also keeps the lubricants distributed on all bearing surfaces. A Tartar training missile is used. A safety watch is posted topside near the launcher, and the EP2 control panel is manned. The system is cycled three or four times in step control and then in automatic control. The warmup supply switch on the EP1 panel is set to EXERCISE so the missiles do not receive actual warmup; the system then acts as if the missiles were on warmup.

The launcher captain stationed at the EP2 control panel watches the cycling of the launching system. If any part does not perform in the cycle, it is rechecked in step control. The action may be too slow, or it may not take place at all. Then a careful check must be made to locate the cause of the trouble. You may need to get the wiring diagrams.

Part of the daily testing is testing the firing circuits. Four tests are involved: (1) normal firing and misfire, (2) normal firing resulting in a dud, (3) firing of a dud, and (4) emergency firing. During practice operation, tests 1 and 2 are performed daily; tests 3 and 4 are performed periodically.

The Tartar training missile is hoisted onto the launcher. Be sure the warmup supply switch on the EP1 panel is set to EXERCISE during the entire checking of the firing circuits. The checking is done in cooperation with weapons control. All panel operators should have a checklist that tells them the things to do (buttons to push, etc.) at their stations, and what indicating lights to observe for each step.

### Weekly

A weekly schedule for testing, checking, servicing, and lubricating launcher components is listed on the maintenance index page for a designated GMLS and contains a list of all MRCs applicable to the system. Each MRC contains step-by-step instructions for performing the weekly task and, where applicable, shows a lubrication chart for the component scheduled for maintenance. A lubrication chart shows the points requiring lubrication, the frequency, and the type of lubrication. If the recommended lubricant is not available, a tested substitute with the same characteristics may be used. Substitute oils and lubricants are listed in OD 3000. Local environmental conditions may require the use of special lubricants such as cold-weather lubricant. Always cycle the equipment after lubrication. This distributes the lubricant and forces out any excess. Clean up any excess, and clean up your grease guns and other applicators after use—before you stow them.

HYDRAULIC CHECKS.—Checking hydraulic fluid levels can be a daily or weekly maintenance requirement. Fluid levels may be checked by a sight gage or a dipstick. Most header tanks have some type of sight gage for quick easy fluid checking where main supply tanks contain a dipstick mechanism. In most GMLSs, a main supply tank contains an oil fill and drain valve while most header tanks contain only a fill valve or fill cap.

CHECKING ACCUMULATOR PRESSURE.—In most missile jettison systems air and/or nitrogen pressures are checked daily. In other accumulator assemblies weekly checking is the rule. The correct pressure of each nitrogen accumulator system varies with the ambient temperature. A table of temperature-pressure requirements may be mounted near a nitrogen accumulator charging assembly. This table lists the required nitrogen pressure for a given temperature recorded on a centigrade thermometer attached to the nitrogen charging valve block assembly. If a table is not attached to the charging asembly, a temperature pressure tabulation chart will be included on the MRC for the system being maintained. If a launching system has been in operation prior to a maintenance requirement, wait about 2 hours before checking an accumulator system. The waiting period should allow the system to cool so that thermometer gage readings represent normal ambient temperatures.

### Monthly

All missile magazines have either a saltwater or an oil operated hydraulic sprinkling system actuated by an automatic (thermopneumatic) control system. Sprinkling systems are tested monthly to ensure proper operation. Tartar missile magazines also have a water injection system which is used to diffuse the exhaust flame resulting from rocket motor ignition in the magazine. Water injection systems are not tested but are checked periodically to ensure that the freshwater and air pressure used within the system are at their required levels. Built-in carbon dioxide systems installed in missile magazines are tested in accordance with the ship's current policy.

## Chapter 9—REPAIR AND TEST

Certain tests and checks of the weapons system are regularly scheduled to be performed every 30 days. A number of the monthly tests are performed on the operation of related parts of the weapons system, with a GMM1 or C cooperating. You may be placed at the launcher captain's control panel to observe and report the indications at your panel. Monthly tests on the Mk 13 system repeat the daily and weekly tests, but with additional procedures. Testing the sprinkling system adds several procedures, including air-testing of pneumatic lines for tightness and operability of the heat sensing devices, air-testing for unobstructed flow between the sprinkling heads, flushing the associated firemains, cleaning the saltwater strainers and the drain hole, giving the system an operational test, and checking the operation of all the valves. Review chapter 7 for these operations.

Testing the carbon dioxide system in the magazine was described in chapter 7. The monthly testing includes inspection to discover any breaks in the tubing or other leakage. The supply piping is air-tested for tightness, and the operation of the system activation alarms are tested. The supply of $CO_2$ is cut off during the tests by disconnecting the control and discharge heads and capping the connections to the supply tanks. These are precautions to prevent the escape of $CO_2$ while working in the launching system. Compressed air is used to test the operation of the system, connecting the ship's air supply to the carbon dioxide lines with an adapter. If the air pressure gage shows even a slight drop in pressure, the leak must be found and repaired. A drop in pressure could prevent the operation of the alarm system when it is needed to warn of a fire.

The control heads are checked with the use of a pneumatic hand pump and an air gage. Remember the warnings about the suffocation danger from $CO_2$, and see that the supply cannot be turned on accidentally. Connect or disconnect lines, control heads, and discharge heads in the order given on the MRC or your checksheet so there is no outflow of carbon dioxide at any time while anyone is working in the launching system.

Other monthly checks and maintenance procedures include the cleaning of the steam strainer and fluid strainer, checking the bladder pressure in the anti-icing system, and checking the operation of interlock switches. Solenoid switch operation is checked quarterly. Among the interlock switches that are tested (whether for actuation or continuity) are numerous sensitive switches, single-element switches, paired switch elements, microswitches, and rotary switches.

### Quarterly Tests

The checking of timing operations usually is done through a series of tests performed quarterly. For the launching system to operate properly, the components must act within the time limits set for each—a matter of seconds. A stopwatch is needed for testing certain operations. An electric timer is used at the EP2 panel. On the Tartar Mk 13 GMLS, use a 60-second, 115-volt a.c. timer with a special, self-contained, 24-volt d.c. rectifier. You will need the wiring diagram to make the proper connections. Each timing operation is repeated two or three times and the time is recorded. If the average is not within the limits needed for that action, you must search for the cause of the trouble and correct it. The tests are conducted with all the motors running (except where indicated otherwise). Figure 9-1 illustrates a test

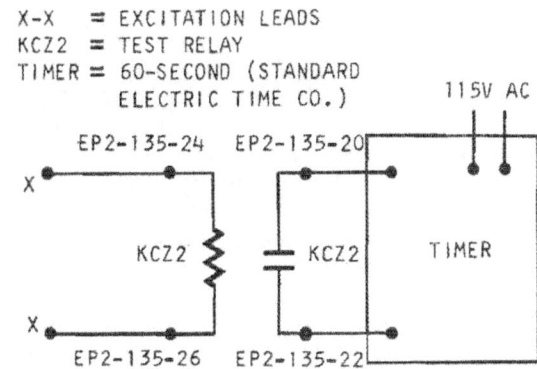

94.161
Figure 9-1.—KCZ2 and timer circuit, used in timing of operations.

circuit used to time the operation of a Tartar system. Timing relay KCZ2 is connected into the circuit being tested by two excitation leads. The contacts of KCZ2 control the start and stop action of the timing mechanism. The following are some of the timing tests made on a Tartar launching system; similar tests are made on other GMLSs.

1. Check and record the time it takes to index between stations on both inner and outer rings of the ready service ring.

2. Check and record the time it takes to extend and retract both the inner and outer hoist retractable rails.

3. Check and record the time it takes to open and close the blast door.

4. Check and record the time it takes to raise and lower the hoist chain under the designated conditions with the chain shifter at either the inner or outer ring (depending on which ring the Tartar training missile is stowed in).

5. Check and record the time it takes to extend and retract the electrical contactor and fin opener housings.

6. Check and record the time it takes to extend and retract the fin opener cranks.

7. Check and record the time it takes to arm and disarm a Tartar training missile.

8. Check and record the time it takes to extend the launcher retractable rail when it is empty.

9. Using a stopwatch, check and record the time it takes to retract the jettison piston (test circuitry and electric timer not used).

10. Check and record the time it takes to load a missile (1) from the initiation of a single load order until the fins are unfolded and the fin opener cranks are retracted, (2) from the initiation of a single load order until the next missile in the magazine is ready to hoist, (3) from activation of the system to automatic until the launcher is loaded and synchronized to the load order, and (4) from activation of the system in automatic until the launcher is loaded, synchronized, and ignition of a simulated missile is indicated.

11. Check and record the time it takes to unload a missile from the launcher.

12. Check and record the effective time delays imposed by the time delay relays that are mounted within the EP1 and EP2 panels. In addition to the electric timer and a stopwatch, a jumper is needed. Figure 9-2 shows how to wire up a time delay relay for testing purposes.

All these tests require careful attention to detail—connecting to the right circuit, exact time, particular sequence of steps, and careful recordkeeping. They cannot be done hurriedly. Schedule them for a time when they are not likely to be interrupted.

**Periodic Tests**

Some tests are scheduled to be made every 3 months, every 6 months, or yearly. Refer to the maintenance index pages for a listing and basic description of the scheduled maintenance and situation requirements covered by the PMS. These requirements are tailored for your particular system and appear on your work

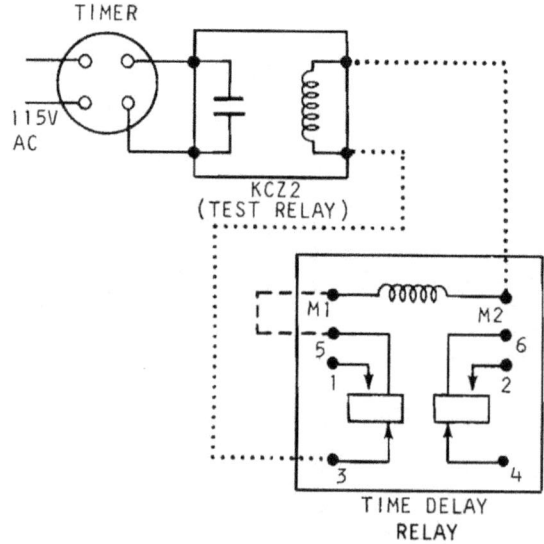

94.162
Figure 9-2.—Circuit for testing time delay relay.

## Chapter 9—REPAIR AND TEST

center cycle and weekly PMS charts. Corrective MRCs are listed on index pages maintained in a separate binder and do not appear as scheduled maintenance. Unscheduled maintenance actions are made when there are indications that maintenance or testing is needed. Your experience and judgement are needed to decide.

## LAUNCHER SHIPBOARD PERFORMANCE TEST

Launcher performance tests determine whether launcher equipment functions satsifactorily under various operating conditions. Most launcher performance tests are conducted quarterly, but an existing condition (a suspected malfunction) may require that certain performance tests be conducted more frequently. Regardless of the frequency of performance tests, personnel conducting these tests must be familiar with both the testing procedures and the test equipment used. The test equipment is used to control and record the performance of the launcher power drives while they are being controlled by the launcher test panel. The responses that are recorded represent instantaneously the error position and velocity of the power drive unit under test. Most shipboard launcher systems have a test panel (EP3 panel) which contains all the necessary test receptacles and test jacks for connecting the different test cables, leads, and jumpers to the test equipment required for a given performance test. Switches, pushbuttons, and control knobs used to control the launcher test equipment are located on the face of the test panel. A local director (internal director) is mounted within the test panel which enables the launcher to be positioned in train and elevation in local control or in a test mode of operation. A dummy director (external director), an error recorder, two limiter and demodulator units, and a frequency generator are used in conjunction with the EP3 test panel when conducting launcher performance tests.

Some of the older GMLSs do not have an EP3 test panel. These systems have separate local control and test panels.

## DUMMY DIRECTOR AND ERROR RECORDER

Dummy directors and error recorders are used for routine shipboard dynamic testing of synchro controlled followup systems such as train and elevation power drives for missile launchers. Their purposes are to simulate a command synchro signal, normally sent to the missile launcher power drive servosystem by the missile fire control system, and to record the error between the selected input signal and the actual output response of the power drive servosystem under test. The information obtained is used for analysis of a launcher's electrohydraulic control system and launcher power drives.

The dummy directors predominantly used with the Tartar and Terrier missile systems are the Mk 6 Mods 0 and 1. Figure 9-3 shows the control panel of the Mod 0. It was designed primarily for shipboard testing of GMLSs. It does not replace test equipment used for laboratory, factory, shipyard, and installation tests. The records of installation tests are retained on board the ship, and the first record made with the dummy director and error recorder aboard the ship is kept for comparison with subsequent tests by the same test equipment. Each major unit of a ship's weapon system controlled by a synchro system and positioned by an electrohydraulic power drive has a separate weapons system publication (OD) which lists all the individual shipboard tests conducted at the time of initial installation. Each major unit, a launcher, a gun mount, a rocket launcher, etc., is subject to a complete set of performance tests as detailed in an installation test instruction OD for each unit installed. The results of all tests are recorded or added as an appendix in a shipboard copy of the OD for future reference.

Any information obtained from a routine shipboard performance test could result in detection of a significant performance deterioration of a launcher component and can warn of impending failure which could result in the need for corrective maintenance, repair, or overhaul.

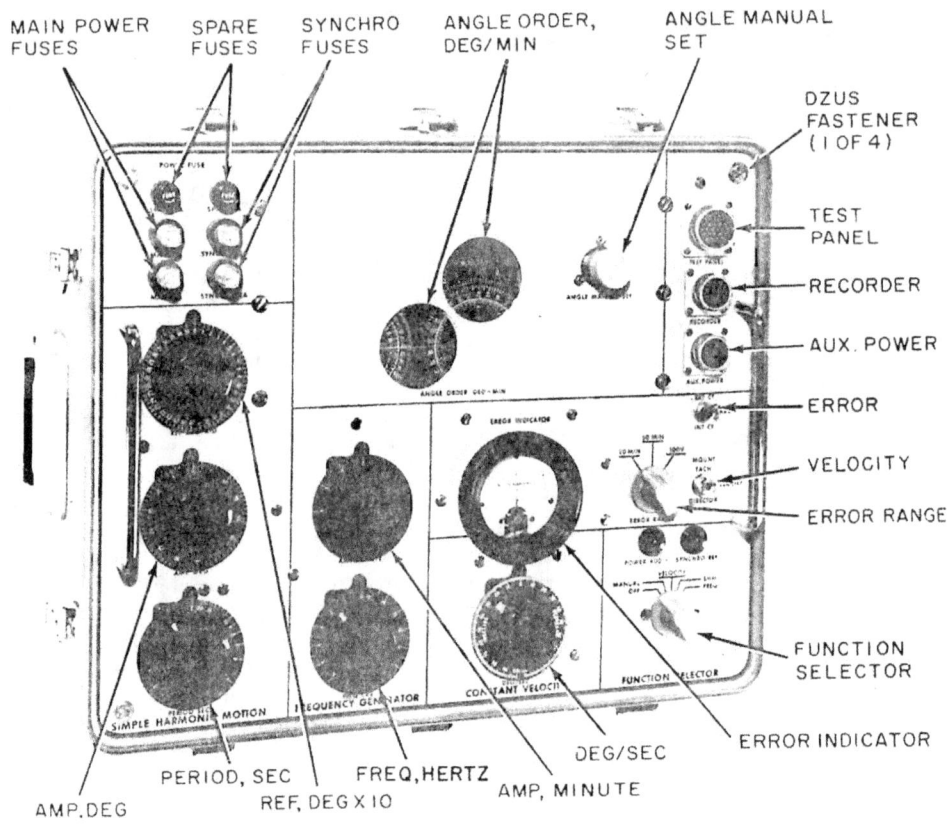

Figure 9-3.—Mk 6 Mod 0 dummy director control panel.

## Mk 6 Dummy Director

The Mk 6 Mod 0 dummy director weighs 78 pounds. It is housed in a portable, aluminum case approximately 19 inches long, 16 inches wide, and 12 inches high. It requires 1-ampere, 115-volt, 400-hertz a.c. power for operation, and it accommodates either 400-hertz or 60-hertz synchro signals. All power is supplied from the launching system test panel through cable stored in the case. An auxiliary power conversion unit is required for 60- to 400-hertz conversion if the system to be tested is limited to a 60-hertz power supply. The principal components of the dummy director are:

1. A main servodrive, with transistor and potentiometer control. It drives two synchro transmitters for 1- and 36-speed order signals, together with a 36-speed synchro control transformer for error detection purposes. It also drives a d.c. tachometer for generation of a signal velocity order required for certain power drives.

2. An auxiliary servodrive, with similar transistor and potentiometer control, for generating oscillating signals for simple harmonic motion control of the main servodrive for frequency generation.

All manual controls, indicators, and connectors are located on the control panel (figure 9-3) for the following operations. These tests are made for train and for elevation.

## Chapter 9—REPAIR AND TEST

MANUAL OPERATION.—During manual operation, the servodrives are disengaged from the synchro gear train which permits the operator to position the output synchro rotors to any selected angle from where the particular test may begin. The manually set output signal is stationary, enabling the power drive under test to synchronize with the dummy director output. Do not, under any conditions, turn the knob on the 1X dial by hand.

CONSTANT VELOCITY OPERATION.—Constant velocity test signals are used to drive the unit under test at a constant speed. The velocity may be adjusted from 0- to 100-degrees-per-second in either direction. An oscillograph is used to make the test traces.

SIMPLE HARMONIC MOTION OPERATION.—Simple harmonic motion test signals provide a sinusoidal input signal to the power drive under test. The sinusoidal signal causes the driven unit to oscillate about a present reference point which is adjustable through 360°. The period of oscillation is adjustable from 4.5 to 18 seconds, and the amplitude of the oscillation is adjustable to a maximum of 60° at a period of 9 seconds and longer and up to 10° at a period of 2 seconds.

FREQUENCY GENERATION OPERATION.—Frequency generation operation of the dummy director produces a sinusoidal signal of 0 to 12 minutes' amplitude which is superimposed on the 36-speed synchro output signal. During frequency generation operation, the output command synchro transmitters of the dummy director are positioned automatically at a 10-degree electrical zero position. The frequency of the frequency generation signal is adjustable from approximately 0 to 18 hertz.

The frequency generator components are mounted to the frequency generator chassis, which has a removable cover. A connecting cable is provided for connecting the generator to the EP3 panel. Open and closed views of the frequency generator are shown in figure 9-4. It is used to test the frequency response characteristics of the launcher train and elevation systems.

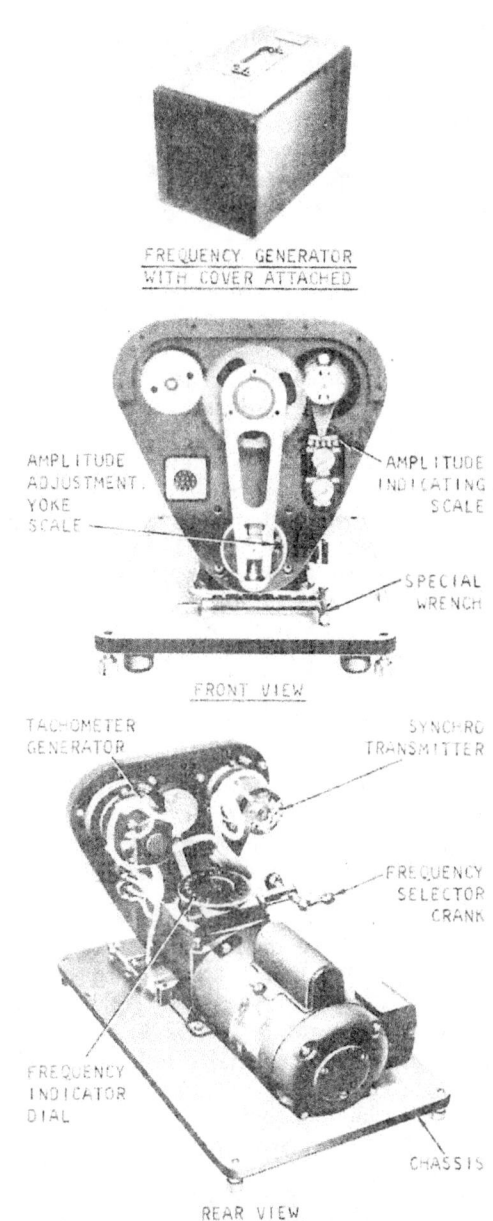

94.163
Figure 9-4.—Frequency generator, enclosed and interior views.

289

# GUNNER'S MATE M 1 & C

## Dual-Channel Oscillograph

Since the oscillograph has two channels, two different traces may be taken at the same time. This allows corresponding trace results to be compared to learn more about the launcher operation. Normally, three types of test traces are taken: B-end error traces, velocity traces, and position traces (figure 9-5).

NOTE: Always calibrate the Brush oscillograph before recording any traces.

The voltages for the B-end error trace are obtained from the 36-speed CT in the receiver-regulator that is geared to B-end response. The synchro rotor is geared to rotate at 36-speed while the stator is electrically connected to the 36-speed CX in the controlling test director. The 400-hertz rotor output voltage indicates the error between the generating director and the B-end response shaft. The CT rotor output voltages are circuited through the control test panel to the limiter and demodulator unit and then to the oscillograph.

The B-end position trace voltages also are obtained from the 36-speed CT. Through the proper switching on the control test panel, the output voltage produced will indicate the B-end position, and not error. The position output voltage also goes through the limiter and demodulator unit, and is recorded by the oscillograph.

The B-end volocity trace voltages are obtained from the d.c. tachometer generators located in the receiver-regulator. The tachometer generators are geared directly to the regulator B-end response input shafts and furnish a d.c. voltage which is proportional to the B-end velocity. The tachometer output is circuited through test instrumentation to the oscillograph.

READING TEST TRACES.—Test traces are read like ordinary graph curves. They illustrate the error, position, or velocity of the launcher at the time the tests were made. Traces below the zero reference line are of the opposite phase from traces above the zero reference line. Be certain to check the following when reading test traces: (1) type of test being checked, (2) type of trace being used, (3) test conditions, (4) calibration on the left margin of the graph, and (5) the time allotted for each of the vertical graph divisions.

Use the calibration curve shown in figure 9-6 to determine the exact B-end positions when they are less than 5°.

As the error and position trace voltages are generated by the 36-speed synchros, difficulty may arise in reading test traces if the error or position reading is greater than 2.5°. The 36-speed synchro is geared to rotate 36° for each 1-degree movement of the launcher. A launcher movement of 2.5°, therefore, corresponds to 90° rotation of the synchro. Since a synchro generates maximum output with a 90-degree rotor or stator displacement, the maximum trace indication occurs at an error or position displacement of 2.5°. Error or position traces greater than 2.5° require a special method of indication.

Since a complete revolution of a 36-speed synchro corresponds to 10° of launcher movement, one complete cycle of a position or error trace corresponds to 10° of launcher movement. For example, if the error or trace position consists of 6-1/2 cycles, the trace will measure 65° of position or error (10 x 6.5).

CALIBRATION OF TRACE RECORDER.—The missile launching system control and its test panels may differ in switch arrangement, identification, and circuitry not only for different missile systems, but for different installations of the same missile system. You will need the elementary wiring diagrams to determine actual identification of switches and positions, and what each switch controls. Use only the special cables supplied for interconnections of test instruments and the test panels of the missile launching system control.

The oscillograph is calibrated during launcher testing procedures; and before any launcher shipboard tests are made, the following general checkoffs must be performed.

1. Check the oil level at the main supply tank.

2. Check the oil level of all gear housings associated with the train and elevation power drives.

## Chapter 9—REPAIR AND TEST

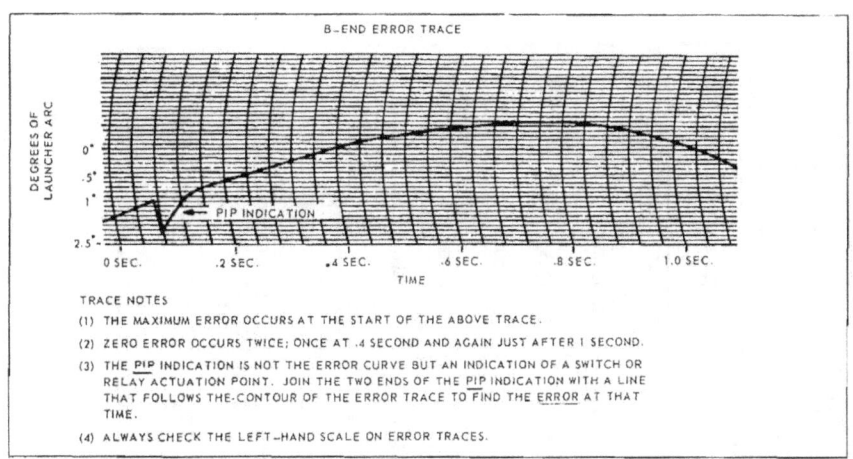

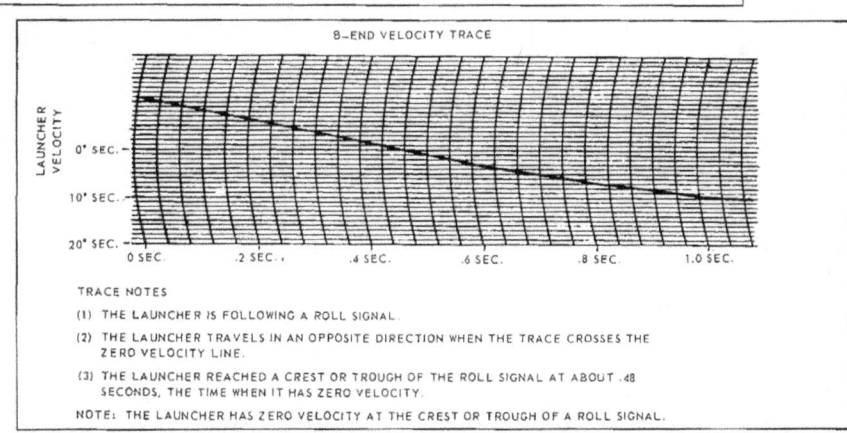

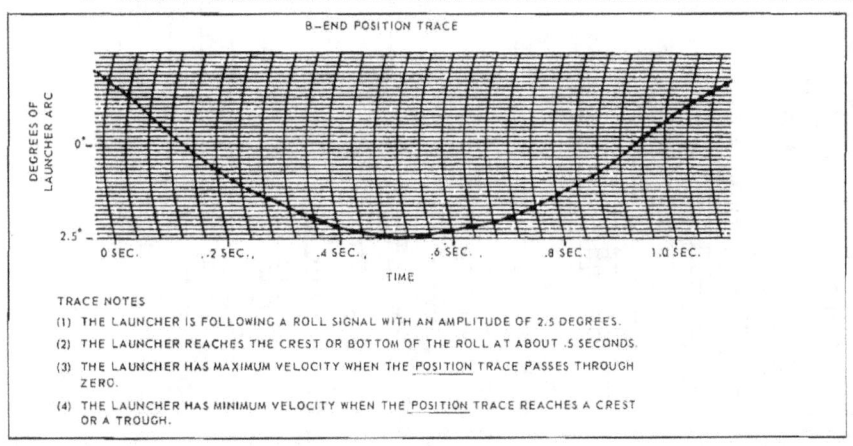

Figure 9-5.—Oscillograph traces of launcher response: A. Sample error trace; B. Sample position trace; C. Sample velocity trace.

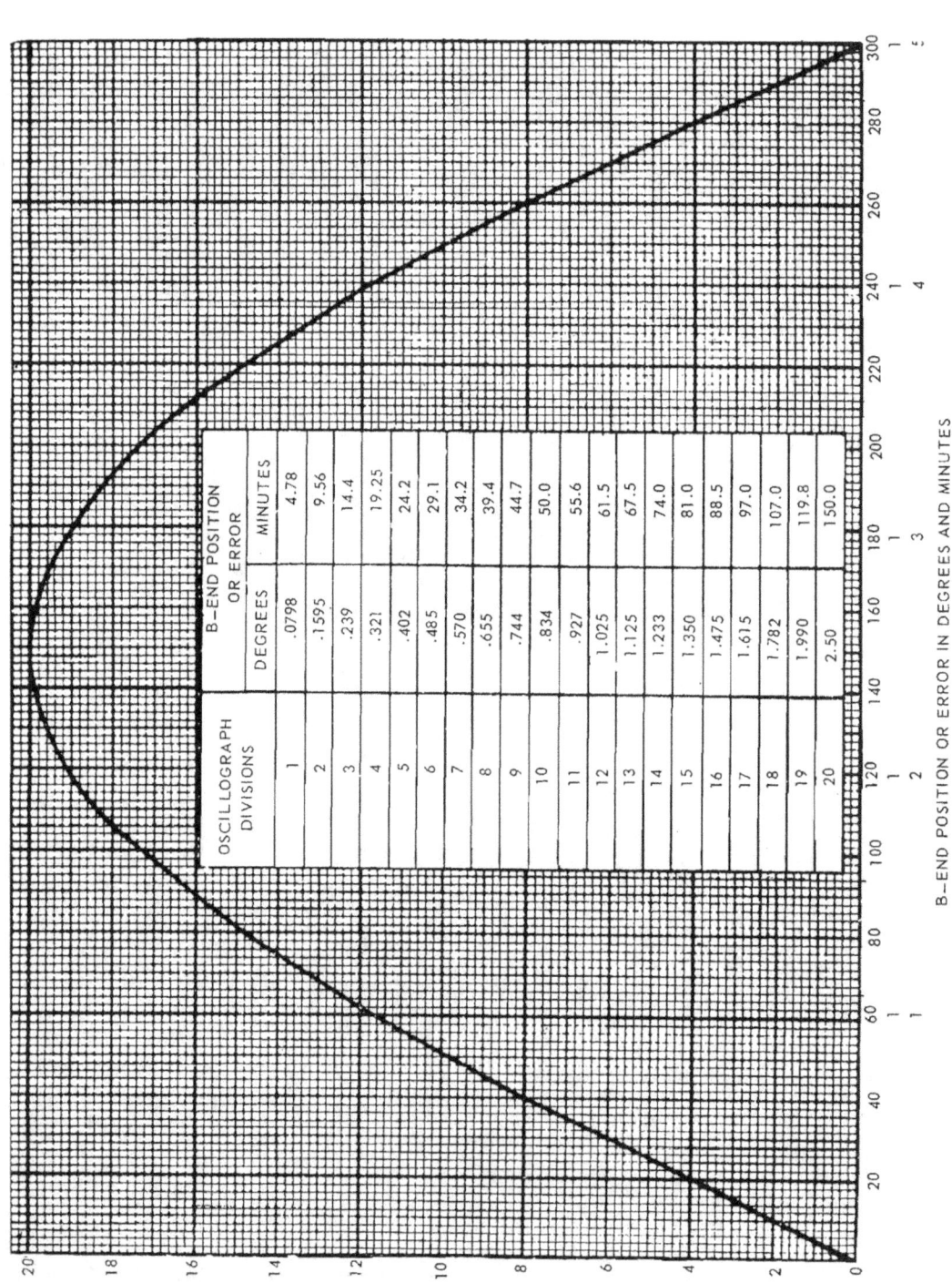

Figure 9-6.—Position and error trace calibration curve.

## Chapter 9—REPAIR AND TEST

3. Lubricate the launcher components properly.

4. Charge the launcher accumulators properly.

5. Vent all hydraulic units properly.

6. Check the train warning bell operation.

7. Train the launcher through its maximum limits to verify free and unobstructed launcher train movements.

8. Elevate and depress the launcher guide arms to their maximum limits of travel to verify free and unrestricted guide arm movements.

Train and elevation air motors are used for items 7 and 8 on systems so equipped. Power drives are not activated for these checks.

CAUTION: Do not move the guide arms or start power drives unless it is known that the firing cutout mechanism is adjusted properly. Failure to do so may result in extensive damage to the firing cutout mechanism.

9. Load the launcher rails with standard inert missiles or equivalent unless specified otherwise for the individual test being performed.

10. Check general condition of test instrumentation and service as required. Use black ink in the oscillograph so that test traces can be reproduced clearly.

After all these preliminary checks are made, activate the launcher by switching on the EP1 power panel, and start the train and elevation motors. The EP2 panel should be switched to STEP control, and control of the launcher power drives switched to the EP3 panel. The test cables are connected to the EP3 panel. Allow the train and elevation power drives to operate at least 30 minutes before making test traces.

Two different methods of calibration are used. Error and position traces are calibrated by one method and velocity traces are calibrated by a second method.

The error trace uses three possible calibration scales: a 10-minute full-scale calibration, a 20-minute full-scale calibration, and a 2.5-degree full-scale calibration.

The position trace normally is calibrated with only one scale, a 2.5-degree full-scale calibration.

The velocity trace uses one calibration scale for elevation and train tests. The train velocity trace is calibrated with a 40-degree-per-second full-scale calibration.

Allow the test equipment at least 10 minutes to warm up before attempting any calibration procedures. (The time requirement varies with different systems; check your OP and the MRC.)

Obtain the instructions for calibrating the oscillograph (error recorder) used with your missile launching system and proceed with the calibration. After you have completed the calibration of the oscillograph, it is ready to be used in testing the accuracy of the launcher. There are many similarities between the train and elevation tests, but each power drive must be tested separately. For example:

Test 1. Simple harmonic motion test.
Test 2. Static test.
Test 3. Five-degree-per-second constant velocity test.
Test 4. Ten-degree-per-second constant velocity test.
Test 5. Fifteen-degree-per-second constant velocity test.
Test 6. Elevation (train) velocity and acceleration test.
Tests 6A and 6B. Launcher elevation (train) synchronized indicator tests.

The train power drive requires an additional test in this series—25-degrees-per-second constant velocity test.

Other tests in this group are elevation (train) synchronizing tests, fixed displacement; elevation (train) harmonic motion synchronizing tests; elevation (train) synchro power failure tests; elevation (train) main power failure tests; elevation (train) return to load tests; and elevation (train) frequency response tests.

The error recorder is used to make traces in each of the tests, the maximum operating errors are calculated, and the traces are compared with typical traces. Copies of typical traces are included in the OP or OD. The traces made at installation of the launching system on the ship are kept aboard for comparison.

Elevation accuracy tests on shipboard include a simple harmonic motion test, a static operation test, and constant velocity tests. The same types of tests are made for train accuracy. Constant velocity and synchronizing tests are performed at different speeds and at different angles of train and elevation, each performed according to specific instructions in the MRC.

These tests are performed quarterly unless circumstances require otherwise. A suspected malfunction may require certain tests to be performed more frequently. Operational tests may be needed to determine if the launcher follows order signals accurately, or to check some other function of the launcher. All personnel who perform the test must be familiar with the equipment and the procedure. Although you follow the steps according to the MRCs studying the procedure beforehand will do much for a smooth operation. If you are the leading petty officer, you will check the work of the personnel who work for you.

The launcher test equipment is stowed in the shipboard instrument storage cabinet when not in use.

## Mk12 Mods 0 and 1 Error Recorder

The Mk 12 Mods 0 and 1 error recorder (figure 9-7) is housed in a portable, aluminum case approximately 21 inches long, 15 inches wide, and 21 inches high. It weighs 76 pounds. It requires 1.6-ampere, 115-volt, 60-hertz a.c. power which normally is supplied from the launching system test panel through the dummy director. The principal component of the error recorder is a modified commerical Brush Instrument Company recorder, Mark III, which provides the immediately visible, permanent chart recordings on two channels. It includes integral amplifiers for a pen deflection of 1-mm per 10-millivolts of input signal, up to 100 hertz. Simple adjustment and chart speed controls are located on the front panel (figure 9-7), which also permit convenient change of chart paper. In addition to the two chart-recording pens, the recorder includes two event-marker pens, individually operated through remote control circuits. The recorder may also be used for time recordings of various launching system sequence operations.

The error recorder must be calibrated to zero position in relation to launcher zero train and elevation, and requires a warmup time of 15 minutes to provide an accurate error trace. (The time requirement may vary with the mod; observe the requirement stated in your instructions.) An error trace may be recorded of the launcher velocity, acceleration and deceleration, hunt, and ability of the launcher to follow static, constant velocity, or simple harmonic motion signals. These traces can be compared to those at installation.

When in use, the error recorder is connected by cable to the dummy director, and the dummy director is connected to the launcher EP3 panel. The receptacles for connecting the dummy director, the two limiter and demodulator units, and frequency generator to the EP3 panel are on the lower part of the EP3 panel, adjacent to the test jacks.

All required cabling for electrical interconnection of test instrumentation and connection of instrumentation to the test panel (EP3) of missile launching systems control is provided with the test equipment. The special cables are designed with proper conductors, length, insulation, and connectors for optimum performance of equipment. Only the approved cabling should be used in the test instrumentation setup. Only one dummy director is used at any one time, either for train or for elevation system testing.

## LIMITER AND DEMODULATOR

The limiter and demodulator unit is a portable electronic test instrument used with a standard oscillograph or other galvanometer

Chapter 9—REPAIR AND TEST

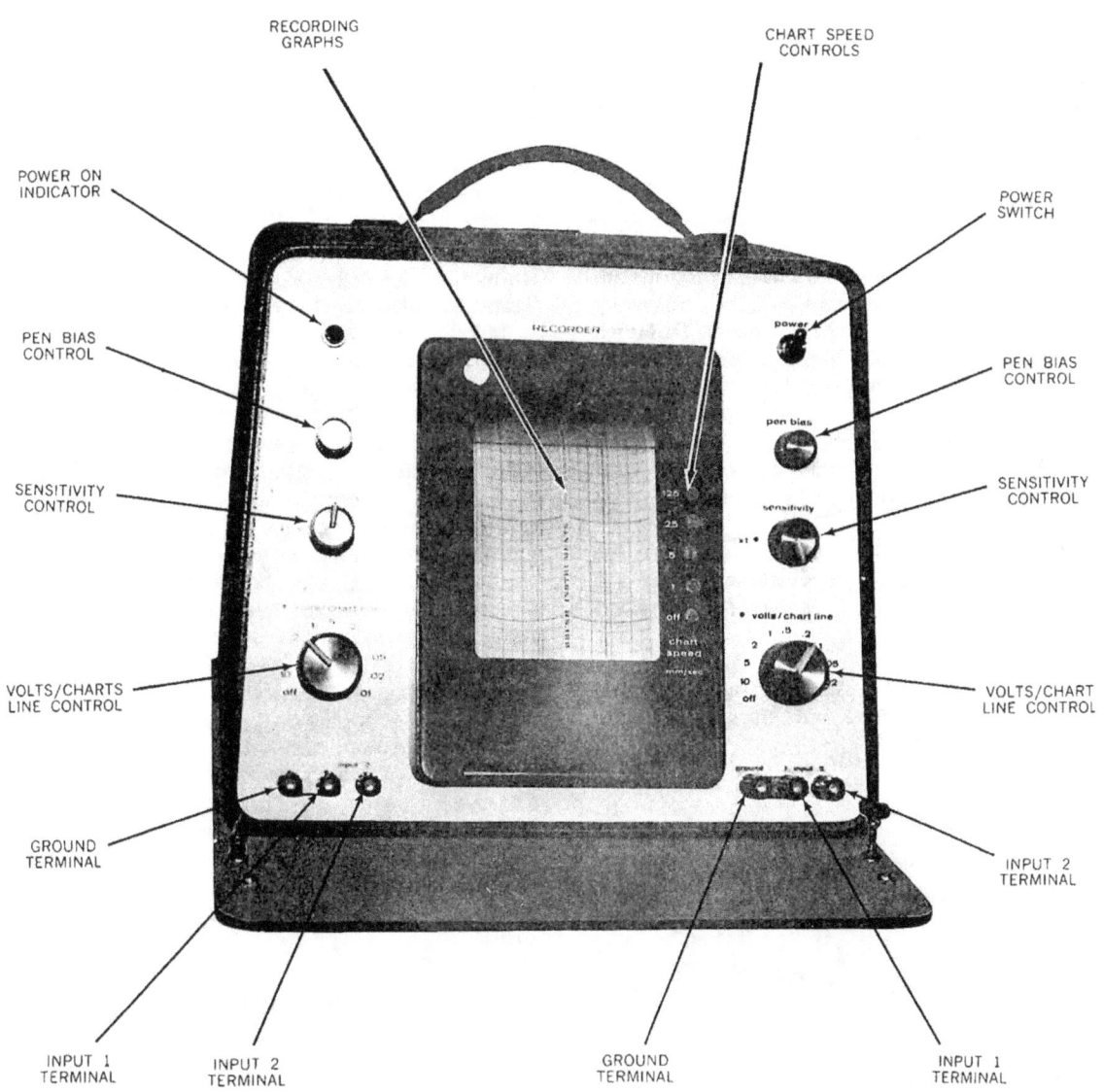

Figure 9-7.—Mk 12 error recorder control panel.

type recording instruments which performs (1) a.c. amplification with demodulation and filtering, and (2) d.c. amplification.

The limiter and demodulator contains an RC low pass filter which can be used in all modes of operation. In addition, provisions are made within the test circuit to superimpose an auxiliary pip onto a position test trace. Pips indicate switch or relay actuation points on a trace to simulate power failure circuitry.

The limiter and demodulator consists essentially of an a.c. amplifier, an attenuator, a demodulator, limiters, filters, and a d.c. amplifier. The output of the d.c. amplifier drives the recording instrument. Control switches are located on the front of each limiter and demodulator which allows three modes of operation (figure 9-8).

The mode of operation switch is used to turn the power on and to obtain the three basic functions of the instrument. These functions are obtained by selecting one of the following switch positions: DC (direct current), DEMOD. FILT. (demodulation filtering), and DEMOD. CAL. (demodulation calibration).

At the DC setting, the limiter and demodulator power amplifies d.c. input signals. This setting is used for velocity test traces.

At the DEMOD. FILT. setting, the instrument amplifiers demodulate and filter a.c. signal inputs. This setting is used for error and position test traces.

At the DEMOD. CAL. setting, the signal input circuit is opened and an a.c. calibrate voltage is introduced.

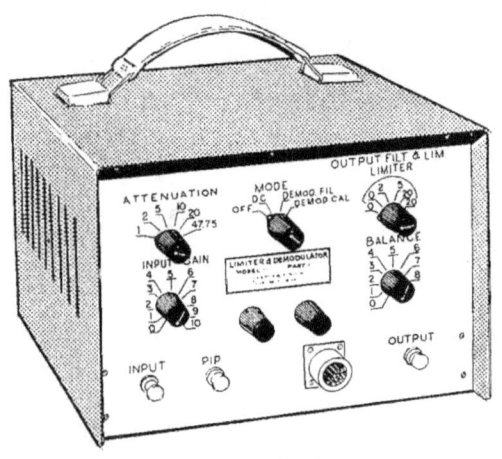

53.298
Figure 9-8.—Model G limiter and demodulator.

## MISSILE SIMULATORS

The use of simulators for training in all phases of missile handling, launcher operation, firing, and securing has been developed to a high degree. The Tartar missile simulator is placed in the training missile, as is the Terrier simulator. Since you will be training lower rated GMMs in the operation of the launching system, you need to become familiar with the operation of the missile simulator you have aboard ship. Missile simulator SM-161/DSM, used to check the Tartar missile launching system aboard ship, is installed in the Tartar training missile (TSAM), occupying the space normally occupied by the auxiliary power supply (APS). The missile simulator test panel is located just under the missile skin and is accessible by a hinged cover. All electrical connections from the launcher to the missile simulator are made through the DTRM firing contacts on the skin of the DTRM, and through the missile-to-launcher contactor on the aft end of the tail cone. The missile simulator furnishes electrical loads equivalent to those in a tactical missile and provides indicators and test jacks as a means of checking the launcher firing control circuits.

Guided missile simulator SM-159D/DSM is used with training rounds Mk 43 Mod 0 (BT-3A/F/ and BT-3B/F/), Mk 44 Mod 0 (BT-3A/N/ and BT-3B/N/), and Mk 45 Mod 0 (HT-3 and HT-3A). The simulator provides the electrical responses and loads equivalent to the missiles, thus allowing realistic loading and firing drills.

## MAINTENANCE OF SIMULATOR

It is part of your job to keep the simulator in operating condition. Periodically, the filter should be cleaned and the burned out panel lamps replaced. The only other authorized maintenance is periodic cleaning of the air filter in accordance with the applicable MRC.

The operation of the simulator can be checked by conducting a dynamic test with the simulator installed in the TSAM and the TSAM on the launcher. Signal voltages from the weapons system can be monitored from jacks located on the front of the simulator. A correct voltage reading on the front panel of the simulator indicates that the simulator and the weapons system are functioning properly. The voltages for each test point are listed on the MRC for each test. Also, there is an instruction plate attached to the front of the simulator that will aid you in performing the tests.

An incorrect voltage reading at the test points presents a problem because it means there is a defect in the simulator, or the signal voltage from the weapons system is either incorrect or has failed to reach the simulator. In the case of an incorrect signal voltage or failure of the signal voltage to reach the simulator, normally it is simpler to begin troubleshooting at the source of the signal voltages, rather than starting at the launcher.

The connectors between the simulator and the TSAM are a problem area on the simulator. These connectors receive a lot of abuse because the simulator normally is removed from the TSAM for in-house DSOTs and system tests.

## TRAINING MISSILES

Training missiles, with or without simulators, are used for practice in the handling of missiles; however, when any feedback or readout is involved, a simulator must be installed in the TSAM. The simulator is electrically connected to the trainer through the booster and missile power receptacles, shown on the left end of the simulator.

Two Terrier training rounds and one ASROC training round are provided for each Mk 10 Mod 7 or 8 launching system. The guided missile simulator used in the Terrier training rounds is described in OP 4093.

The launching system is cycled through all its operations, using a training missile in place of an active missile. For testing the firing circuits, a Terrier training missile with the simulator installed is used, loaded on the launcher rail. It is tested for normal firing, A- and B-side, misfire, and dud firing. The launcher electrical contactor is mated to the trainer electrical contactor. During the warmup, power changeover, and firing phases of the training drill, circuits in the simulator represent or monitor the missile operations. All the input and output signals, except the booster squib ignition voltages, are conducted through the electrical contactor. Automatic control is used for training drills; the observer of the simulator panel has a dangerous position because of launcher movement, and must remain constantly alert. All other personnel must be cleared from the launcher area.

The steps of operation will vary with the missile being tested and type of firing being simulated. Follow the steps exactly as given in the OP or the checksheets for the test you are making.

## OWN SHIP'S MAINTENANCE PROGRAM

The development of planned systems for maintenance of ordnance and ship's equipment was covered in the *Gunner's Mate M 3 & 2* manual, in earlier chapters of this text, and in your military requirements courses. Each ship develops its own current ship's maintenance project (CSMP) file which is used for planning and coordinating the ship's maintenance workload. The CSMP file is made up of deferred action report forms 4790/2K for those maintenance actions which have been deferred because of the unavailability of repair parts, special equipment, or technically skilled personnel. This information is also used for analyzing maintenance and logistic support problems; in addition, the CSMP makes it possible to record and report the need to delay an accomplishment of a required maintenance and can indicate the principal reason for the delay. The 3-M Systems, their purpose, organization, and procedures for shipboard use, and the forms which make up the CSMP are explained in OPNAVINST 4790.4. Reference is made to the 3-M Systems as they apply to missile systems in chapter 10 of this text.

## SLOW RUN-THROUGH (SRT)

Because all parts of the missile system must work together, testing of the complete system must be done before it is put into use, and at intervals thereafter. Shipbuilders and naval shipyards are given general and detailed specifications for installing the equipments and for checkout procedures after installation. The systems test after installation on shipboard is called the slow run-through (SRT). Detailed requirements for the SRT are established by the NAVSEASYSCOM. The SRT must demonstrate the satisfactory operation of the complete shipboard weapons installation, including supporting and auxiliary subsystems.

An SRT may also be necessary after a ship has undergone overhaul or conversion and after new or major alterations to the weapons system. Sometimes it is necessary after a minor alteration because of alignment problems. In that case the necessity for an SRT is determined by the type commander, NAVSEASYSCOM, or the shipyard. Any deficiencies revealed during the SRT must be corrected by the installing activity. The ship's personnel assist in conducting the SRT.

Testing of advanced ASW system installations, surface-to-surface, and surface-to-air missile installations begins with replenishment-at-sea and proceeds through all phases of strikedown, stowage, checkout, disassembly, servicing, checkout servicing, assembly, handling, and simulated launch. For ASW systems (ASROC), an actual or simulated sonar contact is introduced into the overall weapons system and checked through the underwater fire control system. On other systems, an actual or simulated output from the ship's search radars is introduced into the overall missile system and checked through the weapons direction equipment, the missile fire control system, and the ready service feeder and launcher systems. The feeder and launcher systems thereby receive orders that result in a missile being rammed on the launcher and the launcher trained and elevated to the position indicated by the initial input. Each magazine, launcher feeder, checkout, and strikedown system must be tested.

In addition to testing the weapons system, all supporting services and auxiliary subsystems must be tested. These include all associated lighting, air-conditioning, humidity control, security alarms, sprinkling, damage control facilities, communications, and other utilities contributing to the effectiveness of the weapons system installation. Accurate time cycles must be recorded for parts of the system where speed is part of the operational effectiveness.

When an SRT is being conducted on your ship, as a petty officer you will be assigned responsibility for checking the operation of parts of the ship's equipment and keeping the records of the operation. As a GMM1 or C you will supervise the operation of parts of the launching system and the recording of results of the tests. The response of the launcher to train and elevation orders must be noted with care, and adjustment made if necessary.

Since there are differences in each installation, detailed instructions are prepared for each ship. From these instructions the tasks are apportioned among the ship's and contractor's personnel.

# CHAPTER 10

# ADMINISTRATION AND SUPPLY

Now that you are advancing to first class or chief, you can expect more administrative duties than you have had in the past. For example, there are inspections, tests, and checks to be made on equipment; logs and records to be kept; inventories to be taken; spare parts to be ordered; training schedules to be drawn up; reports to be made; and various other "paperwork" jobs to perform. These duties will be as much a part of your job as is supervising the personnel in your division. As you can see, increasing responsibility goes along with that third stripe—or "hat."

## ADMINISTRATION

Under administration we will cover the Maintenance and Material Management (3-M) Systems, the ships' equipment configuration accounting system (SECAS) the ship armament inventory list (SAIL), and where to obtain technical assistance when it is needed.

### 3-M SYSTEMS

The primary objective of the Ships' 3-M Systems is to provide for managing maintenance and maintenance support in a manner which will ensure maximum equipment operational readiness. To this end, the intermediate objectives of the 3-M Systems are as follows:

1. Achievement of uniform maintenance standards and criteria.

2. Effective use of available manpower and material resources in maintenance and maintenance support efforts.

3. Documenting information relating to maintenance and maintenance support actions.

4. Improvement of maintainability and reliability of systems and equipment through analysis of documented maintenance information.

5. Provision of the means for reporting ship configuration changes.

6. Identification and reduction of the cost of maintenance and maintenance support in terms of manpower and material resources.

7. Reduction of the cost of accidental material damage by means of accurate identification and analysis of the cost.

The 3-M Systems are the nucleus for managing maintenance aboard all ships of the Navy. Their purpose is to provide all maintenance and material managers throughout the Navy with the means to plan, acquire, organize, direct, control, and evaluate manpower and material resources expended or planned for expenditure in support of maintenance. Thus, it is essential that all hands recognize the importance of the systems, and understand the role they play in assisting management to improve the material readiness of equipment in the fleet. In referring to "management," the term is used in its broadest sense, including the work center on the ship as well as Navy Headquaters in Washington.

### Planned Maintenance System (PMS)

PMS provides each user with a simple and standard means for planning, scheduling, controlling, and performing planned maintenance of all equipment. PMS is the most efficient means developed to date for using available maintenance resources.

PMS maintenance actions are the minimum required to maintain the equipment in a fully operable condition, within specifications. If performed according to schedule, these maintenance actions will provide the means to identify parts requiring replacement prior to failure. PMS procedures are, therefore, preventive in nature in that they are designed to prevent future equipment failures which might otherwise result in repeated corrective maintenance actions.

These PMS procedures and the periodicities in which they are to be accomplished are developed for each piece of equipment based on good engineering practice, practical experience, and technical standards. These procedures are contained on cards designated maintenance requirement cards (MRCs). MRCs provide the detailed procedures for performing the preventive maintenance and state who, what, when, how, and with what resources a specific requirement is to be accomplished. Some MRCs have equipment guide lists (EGLs) accompanying them to serve as location guides for identical equipment; i.e., gages, valves, $CO_2$ bottles, etc., which are impractical to schedule individually for routine, periodic preventive maintenance.

PMS procedures are developed by the activities and offices of the Naval Material Command that are responsible for the development and procurement of the systems/equipment for active, new construction, major conversion and activation ships, boats, and craft. PMS documentation (maintenance index pages (MIPs) and MRCs) is developed as part of the integrated logistics support (ILS) effort for all new procurements, reprocurements, alterations, and modifications of systems and equipment. Primary requirements for the PMS development activities are as follows:

1. PMS documentation shall be developed in accordance with the current military specification (MIL-P-24534(NAVY)) and critically reviewed in the sponsoring Navy organization prior to release to ensure that the planned maintenance requirements are current, technically correct, not excessive, and practical for fleet use.

2. PMS documentation approved by the Navy organization responsible for the equipment development and procurement shall be delivered concurrently with the installation of the applicable systems/equipments or alterations.

A MIP contains a brief description of the requirements of the MRC(s) for each item of equipment, including the periodicity code, the estimated man-hours involved, the recommended rates and, if applicable, the related maintenance requirements. The MIPs for all equipment in a department are contained in a departmental master PMS record. This record also contains an index of the effective MIPs, called a list of effective pages (LOEP). The department master PMS record is used by the department head as a scheduling tool when scheduling maintenance on the PMS schedule forms, and also as a cross-reference guide. Additionally, each work center has a work center PMS record which is identical to the departmental master PMS record, except that it contains only those MIPs and LOEPs applicable to the work center. The division officers, work center supervisors, and maintenance personnel use these records for cross-reference purposes.

The planning and scheduling of maintenance requirements are accomplished on the cycle, quarterly, and weekly schedules. Transferring maintenance requirements from the MIPs for each work center contained in the departmental master PMS record and scheduling them on the cycle schedule creates the ship's overhaul-to-overhaul maintenance schedule. Quarterly and weekly schedules are prepared, using the cycle schedule as a guide. Maintenance requirements indicated on the weekly schedule are assigned to specific personnel for accomplishment. Scheduled maintenance actions are crossed over with an X when they are completed, and actions not completed are circled and arrowed to a new schedule date. Quarterly schedules are updated in the same manner.

## Chapter 10—ADMINISTRATION AND SUPPLY

Changes to PMS are issued by the Naval Sea Support Centers (NAVSEACENs), Atlantic and Pacific.

The PMS Feedback Report (FBR) form, OPNAV 4790/7B, provides fleet maintenance personnel with the means to report discrepancies, problems, partial source data automation (PSDA) requirements, and to request PMS software. All feedback reports are sent to NAVSEACENs or TYCOMs (type commanders), based on the category of the FBR.

**Maintenance Data System (MDS)**

The following qualifications apply to the scope of MDS:

1. Maintenance actions deferred for outside assistance for all ships for all such maintenance actions.

2. Ships' force maintenance actions on certain selected equipment.

3. All other maintenance actions as directed by Fleet CINC/TYCOMs.

4. MDS includes the following:

   a. Documentation provided by shipboard personnel incident to certain shipboard maintenance actions. This documentation describes what was done or needs to be done, why it was done or why it needs to be done, who did it or who needs to do it, and what resources were used or are needed.

   b. The means for producing an automated current ships maintenance project (CSMP).

   c. The means for producing automated ship work requests for intermediate maintenance activity and shipyard use.

   d. The means for producing automated pre-inspection and survey (PRE-INSURV) deficiency listings.

   e. The means for producing automated reports tailored to meet the needs of all types and levels of management through the Navy.

   f. The tools necessary to effectively manage and control intermediate maintenance activity workloads.

   g. The means for the fleet to report changes to the configuration of equipment installed in ships. Incident to such reporting is the development of the capability to automatically update a ship's PMS coverage.

   h. The means for depot level activities to inform the fleet of the estimated and actual resource expenditures.

   i. The means for managing alterations.

MDS is the means by which maintenance personnel report corrective maintenance actions on specific categories of equipment. Information is retrievable from the TYCOM and the maintenance support office department (MSOD) data banks for analyzing maintenance and logistic support problems, for the development of the CSMP, and for generation of automated work requests for maintenance actions deferred for outside assistance.

It is a basic premise of the MDS that maintenance data will be recorded once and only once by fleet personnel and that the MDS data bank (not the maintenance activity) will thereafter provide information to all who have need for it in such form as may be required. In this connection, the MSOD data bank is designed to be the focal point for receipt and distribution of maintenance and material data. Direct requests to the fleet for data which is available from the MSOD imposes an unnecessary burden on the operating forces. It is the policy of the Chief of Naval Material (CNM) that the Naval Material Command minimize requests to the fleet for special data. However, if some such requests are deemed essential, special requests for data will include the phrase "The MSOD data bank has been queried and the data is not available."

From the 3-M Systems' central data bank maintained at MSOD, numerous reports are already programmed and available upon request by any command. These reports yield data concerning equipment maintainability and reliability, man-hour usage, equipment alteration status, material usage and costs, and fleet material condition. Many reports are produced periodically for users in both the Navy Shore Establishment and Operating Forces.

From the deferred maintenance that is reported, a CSMP file is developed by the automatic data processing (ADP) facility designated by the TYCOM. From the CSMP file a series of computer reports is provided the ship and/or unit commander. These reports are also used by the TYCOM. The reports provide either a detailed or summary listing of deferred maintenance information in various format options. By-products of the CSMP include automated work packages, PRE-INSURV packages, etc. A package of automated work requests (AWRs) is generated by the ADP facility for each ship prior to overhaul and availability periods. These packages contain work requests for CSMP items appropriate to the designated repair activity as well as standard work requests for routine jobs performed during intermediate maintenance activity (IMA) and shipyard availabilities.

Prior to an inspection by the Board of Inspection and Survey, a package of automated INSURV items is generated by the TYCOM from all deferrals listed in a CSMP file. These items are in a format similar to AWRs. After the INSURV board has screened these items and assigned priority numbers, if appropriate, the package is used to update the CSMP.

The usefulness of MDS is dependent upon the accuracy, adequacy, and timeliness of the information reported into the system. It is a system in which potential benefits are directly proportional to the efforts applied. Present programs for improving reliability, maintainability, and logistic support of fleet equipment are dependent upon conscientious adherence to reporting procedures.

Much of the fleet support effort resulting from use of 3-M Systems' data is not always visible immediately to the fleet because of the timespan required to test, evaluate, and implement engineering and design changes. Corrections to malfunctioning equipment through improvements in design often occur after the personnel who originally provided the information have been reassigned.

**Intermediate Maintenance Activity (IMA); Intermediate Maintenance Management System (IMMS)**

IMMS comprises computerized procedures used aboard tenders, repair ships, and repair bases/activities. These mechanized procedures are used to manage the planning, scheduling, production, and monitoring of the maintenance workloads of tended ships.

**Summary**

In summary, the purpose of the 3-M Systems is (1) to provide a tool to the fleet which can be used to manage, schedule, and perform maintenance, and (2) to provide information concerning fleet maintenance and maintenance support experience to organizations responsible for logistic support of the fleet. The Ships' 3-M Systems will operate in spite of meager resources and human error. These deficiencies are correctable. These systems will definitely not work or survive in an atmosphere of indifference, especially an atmosphere created by command indifference.

As a leading petty officer, you should keep abreast of new developments and changes to the 3-M Systems. Details on the systems and changes related to them are available in the Maintenance Material Management Manual, OPNAV-INST 4790.4 (latest revision).

## SHIP EQUIPMENT CONFIGURATION ACCOUNTING SYSTEM (SECAS)

The SECAS has replaced the ordnance logistics information system (ORDLIS) and the ship electronic installation record component index as NAVSEA's equipment accounting system. SECAS is a data collecting program which identifies shipboard equipment/components, locates the identified items, records all associated data, verifies accomplishment of specific engineering changes (ORDALTs, SHIPALTs, etc.), and provides this information to all concerned engineering activities, supply support activities, and operational managers. In addition, SECAS provides data in support of technical

## Chapter 10—ADMINISTRATION AND SUPPLY

publications, the weapons system file (equipment-to-ship and parts-to-equipment record), PMS, test equipment requirement lists, and training course requirements.

The weapons system file located at the Navy Ships Parts Control Center (SPCC), Mechanicsburg, Pa., has been designated as the Navy's control data bank for the equipment configuration information. To obtain this information, SECAS will receive input from the ship and equipment acquisition programs, shipboard validations (SAIL, etc.), shipyard/SUPSHIP postoverhaul reports, and configuration change reporting through the MDS on form OPNAV 4790/2K, in accordance with OPNAVINST 4790.4. Changes made during overhaul will be reported on NAVSEA form 4720/3 by the overhauling activity (Navy yards and SUPSHIP), in accordance with NAVSEAINST 4720.3.

### ORDNANCE ALTERATION (ORDALT)

The configuration control procedures for preparation and implementation of ordnance alterations (ORDALTs) to nonexpendable ordnance equipment is contained in NAVSEAINST 4130.9.

### SHIP ARMAMENT INVENTORY LIST (SAIL)

The SAIL is a reporting system that furnishes a listing of installed shipboard ordnance equipment, including the equipment's ORDALT status, to NAVSEA, SECAS, and all commands concerned with armament configuration. The SAIL is produced on data processing equipment in the format of the sample shown in figure 10-1.

Each ship is supplied with two copies of SAIL. Prior to a scheduled overhaul, one copy of SAIL should be marked up to show any change in the ORDALT status of the equipment since the list was published. Approximately 7 months prior to the scheduled overhaul, the updated list is sent to NAVSEACEN (Atlantic or Pacific). After the overhaul is completed, the SAIL is again marked up and forwarded to the appropriate NAVSEACEN, indicating all deletions or additions to the equipment's ORDALT status that occurred during the overhaul. Changes to the equipment or to the ORDALT status of the equipment made at times other than during an overhaul are reported to the appropriate NAVSEACEN on NAVORD form 8000/2. (See figure 10-2.)

### TECHNICAL ASSISTANCE

Assistance for field services may be obtained when technical difficulties are beyond the repair capabilities of a ship. Certain field activities of the Navy have technical personnel assigned to them and some service forces have system engineering specialists assigned directly to force commanders. These engineers are specialists in particular equipment which makes their availability limited. Emergency requests for technical assistance may be made by message or telephone to the nearest activity having a specialist, but a written request for assistance must follow. Maintenance assistance received from a mobile technical unit (MOTU), or Naval Sea Support Center (NAVSEACEN), Atlantic or Pacific, will be documented on a form 4790/2K and submitted as a work request. The document is originated by the requesting work center in accordance with procedures outlined in OPNAVINST 4790.4.

The objective of NAVSEASYSCOM is to ensure that all ordnance units attain the maximum degree of self-reliance in operation, utilization, and maintenance. Special assistance is provided to fleet personnel who serve these objectives by providing special assistance through different levels of maintenance by contract service engineers, MOTUs, and NAVSEACENs.

Contract service engineers are contractor engineers and technicians specially trained to handle specific equipment. These personnel are available on an "as required" basis either by special assignement to activities or service commands, or by a specific request to NAVSEASYSCOM. Special NAVSEA and type commander (TYCOM) instructions are issued periodically to cover these services.

LSR / EXPANDED SAIL
LOGISTICS SUPPORT REQUIREMENTS (CATALOG NO OD0040)
VESSEL NAME: KNOX SQDRN/DIV: 112

REASON FOR ISSUE: SPECIAL REQUEST TYCOM: COMSURFPAC TYPE/HULL: DE 1C52 CIP: 1052 UIC: 54047

AVAILABILITY DATE: 10/08/79 TO 01/18/80 O/H YARD: LBECH REPORT DATE: 09/10/80 NAVSHIP TYPE DESK 423

| CAN | SYS/EQ NOMENCLATURE | CODE | MARK | MOD | SERIAL | P LOC | EIC | APL | LD/SK | DWG NO | FSN | ORDALT DATA NUMBR | R C S K | COMPL DATE |
|---|---|---|---|---|---|---|---|---|---|---|---|---|---|---|
| 870 | | | | | | | | | | | | 6888 | 00 * | 70-07 |
| 880 | TELESCOPE | 8185 | 100 | 1 | 51 | | G11K700 | 49401989 | | | | | | |
| 880 | | | | | | | | | | | 4N12405937041 | 3867 | 00 * | 69-02 |
| 900 | DYNAMIC TESTER | 2595 | 2 | 3 | 54 | | G1R3000 | 49402528 | 2324850 | | | | | |
| 910 | ERROR RECORDER | 3095 | 7 | 1 | 192 | | G1RA000 | 49402010 | LD412565 | | | | | |
| 920 | TEST SET | 8195 | 346 | 3 | 147 | | | 49402717 | | 2438177 | | | | |
| 930 | | 612 | 14 | 6 | 02-01 | | | | | | | | | |
| 930 | | | | | | | | | | | | 6653 | 00 P | 70-02 |
| 930 | | | | | | | | | | | | 6654 | 00 P | 70-02 |
| 930 | | | | | | | | | | | 2A49251347728 | 6791 | 00 P | |
| 930 | | | | | | | | | | | | 7024 | 00 P | 69-03 |
| 930 | | | | | | | | | | | | 7063 | 00 P | 69-03 |
| 930 | | | | | | | | | | | | 7115 | 00 P | |
| 930 | | | | | | | | | | | 2A12901685649 | 7200 | 00 P | |
| 1000 | MISC FIRE CONTROL EQ | 678 | - | - | - | | | | | | | | | |
| 1010 | BEARING/RANGE INDICA | 1645 | 7 | 4 | 165 | | | | | | | | | |
| 1020 | DUMMY DIRECTOR | 2585 | 3 | 8 | 132 | | JY41000 | 49402156 | LD272568 | | 2J49317708439 | | | |
| 1030 | ERROR RECORDER | 3095 | 6 | 5 | 173 | | | 49400582 | | | | | | |
| 1040 | INDICATOR PANEL | 4925 | 5 | 29 | 653 | | | 49401604 | LD281226 | | | | | |

PAGE
6

TOTALS: SYSTEMS: 10 EQUIP: 69 ORDALTS: 125 COMPLETE: 96 INCOMPLETE: 29 ITEMS: 204

63.12

Figure 10-1.—Ship armament inventory list (SAIL).

Chapter 10—ADMINISTRATION AND SUPPLY

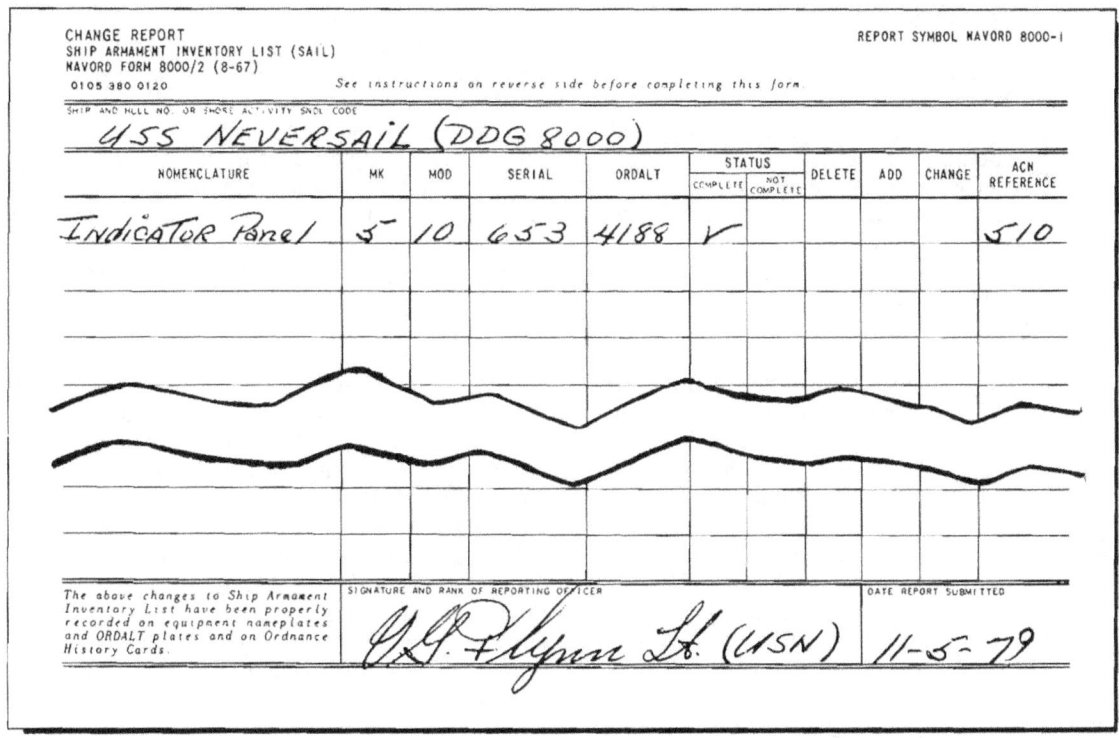

Figure 10-2.—SAIL change report, NAVORD Form 8000/2.

MOTUs comprise specifically selected and trained groups of military personnel assigned to and made available by Commander Service Force Pacific and Commander Service Force Atlantic. These groups normally are made available for assistance in the maintenance of newly installed equipment, but may be called upon in an emergency.

There are other forms of external assistance given to ships when requested, such as engineers and technicians assigned by NAVSEACENs for special projects.

Once the assistance has been rendered, or it has been found that the equipment cannot be repaired or deficiencies in its operation cannot be eliminated without extensive work and time, a maintenance action report (4790/2K) is prepared and used by the ship as a deferred maintenance action. Requests for technical assistance during regular overhaul to augment the technical capability of an activity will be specifically directed to NAVSEACEN Atlantic or Pacific. In the event of an emergency request, the appropriate NAVSEACEN will be advised, by telephone or message, of the scope and nature of the request. Listed below are the major services provided by NAVSEACEN organizations.

1. Liaison, single point of contact for ordnance.
2. Fleet support engineering.
3. Technical assistance (MOTU backup).
4. Predeployment reviews (PDR).
5. Configuration (verifications and update).

6. Special assistance teams.
    a. Technical (system and/or equipment).
    b. Torpedo.
    c. Logistics.
    d. Test equipment.
    e. Publications (technical data and documentation).
7. Ordnance technical training support.
8. Tests and trials on new construction/modernized surface ships.

## SHIPS QUALIFICATION ASSISTANCE TEAM (SQAT)

Naval Ship Weapon Systems Engineering Station (NSWSES) provides a program of timely and competent assistance to the commanding officer of each applicable ship in achievement of a stable level of weapon system operational readiness on a continuing bases. The ship qualification trials (SQTS) are conducted so surface weapons/underway replenishment (UNREP) systems may be adequately checked out and proofed after completion of new construction, conversion, or overhaul, and prior to overseas deployment. A ships qualification assistance team (SQAT) is provided by NSWSES to assist a ship's crew in achieving or demonstrating a set of program objectives for a given weapons/UPREP system during an SQT period. A SQAT normally will report to the ship's commanding officer after completion of fitting-out of new construction and upon completion of conversion/overhaul. Assistance is provided to each ship in direct support of this objective as follows:

1. A test program to demonstrate, on a step-by-step basis, the operability and capability of the weapon system(s) in the at-sea environment and the competence of the ship's force to maintain and operate them.
2. A team to assist and instruct the ship's force in the conduct of the test program. The conduct of system maintenance tests and equipment level tests must conform to the PMS/SMS discipline.
3. Training materials, documentation, logistic support, and special technical assistance required in support of the test program.

During an SQT period, the leading GMM of a system is responsible for following all of the training requirements provided by a SQAT. The GMM also helps in the identification of system problems and the collection of technical information required for corrective action. An SQT plan provided prior to the beginning of the SQT lists the sequence of events for an orderly accomplishment of the program and indicates action responsibility for each event. The listed objectives of the SQT program may be varied, as required, to meet the constraints imposed by each ship's material, personnel, logistics, and operational situation.

Prior to an SQT period, each GMM in charge of a missile launcher should look over the checklist provided in SQT test plan for the SQT program to ensure that all maintenance and training requirements have been accomplished. One of the maintenance management requirements for SQT is to ensure that each launcher has an updated MRC deck and if not, a feedback report (OPNAV 4790/7B) has been submitted. The SQT period for each ship is scheduled by the ship's type commander and terminates upon satisfactory completion of an SQT program.

## SUPPLY RESPONSIBILITIES

You have dealt with supplies in one capacity or another since you first joined the Navy. As you advance in your rating, you will take a more active part in assisting your division officer and the supply officer in estimating material needs and planning for replenishment. A review of the supply chapter in *Military Requirements for Petty Officer 3 & 2*, NAVEDTRA 10056 series, and study of the supply chapter in *Military Requirements for Petty Officer 1 & C*, NAVEDTRA 10057 series, should be helpful. A word of caution is in order here. The spread of automation in many phases of supply, the consolidation of all Armed Forces supply systems into one Defense Logistics Agency (DLA), and cost reduction programs all work together to cause rapid changes in the supply system. Members of the supply department can help you keep abreast of the changes as they affect you. Navy training courses are revised every 4 or 5 years, but the supply system changes cannot wait that long; the supply department receives the directives on the changes as they occur.

## SOURCES OF SUPPLY

The original source of supply is, of course, the manufacturer of the item, but you seldom have direct contact with the original source. Supplies are purchased in huge quantities and stockpiled at various locations where they will be more readily available to the ultimate users.

### Defense Logistics Agency (DLA)

The consolidation of supply and service functions between and among the military departments began several years ago. The DLA acts as the centralized policymaking and monitoring agency sitting on top of the whole single manager concept. It has been assigned management and operational control of a number of defense supply agencies and centers. Supply agencies will be transferred to DLA as the consolidation is carried forward.

DLA publications and directives, under the overall policies of the Secretary of Defense, provide information and direction to all defense agencies that receive support from, or supply support to DLA. These publications include catalogs, stocklists, pricelists, manuals, handbooks, and information bulletins. The numbering system, along with other information on these publications, is given in SECNAVINST 5215.13 series.

Having one supply system for all defense agencies is expected to save the Government many millions of dollars each year. As a taxpayer, that should interest you. As a PO you have a part in implementing the system. Learn enough about it so you can order supplies intelligently. Every incorrect or ambiguous order causes extra correspondence, incorrect shipments, aggravating delays, confusion, and annoyance; and it piles up costs. Even more important, repair of weapons and weapon components is delayed, which could be critical.

### Ships Parts Control Center

The supply of ordnance material is coordinated through the Navy Ships Parts Control Center (SPCC), Mechanicsburg, Pa. This office is operated under management control of NAVSUP. SPCC is responsible for control inventories, procurement, and distribution. It also maintains records, catalogs (NAVSEA lists), and stock levels, and regulates the flow of material throughout the Navy distribution system. The limited application of guided missiles and related test and handling equipment does not justify the positioning of spare parts throughout the distributive system to the same extent as other ordnance material. For the same reason, range and depth of guided missile material has generally been restricted to cover specific program requirements.

### Guided Missile Support Facilities

The Naval Seas Systems Command (NAVSEASYSCOM) has established guided missile support facilities at various coastal and tidewater naval weapon stations. These facilities provide the necessary special test equipment and skilled personnel to assemble, checkout, modify, maintain, and prepare for issue various NAVSEASYSCOM guided missile material scheduled for service use. This material includes requirements for the U.S. Marine Corps, as well as for the fleet and naval air stations. In addition, these facilities will support ammunition components such as warheads, fuzes, and rocket motors for Sparrow and Bullpup missiles. Guided missile facilities, unlike conventional ammunition facilities, require considerable special-purpose electronic and hydraulic test equipment, and special tools and handling equipment. These guided missile support facilities are specially designed to maintain substantial numbers of specific types of guided missiles in a ready-for-issue condition for ship allowances, fleet training, and NAVSEASYSCOM evaluation. The facilities also have capabilities for offloading fleet missiles in conjunction with vessel shipyard overhauls, for accomplishing minor maintenance, and for performing ORDALTs on missiles in stock.

The primary distribution points for guided missile material in the ordnance segment of the Navy supply system are Naval Supply Center (NSC), Norfolk, Va., for all guided missile material on the East Coast, and NSC Oakland, Calif., for all guided missile material on the West Coast.

## COORDINATED SHIPBOARD ALLOWANCE LIST (COSAL)

The COSAL was explained and illustrated in your military requirements text. As a GMM in the weapons department, the part of the COSAL that concerns your work is the section that lists the guided missile launching systems and the tools and repair parts allowed for their maintenance. The type and quantity of repair parts allotted a ship were determined by studies of requirements in the past. Each ship has a COSAL prepared just for it and the ship is stocked accordingly.

The parts of the COSAL prepared by SPCC cover hull, mechanical, electrical, electronics, and ordnance equipment. The Avaition Supply Office (ASO) deals with aviation equipment, and the Naval Ship Engineering Center (NAVSEC) prepares the COSAL for portable communication, radiac and sonar equipment, and electronic and electric test equipment allowance.

Part I of COSAL is basically an index of all GMM equipment and equipage. The equipment and equipage are listed in alpha sequence by noun name and by service application. The index provides each equipment's allowance parts list (APL) identification number, and each equipage item's allowance equipage list (AEL) identification number. There is also a numerical listing of all APL and AEL identification numbers.

Part II of COSAL contains all of the APLs and AELs issued for your weapons equipment. An equipment APL provides a detailed description of the equipment and lists its repair parts, including each part's national stock number (NSN). The AEL provides the same information for each item of equipage.

Part III of COSAL contains a stock number sequence list (SNSL) for the storeroom items and another one for the operating spare items. These lists contain the repair parts for all equipment aboard ship. Since they include the total parts for the ship, this section permits you to locate parts for your equipment that are also used by other departments of the ship. Another section of part III contains the alternate number cross-reference to stock number. This cross-reference section provides a convenient tool for material identification and for obtaining the NSN from a manufacturer's identification number or part reference number.

The stock numbers given in COSAL listings are NSNs. The Federal Supply Classification system and the numbering system were explained in *Military Requirements for Petty Officer 3 & 2*, NAVEDTRA 10056 series. Review it to refresh your memory.

The COSAL does not include ship's store stocks, resale clothing, bulk fuels, subsistence items, expendable ordnance, or repair parts for aircraft. These items are covered by separate outfitting and load lists.

### Ordnance Segment of COSAL

The ordnance segment of the COSAL is an allowance list of ordnance equipment, equipage, and supporting repair parts and other materials, tailored to the individual ship. As each active ship with ordnance installations undergoes overhaul, it is supplied with a new ordnance segment of the COSAL. All active fleet, new construction, and major conversion ships with installed armament receive an ordnance segment of the COSAL. Only the items listed will be placed on board, in the quantity listed.

The ordnance section of the COSAL is made up of an introduction and three separate parts. The introduction is prepared by the Navy Ships Parts Control Center (SPCC) and gives complete information on the use of the COSAL. Study this introduction carefully before using the COSAL.

Allowance requirements for nuclear weapons, guided missiles, and certain FBM

## Chapter 10—ADMINISTRATION AND SUPPLY

equipment are included in special supplements to the COSAL. The supplement consists of an index of all major training items, test and handling equipment, tools, and consumables within the nuclear weapons program; an APL of all authorized repair parts within the equivalent war reserve weapons or components, and test equipment listed in the index (above); and a stock number sequence list of authorized on-board allowances for all equipment and repair parts listed. The supply department does not have cognizance over war reserve nuclear components; the weapons officer must take the responsibility for those. The tools, test equipment, etc., are obtained through the supply department.

The page numbers of the ordnance part of the COSAL are preceded by Z.

Sections A and B of the index are cross-referenced. Knowing how to use the index is an important part of knowing how to use the COSAL to locate an item. Practice is the best way to become familiar with using the index.

The COSAL does not generate any additional report or records but simplifies your recordkeeping. It gives you the national stock number for most of the items you have to order, and saves much looking up of those numbers. It lists the items your ship is allotted, so you will not order something you may not have. When you need a repair part, the COSAL gives you the correct stock number for ordering it, so you will get the correct part. It also give you the correct name for the part and tells where it is used, which gives you another method for checking that you are ordering the right part. If the supply department on the ship has it in stock, the COSAL also gives you that information.

The COSAL for each ship is prepared by the inventory control point (ICP), which also stocks the ship with the listed equipment and repair parts, either at outfitting of a new or a converted ship or just before overhaul. Minor revisions to the COSAL are made as pen-and-ink changes on the ship's copy. More extensive revisions or additions are distributed as changes.

**Procurement of Material**

Procurement is the act of getting or obtaining something. As a rule your supply officer will procure the material for you. The COSAL tells you what material is authorized. You must know what forms to use and the procedure for initiating procurement action.

Material may be procured by: (1) requisition, (2) purchase, (3) transfer, and (4) manufacture.

The requisition method is the one you will use most often. Use NAVSUP Form 1250 or DD Form 1348 to obtain your supplies and repair parts from the ship's storeroom. A supply of the forms is usually kept in the weapons office. While you are in the weapons office, check the COSAL to find out if the part you want is stocked, and copy the national stock number on the form. If it is customary on your ship to list the price, look it up and enter it on the form.

When you have filled in the information on the supply form, you must ask your department head to sign the form before you take it to the storekeeper. The storekeeper will doublecheck your information, and if it is in stock, the item requested will be issued to you. The storekeeper retains the form as authorization for the supply department to requisition the same item from supply ashore to bring the ship's allowance up to full condition again.

It is good practice to maintain a file in the weapons office of all material requisitioned from supply. Record the request or requisition numbers. These numbers are assigned by the supply department and are entered on the supply form. The number is especially helpful when tracing a part that was requisitioned from an activity other than your ship, such as a shore activity. The requisition number is necessary to start a tracer through the supply system to locate your material if it is unnecessarily long in coming.

If you are unable to determine the national stock number of a repair part that you need, use DD Form 1348-6, along with DD Form 1348 to requisition it. Only the Identification Data section of Form 1348-6 must be completed. This data will include nomenclature, identification taken from the nameplate, drawings, or any source that will help identify the item.

Forms 1250, 1348, and 1348-6 are illustrated, and the instructions on their completion are contained in volume II of OPNAVINST 4790.4 (3-M Manual).

When you have used a repair part that was stocked in your department, do not put off ordering a replacement for the part.

IN-EXCESS REQUISITIONS.—Sometimes it is necessary to request material above the quantity shown for the item on the COSAL. Your department must give reasons sufficient to justify the need in excess of the allowance and the supply department prepares the in-excess requisition.

In-excess requisitions are required for all of the following materials:

1. Equipage not on the ship's COSAL.
2. Equipage on the COSAL but requested in greater quantities that allowed.
3. Nonstandard consumable supplies.
4. Repair parts not listed with quantities on the ship's allowance, for which a requirement can be justified.

Approval of an in-excess requisition does not constitute a change of or addition to the COSAL. If replacement of the in-excess articles is required, additional approval is required.

ISSUING PROCEDURES.—Procedures for issuing supplies vary on different ships. Approval by the person maintaining the departmental budget record may be required on some ships. Clearance with the supply department is required before proceeding to the stockroom with the approved requisition form. Be sure that each individual drawing material is instructed to check that each item is received in correct quantity and the price is listed correctly for the material actually received.

DISPOSITION OF REPAIR ITEMS.—In the basic repair cycle of the distribution system, all items fall within two general classifications. These classifications are expendable and nonexpendable items. The expendable items are disposed of in accordance with applicable current instructions. The nonexpendable items, which are classified by material control codes (MCCs) E, G, H, Q, or X, on the allowance list are rated as repairable. The user returns them to the supply system when replacements are requisitioned. Vessels operating outside the continental United States normally will offload such failed components at the first opportunity. Once back in the supply system, these failed or damaged items are reported to SPCC as available for repair.

The primary distribution points have a list of items requiring fast repair. Immediately upon being turned in, these items are returned to the contractor or a qualified repair center. Repairable items not appearing on this list will be accumulated at the distribution point pending disposition instructions from SPCC.

## IDENTIFICATION OF ORDNANCE PARTS

In addition to the COSAL, there are several other places to search when trying to identify an item (which may be an old model not given in new lists, or for some other reason is difficult to identify exactly.

One of the most important sources of identification is the information on nameplates. This may include the manufacturer's name, make or model number, size, voltage, and the like. Identification publications such as manufacturers' technical manuals may help you in identifying an item.

The two-part cognizance symbol that precedes the national stock number indicates the command, agency, or office that has control over the supply and distribution of the material.

Cognizance symbols are assigned to different groups of material. Cognizance symbol 2J material includes, among other things, guided missile launchers (less airborne), torpedo launchers, rocket launchers, selected fire control and optical equipment under the design control of NAVSEASYSCOM, and other major ordnance equipment.

## Chapter 10—ADMINISTRATION AND SUPPLY

Cognizance symbols 2T, 4T, and 8T indicate expendable ordnance and include missiles; signals, underwater sound; related inert and explosive components; and selected support or test equipment for the above items. Torpedoes, mines, and underwater ordnance are designated 6T.

The condition of ordnance material is also designated by code letters which are used in stock recording and reporting procedures for the ammunition segment of the ordnance supply system. A numeric code was formerly used. The changeover calls for the use of a new ammunition class X, new alphabetic condition codes, and MILSTRIP routing identifiers to be used in place of station reporting numbers (e.g., N35 for SPCC, N24 for NAVSEASYSCOM, etc.). The new designation ammunition class W identifies items for which end action disposal has been authorized.

Illustrated Parts Breakdown of Ordnance Equipment (IPB) is a publication prepared by SPCC. Each IPB is published for one particular type or piece of equipment, and describes and illustrates the relationship of all assemblies and parts comprising the equipment. IPB 0000 is an idex of all IPBs.

## NAVY MANAGEMENT DATA LIST (NMDL)

The Navy management data list (NMDL) provides information necessary for good management of the item. It is not practical to include such items as price, unit of issue, and cognizance symbol in the Federal Supply Catalog or in the COSAL since these items are subject to frequent change. Therefore, these items, along with other information, are listed in the NMDL. Basically, you will use this publication to determine the price, unit of issue, cognizance symbol, and material control code (if applicable) for NSNs you have located in the COSAL. All Navy-interest NSNs are listed in the NMDL.

You will continue to use the SPCC Ammunition Index of Navy Ammunition, Navy Stock List of Forms and Publications, and certain ICP specialized supplements.

The NMDL is expected to eliminate a great deal of searching for correct stock numbers, to identify items, and to simplify requisitioning. It will extend the utility of the COSAL by providing updated stock numbers and reference numbers.

The NMDL is being produced in microfiche form. Quarterly, a completely updated NMDL is issued to all activities. This eliminates the requirement for change bulletins to the NMDL.

## OTHER SUPPLY REFERENCES

General purpose items are described and illustrated in the afloat shopping guide (ASG).

Part numbers are cross-referenced in the consolidated master cross-reference list (C-MCRL). Another publication frequency used with the C-MCRL is the Federal Supply Code for Manufacturers (FSCM). It consists of two volumes and lists all commercial firms manufacturing material for DOD. Each manufactuer is identified by a 5-digit number. One volume lists the manufacturers in alphabetical order and identifies them to the code number. The other volume is a numerical listing by code number, and identifies the manufacturer. Frequently, similar parts manufactured by different manufacturers will be identified by the same part number. When this occurs, the FSCM helps you identify and obtain the correct part. If you have any question on supply, the answer can probably be found in the NAVSUP manual, and someone in the supply department will know about it or can find it.

## ORDERING PUBLICATIONS

Cognizance symbol I designates printed material such as standardized forms, handbooks, instructions, and training publications. They are listed in NAVSUP Publication 2002, Navy Stock List of Forms and Publications, Cognizance Symbol I. The initial commissioning allowance is sent without requisitioning. Other recommended publications, classified as category II, may be requisitioned from Forms and Publications Stock Point, U.S. Naval Supply Center, Norfolk, Virginia, or Oakland, California.

Except where indicated otherwise, order OPs and changes from the Naval Publications and Forms Center, Philadelphia, Pa. Use a MILSTRIP form. Changes are automatically supplied, but if you are missing any, write, "include changes 1, 2, and 3" in the Remarks section of MILSTRIP on which you order the OP by stock number.

The publications custodian has the responsibility for ordering publications that are needed and keeping changes inserted. The individual also is responsible for the security of the publications.

## WEAPONS DEPARTMENT REPORTS, RECORDS, AND INSPECTIONS

Various reports are required by the Naval Sea Systems Command (NAVSEASYSCOM), type commanders, and other interested commands or offices to keep informed of the status of ordnance equipment and whether it is functioning efficiently. Most of these reports will be made by the weapons officer, or administrative assistant; however, much of the information required must be supplied by the leading petty officers of the department.

All reports currently required by OPNAV and NAVSEASYSCOM directives are listed in the latest revision of OPNAVINST 5214.1 (Consolidated List of Recurring Reports Required by the Bureaus and Offices, Navy Department, from the Operating Forces of the Navy). This listing does not include reports required by fleet, force, and type commanders.

All reports fall into one of the following three general categories:

1. Periodic reports, which give the same type of information at regular intervals—such as monthly or quarterly.
2. Situation, or performance, reports which are submitted whenever the action being reported on occurs, such as a loss of small arms report.
3. Special reports, which are one-time reports such as a report of inventory of a specific class or type of material or equipment held by the activities of the operating forces at a specific time. These reports are requested by the bureau, systems command, or office desiring the information, and instructions as to the form of the report, report date, symbol, etc., usually are set forth in the form of a speedletter and distributed to all activities concerned. In any case, the requiring directive for each report normally will give you the instructions for preparing and submitting the report.

OPNAVINST 5214.1 series contains a consolidated list of required recurring reports from operating forces of the Navy to Navy headquarters organizations. These reports are made by ship's department heads through commanding officers, from information received from petty officers of different divisions of a ship. The information necessary for these reports comes from equipment logs, missile logs, supply records and the ship's CSMP. An example of a required report is the surface launched missile firing reports which are submitted after each test firing of guided missiles against surface or air targets. A firing report is sent by the ship within 4 working days following the firing to Officer in Charge, Fleet Analysis Center, Naval Weapons Station, Seal Beach, Corona Annex, Corona, Calif., 91720.

## SURFACE-LAUNCHED MISSILE FIRING REPORTS

Reporting requirements for surface-launched missile firing reports are contained in NAVSEAINST 8810.3. Report symbols assigned to the reporting requirements are NAVSEA 8810-2, Surface-Launched Missile Firing Report, and NAVSEA 8810-3, Electromagnetic Interference (EMI) Report. The data required on firing reports is of extreme importance to the Commander Naval Sea Systems Command (COMNAVSEASYSCOM), and the information provided by the fleet will be under constant review to ensure that the maximum degree of missile system performance is achieved. To be successful in this endeavor, the reporting and evaluating of data must be a joint effort. The information supplied to a department head by leading petty officers must be accurate and up to date.

## Chapter 10—ADMINISTRATION AND SUPPLY

The following firing reports are of concern to a GMM:

1. NAVSEA 8810/1C—Improved Point Defense (NATO Seasparrow) Missile Firing Report.
2. NAVSEA 8810/1D—Terrier/Standard (ER) Missile Firing Report for non-DFCS Ships.
3. NAVSEA 8810/1E—Terrier/Standard (ER) Missile Firing Report for DFCS Ships.
4. NAVSEA 8810/1F—Tartar/Standard (MR)/Standard (ARM) Missile Firing Report for DDG, FFG, CTN, and CG Class Ships.
5. NAVSEA 8810/1G—Standard (MR) Missile Firing Report for FF-1052 Class Ships.

All data indicating performance of the weapon system during the missile firing, including film, magnetic tapes, paper records, operations recorder, target plots, and logs for the expended missile and their components are sent with the missile firing reports. Instructions for completion of forms 8810 series are contained in NAVSEAINST 8810.3. Weapons exercises, conducted aboard ship for training purposes, checking out the weapons system, and qualifying and grading the ship and an operating unit, are described in FXP-3 series (ship exercises). When your ship is preparing to engage in firing exercises, become familiar with the plan and your place in it. Teamwork and cooperation are essential to successful performance in firing exercises.

### NONEXPENDABLE SMS EQUIPMENT STATUS REPORTS

All ships of the operating fleets equipped with surface missile systems shall complete the original and one copy of the Nonexpendable SMS Equipment Status Log, NAVORD Form 8810/2 (figure 10-3). The originals of each week's data shall be forwarded within 7 working days after completion to NWS, Seal Beach, Corona, Calif. The log may be filled in by hand. The yellow copy of 8810/2 will be retained by the ship as its equipment rough log; no other log is required. The information recorded on 8810/2 shall not repeat any maintenance action details that are normally reported on Maintenance Action Form 4790/2K. Use as many 8810/2 forms as necessary to record all information for a 7-day period. Continue entries for the next day on the same sheet if space is available. Make at least one entry at the beginning of every day. This entry should show at least the date, time, and code in appropriate columns (figure 10-3). A final daily entry should show at least the date, time, status code, and signature of the person completing the form. Procedures for completing 8810/2 are explained in NAVORDINST 8810.2A. The security classification of 8810/2 when filled in will depend upon the content in the Remarks section. If there are no classified remarks, the form is unclassified and does not need to be marked. Examples of entries which may or may not classify an 8810/2 are:

1. Statement that a missile was fired is unclassified.
2. Statement of a missile firing with results of firing is Confidential.
3. Specific missile frequencies are Secret.
4. Routine operations of the equipment, such as DSOTs, tracking operations, loading operations, or system testing are unclassified.

For the appropriate downgrading statement when 8810/2 is classified, refer to the Information Security Program Regulation, OPNAVINST 5510.1 series.

### COMMANDING OFFICER'S NARRATIVE REPORT (CONAR)

Despite the quantity of statistical data accumulated from the MDC system, there remains a vital need for the comprehensive assessment of the surface missile systems and associated equipment as an integrated part of a ship's defense capability. The commanding officer's narrative report (CONAR) provides an opportunity to comment on the missile system as a whole, and to make such recommendations as desired for improved operation of the system. Problems of quality control or poor design of parts, suggestions for better utilization of equipment or manpower, and changes in test design should be reported routinely on the PMS Feedback Report, OPNAV 4790/7B or, if desired, in a narrative report. To the extent that

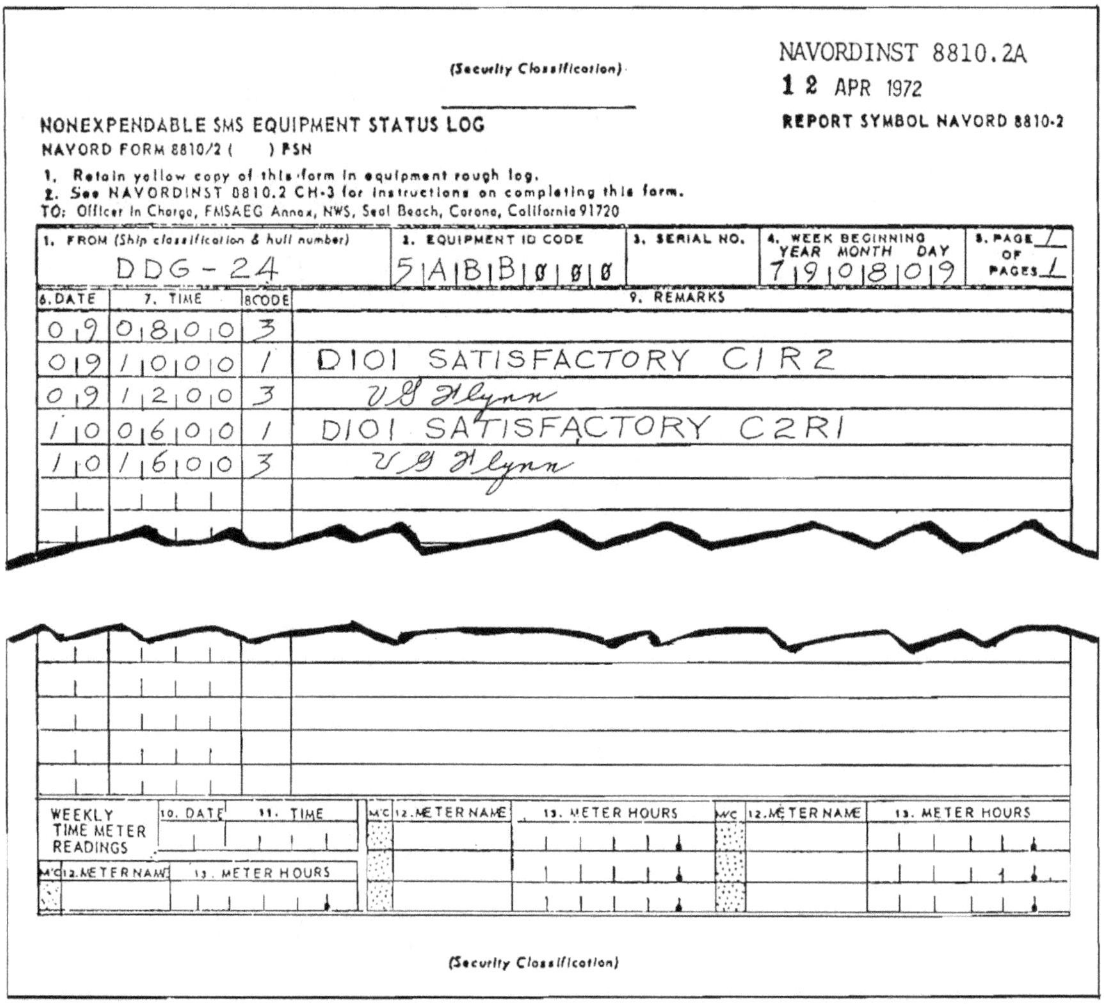

(Security Classification)

NAVORDINST 8810.2A
12 APR 1972
REPORT SYMBOL NAVORD 8810-2

NONEXPENDABLE SMS EQUIPMENT STATUS LOG
NAVORD FORM 8810/2 ( ) FSN

1. Retain yellow copy of this form in equipment rough log.
2. See NAVORDINST 8810.2 CH-3 for instructions on completing this form.

TO: Officer In Charge, FMSAEG Annex, NWS, Seal Beach, Corona, California 91720

| 1. FROM (Ship classification & hull number) | 2. EQUIPMENT ID CODE | 3. SERIAL NO. | 4. WEEK BEGINNING YEAR MONTH DAY | 5. PAGE OF PAGES |
|---|---|---|---|---|
| DDG-24 | 5AB B Ø Ø Ø | | 7 9 0 8 0 9 | 1 of 1 |

| 6. DATE | 7. TIME | 8. CODE | 9. REMARKS |
|---|---|---|---|
| 09 | 0800 | 3 | |
| 09 | 1000 | 1 | D101 SATISFACTORY C1R2 |
| 09 | 1200 | 3 | V.G. Flynn |
| 10 | 0600 | 1 | D101 SATISFACTORY C2R1 |
| 10 | 1600 | 3 | V.G. Flynn |

WEEKLY TIME METER READINGS

| 10. DATE | 11. TIME | M/C | 12. METER NAME | 13. METER HOURS | M/C | 12. METER NAME | 13. METER HOURS |
|---|---|---|---|---|---|---|---|
| | | | | | | | |

(Security Classification)

94.166

Figure 10-3.—Nonexpendable SMS equipment status log report, NAVORD Form 8810/2.

this report is used to define problems, it is an "exception" report intended to uncover problems of more than a routine nature. It may be used to evaluate, compare, or to simply report problems encountered. The CONAR is submitted to Naval Ship Weapon Systems Engineering Station (NSWSES) within 15 working days after each calendar quarter. A guide for preparing the CONAR is contained in NAVSEAINST 8000.1.

A CONAR is divided into sections covering equipment which affects a guided missile or an SMS weapon system's operational capability. A section of this report is used to facilitate the exchange of information among missile systems which may have the same or similar problems. This section can be used by GMLS personnel as a reference to solve a maintenance problem. Each problem area of a missile system is divided into separate sections which explain the

## Chapter 10—ADMINISTRATION AND SUPPLY

casualty, the findings, the remedy, and the probable cause. Since most types of missile weapon systems installed aboard a particular class of ship are physically and functionally the same, the remedy of a casualty of one ship could also be the remedy of a casualty of a ship in the same class. For this reason, all inputs related to casualties and maintenance problems for a given GMLS should originate from a GMLS petty officer whose knowledge and personal judgment are essential for improving and maintaining surface missile launching systems.

Upon receipt of a CONAR, the weapon systems engineering station takes the following actions:

1. Distributes copies of CONAR to prescribed NAVSEA activities.

2. Enters in-service engineering problems into the deficiency corrective action program (DCAP) system in accordance with SMSINST 8810.1 series.

3. Reviews those areas of responsibility of the CONAR and provides a CONAR reply wherein those problems, other than in-service engineering problems, are discussed, indicating action intended, when appropriate. Problems relating to personnel are referred to the Naval Military Personnel Center (NMPC) for possible comment.

### DEFICIENCY CORRECTIVE ACTION PROGRAM (DCAP)

The DCAP is the vehicle for monitoring and controlling the actions necessary to respond to feedback information received from fleet and shore activities. All data elements and information reported through the formalized 3-M reporting system and by CONAR from all sources is processed to establish the in-service engineering responsibilities and authority necessary to solve and correct a problem. The CONAR provides a primary periodic feedback channel for all SMS problems and deficiencies. Acknowledgement of all reported problems and deficiencies is vital to the success of the CONAR and will be continued; however, DCAP reporting will be restricted to progress on in-service engineering problems only. The DCAP system will:

1. Provide one central clearing house for SMS in-service engineering problems and establish communication channels for the expeditious flow of problem information.

2. Provide continually updated information on the status of in-service engineering problems and provide followup action. Problems will not be terminated until the final solution has been accomplished in all affected ships or stations, incorporated into production as required, and all technical data and records have been made.

3. Inform all levels of management, through a series of reports, of all SMS in-service engineering problems, actions planned or underway, and interim or final solutions. An in-service engineering problem is concerned with all engineering actions that are required to ensure the SMS equipment continues to be suitable for its intended service use. In-service engineering does not include within its scope actions which would significantly alter specified operational or performance characteristics, the configuration or interface requirements, or would compromise the reliability or safety of SMS equipment.

### GUIDED MISSILE SERVICE REPORT

The Guided Missile Service Report, form NAVSEA 4855/1, in accordance with NAVSEAINST 4855.19, should be prepared in duplicate. Upon recording missile testing, handling, servicing, and other reportable operations, NAVSEA form 4855/1 shall be forwarded to Officer in Charge, Fleet Analysis Center (Code 845), Naval Weapons Station, Seal Beach, Corona Annex, Corona, Calif., 91720. Negative reports are not required. A copy of the form shall be inserted in the missile log for unique operations; i.e., test, damage. Batch reports (operations that are applicable to several missiless or boosters; i.e., onloading or offloading) shall not be inserted in the missile logs, but shall be forwarded to the Fleet Analysis Center. Examples for filling out the forms are shown in figure 10-4 thru 10-7.

GUIDED MISSILE SERVICE REPORT
NAVSEA 4855/1 (REV. 4-77) (FORMERLY NAVORD 4855/22)

Mail Original to: Officer in Charge
Fleet Analysis Center
Naval Weapons Station, Seal Beach
Corona Annex
Corona, California 91720

RPT. NAVSEA 4855-1

SECURITY CLASSIFICATION: UNCLASSIFIED

(Please Print)

1. ORIGINATING ACTIVITY: (HULL NO.)

2. MISSILE OR ASSY (Rim/AIM or Mk–Mod & S/N)

| 3 DATE Mo Day Yr | 4 OPERATION OR DEFECT | 5 ITEM Name, Item Identifying No., and Serial No. | 6 OP TIME Minutes | 7 TEST RESULTS Check ✓ one: GO | NO GO: ITEM | NO GO: T.E. | NO GO: PERS. | 8 REMARKS |
|---|---|---|---|---|---|---|---|---|
| | | EXAMPLE 1. MISSILE ONLOAD AT | | | | | | |
| | | NAVWPNSTA, CHARLESTON | | | | | | |
| 6-24-79 | ONLOAD | NALC 1949 S/N 2084M | | | | | | FROM NAVWPNSTA, CHARLESTON |
| | | 1910 S/N 2809M | | | | | | |
| | | 1910 S/N 2810M | | | | | | |
| | | 1906 S/N 2812M | | | | | | |
| | | 1934 S/N 2820M | | | | | | |
| | | EXAMPLE 2. MISSILE OFFLOAD | | | | | | |
| | | AT NAVWPNSTA, CHARLESTON | | | | | | |
| 6-24-79 | OFFLOAD | NALC 1960 S/N 2821M | | | | | | TO NAVWPNSTA, CHARLESTON |
| | | S/N 2811M | | | | | | |
| | | S/N 2808M | | | | | | |
| | | S/N 2087M | | | | | | |
| | | | | | | | | |
| | | | | | | | | |
| | | | | | | | | |

94.245

Figure 10-4.—Missile onload/offload, batch report examples.

Detailed instructions for recording data in each block follow:

Block 1—Originating Activity. Enter ship hull number.

Block 2—Missile or Assembly. Enter RIM, RGM, or AIM–Mk/Mod, and serial number. Leave blank for batch reporting.

Block 3—Date. Enter month, day, and year for each entry.

Block 4—Operation or Defect. Operations or conditions that ship's personnel might desire to record for logging purposes may be recorded; however, recording of the following operations or defect is required:

DEFECT (describe defective condition and corrective action, if any.)

FIRED (identify booster, Mk/Mod, and serial number when applicable.)

MISFIRE (identify booster, Mk/Mod, and serial number when applicable.)

OFFLOAD (identify activity where offloaded; i.e., another ship or shore activity.)

ONLOAD (identify activity where onloaded; i.e., another ship or shore activity.)

## Chapter 10—ADMINISTRATION AND SUPPLY

| DATE<br>Mo Day Yr | OPERATION OR DEFECT | ITEM<br>Name, Item Identifying No., and Serial No. | OP TIME<br>Minutes | TEST RESULTS (Check one)<br>GO / NO GO (ITEM / T.E. / PERS) | REMARKS |
|---|---|---|---|---|---|
| | | EXAMPLE 1. MISSILE FIRED | | | |
| 9-30-79 | FIRED | RIM 66A-3  S/N 155M | | | |
| | | EXAMPLE 2. MISSILE MISFIRE | | | |
| 10-1-79 | MISFIRE | RIM 2F  S/N R-1635-B-5 | | | W/BOOSTER MK 12 MOD 1 S/N C1430 |
| 10-8-79 | OFFLOAD | RIM 2F  S/N R-1635-B-5 | | | TO NAVWPNSTA, SEAL BEACH |
| 10-8-79 | OFFLOAD | BOOSTER MK 12 MOD 1 | | | TO NAVWPNSTA, SEAL BEACH |
| | | EXAMPLE 3. MISSILE TRANSFER AT SEA | | | |
| 4-3-79 | ONLOAD | RIM 67A-3  S/N 6734E | | | RECEIVED FROM AE-15 |

Figure 10-5.—Missile, fired/misfire/transfer examples.

WETDOWN (and CLEANUP.)
DSOT AUTO TUNE SEASPARROW (NO-GO on missile only.)

Block 5—Item. Enter RIM, RGM, or AIM, and serial number of the missile; NALC and serial number, or name-Mk/Mod, and serial number.

Block 6—Op Time. Enter operating (run) time to the nearest tenth of a minute. (Not required for Tartar, Terrier, or Standard Missiles.)

Block 7—Test Results. All unsatisfactory (NO-GO) tests must be entered. (Not required for Tartar, Terrier, or Standard missiles.)

Block 8—Remarks. Explain the defect, corrective action, etc.

To obtain the correct security classification concerning guided missiles when filling out NAVSEA form 4855/1, refer to NAVSEAINST 5511.28, Security Classification Guide for Guided Missiles.

## ASROC REPORTS

In the Mk 10 Mod 7/8 and Mk 26 guided missile launching systems that have ASROC capability, an additional log is used for

GUIDED MISSILE SERVICE REPORT
NAVSEA 4855/1 (REV. 4-77) (FORMERLY NAVORD 4855/22)

Mail Original to: Officer in Charge
Fleet Analysis Center
Naval Weapons Station, Seal Beach
Corona Annex
Corona, California 91720 (Please Print)

RPT. NAVSEA 4855.1

SECURITY CLASSIFICATION: UNCLASSIFIED

1. ORIGINATING ACTIVITY: (HULL NO.)

2. MISSILE OR ASSY (Rim/AIM or Mk-Mod & S/N)

| 3 DATE Mo Day Yr | 4 OPERATION OR DEFECT | 5 ITEM Name, Item Identifying No., and Serial No. | 6 OP TIME Minutes | 7 TEST RESULTS Check ✓ one: GO | NO GO: ITEM | NO GO: T.E. | NO GO: PERS | 8 REMARKS |
|---|---|---|---|---|---|---|---|---|
| 5-2-79 | WETDOWN | RIM 24C-2 S/N 4339 | | | | | | MAGAZINE SPRINKLER |
| | | | | | | | | SPRAYED MISSILE WITH SALT |
| | | | | | | | | WATER APPROXIMATELY 2 MINUTES. |

94.247

Figure 10-6.—Missile wetdown example.

identification and transactions of ASROC weapons. NAVSEA form 8830/1 in accordance with NAVSEAINST 8830.1 is used to record all information about ASROC weapons, including mark, modification, and serial numbers, and ammunition lot numbers (if required) for each component that makes up an assembled weapon. Each ASROC is identified by an assembly identification number (AIN) for either a rocket thrown torpedo (RTT) or rocket thrown depth charge (RTDC). When a weapon is expended by firing, the log is sent to NWS, Seal Beach, Calif.

## NUCLEAR REPORTS AND INSPECTIONS

All nuclear material is subject to the control of Defense Nuclear Agency (DNA). Personnel assigned to work with nuclear weapons must receive special training in the handling, stowage, and accounting methods peculiar to nuclear weapons. Prior to such training they must possess at least a Secret clearance based on a background investigation. If you work in a GMLS that has nuclear weapons, a series of

## Chapter 10—ADMINISTRATION AND SUPPLY

Figure 10-7.—Missile, damage/offload, fired and test examples.

special publications is issued by the Joint Atomic Weapons Publication System and are used as reference publications for matenance, tests, storage requirements, and identifications of all nuclear weapons used by all services. In addition, a series of Navy special weapons ordnance publication (SWOPs) for Navy nuclear weapons and their related equipment for each type of nuclear weapon used with shipboard GMLS is also issued. The SWOPs take precedence over all other technical publications where conflicting information is present. Since all the procedures in SWOPs are mandatory, it is important that all shipboard activities expedite the processing and routing of message/speedletcr changes, and interim and permanent changes to SWOPs to ensure prompt updating of affected publications. Each type of nuclear weapon and its related components and their test equipment is assigned a separate publication number series applicable to the Navy's nuclear weapons program. For example SWOP 45.21 series is applicable to Terrier. The requirements for the SWOP documents are determined by the individual commands in accordance with their nuclear weapons capabilities. An index to joint atomic weapons publications used for multiple service purposes is listed in SWOP 0-1; a Navy supplement, SWOP 0-1B, lists all publications applicable for Navy use only.

All nuclear weapons, handling equipment, and reports are subject to special inspection in accordance with OPNAVINST 5040.6 series. The type of inspection required for any activity having nuclear capabilities depends upon the type of installation and type of weapons.

**Navy Technical Proficiency Inspection (NTPI)**

This inspection is performed once a year by naval inspectors. The inspection teams are sent aboard ship from nuclear weapons training centers. The NTPI determines the capabilities of the naval activities for storing, testing, assembling, maintaining, handling, and loading nuclear weapons.

Inspectors are provided from the nuclear weapons training centers, Atlantic or Pacific, by the appropriate fleet training commands for all the respective fleet inspections. The inspectors are usually officers and chiefs who are skilled in nuclear weapons handling. When practicable, NTPIs of fleet units are conducted in conjunctions with, and as part of, operational readiness inspections (ORI) or other major readiness exercises.

NTPIs are graded in accordance with SWOP 25-1, Nuclear Weapons Technical Inspection System. Inspection reports are made in accordance with applicable directives of fleet commanders and NAVSEA, and include all the inspector's comments, recommendations, and discrepancies noted. They are appended to the chief inspector's report.

NTPI GUIDES.—The Nuclear Weapons Management Manual, COMNAVSURFLANTINST C8120.1 series or COMNAVSURFPACINST C8120.1 series, is a management manual, not a technical manual. Some technical items are included only to emphasize the importance of safety, quality, or readiness. Virtually everything in the management area required or expected of a nuclear capable ship/unit, under routine conditions, has been included. The Nuclear Weapons Management Manual is the prime source of information in its subject area, and only under unusual circumstances will it be necessary to consult another source.

The intent in developing this manual is to—

1. Standardize procedures and requirements to the maximum extent.
2. Minimize the need for additional reference to other directives.
3. Reduce shipboard administrative efforts.
4. Remove the responsibility for the accuracy and currency of shipboard documentation from the user.

All changes by higher authority to the nuclear weapons program that are suitable for inclusion in the manual will be so done expeditiously. Consult your Nuclear Weapons Management Manual often to make sure you have not overlooked any management items.

The scope of coverage of an NTPI differs between activities. All matters directly related to the processing, handling, inspecting, maintaining, and storing of nuclear weapons, and all matters and procedures involved in the administration of a nuclear unit are included in every NTPI. Some of the items checked by an NTPI follow:

1. Is the area clean and free from tripping or slipping hazards? Is the lighting sufficient? Is the noise level low? Is the fire equipment inspected and maintained periodically? Are first aid and decontamination facilities properly identified and readily available?
2. Are the safety precautions posted and are personnel familiar with them?
3. Is the emergency equipment readily available, properly stowed, and in good condition?
4. Are the checksheets up to date and do they include all manual changes and contain all applicable notes, cautions, and warnings?
5. Are all tools clean, in good condition, and used properly?

## Chapter 10—ADMINISTRATION AND SUPPLY

6. Is the two-man rule enforced with personnel who are equally knowledgeable with respect to the task being performed?

7. Are containers holding toxic and hazardous liquids or materials properly labeled for identification? Are they stowed properly?

8. Is the handling equipment being properly maintained and weight tested at the appropriate intervals? Are there weight test inspection tags for each piece of handling equipment in accordance with the Nuclear Weapons Management Manual, COMNAVSURFLANT or PACINST C8120.1 series? Does the weight test log reflect the current handling equipment's weight test status?

9. Is the personnel reliability program being implemented? Are the records of such personnel maintained properly and are medical and supervisory personnel provided observation of such personnel?

10. Is the intrusion alarm system being operated and maintained in accordance with existing instructions?

11. Are the local instructions current and accurate? Are personnel aware of them and are they followed? (Many discrepancies are noted in this area when these instructions are not reviewed and updated periodically.)

12. Are Navy SWOPs, OPs, and ODs up to date in accordance with current allowance lists? Are subcustody procedures in effect sufficient to allow a rapid, complete inventory?

13. Is the emergency destruction bill current and are personnel assignments reviewed to ensure accuracy? Is the emergency recall bill exercised periodically?

14. Are personnel allowances adequate in view of the responsibilities and workload imposed?

15. Does the pass and badge system comply with current directives?

16. Are reports properlyy submitted?

In addition to the preceding checks, are accident/incident drills performed by the inspected organization and observed by the NTPI team. Accident/incident drills demonstrate initial procedures performed by local station personnel. These drills are designed to check the following:

1. Medical, firefighting, and guard force procedures.
2. On-scene survivors and on-scene commander's procedures.
3. Hot line procedures.
4. Explosive disposal procedures.

COMPLETION INSPECTION REPORT.—At the completion of the NTPI, a rough draft report is prepared. A copy of this report is given to the activity and a critique is conducted. All discrepancies and comments are read to the technicians and supervisors and they are given an opportunity to dispute them. The final smooth report is issued about 2 weeks later, after it is thoroughly checked for accuracy. This report is submitted to the appropriate activities in the chain of command.

Upon completion of the inspection, the Commander Naval Sea Systems Command makes a report directly to the Chief of Naval Operations, with a copy to the fleet commander in chief, certifying the readiness of the facility. This report contains (1) a statement that all safety (nuclear and explosive), technical, and security criteria have been met, or a statement listing the deviations from those criteria with the justification for waiver; and (2) a statement as to whether the facility is ready to perform its assigned mission.

**Nuclear Weapons Acceptance Inspection**

A nuclear weapons acceptance inspection (NWAI) is conducted on all prospective naval nuclear weapons activities by Navy inspectors. This inspection determines the readiness of the activity to perform technical, administrative, and logistical procedures directly related to their nuclear weapons capabilities. This could be a newly constructed ship, a ship just recently operational after an extended yard period during which a large number of personnel were transferred, or a newly constructed shore site.

Each activity is inspected for each of its capabilities and must receive a grade of satisfactory before it is considered operational for any capability. A regularly scheduled NTPI can serve as an NWAI for an operational activity achieving a new capability in addition to those it already holds.

## Technical Standardization Inspection (TSI)

This inspection is performed by personnel from Field Command, Defense Nuclear Agency (DNA). The schedule of inspections is published, so you will be aware of the approximate date your activity will be inspected. In addition to judging your capability in the handling and care of missiles with nuclear warheads, all the records, reports, publications, training programs, and team orgainzations will be inspected. As a guide to the inspectors, SWOP 25-1 lists the points to be inspected. This list can also be a valuable guide for checking yourself and for training your personnel. The TSI also determines the status of maintenance of your nuclear missiles.

## Personnel Injuries

Any time an individual sustains a personal injury, a report must be made. The medical department prepares the medical report, but the weapons department must explain the circumstances under which the injury occurred. If a casualty is caused by an explosive accident or incident (nonnuclear), it is reported according to the message format given in OPNAVINST 5102.1. The same format is used for reporting explosive accidents and incidents without personnel casualties. The messages may be supplemented with written descriptions, photographs, diagrams, etc. The purpose is to determine by what means the accident could have been avoided and to issue instructions to correct the conditions that caused the accident or incident. Analysis of the reports supplies much of the information included in OP 3565, Technical Manual Radio Frequency Hazards to Ordnance, Personnel, and Fuel.

Supervisor's Report of Injury, NAVEXOS Form 180, is an administrative report about any type of injury sustained by military personnel. The information given on these reports is analyzed by the coordinator of safety programs to discover the causes and frequency of different type of accidents.

## Security

Commanding officers are responsible for establishing security orientation, education, and training programs for all personnel assigned. OPNAVINST 5510.1 series delineates the necessary minimum requirements for implementing these programs.

The central aim of the security education program is to make all personnel secuirty-minded. To achieve this, a continuing training program must be conducted at all levels of the command. Senior petty officer should emphasize the importance of this program to their subordinates.

Effective security also requires that the handling, accountability, procurement, stowage, and transmission of classified records, reports, and publications be managed in an efficient manner. Good classification management practices are important in obtaining effective security control. If security needs are to be met, constant attention must be given to the way in which classified material is handled. With Confidential and Secret publications and material in daily use in a GMLS, the responsibility for safeguarding classified material lies with each person who handles the material. In addition to this general responsibility, each leading petty officer should establish organizational responsibilities for the control of classified material. A simple, effective control system can provide a readily available accurate accounting system for classified documents.

# INDEX

## A

Actuator units, 212
Adjustment and repair of hydraulic systems, 196-202
    at a Navy yard or repair tender, 201, 202
    shipboard maintenance, 196-200
    troubleshooting, 200, 201
Administration and supply, 299-322
    Coordinated Shipboard Allowance List (COSAL), 308-310
    identification of ordnance parts, 310, 311
    Navy management data list (NMDL), 311
    ordering publications, 311, 312
    ordnance alteration (ORDALT), 303
    ship armament inventory list (SAIL), 303
    Ship Equipment Configuration Accounting System (SECAS), 302, 303
    sources of supply, 307, 308
    supply responsibilities, 306, 307
    technical assistance, 303-306
    3-M systems, 299-303
    weapons department reports, records, and inspections, 312-322
Air driven missile handling equipment, 213-218
Aligning the missile batteries, 267-277
    alignment of launcher, 270-274
    final alignment adjustment procedures, 276, 277
    firing cutout mechanisms, 274, 275
    radar alignment, 275
    sonar-to-radar alignment checkout, 275, 276
    sources of alignment information, 267
    system alignment, 267-270
Alignment and fire control, 250-278
Ammunition records and reports, 248

ASROC weapons, 86-92
    duds and misfires, 90-92
    loading, 86-88
    unloading, 88-90

## C

Classification codes, 4-6
    assignment priorities, 5, 6
    entry series, 4
    rating series, 5
    special series, 5
Components of missile control systems, 210
Control circuits, 107-128
    conventional type, 107
    logic circuits, 120
    solid-state type, 107-120
Controller units, 210, 211
Control system, GMLS, 106-148
    control circuits, 107-128
    digital circuits and devices, 128-136
    power and control panels, 106, 107
    safety, 147-149
    troubleshooting control and power circuits, 136-147
Coordinated Shipboard Allowance List (COSAL), 308-310

## D

Digital circuits and devices, 128-136
    binary numbering system, 131
    computers, 130
    converters, 134-136
    data transmission, 132-134
    position notation, 130, 131

Dummy director and error recorder, 287-294

Duty assignments, GMM, 6
    out-of-rating assignments, 6
    tour lengths, 6

## E

Electricity and electronics, 106-163
    control system, 106-148
    references, 163
    servoamplifier, 157-163
    servomechanisms, 149-151
    synchro systems, 151-156

Explosive hazard classification, 232, 233

## F

Fire control and alignment, 250-278
    aligning the missile batteries, 267-277
        alignment of launcher, 270-274
        final alignment adjustment procedures, 276, 277
        firing cutout mechanisms, 274-275
        radar alignment, 275
        sonar-to-radar alignment checkout, 275, 276
        sources of alignment information, 267
        system alignment, 267, 270
    Terrier weapon system, 253-267
        Mk 11 NTDS/WDS functions in AAW operations, 262-267
        Mk 76 Mod 5 guided missile fire control system, 254-261
        Mk 76 Mods 6, 7, and 8 guided missile fire control systems, 261, 262
        NTDS/WDS, 254
        search radars, 253, 254
    weapon system configuration and data flow, 250-253
        command and control, 251, 252
        delivery, 253
        destruction, 253
        fire control system (FCS), 252, 253
        surveillance, 250, 251

Firing cutout mechanisms, 274, 275

Firing exercises, 65-77
    Harpoon GMLSs, 77
    Sparrow III, Mk 29 GMLS, 76, 77
    Standard GMLSs, 75-76
    Tartar GMLSs, 67-75
        Mk 11, 67-69
        Mk 13, 69-74
        Mk 22, 74, 75
    Terrier GMLSs, 65-67
        Mk 10 mods 0 through 6, 65-67
        Mk 10 mods 7 and 8 launching systems, 67
    Tomahawk GMLSs, 77

## G

GMLS, pneumatics in, 212-229
    air driven missile handling equipment, 213-218
    launcher air drive systems, 218-222
    other uses of pneumatics, 222-228
    pneumatics in missiles, 228, 229

GMLSs, 65-105
    Harpoon, 77
    Mk 11, 95-96
    Mk 13, 97-101
    Mk 22, 101, 102
    Mk 26, 102-105
    Sparrow III, Mk 29, 76, 77
    Standard, 75-76
    Tartar, 67-75, 93
    Terrier, 65-67, 79-92
    Tomahawk, 77

GMLSs, hydraulics in, 164-207
    adjustment and repair of hydraulic systems, 196-202
    handling and strikedown equipment, 164-168
    hydraulically operated launcher components, 183-194
    hydraulic schematics, 194-196
    loader components, 179-183
    magazines, 169-178
    receiver-regulators, 202-207

Gunner's Mate M, the, 1-16
    Gunner's Mate rating, 3-6
        classification codes, 4-6
        duty assignments, 6
        missile ratings, 3

# INDEX

Gunner's Mate M, the—Continued
    personnel assignment system, 7-16
        advancement, 10
        advancement opportunities for senior
            petty officers, 15-16
        detailing, 8-9
        duty preferences, 8
        how to prepare for advancement, 11-15
        qualifying for advancement, 10
        Personnal Requisition System, 7
        types of duty, 7
        who will be advanced?, 10, 11
    senior petty officer responsibilities, 1-3
        effective communication, 1
        training, 2, 3

## H

Handling and stowage of other missiles, 49, 50
    aircraft missiles, 50
    standard missiles (MR/ER), 49, 50
Handling and strikedown equipment, 164-168
Harpoon launching group, 36-42
    canister, 39
    canister frame assemblies, 39
    grade B canister launcher, 39
    Harpoon casualty panel (HCP), 39
    Harpoon lightweight (LTWT) canister launcher, 39
    Harpoon trainer module (HTM), 39
    Harpoon weapon control console (HWCC), 36-38
    Harpoon weapon control indicator panel (HWCIP), 39
    launcher switching unit (LSU), 39
    onloading operation, 40-42
Hydraulics and pneumatics, 164-229
    hydraulics in GMLSs, 164-207
        adjustment and repair of hydraulic systems, 196-202
        handling and strikedown equipment, 164-168
        hydraulically operated launcher components, 183-194
        hydraulic schematics, 194-196
        loader components, 179-183
        magazines, 169-178
        receiver-regulators, 202-207

Hydraulics and pneumatics—Continued
    hydraulics in missiles, 208-212
        actuator units, 212
        components of missile control systems, 210
        controller units, 210, 211
        overall operation, 209, 210
    pneumatics in a GMLS, 212-229
        air driven missile handling equipment, 213-218
        launcher air drive systems, 218-222
        other uses of pneumatics, 222-228
        pneumatics in missiles, 228, 229

## I

Identification, missile component, 236-238
Information, sources of, advancement, 11-15
Inspection and disposal of explosive components, 235, 236
Inspections, reports, and records, weapon department, 312-322
Inspections, weapon, 231, 232
Inventory record of small arms and pyrotechnics, 247, 248

## L

Launcher air drive systems, 218-222
Launcher, alignment of, 270-274
Launcher components, hydraulically operated, 183-194
    carriage-mounted hydraulic parts, 184-188
    guide and guide arms, 184
    power drives, 188-194
Launcher power drive servoamplifiers, 157-162
Launcher shipboard performance test, 287-296
    dummy director and error recorder, 287-294
    limiter and demodulator, 294-296
Limiter and demodulator, 294-296
Loader components, 179-183
    floating tracks, 181
    power drives, 179
    retractable rails, 181, 182
    spanning rails and blast doors, 183
    tilting rail, 179-181
    trunk, 179

Loading and stowage plans, 17-20
    knowledge required for planning work, 17, 18
    planning sequence of operation, 19
    scheduling of work, 18
    security, 19
    stowage areas, 19, 20

Loading, unloading, and dud jettisoning, 79-105
    Tartar GMLSs, 93-105
        dud jettisoning, 95
        loading, 93, 94
        step control, 95
        unloading, 94, 95
    Terrier GMLSs, 79-92
        ASROC duds and misfires, 90-92
        ASROC weapons, loading of, 86-88
        ASROC weapons, unloading of, 88-90
        dud jettisoning, 81-86
        loading, 79
        step control, 80, 81
        unloading, 80

## M

Magazine and weapon inspection, records, and reports, 246-249
    ammunition records and reports, 248
    inventory record of small arms and pyrotechnics, 247, 248
    nuclear warhead reports, 248, 249
    visual magazine inspection and recordkeeping, 246, 247

Magazines, 169-178
    hoist mechanism, 176
    magazine doors, 176-178
    power drive, 169-172
    ready service ring drive, 172, 173
    tray shift mechanism, 173-176

Magazines, missile, 238-246
    construction and arrangement, 238
    fire suppresion equipment, 241-243
    other magazine protective devices, 244-246
    safety requirements, magazine, 238-241

Maintenance procedure, steps in, 281-287
    cleaning of parts, 281, 282
    lubrication, 282, 283
    tests and checks, 283-287
    visual inspection, 281

Manning the system, 55-64
    GMLS control panels, 57-61
        launcher captain's panel, 57
        power panels, 57
    GMLS stations, 61, 62
        checkout area, 62
        magazines, 62
        wing and fin assembly areas, 62
    safety, 62-64
        provisions and devices, 64
        types of danger, 63
    weapons system operation, 55-58

Missile component identification, 236-238
    color code interpretation, 236
    miscellaneous explosive devices, 237, 238
    missile and component markings, 236, 237

Missile Handling, 17-54
    handling and stowage of other missiles, 49, 50
        aircraft missiles, 50
        Standard missiles (MR/ER), 49, 50
    Harpoon launching group, 36-42
        canister, 39
        canister frame assemblies, 39
        grade B canister launcher, 39
        Harpoon casualty panel (HCP), 39
        Harpoon lightweight (LTWT) canister launcher, 39
        Harpoon trainer module (HTM), 39
        Harpoon weapon control console (HWCC), 36-38
        Harpoon weapon control indicator panel (HWCIP), 39
        launcher switching unit (LSU), 39
        onloading operation, 40-42
    loading and stowage plans, 17-20
        knowledge required for planning work, 17, 18
        planning sequence of operation, 19
        scheduling of work, 18
        security, 19
        stowage areas, 19, 20
    NATO Seasparrow surface missile system, (NSSMS), 33-36
        deck station launcher control, 34
        launcher, 33
        loader, 33
        onloading operation, 34-36

# INDEX

Missile Handling—Continued
    nuclear weapons handling and stowage, 50-54
        alarm and warning systems, 53
        entry and access control, 51, 52
        handling, 52
        personnel reliability program, 53
        safeguarding, 51
        safety requirements, 52
        stowage requirements, 53
        training, 53
        ventilation in nuclear weapons spaces, 53, 54
    references, 54
    Tartar missile handling and stowage, 42-49
        handling, 49
        no-test program, 48
        strikedown operations on the Mk 13 and Mk 22 GMLSs, 47, 48
    Terrier missile handling and stowage, 20-33
        depot handling and stowage, 30-33
        initial receipt, 21
        shipboard assembly and disassembly, 30
        special problems with Terrier, 20
        tests and inspections, 28, 29
    Tomahawk weapon system, 42
Missiles, hydraulics in, 208-212
    acutator units, 212
    components of missile control systems, 210
    controller units, 210, 211
    overall operation, 209, 210
Missile simulators, 296, 297
Mk 11 NTDS/WDS functions in AAW operations, 262-267
Mk 76 Mod 5 guided missile fire control system, 254-261
Mk 76 Mods 6, 7, and 8 guided missile fire control systems, 261, 262

## N

NATO Seasparrow surface missile system, (NSSMS), 33-36
    deck station launcher control, 34
    launcher, 33
    loader, 33
    onloading operation, 34-36
Navy maintenance program, 279-281

Navy Management Data List (NMDL), 311
Nuclear warhead reports, 248, 249
Nuclear weapons handling and stowage, 50-54
    alarm and warning systems, 53
    entry and access control, 51, 52
    handling, 52
    personnel reliability program, 53
    safeguarding, 51
    safety requirements, 52
    stowage requirements, 53
    training, 53
    ventilation in nuclear weapons spaces, 53, 54
Nuclear weapons hazards, 233-235

## O

Ordnance alteration (ORDALT), 303
Ordnance parts, identification of, 310, 311
Own ship's maintenance program, 297, 298

## P

Personnel assignment system, 7-16
    advancement, 10
    advancement opportunities for senior petty officers, 15-16
    detailing, 8-9
    duty preferences, 8
    how to prepare for advancement, 11-15
        how to study, 15
        sources of information, 11-15
    Personnel Requisition System, 7
    qualifying for advancement, 10
    types of duty, 7
    who will be advanced?, 10, 11
Pneumatics and hydraulics, 164-229
PQS, Personnel Qualification Standards Program, 3
Preparing for a firing exercise, 55-78
    firing exercises, 65-77
        Harpoon GMLSs, 77
        Sparrow III, Mk 29 GMLS, 76, 77
        Standard GMLSs, 75-76
        Tartar GMLSs, 67-75
        Terrier GMLSs, 65-67
        Tomahawk GMLSs, 77

Preparing for a firing exercise—Continued
　　manning the system, 55-64
　　　　GMLS control panels, 57-61
　　　　GMLS stations, 61, 62
　　　　safety, 62-64
　　　　weapon system operation, 55-58
　　references, 77, 78

## R

Radar alignment, 275
Radio frequency (RF) hazards, 230, 231
Receiver-regulators, 202-207
References, missile handling, 54
Repair and test, 279-298
　　launcher shipboard performance test, 287-296
　　missile simulators, 296, 297
　　Navy maintenance program, 279-281
　　own ship's maintenance program, 297, 298
　　steps in maintenance procedure, 281-287
　　training missiles, 297
Reports, records, and inspections, weapons department, 312-322

## S

Schematics, hydraulic, 194-196
Search radars, Terrier weapon system, 253, 254
Senior petty officer responsibilities, 1-3
　　effective communication, 1
　　training, 2, 3
　　　　PQS, Personnel Qualification Standards Program, 3
Servoamplifier, 157-163
　　launcher power drive servoamplifiers, 157-162
　　troubleshooting and adjustments, 162, 163
Servomechanisms, 149-151
Ship Armament Inventory List (SAIL), 303
Ship Equipment Configuration Accounting System (SECAS), 302, 303
Ship's Qualification Assistance Team (SQAT), 306
Simulator, missile, 296, 297
Sonar-to-radar alignment checkout, 275, 276

Sources of supply, 307, 308
　　Defense Logistics Agency, 307
　　Guided Missile Support Facilities, 307, 308
　　Ship's Parts Control Center, 307
Strikedown operations on the Mk 13 and Mk 22 GMLSs, 47, 48
Strikedown, terrier missile handling and stowage, 24-27
Supply and administration, 299-322
Synchro systems, 151-156
　　alignment, 151-155
　　maintaining a synchro system, 155
　　troubleshooting synchro systems, 155, 156
System alignment, 267-270

## T

Tartar GMLSs, 93-105
　　dud jettisoning, 95
　　　　Mk 11 GMLS, 95-96
　　　　Mk 13 GMLS, 97-101
　　　　Mk 22 GMLS, 101, 102
　　　　Mk 26 GMLS, 102-105
　　loading, 93, 94
　　step control, 95
　　unloading, 94, 95
Tartar missile handling and stowage, 42-49
　　handling, 49
　　no-test program, 48
　　strikedown operations on the Mk 13 and Mk 22 GMLSs, 47, 48
Technical assistance, 303-306
Terrier GMLSs, 79-92
　　ASROC duds and misfires, 90-92
　　ASROC weapons, loading of, 86-88
　　　　care of cable assemblies, 87
　　　　depth charge configuration, 87, 88
　　ASROC weapons, unloading of, 88-90
　　　　dud jettisoning, 81-86
　　　　dud or misfire, 85
　　　　maintenance, 85, 86
　　　　operation, 82-84
　　loading, 79
　　step control, 80, 81
　　unloading, 80

# INDEX

Terrier missile handling and stowage, 20-33
    depot handling and stowage, 30-33
    initial receipt, 21-27
        handling equipment needed, 21-23
        safety precautions in handling, 23, 24
        strikedown, 24
    shipboard assembly and disassembly, 30
    special problems with terrier, 20
    tests and inspections, 28, 29
        radiation and monitoring, 29, 30
Terrier weapon system, 253-267
    Mk 11 NTDS/WDS functions in AAW operations, 262-267
    Mk 76 Mod 5 guided missile fire control system, 254-261
    Mk 76 Mods 6, 7, and 8 guided missile fire control systems, 261, 262
    NTDS/WDS, 254
    search radars, 253, 254
Test and repair, 279-298
Tests and checks, maintenance, 283-287
    daily, 283, 284
    monthly, 284, 285
    periodic tests, 286, 287
    quarterly tests, 285, 286
    weekly, 284
3-M Systems, 299-302
    Intermediate Maintenance Activity (IMS); Intermediate Maintenance Management System (IMMS), 302
    Maintenance Data System (MDS), 301, 302
    Planned Maintenance System (PMS), 299-301
Tomahawk weapon system, 42
Training missiles, 297
Troubleshooting control and power circuits, 136-147
    component isolation, 142-147
    preliminary isolation, 136-140

## V

Visual magazine inspection and recordkeeping, 246, 247

## W

Weapons department reports, records, and inspections, 312-322
    ASROC reports, 317, 318
    Commanding Officer's Narrative Report (CONAR), 313-315
    Deficiency Corrective Action Program (DCAP), 315
    Guided Missile Service Report, 315-317
    Nonexpendable SMS Equipment Status Reports, 313
    nuclear reports and inspections, 318-322
    Surface-Launched Missile Firing Reports, 312, 313
Weapon surveillance and stowage, 230-249
    inspection and disposal of explosive components, 235, 236
    magazine and weapon inspections, records, and reports, 246-249
    missile component identification, 236-238
    missile magazines, 238-246
    safety considerations, 230-235
Weapon system configuration and data flow, 250-253
    command and control, 251, 252
    delivery, 253
    destruction, 253
    fire control system (FCS), 252, 253
    surveillance, 250, 251

☆U.S. GOVERNMENT PRINTING OFFICE:1983 -651 -126/ 4904

# GUNNER'S MATE M 1 & C

## NAVEDTRA 10200-D

Prepared by the Naval Education and Training Program Development Center, Pensacola, Florida

Your NRCC contains a set of assignments and perforated answer sheets. The Rate Training Manual, Gunner's Mate M 1&C, NAVEDTRA 10200-D, is your textbook for the NRCC. If an errata sheet comes with the NRCC, make all indicated changes or corrections. Do not change or correct the textbook or assignments in any other way.

## HOW TO COMPLETE THIS COURSE SUCCESSFULLY

Study the textbook pages given at the beginning of each assignment before trying to answer the items. Pay attention to tables and illustrations as they contain a lot of information. Making your own drawings can help you understand the subject matter. Also, read the learning objectives that precede the sets of items. The learning objectives and items are based on the subject matter or study material in the textbook. The objectives tell you what you should be able to do by studying assigned textual material and answering the items.

At this point you should be ready to answer the items in the assignment. Read each item carefully. Select the BEST ANSWER for each item, consulting your textbook when necessary. Be sure to select the BEST ANSWER from the subject matter in the textbook. You may discuss difficult points in the course with others. However, the answer you select must be your own. Remove a perforated answer sheet from the back of this text, write in the proper assignment number, and enter your answer for each item.

Your NRCC will be administered by your command or, in the case of small commands, by the Naval Education and Training Program Development Center. No matter who administers your course you can complete it successfully by earning a 3.2 for each assignment. The unit breakdown of the course, if any, is shown later under Naval Reserve Retirement Credit.

## WHEN YOUR COURSE IS ADMINISTERED BY LOCAL COMMAND

As soon as you have finished an assignment, submit the completed answer sheet to the officer designated to grade it. The graded answer sheet will not be returned to you.

If you are completing this NRCC to become eligible to take the fleetwide advancement examination, follow a schedule that will enable you to complete all assignments in time. Your schedule should call for the completion of at least one assignment per month.

Although you complete the course successfully, the Naval Education and Training Program Development Center will not issue you a letter of satisfactory completion. Your command will make an entry in your service record, giving you credit for your work.

## WHEN YOUR COURSE IS ADMINISTERED BY THE NAVAL EDUCATION AND TRAINING PROGRAM DEVELOPMENT CENTER

After finishing an assignment, go on to the next. Retain each completed answer sheet until you finish all the assignments in a unit (or in the course if it is not divided into units). Using the envelopes provided, mail your completed answer sheets to the Naval Education and Training Program Development Center where they will be graded and the score recorded. Make sure all blanks at the top of each answer sheet are filled in. Unless you furnish all the information required, it will be impossible to give you credit for your work. The graded answer sheets will not be returned.

The Naval Education and Training Program Development Center will issue a letter of satisfactory completion to certify successful completion of the course (or a creditable unit of the course). To receive a course-completion letter, follow the directions given on the course-completion form in the back of this NRCC.

You may keep the textbook and assignments for this course. Return them only in the event you disenroll from the course or otherwise fail to complete the course. Directions for returning the textbook and assignments are given on the book-return form in the back of this NRCC.

## PREPARING FOR YOUR ADVANCEMENT EXAMINATION

Your examination for advancement is based on the Occupational Standards for your rating as found in the MANUAL OF NAVY ENLISTED MANPOWER AND PERSONNEL CLASSIFICATIONS AND OCCUPATIONAL STANDARDS (NAVPERS 18068). These Occupational Standards define the minimum tasks required of your rating. The sources of questions in your advancement examination are listed in the BIBLIOGRAPHY FOR ADVANCEMENT STUDY (NAVEDTRA 10052). For your convenience, the Occupational Standards and the sources of questions for your rating are combined in a single pamphlet for the series of examinations for each year. These OCCUPATIONAL STANDARDS AND BIBLIOGRAPHY SHEETS (called Bib Sheets), are available from your ESO. Since your textbook and NRCC are among the sources listed in the bibliography, be sure to study both as you take the course. The qualifications for your rating may have changed since your course and textbook were printed, so refer to the latest edition of the Bib Sheets.

## NAVAL RESERVE RETIREMENT CREDIT

The course is evaluated at 16 Naval Reserve retirement points, which will be credited as follows: 12 points upon completion of assignments 1 through 8; 4 points upon completion of assignments 9 through 10. These points are creditable to personnel eligible to receive them under current directives governing the retirement of Naval Reserve personnel.

## COURSE OBJECTIVE

While completing this course, the student will demonstrate understanding of course material by correctly answering items on the following: responsibilities and skills of a GMM1 or GMMC; basic fundamentals of hydraulic, pneumatic, and electrical devices used in missile launching systems; characteristics, classification, and types of explosives and pyrotechnics; the related safety precautions which must be observed when handling explosives and pyrotechnics; magazines and their associated sprinkling and alarm systems; basic principles of missile flight and guidance control equipment; types of guided missile launching systems, their function and component identification and related safety precautions; application of fire control units that make up a missile weapons system; function and use of servosystems in battery alignment; maintenance, overhaul, repair, adjustment and testing of a missile launching system; administrative duties of a GMM; logs, records, and reports.

---

While working on this correspondence course, you may refer freely to the text. You may seek advice and instruction from others on problems arising in the course, but the solutions submitted must be the result of your own work and decisions. You are prohibited from referring to or copying the solutions of others, or giving completed solutions to anyone else taking the same course.

Naval courses may include a variety of questions -- multiple-choice, true-false, matching, etc. The questions are not grouped by type; regardless of type, they are presented in the same general sequence as the textbook material upon which they are based. This presentation is designed to preserve continuity of thought, permitting step-by-step development of ideas. Some courses use many types of questions, others only a few. The student can readily identify the type of each question (and the action required) through inspection of the samples given below.

## MULTIPLE-CHOICE QUESTIONS

Each question contains several alternatives, one of which provides the best answer to the question. Select the best alternative, and blacken the appropriate box on the answer sheet.

### SAMPLE

s-1. The first person to be appointed Secretary of Defense under the National Security Act of 1947 was
1. George Marshall
2. James Forrestal
3. Chester Nimitz
4. William Halsey

Indicate in this way on the answer sheet:

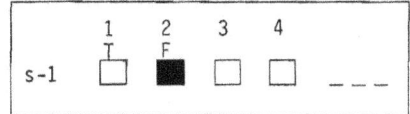

## TRUE-FALSE QUESTIONS

Mark each statement true or false as indicated below. If any part of the statement is false the statement is to be considered false. Make the decision, and blacken the appropriate box on the answer sheet.

### SAMPLE

s-2. Any naval officer is authorized to correspond officially with any systems command of the Department of the Navy without his commanding officer's endorsement.

Indicate in this way on the answer sheet:

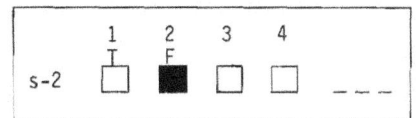

## MATCHING QUESTIONS

Each set of questions consists of two columns, each listing words, phrases or sentences. The task is to select the item in column B which is the best match for the item in column A that is being considered. Items in column B may be used once, more than once, or not at all. Specific instructions are given with each set of questions. Select the numbers identifying the answers and blacken the appropriate boxes on the answer sheet.

### SAMPLE

In questions s-3 through s-6, match the name of the shipboard officer in column A by selecting from column B the name of the department in which the officer functions.

| A | B |
|---|---|
| s-3. Damage Control Assistant | 1. Operations Department |
| s-4. CIC Officer | 2. Engineering Department |
| s-5. Disbursing Officer | 3. Supply Department |
| s-6. Communications Officer | |

Indicate in this way on the answer sheet:

```
        1   2   3   4
           T   F
s-3    [ ] [■] [ ] [ ]   _ _ _
s-4    [■] [ ] [ ] [ ]   _ _ _
s-5    [ ] [ ] [■] [ ]   _ _ _
s-6    [■] [ ] [ ] [ ]   _ _ _
```

# Assignment 1

The Gunner's Mate M 1&C; Missile Handling

Textbook, NAVEDTRA 10200-D: Pages 1 through 23

---

Learning Objective: Define the purpose of the *Gunner's Mate M 1&C* rate training manual (RTM) and nonresident career course (NRCC).

---

1-1. Which of the following is the basis for the information contained in the GMM 1&C rate training manual?

1. Occupational standards for advancement
2. Navy enlisted classifications (NECs)
3. Enlisted manpower requirements
4. All of the above

1-2. Which of the following statements best describes a nonresident career course?

1. It is a self-study enlisted training course based on material contained in an associated rate training manual
2. It is a correspondence course designed primarily for use by personnel attending a formal Navy school
3. It is a self-study training course recommended for completion prior to advancement
4. It is a self-study training course for officer and enlisted personnel

---

Learning Objective: Describe the responsibilities of the senior Gunner's Mate while working with subordinates and superiors.

---

1-3. Which of the following actions are essential for improving your leadership abilities and technical knowledge?

1. Study, evaluation, and observation
2. Evaluation, observation, and practical application
3. Observation, practical application, and study
4. Practical application, study, and evaluation

1-4. Which of the following functions is/are included in your duties as a GMM1 or GMMC?

1. Translating orders given by higher authority into practical on-the-job language that can be understood by relatively inexperienced personnel
2. Explaining the problems and needs of enlisted personnel to higher authority
3. Both 1 and 2 above
4. Administering proper but fair discipline when a subordinate commits a serious wrong

1

1-5. To properly communicate with and train others, you should make it a practice to continually develop your ability, and that of your trainees to

1. avoid criticism from others having higher formal education
2. make use of every opportunity to impress others with your superior command of language
3. learn a foreign language
4. use correct terminology

1-6. Which of the following practices is recommended for keeping up with new developments and changes?

1. Complete all officer correspondence courses
2. Remain alert for new information from sources such as technical data publications
3. Complete all enlisted non-resident career courses
4. Collect personal copies of all pertinent technical manuals

1-7. If your missile gang is composed solely of highly skilled, well-trained, and experienced Gunner's Mates, there is no need to continually train them.

1-8. The Navy is constantly working toward improvements in training. The present trend is to

1. provide all the training needed but at the same time eliminate irrelevant training
2. provide more A schools and less training aboard ship or station
3. provide more group training and less individualized training
4. do all training through the use of correspondence courses, where possible, and eliminate all but a few resident schools

1-9. The personnel qualification standards (PQS) program is an operator-oriented program consisting of a compilation of knowledge and skill requirements designed to qualify personnel for specific watch stations.

---

Learning Objective: Point out the history of the GM rating with emphasis upon the GMM, weapons, and technical requirements.

---

1-10. In ___A___, the Gunner's Mate
      1797, 1894
      rating was established and pay grades third through first class petty officer were established by General Order ___B___.
                                37, 409

1. A - 1797, B - 37
2. A - 1894, B - 37
3. A - 1894, B - 409
4. A - 1797, B - 409

1-11. When was the present GM rate structure established?

1. 1969
2. 1961
3. 1958
4. 1948

1-12. When did the first surface-to-air missile system become operational in the U.S. Fleet?

1. 1961
2. 1958
3. 1955
4. 1944

1-13. Which of the following types of Navy ships was the first to have an operational GMLS?

1. CLG
2. CA
3. DDG
4. DLG

1-14. Which of the following technical skills is NOT one of those normally considered necessary for maintaining a GMLS?

1. Hydraulics
2. Mechanical
3. Metallurgy
4. Electrical/electronic

1-15. At what rate level are the qualifications for the GMM and GMG ratings combined?

1. E-6
2. E-7
3. E-8
4. E-9

Learning Objective: Explain how GMMs are classified according to their skills and how this classification affects the duty to which they may be assigned.

1-16. A system that is designed to identify special skills within a general rating, when the general rating structure itself is insufficient to otherwise identify those skills, is called

1. Navy enlisted classifications
2. personnel qualification standards
3. occupational standards
4. military occupational specialties

1-17. Of the following NEC series, which one identifies skills that are NOT related to any particular rating?

1. Special
2. Rating
3. Entry
4. Conversion

1-18. As a senior Gunner's Mate, your duty assignments are governed primarily by your

1. experience
2. desires
3. capabilities
4. job code

1-19. Tour lengths are influenced by which of the following considerations?

1. Distribution of billets
2. Retention rate
3. Manning level projections
4. All of the above

Learning Objective: Describe the Navy's process of detailing Gunner's Mates.

1-20. Which of the following naval personnel are assigned to duty stations by the Enlisted Personnel Management Center (EPMAC) in New Orleans?

1. Designated Seamen
2. Nondesignated Seamen
3. All petty officers
4. All personnel in pay grades E-1 thru E-3

To answer questions 1-21 thru 1-26, select from column B the type of duty that is described in column A.

| A. Descriptions of Duty | B. Types of Duty |
|---|---|
| 1-21. Duty aboard the USS *Orion* (AS-18), which goes to sea for about 2 weeks each year | 1. Sea duty |
| | 2. Non-rotated sea duty |
| 1-22. Duty aboard the USS *Midway* (CV-41), homeported in Yokosuka, Japan | 3. Preferred overseas shore duty |
| 1-23. Duty aboard the USS *Preble* (DDG-46), homeported in Hawaii | 4. Neutral duty |
| 1-24. Duty at the Ship Repair Facility (SRF), Yokosuka, Japan, for a 42-month tour | |
| 1-25. Duty aboard an ATF, homeported in Adak, Alaska | |
| 1-26. Duty for a normal 13-month tour on the staff of the Commander, Naval Forces Korea, in Seoul | |

1-27. When filling out an Enlisted Duty Preference form (NAVPERS 1306/63), you will improve your chances of getting the duty you desire if you list broad preferences instead of specific ones.

Learning Objective: Specify the basic principles of the Navy advancement system.

1-28. Advancements in rate are profitable to both the persons being advanced and to the Navy. In which of the following ways do you become more valuable to the Navy when you are advanced?

1. By being able to maintain a higher standard of living due to increased pay
2. By being able to train, supervise, and lead others
3. By being able to demand more respect from subordinates
4. By the good example your advancement provides to others

1-29. Changes to the Navy's advancement system are first promulgated in which of the following documents?

1. NAVPERS 18068 series
2. NAVPERS 1414/4
3. BUPERSINST 1418.10
4. BUPERS Notice 1418

1-30. Once you meet all the requirements for advancement, you will be advanced.

1-31. To determine the best qualified personnel for advancement, the Navy follows a controlled procedure based on which of the following factors?

1. Tests administered to qualified personnel
2. Annual evaluations and CO's recommendations
3. Time in rate and time in service
4. All of the above

1-32. The final multiple score which determines your place on the list of qualified candidates for advancement is determined by your

1. performance and knowledge only
2. performance and seniority only
3. knowledge and seniority only
4. performance, knowledge, and seniority

1-33. Assume there are five E-7 vacancies and six candidates with final multiple scores high enough for advancement. Of the six candidates, who will be advanced?

1. All six
2. The five with the highest score
3. The five senior candidates
4. The five selected by a CPO selection board

1-34. What is the total number of high quality bonus points a person can accumulate as part of the final multiple for advancement?

1. 3
2. 6
3. 15
4. 18

1-35. Which of the following publications delineates the minimum skill requirements for each rate and rating?

1. NAVEDTRA 10052
2. NAVPERS 18068
3. Rate training manuals
4. BUPERS Manual

1-36. Minimum skill requirements which are universal to most rates and ratings are listed in the *Manual of Navy Enlisted Manpower and Personnel Classifications and Occupational Standards* under the category of

1. practical factors
2. professional factors
3. naval standards
4. occupational standards

1-37. The format of occupational standards is not readily adaptable to the practical factor checkoff list concept because the standards are presented as

1. knowledge factors
2. task statements
3. skill requirements
4. aptitude factors

1-38. The PAR program enables a command to evaluate the overall abilities of personnel preparing for advancement to which of the following pay grades?

1. E-1 thru E-9
2. E-8 and E-9
3. E-1 thru E-7
4. E-4 thru E-7

1-39. In what section of the PAR pamphlet should you make the entry indicating that an individual has passed the E-4/E-5 military leadership examination?

1. I
2. II
3. III
4. IV

1-40. To satisfy the requirements of the PAR program, an individual's ability to perform a specific task should be evaluated by which of the following methods?

1. Observation of ability to perform a related task
2. Training received
3. Demonstration
4. All of the above

To answer questions 1-41 thru 1-46, select from column B the source of information that is described in column A.

| A. Information Source Descriptions | B. Sources of Information |
|---|---|

1-41. A total of four of these information sources has been written to present information on naval standards for advancement

1. NAVEDTRA 10052
2. Rate training manuals
3. NAVPERS 18068
4. NAVEDTRA 10061

1-42. It is the most important single source of information when studying for advancement

1-43. These sources may be written to cover general information to be used by personnel of more than one rating, or may be written specifically for one rating

1-44. This source of information provides a list of available training courses

1-45. This source of information contains a complete listing of all naval enlisted classification codes

1-46. Publications listed in this source of information are used by the examination writer in preparing the advancement examination

1-47. To make the best use of this course as you prepare for advancement, it is recommended that you

1. vary your study time from day to day
2. refer to the occupational standards frequently as you study
3. avoid following a regular study plan
4. study the material in the same order as given in the text

1-48. Which of the following hints is recommended for improving your study habits?

1. Select a time to study when you are completely relaxed
2. Memorize all important information
3. Make notes and write down questions as they occur
4. Disregard the time-consuming introductions in the text

1-49. Which of the following is the best indication that you have mastered the material in a chapter or manual?

1. You can recite word-for-word passages of the chapter or manual
2. You can recall exact dates, weights, and measures
3. You can quickly locate important information in the manual
4. You can explain in your own words the main ideas covered in the text

Learning Objective: Specify the knowledge required for planning work relating to the handling and stowage of shipboard missiles.

1-50. Which of the following duties should you delegate to a lower rated individual?

1. Routine preventive maintenance
2. Planning a repair program
3. Enforcement of special rules of safety for the stowage of missiles
4. Troubleshooting unusual problems

1-51. What must you know to plan the handling of missiles being received aboard ship?

1. The location of strikedown hatches
2. The availability of stowage spaces
3. The mode of delivery
4. All of the above

1-52. Which of the following missiles is/are received on board in the desired configuration?

1. Terrier
2. Standard MR
3. Tartar
4. All of the above

1-53. Safety rules for the handling of explosives are applicable to all guided missiles, but there are some additional rules for handling particular missiles.

Learning Objective: Indicate the work schedule, operation sequence, and security procedures pertaining to loading and stowing missiles aboard ship.

1-54. The best time of day for handling explosives aboard ship is during daylight hours.

1-55. GMMs are responsible for rigging the ship's cargo handling gear.

1-56. The GMMs handling containers of classified ordnance material should know the explosive nature of such material because

1. stenciled information reveals the classified nature
2. only GMMs are assigned to handle nuclear warheads
3. such awareness encourages careful handling
4. they are responsible for acknowledging its receipt

1-57. Planning the operation sequences of loading and unloading missiles should follow the

1. stowage area checks
2. working party assignment
3. loading schedule arrival
4. special handling equipment checks

1-58. Which of the following spaces require continuous operation of a radiation detection device?

1. Submarine nuclear missile stowage spaces
2. Surface ship nuclear missile stowage spaces
3. All shipboard missile stowage spaces
4. All shipboard nuclear missile stowage spaces not equipped with automatic sprinkler systems

1-59. To determine the areas designated for stowage of ammunition on your ship, refer to

1. NWP 14
2. the ship's plans
3. OP 4
4. OP 3347

1-60. Which of the following references do you consult to determine the best place to stow primers when the small arms magazine must undergo repairs?

1. Naval Ships' Technical Manual
2. NAVSEA Instruction 8650.5
3. NWP 14
4. OP 4

---

Learning Objective: Identify the special problems concerning the initial receipt of Terrier for stowage, including the handling equipment needed for handling and transferring missiles at sea and in port.

---

1-61. In what sequence are the components of Terrier rounds taken aboard ship?

1. A booster, a missile, complementary parts
2. All the boosters, all the missiles, all the complementary parts
3. All the missiles, all the boosters, all the complementary parts
4. A missile, a booster, complementary parts

1-62. The Mk 6 Mod 1 and/or Mod 3 transfer dolly(ies) is/are used for the transfer of Tartar and Standard MR missiles.

1-63. What is the preferred method for underway replenishment from an ammunition ship (AE)?

1. Modified housefall
2. STREAM rig
3. Double burton highline
4. Constant tension highline

1-64. A cargo drop reel is attached to the trolley when a STREAM rig is rigged to a fixed pad eye.

1-65. What is the purpose of the strongback adapter during at-sea replenishment using the STREAM system?

1. To provide faster hookup time
2. To decrease the number of personnel required on the sending ship
3. To provide more control of the dolly
4. To eliminate the need for tag lines on the dolly

1-66. What is the minimum number of personnel required to maintain positive control of a Mk 6 dolly, while it is in motion?

   1. Three
   2. Four
   3. Five
   4. Six

1-67. What is the approximate weight of an empty Mk 6 dolly?

   1. 700 pounds
   2. 900 pounds
   3. 1100 pounds
   4. 1300 pounds

1-68. Completely assembled Terrier missiles and boosters are stored in Mk 199 and Mk 200 containers, respectively.

1-69. Shipboard personnel assigned duties on a pier while transferring missiles to or from an ordnance facility are under the authority of the

   1. ship's commanding officer
   2. ship's safety officer
   3. facility's commanding officer
   4. facility's safety officer

# Assignment 2

Missile Handling

Textbook, NAVEDTRA 10200-D: Pages 23 through 54

---

Learning Objective: Recognize the safety precautions that must be observed during the handling of Terrier missiles.

2-1. When a loaded Terrier dolly is dropped 18 inches or more it must be returned to a weapons station.

2-2. In a magazine, the smell of ether from double-base propellants indicates a normal condition.

Learning Objective: Specify the procedures for striking down Terrier and ASROC weapons and their complementary items.

2-3. The shipping band (third handling band) is used only on the missile.

2-4. How are the handling bands released from the dolly prior to lowering a missile or booster on the strike-down elevator?

1. Pneumatically
2. Hydraulically
3. Manually
4. Mechanically

2-5. The correct sequence for striking down Terrier missiles and boosters is for a booster to precede a missile.

2-6. Which of the following missiles requires code plugs to enable the missile to match up with the ship's assigned guidance codes?

1. Standard ER
2. Standard MR
3. HT
4. BT

2-7. There is considerable variation in the manner in which Terrier and Standard ER missiles are mated to their boosters.

2-8. What is the maximum number of ASROC weapons that can be stowed in the Mk 10 Mod 7 or 8 GMLSs?

1. 10
2. 20
3. 30
4. 40

2-9. The lower ready service ring in the Mk 10 Mods 7 and 8 GMLSs stows only Terrier weapons.

2-10. ASROC rounds can be located next to each other in the ready service rings.

2-11. What piece of equipment is used to make the ASROC weapon compatible with the strikedown elevators and transfer car carriage?

1. A cradle
2. A handling fixture
3. An adapter rail
4. A skid

2-12. What types of containers and handlift trucks are used with the ASROC weapons?

1. Mk 200 containers; Mk 45 handlift trucks
2. Mk 199 containers; Mk 42 handlift trucks
3. Mk 183 containers; Mk 45 handlift trucks
4. Mk 205 containers; Mk 42 handlift trucks

2-13. How is the ASROC decanning hoist released from the weapon, after the weapon is lowered onto the strikedown elevators?

1. Manually
2. Hydraulically
3. Electrically
4. Pneumatically

2-14. At the present time, how many basic configurations are there for ASROC weapons?

1. Five
2. Four
3. Three
4. Two

2-15. When loading an ASROC RTDC into an adapter, it is necessary to remove the inserts.

2-16. How many Terrier booster fins normally are shipped in a Mk 205 container?

1. 16
2. 12
3. 8
4. 4

---

Learning Objective: Recognize the shipboard inspection procedures conducted on Terrier missiles, including the procedures for radiation monitoring.

---

2-17. Where is the SAFE-ARM indicator located on the Mk 30 rocket motor?

1. On the forward end of the rocket motor
2. In close proximity to the delay cartridge
3. On the after end of the rocket motor nozzle
4. Just aft of the clamp ring assembly

2-18. How can you determine that the proper fin support mounting snaprings are being used?

1. An "R" is etched into the face of each snapring
2. The snaprings are painted black or have a black dot
3. The snaprings are painted green
4. The edges of the snaprings are beveled

2-19. The acceptance of a complete round is limited by the expiration dates of the explosive components in the missile or booster.

2-20. What are the three types of radiation monitoring equipment presently used on board surface combatant ships?

1. AN/PDR 27, 43, and IC/T2-PAB-M
2. AN/PDR 27, 43, and 56
3. AN/PDR 27, 56, and T-290
4. IC/T2-PAB-M, AN/PDR 27, and T-290

2-21. What is the prescribed charging time for the internal battery of the IC/T2-PAB-M air monitor?

1. 24 hours
2. 18 hours
3. 12 hours
4. 6 hours

2-22. Which of the following statements best describes the function of the IC/T2-PAB-M portable air monitor?

1. It measures the radiation level
2. It detects the radiation and initially measures the amount
3. It detects the radiation, but does not measure the amount
4. It detects the radiation and continously measures the amount

Learning Objective: Indicate the handling and stowage procedures at depots and naval weapons stations.

2-23. In what type of magazine should flash signals be stowed at a weapons station?

1. High explosive
2. Smokeless powder
3. Pyrotechnic
4. Black powder

2-24. To which of the following publications should you refer when attempting to locate information for ammunition ashore at advanced bases?

1. OP 5, volume 2
2. OP 5, volume 3
3. OP 4, volume 1
4. OP 4, volume 2

2-25. How many passengers, if any, are allowed on trucks carrying explosives?

1. One
2. Two
3. Three
4. None

2-26. Which of the following missiles does NOT contain a liquid propellant?

1. Bullpup
2. Harpoon
3. Sparrow III
4. Tomahawk

2-27. When handling Mk 199 and Mk 200 containers, the Mk 42 or Mk 45 handlift truck requires the installation of a

1. stabilizer beam
2. Mk 93 adapter
3. Mk 28 adapter
4. Mk 26 adapter

2-28. What handling beam is used to lift Terrier missiles and boosters from their containers?

1. The Mk 18
2. The Mk 16
3. The Mk 15
4. The Mk 14

Learning Objective: Identify the Sparrow III and Harpoon systems' components and handling equipment.

2-29. Which of the following items is/are NOT included in a loaded Mk 470 container?

1. Four folding wings
2. A RIM-2D missile
3. One shear-off connector
4. One vibration dampener

2-30. For the NSSMS, what is the purpose of the holdback mechanism and the quick reaction brake mechanism on the launcher?

1. To provide inadvertent firing protection
2. To allow the launcher to pinwheel
3. To enable the missile to be restrained during normal firing
4. To restrain the missile until the proper thrust is developed

2-31. At what location is the Mk 567 MSTS plugged into the NSSMS to ensure that all functions of the programmed test are being supplied?

1. The deck station launcher control
2. The launcher rail umbilical connector
3. The launcher interconnection maintenance cabinet
4. The launcher control cabinet

2-32. Assuming that you are receiving a complete loadout of Sparrow III missiles, which of the following rail removal procedures is it more practical to complete before starting onload operations?

1. Remove one rail at a time
2. Remove two rails at a time
3. Remove four rails at a time
4. Remove all eight rails

2-33. During a Sparrow III missile loading operation, at what point is the shorting plug removed from the rail-to-cell connector?

1. Before removing the rail from the launcher
2. Before disengaging the rail from the loader
3. After the rail/missile combination is driven a short distance into the launcher cell
4. After manually positioning the rail assembly on the missile in the container

2-34. By what means are the rail assemblies locked into the cells?

1. By spring actuated latches
2. By solenoid-actuated latches
3. By rotating the rail release ring using a hand-held rod
4. By turning the rail lock set screws clockwise

2-35. Against which of the following targets is the Harpoon missile designed to be used?

1. Missiles
2. Surface ships
3. Submarines
4. Aircraft

2-36. In the Harpoon weapon system (HWS), where are the digital inputs reformatted from ships systems negative 15-volts to a positive level for use by the data processor computer (DPC)?

1. In the Harpoon casualty panel (HCP)
2. In the weapon control indicator panel (WCIP)
3. In the Harpoon trainer module (HTM)
4. In the data conversion unit (DCU)

2-37. Which of the following HWS tests is NOT performed by the DPC using the built-in-test (BIT) feature?

1. Continuous on-line Harpoon weapon control console (HWCC) test
2. Off-line HWCC test, when selected
3. Continuous off-line HWCC test
4. Complete self-test when power is first applied to the HWCC

2-38. Which of the following data inputs CANNOT be manually introduced by the keyboard on the HWCIP?

1. True target bearing numeric display
2. Multiple air target threat axis
3. Bearing only minimum range numeric display
4. Fuze selection

2-39. What component provides the interface between HWCC and the launcher?

1. The launcher relay assembly (LRA)
2. The HWCIP
3. The HTM
4. The launcher switching unit (LSU)

2-40. One Harpoon grade B canister launcher can accommodate up to how many missile/canisters?

1. Two
2. Four
3. Six
4. Eight

2-41. What is the fixed elevation angle for the Harpoon launcher platform?

1. 15°
2. 25°
3. 35°
4. 45°

2-42. Before hooking up to another canister on the pier, the hoist rotation beam should be set to what position?

1. 0°
2. 25°
3. 35°
4. 45°

2-43. Although minor modifications to the launching systems are required, the Harpoon missile is compatible with which of the following GMLSs?

1. Mk 10
2. Mk 13
3. Mk 29
4. All surface systems

Learning Objective: Describe Tartar handling equipment and precautions to be taken during handling evolutions.

2-44. Which of the following transfer dollies is used for handling the Tartar missile?

1. Mk 6 Mod 1
2. Mk 6 Mod 3
3. Mk 6 Mod 4
4. Mk 8 Mod 0

2-45. How are the chain drive fixtures for the Mk 13 and Mk 22 GMLSs operated?

1. Manually
2. Electrically
3. Hydraulically
4. Pneumatically

2-46. The Tartar DTRM is considered to be what class of explosive?

1. A
2. B
3. C
4. D

2-47. A Tartar DTRM is received aboard ship in an armed condition. After manually moving the arming lever on the DTRM to SAFE, to which of the following activities should a request for disposition of the missile be sent?

1. NAVSEASYSCOM
2. NSWSES
3. DNA
4. NAVSEACEN LANT/PAC

Learning Objective: Point out the special handling requirements for Standard missiles.

2-48. Which of the following missiles is handled and stowed in the same manner as the Standard MR?

1. Terrier
2. Tartar
3. Sparrow III
4. Tomahawk

2-49. For a Standard missile, how many hours should be allowed for an activated battery to return to normal temperature?

1. 16
2. 12
3. 8
4. 4

2-50. The electrolyte in the battery of a Standard missile is composed of which of the following chemicals?

1. Potassium hydroxide
2. Sodium chloride
3. Hydrogen peroxide
4. Dimethyl sulfate

2-51. If battery electrolyte comes in contact with the skin, the contaminated area(s) should be flushed with large quantities of

1. alkaline soap
2. saltwater
3. vinegar and water or freshwater
4. foam

2-52. Which of the following procedures should be followed during a firing exercise of a Standard missile if black smoke appears from the battery vent?

1. Notify NAVSEASYSCOM
2. Jettison the missile immediately
3. Wait 4 hours and restow the missile
4. Notify EOD personnel

---

Learning Objective: Point out the special handling requirements for nuclear weapons, and the training required of assigned personnel. Define the terms dealing with nuclear weapons security.

---

2-53. In accordance with which of the following publications are nuclear weapons handled and stowed?

1. OP 2351 and OP 3114
2. Navy SWOP 50-1 and SWOPs of the 20 series
3. Navy SWOP 50-1 and CG-108
4. GC 108 and SWOP 4-1

2-54. Which of the following instructions is used as the basis for determining the minimum security requirements for all nuclear weapons of the Navy?

1. OPNAVINST C5510.83
2. OPNAVINST 5510.1
3. NAVSEAINST 5510.45
4. OPNAVINST 5100.19

2-55. Which of the following SWOPs is an approved source for definitions used in conjunction with nuclear weapons?

1. H-1
2. 4-1
3. 50-1
4. 44-1

2-56. Which of the following is the correct definition of critical position?

1. One in which the incumbent is performing duties physically associated with nuclear weapons, but has no technical knowledge
2. One in which the incumbent has graduated from a special weapons course
3. One in which the incumbent has technical knowledge of and access to nuclear weapons
4. One in which the incumbent has readily available access

2-57. Which of the following areas is considered an exclusion area?

1. The designated area surrounding one or more controlled areas
2. The designated area containing one or more nuclear weapons
3. The entire ship
4. The designated area surrounding one or more limited areas

2-58. What is the minimum number of months a log for an exclusion area should be kept and maintained locally?

1. 12
2. 18
3. 24
4. 30

2-59. Which of the following statements is NOT correct concerning nuclear weapons spaces occupied by two individuals?

1. All unlocked entrances are guarded
2. The entrance in use shall be unlocked and guarded
3. Unlocked entrances are alarmed and unguarded
4. Entrances not in use have to be locked and alarmed

2-60. Once an exclusion area has been entered by two authorized persons, who has the responsibility for maintaining the two-man rule?

1. Senior person present
2. Guard force officer
3. Sentry guard
4. Security officer

2-61. As it pertains to a missile handling evolution, which of the following statements is NOT correct?

1. A first aid kit shall be available in the area
2. Firehoses shall be laid out and charged
3. All handling equipment shall be inspected prior to use and periodically thereafter
4. Combination wire and fiber slings are authorized

2-62. Concerning their maintenance, the GMMs do NOT have a need to know which of the following alarm circuits?

1. FZ
2. FS
3. FH
4. F

2-63. How often should the air-conditioning unit in a nuclear weapons stowage space be secured and the exhaust system turned on to vent stale air from the stowage area?

1. Twice a day
2. Once a day
3. Once every second day
4. Intermittently, as prescribed by ship's policy

# Assignment 3

Preparing for a Firing Exercise

Textbook, NAVEDTRA 10200-D: Pages 55 through 78

---

Learning Objective: Indicate the basic steps of weapons system operation for conducting a firing exercise.

---

In items 3-1 through 3-3, list the step in column B that corresponds to the guided missile weapons system operation listed in column A.

| A. Operations | B. Steps |
|---|---|
| 3-1. Fire control radar tracking | 1. 1 |
| | 2. 2 |
| 3-2. Search radar detection | 3. 3 |
| 3-3. Missile launching | 4. 4 |

3-4. The information upon which target designation is based is derived largely from the

1. fire control director
2. lookout reports
3. search radar
4. tracking guidance radar

3-5. What stations provide the information upon which the AAW commander bases the selection of weapon to destroy a bogey?

1. CIC and WAC
2. CIC and WCS
3. The director and computer rooms
4. WCS and WAC

---

Learning Objective: Identify the control panels used during, and the orders associated with, a firing exercise.

---

3-6. What panels are common to all launcher control systems?

1. Power panel, launcher captain's panel, test panel
2. Launcher captain's panel, test panel, and jettisoning panel
3. Test panel, dud-jettisoning panel, power panel
4. Dud-jettisoning panel, power panel, launcher captain's panel

3-7. Assume the GMLS is being prepared for operation. When does the launcher captain's panel transmit the ready signal to WCS?

1. Immediately upon initiation of an alert signal by WCS
2. The instant the power is turned on
3. As soon as one of the system's panels is set on READY
4. After all systems' panels have been set on READY

3-8. Which of the following missiles requires manual actuation of the safety switch activator during a firing exercise?

1. SM/ER
2. SM/MR
3. HT-3
4. BT-3A(N)

3-9. What does the operator of the EP2 panel do when a HOLD order is received during the missile order phase?

1. Returns the round to its original stow position and awaits the next order
2. Acknowledges the order and the round is held at the hoist position until the next order
3. Stops the loading operation and prepares for dud-jettisoning
4. Loads the round on the launcher and maintains the GMLS in the ready condition

3-10. When the unloading order is indicated by WCS, automatic unloading operations are initiated by the

1. WCS
2. WAC
3. AAW commander
4. launcher captain

3-11. During remote operation, the launcher is positioned in response to train and elevation signals developed by the

1. director
2. computer
3. search radar
4. test panel

3-12. Which of the following status lights on the WAC/EC console is NOT applicable when firing a Standard ER weapon?

1. Warmup power on
2. Launcher synchronized
3. Ready to fire
4. Launching system ready

3-13. Test panels used during launcher test operations indicate the state of operability of

1. missile components
2. auxiliary equipment connected to the EP3 panel
3. train and elevation systems
4. nuclear warheads

3-14. When a Y-type missile is being loaded in the Mk 10 GMLS, a key-operated switch in the SAFE position prevents the opening of the blast doors. At which of the following panels is the key-operated switch located?

1. EP1
2. EP2
3. EP3
4. EP4

Learning Objective: Specify the stations to be manned during a firing exercise.

3-15. Which of the following GMLSs has checkout areas?

1. Mk 10
2. Mk 11
3. Mk 13
4. Mk 26

3-16. How many assemblymen are required for A- and B-side operation of a Mk 10 GMLS?

1. Two
2. Four
3. Six
4. Eight

3-17. Before returning a Tartar missile to the magazine, the fins must be manually folded.

3-18. In automatic and step operation of a GMLS, the magazines are unmanned.

Learning Objective: Identify the safety check requirements necessary for operating a GMLS, including the requirements for safety devices that are built into the system.

3-19. Where are safety precautions inserted in the checkoff lists that are posted at each station?

1. Where they apply
2. At the beginning of each phase of operation
3. In the safety summary
4. At the end of each phase of operation

3-20. Observance of the safety precautions is mandatory.

3-21. Who is responsible for sounding the warning bell before starting any moving equipment in a GMLS?

1. The safety observer
2. The launcher captain
3. The loader captain
4. The assembler captain

3-22. To which of the following publications should you refer when seeking information on the danger circle around a launcher?

1. OP 3347
2. OP 3565
3. OP 3000
4. OP 1014

3-23. When activating a GMLS, who notifies the launcher captain (EP2 panel operator) that the launcher is all clear?

1. The loader captain
2. The assembler captain
3. The feeder captain
4. The safety observer

3-24. What prevents a missile from being fired at a target whose line of sight is obstructed by a part of the superstructure of your ship?

1. The computer ceases to function when a missile is pointed into a part of the ship
2. Firing cutout cams limit the training arc of the launcher
3. WCS manually controls any train past a certain point
4. Power to the synchro-servo systems of the launcher is reduced in direct proportion to the size of the angle of train

3-25. When is the missile launcher position checked for alignment with the train and elevation orders?

1. During preparation for firing
2. When performing maintenance work
3. During training exercises
4. All of the above

---

Learning Objective: Specify the step-by-step procedures for conducting firing exercises on the Mk 10 GMLS, including the operational differences between mods of the system.

---

3-26. How many weapons can be stowed in each ready service ring of the Mk 10 GMLS?

1. 10
2. 20
3. 40
4. 60

3-27. What must each assemblyman do after attaching wings and fins to the missile so that the missile may be moved to the launcher?

1. Depress a foot switch
2. Stow the missile support bands in the assembly area
3. Select a pushbutton corresponding to the type of missile ordered
4. Lock a hydraulically operated door

3-28. Which of the following Mk 10 GMLSs have tilting rails?

1. Mods 0
2. Mods 1
3. Mods 3
4. Mods 5

Information for item 3-29. Steps in the positioning of a Terrier missile on a launcher rail include:

A. Retracting of the spanning rail
B. Extending of the arming tools
C. Extending of the aft-shoe latches and their receiving of the missile from the loader pawls
D. Closing of the blast doors

3-29. In what order do the steps occur?

1. C, B, A, D
2. C, D, B, A
3. B, C, D, A
4. B, D, A, C

3-30. Which of the following personnel operates the firing safety switch?

1. The weapons control operator
2. The assembly captain
3. The launcher captain
4. The antiaircraft warfare commander

3-31. The GMLS that has provisions for stowing ASROC missiles is the

1. Mk 22 Mod 1
2. Mk 13 Mod 4
3. Mk 10 Mods 7 and 8
4. Mk 10 Mods 5 and 6

---

Learning Objective: Point out the operating procedures for conducting firing exercises on Tartar GMLSs.

---

3-32. The Tartar weapon system constitutes the primary means of defense against air attack on

1. PGs
2. DDGs
3. CGs
4. CVs

3-33. What is the most prominent difference between the Mk 11 and the Mk 13 Tartar GMLSs?

1. The Mk 11 is a higher speed system than the Mk 13
2. The Mk 11 has two launcher arms and the Mk 13 has only one
3. The Mk 11 has only one launcher arm and the Mk 13 has two
4. The Mk 13 has more control panels than the Mk 11

3-34. A Mk 11 Tartar GMLS uses two panels which are designated differently from those in other systems. They are the

1. launcher panel (EP5) and the loader control panel (EP3)
2. launcher panel (EP2) and the feeder control panel (EP1)
3. feeder control panel (EP2) and the loader control panel (EP1)
4. loader control panel (EP2) and the launcher control panel (EP3)

3-35. Which of the following panels in the Mk 11 GMLS is used to train and elevate the launcher in local control?

1. EP5
2. EP1
3. IP5
4. IP1

3-36. Which of the following panels indicates the relation of the blast doors to the launcher on the Mk 11 GMLS?

1. IP1
2. EP2
3. EP3
4. EP5

3-37. Which of the following indications is NOT displayed on the IP5 panel of the Mk 11 GMLS?

1. Magazine mode
2. The type of missile on guide arms
3. Train and elevation amplifiers ready
4. Mixed load warning

3-38. Which of the following panels for the Mk 13 GMLS are unmanned in automatic operation?

1. EP1 and EP2 only
2. EP2 and EP3 only
3. EP1 and EP3 only
4. EP1, EP2, and EP3

3-39. When should the Mk 13 GMLS be energized?

1. As soon as the AAW commander has given the command
2. As soon as a contact has been made with weapons control
3. Once selection of the weapon system to be used has been made
4. After the safety observer has communicated an all clear report to the launcher captain

3-40. If a missile has not been loaded onto the launcher for the Mk 13 GMLS within 14 minutes after the onset of warmup, what event should follow?

1. The missile should be labeled a dud
2. The missile should automatically be replaced on warmup by another missile
3. A red light should flash on the loader control panel
4. The assembly captain should depress the panel's foot switch

3-41. Which of the following missiles does not require warmup?

1. IT
2. ITR
3. Standard MR
4. BT

3-42. Assume that the outer ring of missiles in the Mk 13 GMLS has been initially selected for continuous loading. In what sequence will warmup power be applied to the missiles in the outer and inner rings?

1. Missiles in both the outer and inner rings will receive power at the same time
2. Missiles in the outer ring will receive power at the same time after which missiles in the inner ring will receive power at the same time
3. Missiles in the inner ring will successively receive power followed by successive warming up of missiles in the outer ring
4. Missiles in the outer ring will successively receive power followed by successive warming up on missiles in the inner ring

3-43. In the Mk 13 GMLS, at what stage in the transfer of the missile from the magazine to the loader rail does the warmup contactor break contact?

1. When the missile is moved from station at the hoist position
2. When the blast doors close
3. When the launcher is trained to the load position
4. When the rail-loaded indicator plunger is actuated

3-44. Which of the following components secures the Tartar missile on the launcher rail after the hoist completes its raise cycle?

1. The contactor
2. The forward motion latch
3. The aft motion latch
4. The rail loaded plunger

3-45. To which stations does the rail-loaded indicator plunger send signals in the Mk 13 GMLS?

1. WCS and launcher captain's panel
2. Bridge and feeded control panel
3. CIC and feeder control panel
4. Bridge and launcher captain's panel

3-46. At what point are the Tartar missile fins automatically extended (unfolded) during firing?

1. At target acquisition
2. As the missile clears the rail
3. At launcher assignment
4. At approximately 5 seconds before firing

3-47. The firing cutout system opens the firing circuit when the launcher points into an unsafe firing zone.

20

3-48. What are the conditions that must be satisfied in the Mk 13 GMLS before the ready-to-fire signal can be given?

1. The blast door must be closed
2. The fins on the missile must be unfolded
3. Minimum warmup time must have elapsed
4. All of the above

3-49. Where are the firing safety switches located in the Mk 13 GMLS?

1. At the EP1 and EP2 panels
2. At the EP2 and EP3 panels
3. At the EP2 and IP5 panels
4. At the EP2 panel and the safety observer's position

3-50. In the Mk 13 GMLS, the hot gas generator squibs are fired by a circuit which passes through the

1. launcher contactor
2. hoist contactor
3. blast doors
4. spanning rail

3-51. When are the circuits through the launcher contactor broken?

1. As soon as the blast doors close
2. After the hot gas generator squibs have fired
3. At the arming of the missile
4. At the completion of warmup (ITR only)

3-52. What major igniter modification has been incorporated into the Mk 13 Mods 1, 2, and 3 GMLSs?

1. The igniters have key-operated safety locks
2. The igniters contact the missile at all times
3. The igniter housing has been made smaller
4. The igniters will contact only an armed missile

3-53. Since its modification, the magazine of a Mk 13 GMLS can stow how many types of missiles?

1. Three
2. Four
3. Five
4. Seven

3-54. The missile identification system is capable of identifying how many types of missiles?

1. 64
2. 48
3. 24
4. 16

3-55. Which of the following GMLSs has a single ready service ring which stows missiles vertically, and a mounting ring similar to that of a 5"/54 gun mount?

1. Mk 10
2. Mk 11
3. Mk 13
4. Mk 22

3-56. In the Mk 22 guided missile launching system, the retraction of the launching rail after it has guided the missile for its first 20 inches of travel serves to

1. lessen the possibility of damage to the missile during adverse weather conditions
2. diminish solid fuel requirements
3. permit dud-jettisoning more readily if necessary
4. provide more space for the fins to be extended

3-57. The Mk 22 Mod 0 Tartar GMLS is very similar in its mode of operation and control to which of the following GMLSs?

1. Mk 11
2. Mk 13
3. Mk 26
4. Mk 29

3-58. The function of a plenum chamber in a Tartar missile magazine is to

1. remove water accidentally sprayed from the sprinkler system
2. safely hold nuclear warheads
3. vent gases caused by accidental ignition of a missile
4. stow dud missiles until they can be repaired

---

Learning Objective: Identify the operating procedures for the Mk 26 and Mk 29 GMLSs, and point out the compatibility of Standard, Harpoon, and ASROC weapons with other GMLSs.

---

3-59. Standard missiles SM1 and SM2 are technically equivalent.

3-60. Which of the following components is used to ignite an ASW (ASROC) weapon on the Mk 26 GMLS launcher?

1. The fire-thru-latch
2. The arming tool
3. The ASW contactor
4. The igniter blades

3-61. What is the power rating and the type of drive motor used on the launcher for the Mk 29 GMLS?

1. 40 horsepower; a.c.
2. 10 horsepower; d.c.
3. 5 horsepower; a.c.
4. 1 horsepower; d.c.

3-62. To which of the following positions should the mode select switch on the maintenance interconnecting panel be set for completely automatic operation?

1. AUTO
2. REMOTE
3. LOCAL
4. STEP

3-63. The firing command for the Mk 29 GMLS can originate at which of the following locations?

1. The maintenance interconnecting panel
2. The firing officer's console
3. The fire control computer
4. Both 2 and 3 above

3-64. What is the purpose of the infrared sensor in each launcher cell?

1. To sense ambient temperature
2. To sense inadvertent firing
3. To release the launcher brakes
4. To enable the launcher to pinwheel freely

3-65. During power failure the brakes will automatically set on the Mk 132 launcher.

3-66. What component controls the missile's heater operation for the Harpoon canister-launcher configurations once the power has been applied at the weapons control indicator panel (WCIP)?

1. The potentiometer that is attached to the missile's warm air expansion bellows
2. An infrared sensor
3. The missile's heat exchanger
4. The missile's thermostat

3-67. What is the maximum number of Harpoon missiles that can be supplied warmup and electronic power at the same time on a Harpoon canister-launcher configuration?

1. Four
2. Five
3. Six
4. Seven

3-68. The Harpoon ship command launch control set (HSCLCS) provides its own target data.

# Assignment 4

Loading, Unloading, and Dud Jettisoning

Textbook, NAVEDTRA 10200-D: Pages 79 through 105

---

Learning Objective: Point out the type of control used and the operational requirements for loading and unloading Terrier missiles.

4-1. Which of the following procedures normally requires manual labor during loading of Terrier missiles in the automatic mode?

1. Transferring the round to the assembly area
2. Folding wings and fins
3. Unfolding missile control surfaces and attaching booster fins
4. Placing the complete missile on the launcher

4-2. During the automatic loading of a launcher, the equipment suddenly fails. The launcher captain should immediately stop the launcher movement and report the failure to

1. weapons control and the assembly captain
2. weapons control and the feeder system captain
3. CIC and the feeder system captain
4. CIC and the assembly captain

4-3. How are the safety switches for the assemblymen actuated?

1. By the assemblymen depressing a hand-held switch
2. By the assemblymen stepping on the switches
3. By the assembly captain depressing the switches
4. By photoelectric light guns

4-4. Which of the following GMLSs does NOT have automatic unload capability?

1. Mk 22
2. Mk 13
3. Mk 11
4. Mk 10

4-5. Which of the following stations initiates automatic unload launcher orders?

1. CIC
2. Computer
3. Director
4. WCS

4-6. Which of the following components is/are NOT required to be visually inspected on the weapon once it is unloaded from the launcher and reaches the assembly area?

1. The booster aft closure
2. The booster fins
3. The S&A device
4. The missile control surfaces

4-7. When an S&A device is found in the ARMED position, the missile must be offloaded at the earliest possible time.

4-8. Which of the following missiles has a safety switch actuator?

1. SM/ER
2. SM/MR
3. BTN
4. HTR

23

4-9. When should the emergency wing and fin assembly bypass switch on the EP4 or EP5 panel be used?

1. When a faster reaction time is required
2. When the load cycle to the launcher is completed
3. When more time is needed by the assemblymen to perform their functions
4. During equipment checkout and when personnel are clear

4-10. An unloading cycle is NOT necessary after every firing of an ASROC weapon from a Terrier system.

Information for item 4-11. Some of the steps in the unloading procedure of a Terrier missile include:
A. The loader moves to the assembly area
B. The blast doors are opened
C. The launcher is latched in the load position
D. The arming tool is unwound and retracted

4-11. In what order do the steps occur?

1. D, B, A, C
2. C, D, B, A
3. B, A, C, D
4. A, C, D, B

4-12. The Mk 10 GMLS has interlock switches that prevent lowering of a round into the magazine with missile fins unfolded and booster fins installed.

---

Learning Objective: Point out the operating procedures, safety requirements (including how to differentiate between a dud and a misfire), and maintenance requirements of the Terrier dud-jettisoning unit.

---

4-13. The commanding officer's order to jettison a missile is transmitted to the GMLS from the

1. missile director
2. antiair warfare station
3. combat information center
4. weapons control station

4-14. When the clinometer indicates that the ship's roll exceeds 20°, at what point should you jettison a missile?

1. On the uproll
2. On the downroll
3. On a turn into the side of jettisoning
4. On a turn away from the side of jettisoning

4-15. To prevent having a dud, what is the minimum time that the firing key should be held closed during a missile firing?

1. 1.50 seconds
2. 1.25 seconds
3. 1.00 second
4. .25 second

Information for item 4-16. You are operating the dud-jettison panel of a Mk 10 launching system. Four steps you take at the panel to jettison an SM/ER missile from the A-side include:
A. Rotating the jettison lever from READY to JETTISON
B. Turning the air supply lever to OPEN
C. Positioning the jettison lever to CHARGE until air pressure is 3500 psi
D. Turning the positioner control lever to POSITION I

4-16. In what order do you perform these steps?

1. B, C, A, D
2. D, C, A, B
3. B, D, C, A
4. D, B, A, C

4-17. Assume that a misfire has occurred on the launcher and the situation presents a serious danger to your ship. The emergency firing procedure calls for the depression of the dud firing switch by the WCS operator after the launcher captain has placed the emergency enabling switch to which of the following positions?

1. EMERGENCY
2. ENABLE
3. OPEN
4. FIRE

4-18. As a general rule, when a Terrier booster misfire has occurred, all personnel should be kept clear of the launching area for at least how many minutes?

1. 5
2. 10
3. 20
4. 30

4-19. The main difference between a misfire and a dud is that in a misfire the

1. contactor fails to retract
2. firing key is released quickly
3. explosive system, although activated, is interrupted
4. booster firing relay is not energized

4-20. A misfire is indicated on the control panel by a continuous flashing of the

1. dud lamp
2. misfire lamp, while the missile ready lamp goes out
3. misfire lamp, while the dud lamp remains lighted until KCFA(B)2 energizes
4. dud lamp, while the misfire lamp remains lighted

4-21. Suppose that a missile has misfired and you suspect that the failure was due to a malfunction of the normal firing relay contacts. An occasionally successful means of launching such a misfire is to place the emergency enabling switch to which of the following positions?

1. NORMAL and ACTIVATION
2. ACTIVATION and ENABLE
3. EMERGENCY and ENABLE
4. ARMED and READY

4-22. A major cause of malfunctioning of dud-jettisoning units is ice formation in the valve passages caused by

1. moist air meeting a wave of warm air
2. rapid compression of moist air
3. sudden expansion of moist air under pressure
4. moist cold air meeting a wave of hot dry air

4-23. The only suitable lubricant for the dud-jettison piston head and stem is aircraft instrument oil.

4-24. Which of the following is a possible source of moisture draining from the jettisoning air-charging chamber's drain port?

1. A clogged air breather
2. A defective ejector unit gasket
3. A malfunctioning antiicing system
4. A dirty air filter

4-25. What position of the charge and fire control valve jettison lever is used to vent the ejector pneumatic lines?

1. OPEN
2. CHARGE
3. JETTISON AND OFF
4. READY

4-26. Where are the solenoid assemblies for the Mk 108 dud-jettison unit located?

1. In the jettison control panel
2. Adjacent to the nitrogen booster pump
3. Under the pressure intensifier pump
4. On the jettison shafts

4-27. What action, if any, must you take prior to measuring the insulation resistance of power supply cable?

1. Connect the ground-detection indicator to the EP1 panel
2. Disconnect the umbilical cable from the missile
3. Disconnect the ground-detection indicator from the EP1 panel
4. None; insulation resistance of power supply cables is performed without any preparatory actions

---
Learning Objective: Outline the procedures for handling and safing either configuration of a dud or misfired ASROC weapon.
---

4-28. In the Mk 10 Mod 7 GMLS, the operation of loading Terrier missiles or ASROC weapons is performed in the same mode of control.

4-29. Assume the entire launching system is energized and in automatic operation. At what point during loading of an ASROC will the arming tool be rotated into position on the launcher?

1. When the RSR tray is in position to hoist
2. When the ASROC is at the assembly area
3. When the ASROC is at the launcher
4. Upon selection of ASROC mode

4-30. The Y stop keylock switch is used only when Y-type weapons are being loaded onto the launcher.

4-31. What guide arm component(s) initiate(s) opening of the snubbers on the ASROC adapter?

1. The arming tool
2. The forward motion latch
3. The firing contacts
4. The contactor

4-32. At what location should the snubbers be manually closed, after the snubbers have been opened on the launcher during an ASROC firing exercise?

1. On the launcher
2. In the assembly area
3. In the magazine
4. In the checkout area

4-33. At what location within the GMLS should a new umbilical cable be inserted in the ASROC adapter after a firing exercise?

1. At the strikedown elevator
2. In the assembly area
3. In the magazine area
4. In the checkout area

4-34. How many streamer tapes are normally attached to the ASROC depth charge?

1. Two
2. Four
3. Six
4. Eight

4-35. What is the purpose of red streamers attached to the tapes which cover the hydrostatic ports of the fuze of an ASROC depth charge?

1. To counteract the effects of magnetic material in the surroundings
2. To alert you to the necessity of removing the tapes
3. To counter electronic interference from radar equipment
4. To indicate when the depth charge has been exposed to extreme temperatures

4-36. How many pounds of thrust is required for an ASROC weapon to overcome the forward restraining latch?

1. 1000
2. 1500
3. 2000
4. 2500

4-37. What device on the launcher guide arm prevents the ASROC adapter rail from being fired with the weapon?

1. A positive stop
2. The forward motion latch
3. The aft lug latch
4. The fire thru latch

4-38. What is the reason for attaching inserts to the snubbers on an ASROC adapter rail?

1. To facilitate loading an ASROC depth charge into the adapter rail
2. To ensure compatibility of the ASROC depth charge with the adapter rail
3. Because the torpedo configuration of ASROC is smaller in diameter than the depth charge
4. Because the depth charge configuration of ASROC is smaller in diameter than the torpedo

4-39. Which of the following statements best describes the function of the ASROC adapter rail snubbers?

1. They prevent fore-and-aft motion of the weapon
2. They prevent premature arming of the weapon
3. They prevent weapon overtravel during firing
4. They prevent lateral movement of the weapon

4-40. What are the torque requirements for the ASROC thrust neutralizer?

1. 50 - 75 foot pounds
2. 100 - 125 inch pounds
3. 50 - 75 inch pounds
4. 100 - 125 foot pounds

4-41. During peacetime, where is the power supply for the ASROC depth charge stored?

1. In the weapon
2. In a special safe
3. In a filing cabinet
4. In the commanding officer's stateroom

4-42. At what point during ASROC offload is the container ground wire connected to the thrust neutralizer?

1. As soon as the strikedown elevator is at the fully raised position
2. Before disconnecting the ASROC hoist, after the weapon is in the container
3. As soon as the weapon is on the transfer car
4. As soon as the weapon is placed on the strikedown elevator

4-43. What is the configuration of the ASROC RTDC?

1. Mk 2
2. Mk 17
3. Mk 45
4. Mk 46

4-44. What type of container is used for ASROC weapons?

1. Mk 205
2. Mk 200
3. Mk 199
4. Mk 183

4-45. Which of the following components is/are removed from the ASROC RTDC prior to stowing it in the magazine?

1. The ISA shorting plug
2. The thrust neutralizer
3. Both 1 and 2 above
4. The power supply

4-46. If the exploder bore rod on the Mk 19 ASROC torpedo exploder is tripped, how can the exploder be sterilized?

1. By breaking the foil seal in the top of the exploder and depressing the sterilizer switch
2. By breaking the foil seal in the top of the exploder and rotating the sterilizer switch 90° clockwise using a screwdriver
3. By breaking the foil seal in the top of the exploder and rotating the sterilizer switch 90° counterclockwise with a screwdriver
4. Both 2 and 3 above

4-47. What type of exploder is used in the Mk 46 torpedo?

1. Mk 7
2. Mk 19
3. Mk 20
4. Mk 30

4-48. A screwdriver made of nonferrous material should be used to rotate the sterilizer switch for the Mk 20 exploder.

Learning Objective: Indicate the types of control used in, and the operational requirements for, loading and unloading Tartar missiles and exercising the GMLS equipment.

Information for item 4-49. Four of the many operations performed automatically during the loading procedure in a Tartar missile system include:
- A. The fins are unfolded as launcher warmup power is applied
- B. The hoist pawl comes in contact with the missile aft shoe at the intermediate position
- C. Warmup is applied to Tartar missiles in the ready-service ring
- D. The span track extends to permit connection of the launcher fixed rail to the launcher

4-49. In what order do these steps occur?

1. C, A, D, B
2. A, C, D, B
3. C, B, D, A
4. C, A, B, D

4-50. How close to the ordered position does the launcher have to be before a firing order can be given?

1. 1°
2. 2-1/2°
3. 30 minutes
4. 90 minutes

4-51. Assume that a loaded Tartar missile launcher is trained and elevated to the load position. What precaution should you take before folding the fins during unloading?

1. Cut the power to the launcher by turning the fire safety switch to SAFE and remove the switch lever
2. Depress the fins-manually-folded switch
3. Place the firing safety switch on the EP2 panel at UNLOAD
4. Retrain and re-elevate the launcher to the unload position

Learning Objective: Demonstrate the procedures for dud and misfired Tartar missiles, and explain the operating and maintenance requirements for Tartar dud-jettisoning units.

4-52. What quantity of air pressure generates the hydraulic fluid pressure that retracts the spud after a dud missile is ejected on the Mk 11 GMLS?

1. 100 psi
2. 500 psi
3. 1000 psi
4. 2100 psi

4-53. In the automatic mode of the Mk 13 GMLS dud-jettisoning equipment, the fixed jettison elevation position of the launcher is controlled by the

1. clinometer
2. FCS gyrocompass (stable element)
3. compass
4. nitrogen-actuated piston

4-54. What type of motive force operates the Mk 13 Mod 1 nitrogen booster pump?

1. Pneumatic
2. Manual
3. Hydraulic
4. Electronic

4-55. At what interval should the jettison device be exercised?

1. Once a day
2. Once a week
3. Once every 2 weeks
4. Once a month

4-56. How often should (A) the level of the hydraulic fluid in the jettison booster piston and (B) the nitrogen pressure in the jettison accumulator tank be checked?

1. (A) Daily, (B) daily
2. (A) Daily, (B) weekly
3. (A) Weekly, (B) weekly
4. (A) Weekly, (B) daily

4-57. At what staion in the Tartar Mk 13 GMLS dud-jettisoning system is the dud/emergency firing key pressed?

1. EP2 panel
2. Weapons control
3. Attack console
4. CIC

4-58. Normally, (A) how often are jettison device timing tests performed, and (B) what is an acceptable test time?

1. (A) Every month, (B) 25 seconds
2. (A) Every 3 months, (B) 20 seconds
3. (A) Every month, (B) 28 seconds
4. (A) Every 3 months, (B) 25 seconds

4-59. Which of the following elements should you use to charge a jettison device accumulator flask bladder?

1. Oxygen that comes in a green cylinder
2. Nitrogen that comes in a green cylinder
3. Oxygen that comes in a gray cylinder with black bands near the top
4. Nitrogen that comes in a gray cylinder with black bands near the top

4-60. In the Mk 22 GMLS, the proper jettison train and elevation positions are signaled to the EP2 operator by the

1. dud indicator
2. misfire indicator
3. jettison light
4. launcher synchronized light

4-61. In the Mk 22 GMLS, a missile disposal by means of the emergency firing circuits may result in a damaged

1. contactor pad
2. launcher guide arm
3. emergency igniter latch
4. after handling shoe support

4-62. On the dud-jettison unit for the Mk 26 GMLS, what source of energy is used to fire the expendable piston?

1. Nitrogen accumulator
2. Pneumatic pump
3. Gas generator
4. Spring pressure

4-63. On the dud-jettison unit for the Mk 26 GMLS, what is the purpose of the blow-in plug in the wall of the extender sleeve?

1. To keep the extender sleeve from turning when extending and retracting
2. To let gas escape harmlessly to the atmosphere in case the gas generator accidentally fires when the sleeve is retracted
3. To vent housing chamber pressure to atmosphere when the extendable piston is extended
4. To withstand the explosive force of the gas generator

4-64. What is the weight of the extendable piston assembly on the dud-jettison unit for the Mk 26 GMLS?

1. 25 pounds
2. 50 pounds
3. 75 pounds
4. 100 pounds

4-65. On the dud-jettison unit for the Mk 26 GMLS, besides acting as a gas vent if the gas generator accidentally fires, which of the following functions does the pressure relief safety mechanism serve?

1. As a hydraulic relief valve
2. As a moisture seal
3. As a blow-in plug capture cage
4. Both 2 and 3 above

# Assignment 5

Electricity and Electronics

Textbook, NAVEDTRA 10200-D:  Pages 106 through 163

---

> Learning Objective: Identify the power and control panels used in GMLS control systems.

5-1. In which of the following GMLSs is the power distribution panel NOT designated EP1?

1. The Mk 10
2. The Mk 13
3. The Mk 22
4. The Mk 26

5-2. In which of the following GMLSs does the operational control panel (EP2) contain the strikedown position synchros for the launcher?

1. The Mk 10
2. The Mk 11
3. The Mk 22
4. The Mk 26

5-3. The control panel circuitry for which of the following GMLSs is predominantly solid state?

1. Mk 10
2. Mk 11
3. Mk 26
4. All the above

> Learning Objective: Explain how transistors are used as switching devices in solid-state control circuits.

5-4. The use of transistors as switching elements in the control circuits for GMLSs has greatly reduced the system's requirements for

1. interlock switches
2. relays
3. controllers
4. solenoids

5-5. A switching transistor that is in a nonconducting state is described as being

1. saturated
2. reverse biased
3. closed
4. forward biased

5-6. The bases of both the switching PNP and NPN transistors are connected to the input. Where are the emitters of the switching PNP and NPN transistors connected?

1. PNP to d.c. return; NPN to the high voltage side
2. PNP to the high voltage side; NPN to d.c. return
3. Both the PNP and NPN to the high voltage side
4. Both the PNP and NPN to d.c. return

5-7. The PNP transistor in text figure 5-2(d) will be forward biased when the base becomes more positive than the emitter.

5-8. If a voltmeter is connected from d.c. return to point (1) in textbook figure 5-3(b), what is the meter's approximate reading when Q1 is conducting?

1. 24 volts
2. 12 volts
3. 6 volts
4. 0 volts

---

Learning Objective: Point out how a typical solid-state control circuit operates.

---

5-9. The proximity switch is activated by a permanent magnet located on the ____A____ equipment,
      moving, nonmoving
and the switch's output is applied directly to a(n) _____
                                 matrix,
____B____ circuit.
inverter-buffer

1. A - moving, B - inverter-buffer
2. A - nonmoving, B - matrix
3. A - moving, B - matrix
4. A - nonmoving, B - inverter-buffer

5-10. Which of the following components of the optical limit switch circuitry generates the infrared light beam?

1. Transmitter
2. Phototransistor
3. LED
4. Detector

5-11. If a high output from a diode matrix circuit is applied to the input of the buffer circuit shown in textbook figure 5-7, which of the following buffer circuit conditions exists?

1. Transistor Q1 is reverse biased
2. The left-hand buffer output is high
3. The right-hand buffer output is high
4. Transistor Q2 is forward biased

5-12. If a low output from an optical limit switch is applied to the input of the buffer circuit shown in textbook figure 5-7, which of the following buffer circuit conditions exists?

1. Transistor Q1 is saturated
2. Transistor Q3 is turned on
3. The left-hand buffer output is low
4. The right-hand buffer output is high

5-13. Which of the following groups of transistor switches is turned off (reverse-biased) when a low signal is applied to the input of the inverter-buffer circuit shown in textbook figure 5-8?

1. Q1, Q3, and Q4
2. Q1, Q2, and Q3
3. Q3, Q4, and Q5
4. Q2, Q3, and Q4

5-14. Which of the following outputs of the inverter-buffer circuit shown in textbook figure 5-8 will be high when a high input is applied to the circuit?

1. Unflagged 2 and flagged 1
2. Flagged 2 and unflagged 1
3. Flagged 2 and flagged 1
4. Unflagged 2 and unflagged 1

5-15. The inverter-buffer circuit in textbook figure 5-8 is designed to deactivate all of the GMLS components it controls if its input is shorted to ground or it opens. What are the conditions of the outputs of this inverter-buffer circuit with a shorted or open input?

1. Flagged 1 and unflagged 1, high; flagged 2 and unflagged 2, low
2. Flagged 1 and flagged 2, low; unflagged 1 and unflagged 2, high
3. Flagged 1 and unflagged 1, low; flagged 2 and unflagged 2, high
4. Flagged 1 and flagged 2, high; unflagged 1 and unflagged 2, low

5-16. The input amplifier to a diode matrix circuit is shown in textbook figure 5-9. If a high signal is applied to this amplifier transistor, Q1 is turned on causing transistor Q2 to ___A___ and produce a ___B___ signal at the output of the input amplifier.  
(A: turn on, turn off; B: low, high)

1. A - turn on, B - low
2. A - turn on, B - high
3. A - turn off, B - low
4. A - turn off, B - high

5-17. The symbols that represent interlocks in the diode matrix circuit shown in textbook figure 5-10 are actually

1. transistors
2. relay contacts
3. diodes
4. switch contacts

5-18. To have a high output from the diode matrix output amplifier circuit shown in textbook figure 5-11, which of the following conditions must exist?

1. The circuit is enabled, transistor Q3 is turned on, and transistor Q5 is turned off
2. The circuit is enabled and transistors Q3 and Q5 are turned on
3. The circuit is enabled and transistors Q3 and Q5 are turned off
4. The circuit is disabled and transistors Q3 and Q5 are turned on

5-19. The driver circuits shown in textbook figures 5-12 and 5-13 perform the same basic function. In which of the following ways do they differ?

1. Transistor Q2 must be turned on to have a high output in the driver circuit shown in figure 5-12, while transistor Q2 must be turned off to have a high output from the driver circuit shown in figure 5-13
2. Only the input to the driver circuit shown in figure 5-13 can be provided from a diode matrix circuit
3. Transistor Q2 in figure 5-12 is of the NPN type, while transistor Q2 in figure 5-13 is of the PNP type
4. Only the driver circuit in figure 5-12 must be provided a high input to produce a high output

---

Learning Objective: Show how logic circuits can be used to depict a GMLS control circuit.

---

5-20. Logic circuits are a special type of control circuit used on the newer types of GMLSs.

5-21. The polarity indicator system is also known as

1. negative logic
2. positive logic
3. mixed logic
4. standard logic

5-22. On the Mk 26 GMLS, a logic circuit in which high inputs produce a low output is identified by the location of a

1. polarity indicator (flag) at the input of the logic symbol
2. negation indicator at the input of the logic symbol
3. polarity indicator (flag) at the output of the logic symbol
4. negation indicator at the output of the logic symbol

5-23. Which of the AND gate symbols in textbook figure 5-18 represents a logic circuit in which all inputs must be high before a low output can be produced?

1. (c)
2. (d)
3. (e)
4. (f)

5-24. The basic AND gate logic symbol is depicted in textbook figure 5-18(a). It represents a logic circuit in which all inputs must be high to produce a high output. Which of the other AND symbols in figure 5-18 is used to represent a logic circuit in which all inputs must be low before a high output can be produced?

1. (b)
2. (c)
3. (d)
4. (e)

5-25. Which of the following statements concerning the basic OR gate is/are true?

1. If any of the inputs is high, the output will be high
2. All inputs must be low to produce a low output
3. If all inputs are high, the output will be high
4. All the above

5-26. Which of the following OR gate outputs shown in textbook figure 5-20 will be high only when all inputs are high?

1. (b)
2. (c)
3. (d)
4. (e)

5-27. Which of the following OR gates shown in textbook figure 5-20 produces the identical output for the given inputs as the AND gate in textbook figure 5-18(d)?

1. (b)
2. (c)
3. (d)
4. (e)

5-28. Which of the following OR gates has only two inputs, and its output is low when both inputs are either high or low?

1. Basic OR
2. NOR
3. DOT-OR
4. Exclusive OR

5-29. Since there are no flags or negation symbols at the input or output of the AND gates in textbook figure 5-24, it can be assumed that these AND gates are of the basic logic type and all inputs must be high to produce a high output.

---

Learning Objective: Present an overview of digital circuits, binary numbering systems, and digital devices that are used on the Mk 26 GMLS.

---

5-30. Which of the following is NOT a limitation of the digital computer?

1. Continuous variables cannot be processed
2. Programming must be provided for each operation
3. Accuracy decreases with speed
4. Programming sometimes is more complicated than the problem

5-31. A digital computer performs its calculations by counting and comparing. What number system does it use?

1. Base 10 (decimal)
2. Base 8 (octal)
3. Base 5 (quinary)
4. Base 2 (binary)

33

5-32. What is the decimal equivalent to binary 10110?

1. 20
2. 22
3. 24
4. 26

5-33. What is the binary equivalent to decimal 29?

1. 11111
2. 11101
3. 11011
4. 11001

5-34. In the Mk 26 GMLS, in which of the following digital words is the launcher train velocity signal transmitted?

1. Word 1
2. Word 2
3. Word 3
4. Word 4

5-35. How does weapons control determine that the Mk 26 GMLS is correctly responding to an order?

1. By observing TV monitors
2. By observing hardwired synchro and light indicators
3. By comparing the digital order bits with the response bits
4. By communicating verbally through sound-powered phones

5-36. What component of the Mk 26 GMLS converts the launcher power drive digital order signals to analog signals?

1. Digital serial transceiver (DST)
2. Local control module (LCM)
3. Digital interface module (DIM)
4. Electronic servocontrol unit (ESCU)

---

Learning Objective: Recognize the troubleshooting and correction procedures for repairing electrical and electronic equipment that makes up a launcher system.

---

5-37. When you are testing a weapons control circuit's resistance with a megger, the synchros in the circuit that are being tested should be disconnected to prevent

1. damage to the synchros from the megger high voltage output
2. the possibility of obtaining incorrect readings
3. rotation of the synchros due to the megger output
4. arcing between the synchro windings

5-38. Which circuit breaker on the EP1 panel in textbook figure 5-27 must be reset after correcting a short circuit in the A-side magazine accumulator motor?

1. 3a
2. 3d
3. 5d
4. 7b

5-39. What is the designation for a cable in the EP1 panel that is numbered 327?

1. Power distribution
2. B-side loading control
3. Interlock
4. A-side loading control

5-40. Taking which of the following precautions is essential when replacing a burnt out fuse?

1. Check the circuit for opens
2. Use only one hand to insert the fuse
3. Wear safety shoes with non-conducting soles
4. Turn off the power

5-41. Suppose you are starting the magazine accumulator motor shown in textbook figure 5-29. The motor starts when pushbutton SMXA16A is pressed, but stops when the pushbutton is released. Which of the following components is most likely at fault?

1. KCCA28B
2. KCHA1B
3. KPXA1-D
4. KPXA2-A

5-42. Which of the following circuit troubles is LEAST likely to occur?

1. A shorted relay coil
2. A shorted diode
3. An open relay coil
4. An open diode

5-43. Which of the following "tricks of the trade" is considered the most important when you are circuit tracing control circuits?

1. Begin at the upper left-hand corner of the print and proceed from left to right and top to bottom
2. Disregard all circuits that are unnecessary to the one you are attempting to trace
3. Begin at the return or grounded side of the line and proceed toward the source or high side of the line
4. Begin at the source or high side of the line and proceed toward the return or grounded side of the line

5-44. Where is the relay tester located on the Mk 10 GMLS?

1. In the EP1 panel
2. In the EP2 panel
3. In the EP3 panel
4. In the checkout area

5-45. To properly adjust the air gap for the two microswitches shown in figure 5-32 of your textbook, you must first make sure that

1. solenoids LC1 and LC2 are energized
2. solenoid LC1 is energized and solenoid LC2 is deenergized
3. solenoids LC1 and LC2 are deenergized
4. solenoid LC1 is deenergized and solenoid LC2 is energized

5-46. A solenoid coil with an infinity resistance reading indicates an open circuit.

5-47. When a faulty printed circuit board is discovered, the trend is to attempt to repair it. If it cannot be repaired, turn it in to the supply department and obtain a replacement.

5-48. To check the operation of the Mk 10 GMLS load status recorder when the system is in the strike-down mode, the ready service ring is operated from which of the following panels?

1. EP1
2. EP2
3. EP3
4. EP4

---

Learning Objective: Identify the safety precautions and general safety rules required to maintain and operate electrical and electronic equipment.

---

5-49. Because of their work with high voltages, GMMs should know and be able to apply which of the following recommended methods of artificial respiration?

1. Back pressure/arm lift
2. Mouth-to-mouth resuscitation
3. Back pressure/hip lift
4. Chest pressure/arm lift

5-50. The first step you should take before trying to repair electronic equipment is to

1. turn off the main power at its source and tag the circuit "OFF"
2. pull the fuses to the main power supply
3. turn off the main power supply and remove the defective component from its cabinet
4. study all the wiring diagrams for the equipment to learn where all power to the equipment is to be turned off

5-51. Power supply and cutout switches must be turned off and tagged before a circuit may be repaired. Information on the tags includes the

1. nature of the repairs
2. length of time required for repairs
3. time repairs were begun
4. name of the person directly supervising or making the repairs

5-52. Which of the following devices should be used to discharge capacitors in deenergized equipment?

1. Interlocks
2. Ground straps
3. Shorting sticks
4. Rubber gloves

5-53. Suppose you are working on high-voltage equipment in the presence of a safety observer. If an accident occurs, the observer should take which of the following actions?

1. Secure the power
2. Give artificial respiration
3. Perform first aid for electrical shock
4. All the above

5-54. Assume that you are acting as a safety observer for Jim Hamilton, who is working in an energized power panel. Hamilton slips and falls in the panel in such a position that his body is blocking the power switch. What is the first action that you should take?

1. Call the main switchboard and have the power secured
2. Grasp Hamilton by the foot and pull him clear of the panel
3. Pull Hamilton clear of the panel using any nonconductive material at hand
4. Call sickbay and have a hospital corpsman report to the scene immediately

---

Learning Objective: Describe the function of a servomechanism, define special terms used with servomechanisms, and explain the operation of the closed-loop servosystem.

---

5-55. What do all GMLS servomechanisms have in common?

1. A hydraulic input
2. A feedback principle
3. An electrical output
4. A mechanical error detector

5-56. What is the minimum continuous voltage rating of a power servo?

1. 40
2. 60
3. 100
4. 140

5-57. To perform its specified task, what must the servosystem receive in the way of an order signal?

1. The desired results
2. The existing conditions
3. Both 1 and 2 above
4. The power requirements

---

Learning Objective: Point out how synchro systems are used in GMLSs, how synchro systems are aligned, and how to troubleshoot synchro systems.

---

5-58. The transmitter in a torque synchro system is designated a

1. TX
2. TR
3. CX
4. CT

5-59. What is the main function of a synchro tester of the type shown in textbook figure 5-35?

1. To locate defective synchros
2. To set synchro transmitters on electrical zero
3. To set CTs on electrical zero
4. To test synchro capacitors

5-60. When using a voltmeter to make the fine setting on a control transmitter (CX), between which synchro leads is the voltmeter connected?

1. S2 and R1
2. S3 and R2
3. S1 and S3
4. S1 and S2

● Information for item 5-61: GMM2 Sam Jones is assigned the job of zeroing a synchro control transformer (CT). First, he disconnects all the leads from the CT and connects a voltmeter between the R2 and S3 windings. With the voltmeter set to the 250-volt scale, he applies 115-volts a.c. to the S1 and S3 windings for 2 minutes and adjusts for a minimum voltage reading. Jones secures the power, reconnects the voltmeter between the R1 and R2 windings, and places the jumper between the S1 and S3 windings. He applies 115-volts a.c. power again for 1.5 minutes, while he is using the meter's lowest scale to adjust for minimum voltage. Jones then secures the power and connects the CT back in the system.

5-61. Did Jones zero the CT correctly?

1. Yes, because he performed every operation in the correct order and manner and achieved the desired results
2. No, because the first adjustment should be made for a maximum reading on the voltmeter
3. No, because the second adjustment should be made with the same connections as the first adjustment
4. No, because he used the 115-volt a.c. source instead of a 78-volt a.c. source for the zeroing procedure

5-62. What is the recommended procedure if an elevation regulator CT malfunctions?

1. Adjust and replace the synchro
2. Zero the synchro using an a.c. voltmeter
3. Replace and zero the synchro
4. Zero the synchro using two lamps and a pair of headphones

5-63. The torque of a synchro receiver is weakest when the receiver

1. approaches synchronization
2. is 15° from synchronization
3. is 60° from synchronization
4. is 120° from synchronization

5-64. Other than the improper zeroing of a newly installed synchro unit, what problem is most prevalent in initial malfunctions?

1. Switch shorts or opens
2. Corrosion
3. Wrong connections
4. Water leaking into new unit

Learning Objective: Point out the general characteristics of servoamplifiers used in GMLSs.

5-65. The amplifier of a servosystem is used to amplify the

1. followup voltage
2. synchro stator voltage
3. error voltage
4. fixed field voltage

5-66. What is one of the disadvantages of high gain in the amplifier of a servosystem?

1. Reduced speed of response
2. Reduced operating stability
3. Increased steady-state errors
4. Increased velocity errors

Learning Objective: Describe the principles of operation of the magnetic amplifiers used in the launcher power drives for the Mk 10 GMLS.

5-67. The Mk 5 launcher train and elevation power drives are serviced by a dual-channel magnetic amplifier consisting of

1. four magnetic amplifier stages and two power supplies
2. two magnetic amplifier stages and two power supplies
3. four magnetic amplifier stages and a single power supply
4. two magnetic amplifier stages and a single power supply

5-68. The fluctuations in the ship's electric power are corrected for before they reach the magnetic amplifier stages by the missile launcher's

1. regulated voltage supply
2. primary servoamplifier
3. velocity servoamplifier
4. offset voltage

5-69. A main advantage of magnetic amplifiers over other common amplifier circuits is that magnetic amplifiers provide

1. faster input signal response
2. more dependable service
3. superior high-frequency response
4. less input wave distortion

5-70. What is the purpose of the control winding in the magnetic amplifier voltage regulator shown in textbook figure 5-41?

1. To help smooth out the a.c. ripple
2. To set the amplifier operating point
3. To set the amplifier bias point
4. To eliminate transformer action

Learning Objective: Present the operating principles of the transistorized servoamplifier for the Mk 13 GMLS.

5-71. The error signals for the Mk 13 GMLS launcher power drives are applied to which of the following servoamplifier circuit boards?

1. PC5
2. PC6
3. PC2
4. PC1

5-72. In the Mk 13 GMLS servoamplifier shown in textbook figure 5-43, the B-end tachometer velocity signal is applied to the error signal in which of the following circuit boards?

1. PC2
2. PC3
3. PC4
4. PC6

Learning Objective: List the troubleshooting and adjustment procedures for servoamplifiers used in GMLSs.

5-73. Which of the following test equipment is normally used to check the operation of a rectifier in a servoamplifier?

1. Vacuum tube voltmeter
2. Ohmmeter
3. Signal generator
4. Cathode-ray oscilloscope

5-74. What is the desired setting for the gain adjustment in a launcher power drive servoamplifier?

1. Set it as low as possible, with the servosystem possessing a satisfactory degree of stability
2. Set it as high as possible, with the servosystem possessing a satisfactory degree of stability
3. Turn it in a clockwise direction until the servosystem appears to be stable
4. Turn it in a counterclockwise direction until the servosystem appears to be stable

5-75. When making the phase adjustment on a launcher power drive servoamplifier that also contains gain and balance adjustments, which of the following procedures is considered good practice?

1. Make the balance adjustment first
2. Make the gain adjustment first
3. Check the phase of the current using a cathode-ray oscillograph
4. Check the balance and gain adjustments after making the phase adjustment

# Assignment 6

Hydraulics and Pneumatics

Textbook, NAVEDTRA 10200-D:   Pages 164 through 215

---

Learning Objective: Describe how hydraulics are used in missile handling and strikedown equipment.

6-1.  Which of the following components do NOT receive hydraulic power from the loader accumulators in a Mk 10 GMLS?

1. The spanning rails and the blast doors
2. The retractable rails
3. The loader positioners
4. The launcher's aft motion latches

6-2.  The symbols used on hydraulic schematics for strikedown equipment on the Mk 10 GMLS conform to which of the following standards?

1. American National Standard Institute (ANSI)
2. American Institute for Design and Drafting (AIDD)
3. Joint Industrial Conference (JIC)
4. National Fluid Power Association (NFPA)

6-3.  You must manually remove the mechanical locking mechanism from the strikedown elevators before operating the elevator.

6-4.  Which of the following valves should be closed after securing the strikedown elevator and hatch to prevent hydraulic shock from possibly damaging the equipment?

1. A
2. C
3. D
4. E

6-5.  What switch is actuated by the shaft for the strikedown hatch dogs?

1. K
2. N
3. A1
4. B1

6-6.  If an A1 or B1 switch operator releases a switch prematurely during a raise cycle, what will happen?

1. The strikedown hatch will close
2. The raise cycle will stop
3. The raise cycle will be completed automatically
4. The lower cycle will be automatically initiated

6-7.  Assuming that valves D and E are set and locked to specifications, what other valve should be set and locked at the position which will ensure the correct speed of the strikedown elevator and hatch?

1. T
2. U
3. F
4. Z

39

6-8. What is the designation for the hatch-cover counterbalance valve?

1. J
2. T
3. X
4. Y

6-9. During a raise cycle, which of the following valves is used to decelerate the strikedown elevator?

1. U
2. S
3. Z
4. T

6-10. What switch is tripped when the hatch cover is in the fully closed position?

1. A1
2. N
3. M
4. K

---

Learning Objective: Point out how hydraulics are used to operate Terrier magazines.

---

6-11. Which of the following components is/are supplied hydraulic power by the Mk 10 GMLS magazine accumulator system?

1. The RSR motor
2. The magazine doors
3. The tray-shift mechanism
4. All of the above

6-12. Which of the following pumps is normally used on launching systems that require a large volume of pressurized fluid?

1. Gear
2. Parallel piston
3. Vane
4. Radial piston

6-13. The nitrogen charging pressure for accumulators varies with the temperature.

6-14. How many filter elements are contained in the multi-element filter assembly shown in textbook figure 6-5?

1. 4
2. 6
3. 12
4. 18

6-15. The control valve in the magazine accumulator system for the Mk 10 GMLS maintains fluid pressure at what set limit?

1. 700 to 900 psi
2. 1000 to 1200 psi
3. 1300 to 1500 psi
4. 1600 to 1800 psi

6-16. What is the purpose of the gear pump shown in textbook figure 6-7?

1. To supply fluid under pressure to release the power-off brake
2. To supply servo pressure to the magazine control valve block
3. To provide lubricant to the gears and bearings not submerged in oil
4. To supply supercharge pressure to the magazine control valve block

6-17. The slip clutch for the power-off brake on the Mk 10 GMLS RSR will yield and prevent manual operation when the force exceeds

1. 7 foot-pounds
2. 10 foot-pounds
3. 17 foot-pounds
4. 22 foot-pounds

6-18. What is the purpose of the slip clutch on the Mk 10 GMLS RSR?

1. To improve the mechanical advantage of the handcrank
2. To prevent damaging the drive train in case of an obstruction
3. To restrain the RSR while handcranking
4. To prevent an unbalanced load from rotating the RSR

6-19. Which of the following conditions must be met before the tray, shown in textbook figure 6-8, is positioned to the hoist position?

1. The hoist must be down and latched
2. The RSR positioning latches must be engaged
3. Both 1 and 2 above
4. The magazine doors must be open

6-20. In the Mk 10 GMLS, the RSR positioning latch solenoid valve is hydraulically interlocked so that it cannot be retracted unless the tray is latched at the ready service ring position. Which of the following valves, shown in textbook figure 6-8, performs this function by ensuring that PA pressure is NOT available at UVCA(B)1?

1. UVDA(B)4
2. UVDA(B)5
3. UVDA(B)6
4. UVDA(B)7

Learning Objective: Point out how hydraulics are used to operate the loader and hoist equipment on the Terrier and Tartar GMLSs.

6-21. The purpose of the probe that is mounted on the aft head of the hoist mechanism in figure 6-9 of your textbook is to

1. align the head of the hoist with the loader trunk when the hoist is in a raised position
2. act as a buffer in the event that the hoist overtravels
3. counterbalance the weight of the aft shoe hoist head
4. align the missile in the hoist shoes

6-22. Which of the following components is included in the Mk 10 GMLS magazine door drive bracket assembly?

1. Door operating piston assembly
2. Latch control valve
3. Solenoid valve
4. All of the above

6-23. Which of the following components do NOT receive pressure to operate from the loader accumulator power drive on the Mk 10 GMLS?

1. Blast door latches
2. Magazine door latches
3. Strikeup/strikedown gear
4. Retractable rails

6-24. What assembly within the Mk 10 GMLS loader power drive unit serves to stop the moving equipment in the event of a power failure?

1. Magnetic clutch
2. Loader pawl positioner
3. Tilting rail
4. Power-off brake

6-25. What component distributes the hydraulic fluid from the Mk 10 GMLS loader accumulator to the floating track piston assemblies?

1. A flexible line
2. The rotating joint
3. The loader valve block
4. The transfer pin

6-26. In what manner do the floating tracks on the Mk 10 GMLS align the booster shoes with the track grooves of the fixed loader rail?

1. Pneumatically
2. Hydraulically
3. Mechanically
4. Electrically

6-27. During which of the following operations are the retractable rails used in the handling of a missile on the Mk 10 GMLS?

1. Launcher loading
2. Launcher unloading
3. Assembly area unloading
4. Strikedown operation

6-28. What devices are used to mechanically couple the spanning rail to the blast doors?

1. Sliding wedges
2. Operating links
3. Torsion arms
4. Transfer pins

6-29. What is the purpose of the spud located at the forward end of the Mk 10 GMLS spanning rail?

1. To ensure proper alignment between the spanning rail and the guide arm
2. To ensure proper alignment between the fixed loader and the spanning rail
3. To provide increased strength between the spanning rail and the fixed loader rail
4. To provide increased strength between trunk sections IV and V

6-30. Which of the following GMLSs stows the hoist chains in the launcher guide arm?

1. Mk 10
2. Mk 11
3. Mk 22
4. Mk 26

Learning Objective: Point out how hydraulics are used to operate the various components of the launcher and the guide arm components of the Terrier and Tartar launchers.

6-31. What component(s) on the launcher support(s) the trunnion tube?

1. The guides
2. The stand
3. The carriage
4. The train circle

6-32. The purpose of the guide arm in the GMLS is to

1. guide the missile from the assembly area to the launcher
2. guide the missile while in flight
3. support the missile on the launcher
4. guide the missile to the magazine

6-33. Hydraulic power for operating the components within the Terrier launcher guide is provided by the

1. guide arm accumulator power drive
2. carriage accumulator power drive
3. trunnion tube accumulator power drive
4. train and elevation accumulator power drives

6-34. The purpose of the hydraulically operated steel fork in the elevation latch is to prevent the

1. guide arm from elevating while the launcher is training
2. guide arm from moving in elevation when the launcher is being loaded
3. launcher from training while the guide arm is elevating
4. guide arm from moving when the pump motor is not energized

6-35. The elevation guide arms of the Terrier launcher are caused to elevate or depress by direct action of the

1. elevation positioning valve
2. elevation drive pinion gear
3. elevation reduction gear
4. trunnion tube

6-36. The spring-loaded, mechanically operated valves which ensure that the launcher is in the proper LOAD position before the train and elevation latches may extend are called

1. positioning valves
2. control valves
3. pilot valves
4. floating valves

6-37. The purpose of the train and elevation buffers in a GMLS is to

1. keep the train and elevation gearing polished
2. control the rate of response to a large input signal
3. prevent excessive stress on the missile when the guide arms reach the elevation and depression limits
4. limit launcher movement when nearing dud-jettison positions

6-38. One factor that the piston-type accumulator and the bag-type accumulator have in common is that both types

1. are controlled automatically by a pilot valve
2. use nitrogen gas to maintain a reserve quantity of fluid under pressure
3. have poppet valves in the fluid manifold
4. use movable pistons

6-39. Which of the following valves prevents the bladder in a bag-type accumulator from entering the hydraulic manifold?

1. Spring-loaded relief
2. Nitrogen pressure regulator
3. Hydraulic flow sensitive check
4. Spring-loaded poppet

6-40. The pilot valve in a piston accumulator controls the

1. nigrogen pressure in the gas chamber
2. hydraulic pressure in the fluid chamber
3. fluid flow through the manifold
4. position of the piston within the chamber

6-41. What force holds the control valve open when the accumulator is being charged?

1. Spring tension
2. Accumulator pressure
3. Pump pressure
4. Nitrogen pressure

---

Learning Objective: Indicate the purpose, application, and operating principles of hydraulically operated launcher components.

---

6-42. Where are the train and elevation power drives of the Mk 13 GMLS located?

1. In the right and left trunnion supports, respectively
2. On the right and left sides of the base ring, respectively
3. In the inner structure of the stand
4. In the outer structure of the stand

6-43. Besides providing a head of fluid to keep air from the hydraulic lines, the header tank for the launcher guide power unit in the Mk 22 GMLS is capable of which of the following functions?

1. Acting as a cooling chamber for fluid returning from the system
2. Providing fluid for the train and elevation CAB units
3. Furnishing supercharge pressure for the hoist drive unit
4. All of the above

6-44. Which of the following events will take place if the pressure in the discharge line of the train lubrication pump suddenly drops to zero?

1. The pump drive motor will burn out
2. A red warning light will glow on the EP2 panel
3. An auxiliary pump will be switched on automatically
4. The pump motor will be switched on automatically

6-45. Which of the following components of a CAB unit converts hydraulic energy into mechanical motion?

1. B-end
2. A-end
3. Drive train
4. Control valve block

6-46. What is the purpose of the drive train in a CAB unit?

1. To convert electrical power into hydraulic energy
2. To convert hydraulic energy into mechanical motion
3. To transmit mechanical motion
4. All of the above

---

Learning Objective: Identify the maintenance procedures for repair and adjustment of missile launching system hydraulic equipment and components.

---

6-47. Which of the following cleaning agents is used to clean a bypass oil filter element that is made up of perforated paper discs?

1. Diesel fuel
2. Low-pressure air
3. Soap and water
4. Carbon dioxide

6-48. Which of the following precautions should you take to decrease the frequency of filter cleaning and to reduce the damage to hydraulic components caused by contaminants?

1. Keep containers of hydraulic fluid tightly closed
2. Strain or filter all replenishment hydraulic fluid
3. Keep all openings in the hydraulic system secured (sealed)
4. All of the above

6-49. If you must open the hydraulic system to repair a component, what precaution must you take before you begin the repair?

1. Ensure that the pressure has been released from the system
2. Remove the component from the system
3. Remove the electric power from the system
4. Inform damage control that the system is down

6-50. What lubricant do you use to coat the external threads of the filter head when you install a new micronic filter element?

1. Graphite
2. Molykote
3. Petrolatum
4. Gyro oil

6-51. Which of the following types of valves should normally be disassembled only during an overhaul period?

1. Automatically operated relief
2. Manually operated relief
3. Manually operated gate
4. Cone-type check

6-52. Assume that you are adjusting a relief valve in a 100 psi hydraulic system. At what pressure should the relief valve be set to open?

1. 25 psi
2. 60 psi
3. 100 psi
4. 125 psi

6-53. From which of the following references do you normally obtain the name and identifying stock number for a part you need to repair the Tartar launcher?

1. Ordnance data pamphlet
2. Equipment illustrated parts breakdown (IPB)
3. General stores catalog
4. Equipment schematic drawing

6-54. A low oil level in the reservoir or clogged filters in the supply lines can result in which of the following conditions?

1. Improper output pressure
2. Noisy pump operation
3. Overheated pump
4. All of the above

---

Learning Objective: Point out the component arrangement and the operating principles of launching system receiver-regulators.

---

6-55. Which of the following factors determines the A-end's output?

1. The position of the A-end tilt plate
2. The number of pump pistons
3. The electric motor's speed
4. The electric motor's direction of rotation

6-56. The position signal from the computer controls the

1. position of the A-end tilt plate
2. position of the B-end tilt plate
3. electric motor's speed
4. electric motor's direction of rotation

6-57. If one rotation of the rotor of a 1-speed synchro is equal to 360° of launcher movement, one rotation of the rotor on a 36-speed synchro is equal to

1. 360°
2. 36°
3. 10°
4. 1°

6-58. What component transmits the A-end response to the amplifier (servo-control unit) of the Mk 116 launcher?

1. Control transformer
2. Synchro differential
3. Transformer resolver
4. Potentiometer

6-59. Which of the following devices limits maximum pressure buildup in the A-end high pressure output line of the Mk 13 GMLS train and elevation CAB units?

1. A pair of orifices in the A-end return line
2. A check valve assembly mounted in the output line of the A-end pump
3. A compound relief valve assembly mounted in the output line of the B-end motor
4. A compound relief valve assembly mounted on the valve plate between the A-end pump and the B-end motor

6-60. Which of the following units in the Mk 10 GMLS receiver-regulator are NOT driven either directly or indirectly by the B-end response shaft?

1. Indicator dials
2. Electrohydraulic valve
3. Limit-stop assembly
4. B-end tachometer

6-61. In the Mk 10 GMLS, the difference between the contours of the train rotary piston cam and the elevation rotary piston cam is determined by the difference in the

1. mounting positions of the units
2. nonpointing zone positions
3. acceleration and velocity specifications
4. type of servocontrol valves used in the units

6-62. Which of the following statements correctly describes the output of the electrohydraulic servovalve?

1. Its hydraulic pressure output is proportional to the electrical input signal
2. Its mechanical movement is directly proportional to the electrical input signal
3. Its mechanical movement is inversely proportional to the electrical input signal
4. Its hydraulic pressure is inversely proportional to the electrical input signal

---
Learning Objective: Point out the types, functions, and operating principles of missile control systems and related components.
---

6-63. Which of the following missiles does NOT have a hydraulic sump?

1. HTR
2. BT
3. Sparrow III
4. SM/ER

6-64. Which of the following functions is/are performed by a missile servomechanism?

1. Accepts an order which defines the desired result
2. Evaluates existing conditions
3. Carries out an order
4. All of the above

6-65. Which of the following missile components provides physical (spatial) references from which missile attitude can be determined?

1. Tachometer generators
2. Caged gyroscopes
3. Free gyroscopes
4. Accelerometers

6-66. What does the external followup system shown in textbook figure 6-31 measure?

1. Gyro displacement
2. Accelerometer displacement
3. Control surface displacement
4. Missile attitude

6-67. The controller unit shown in textbook figure 6-32 is made up of two solenoids, a transfer valve, and an actuator. What is the main disadvantage of this type of controller in missile use?

1. The operation of the unit is too slow
2. The control unit does not possess sufficient power to cause actuator movement at high missile speeds
3. The design of the unit does not permit a varying degree of control of the actuator
4. The control of the unit is not positive

6-68. Which of the following is a disadvantage of a hydraulic actuator unit?

1. Complexity of mechanical design
2. Speed of reaction
3. Power available
4. Slight losses due to friction

---
Learning Objective: Identify the missile handling equipment that is air driven, and point out the operating features of the chain drive fixture used for strikedown and offload of missiles on the Mk 13 and Mk 22 GMLSs.
---

6-69. An additional piece of equipment which needs 640 psi of compressed air for its operation is being installed aboard your ship and the existing air supply system is 1000 psi. Which of the following changes must be incorporated into the ship's air supply system to operate the new equipment?

1. A 640 psi compressed air system must be installed on the ship
2. The 1000 psi air supply must be reduced to 640 psi by the installation of a reducing valve
3. The new equipment must be modified to operate on 1000 psi
4. The output pressure of the 1000 psi compressor must be lowered to 640 psi to produce a 640 psi supply system

6-70. What is the maximum amount of air pressure that may be used for drying out an electrical motor having a rating of 50 horsepower or less?

1. 10 psi
2. 15 psi
3. 25 psi
4. 30 psi

6-71. The quick-release pin on the chain drive fixture of the Mk 13 GMLS CANNOT be inserted until which of the following conditions have been met?

1. The latch engages the retractable rail
2. The launcher has been depressed in local test mode
3. The latch lever has been pushed
4. The strikedown chain has fully extended

6-72. What component of the auxiliary strikedown gear used with the Mk 13 and Mk 22 GMLSs controls the speed and direction of the air motor?

1. Pressure regulator
2. Air throttle valve
3. Manual control valve
4. Latch lever

6-73. What is one purpose of the hand control box used with the Mk 13 GMLS?

1. To control the operation of the chain drive fixture
2. To enable the launcher captain to control and see the launcher at the same time
3. To provide control of the missile component handling crane
4. To control the operation of the strikedown gear

6-74. On which of the following panels is the firing safety switch located?

1. EP1
2. EP2
3. EP3
4. EP4

6-75. What publication should you consult for the description of the steps in strikedown, off-loading, checkout, and deactivation of the Mk 13 GMLS?

1. OP 3114
2. OP 3115
3. OP 2665
4. OP 2351

47

# Assignment 7

Hydraulics and Pneumatics (continued); Weapon Surveillance and Stowage

Textbook, NAVEDTRA 10200-D: Pages 215 through 249

---

Learning Objective: Point out the pneumatic operation of the ASROC loading fixture used on the Mk 10 Mods 7 and 8 GMLS.

---

7-1. To ready the ASROC loading fixture for loading missiles, it must be lowered from its stowed position by means of a

1. chain hoist
2. handcrank
3. hydraulic jacking device
4. specially designed pneumatic piston

7-2. The three positions of the manual control valve used in the ASROC loading fixture are

1. OFF, ON, and NEUTRAL
2. OFF, UNLATCHED, and LATCHED
3. NEUTRAL, LATCHED, and UNLATCHED
4. OFF, EXTEND, and RETRACT

7-3. What device is used to close the adapter rail snubbers?

1. A manual pump handle
2. A manual jack screw
3. A pneumatic piston
4. A cantilever attachment

---

Learning Objective: Point out the operating principles of the air drive motors of the launcher and the function of air lubricators.

---

7-4. Air drive motors are used with which of the following systems on the Mk 10 GMLS?

1. Ready service ring
2. Train
3. Elevation
4. Both 2 and 3 above

7-5. You should always use extreme caution when using the air motor in training or elevating the launcher because the normal safety interlocks are bypassed.

7-6. The power drives must be __A__ and the train and elevation
       on, off
latches __B__ before
       extended, retracted
elevating or training the Mk 10 GMLS launcher using the air motors.

1. A - on, B - extended
2. A - off, B - extended
3. A - on, B - retracted
4. A - off, B - retracted

7-7. Air lubricators are of the micro fog type.

---

Learning Objective: Identify the function and basic operation of the dud-jettisoning pneumatic equipment and the anti-icing system of the Mk 10 Mods 7 and 8 GMLS.

---

48

7-8. What position of the charge and fire control valve lever cuts off the 4500 psi air supply to the Mk 10 Mods 7 and 8 GMLS dud-jettisoning equipment when preparing to jettison a missile?

1. READY
2. CHARGE
3. POSITION I
4. JETTISON AND OFF

7-9. The 100 psi air pressure from the ship's supply is used in the dud-jettisoning equipment for the Mk 10 Mods 7 and 8 GMLS to

1. operate the ejector sleeve
2. charge the air chamber
3. perform 1 and 2 above
4. operate the firing valve

7-10. The missile must be jettisoned only on the ___A___ if
uproll, downroll
the roll of the ship is more than ___B___ degrees.
10, 20

1. A - uproll, B - 10
2. A - downroll, B - 10
3. A - downroll, B - 20
4. A - uproll, B - 20

7-11. What assures that the dud-jettison piston for the Mk 10 Mods 7 and 8 GMLS moves at the same speed when being exercised or in training periods as it does when being used with a missile?

1. Check valve
2. Orifices
3. Control valve
4. Both 2 and 3 above

7-12. Anti-icing fluid is maintained at a constant head of pressure to compensate for expansion and contraction of the fluid under varying temperature conditions.

---

Learning Objective: Specify the uses of thermopneumatic devices in missile magazines to activate fire suppression systems.

---

7-13. Automatic thermopneumatic control systems are installed in missile magazines as part of which of the following fire suppression systems?

1. Water injection
2. Carbon dioxide ($CO_2$)
3. Magazine sprinkling
4. Both 2 and 3 above

7-14. Which of the following magazine fire suppression systems uses the heat sensing device and a pneumatic control head?

1. Dry-type sprinkling
2. Wet-type sprinkling
3. $CO_2$
4. Water injection

7-15. The diaphragm in the ___A___
control head,
reacts to a change in
PRP valve
air pressure from a heat sensing device and, through linkage, connects ship's ___B___ to
air, saltwater
the sprinkling system's main control valve.

1. A - control head, B - air
2. A - PRP valve, B - saltwater
3. A - control head, B - saltwater
4. A - PRP valve, B - air

7-16. Which of the following components of the PRP valve prevents activation of the sprinkling system due to normal magazine temperature variations?

1. Compensating vent
2. Diaphragm
3. Spring mechanism
4. Lever

Learning Objective: Recognize the Navy's shipboard surface-to-air missiles' use of air pressure to operate parts of their internal system.

7-17. The purpose of the probe-shield on the Terrier BT-3 missile is to

1. act as an air scoop to bring additional pressure into the nose orifice during flight
2. prevent dust and moisture from entering the nose orifice during the terminal stage of the flight
3. prevent dust and moisture from entering the nose orifice prior to missile firing
4. give the missile a stabilizing roll during its initial stage of flight

7-18. What device in the Terrier BT-3 missile senses the changes in missile velocity and altitude and causes the missile's control surfaces to compensate for these changes?

1. Transducer
2. Potentiometer
3. Probe
4. Guidance computer

7-19. The pressure transducer of the Terrier HT-3 missile supplies the electrical output that directly drives the

1. ganged potentiometers
2. radome section
3. guidance computer
4. servomotor

Learning Objective: Point out the hazards to guided missiles, electroexplosive devices (EEDs), personnel, and fuel from radio frequency (rf) transmission.

7-20. Which of the following publications is the Navy's primary reference source concerning the hazards of electromagnetic radiation to ordnance?

1. OP 4
2. OP 5
3. OP 3347
4. OP 3565

7-21. Which of the following classifications is NOT assigned to EED-contained ordnance?

1. HERO SAFE
2. HERO READY
3. HERO UNSAFE
4. HERO SUSCEPTIBLE

7-22. Which of the following publications requires the development and maintenance of a HERO emission control (EMCON) bill?

1. OP 2173
2. OP 3000
3. OP 3565
4. OP 4093

Learning Objective: Point out weapon inspection guidelines for the Mk 10 GMLS.

7-23. The best method for simplifying the redundant operational steps of a checklist is to memorize the steps.

7-24. Which of the following publications takes precedence over OP 4093 when handling Terrier BTN missiles?

1. The OP for the Standard ER missile
2. The OP for the Mk 10 GMLS
3. The SWOP for the Terrier BTN missile
4. The OP for the Terrier HTR missile

7-25. In what volume of OP 4093 will you find information applicable to all Terrier class ships?

1. 1
2. 2
3. 3
4. 4

---
Learning Objective: Point out the explosive component hazard categories and identify the special handling requirements and hazards associated with explosives and propellants.
---

7-26. Using the Department of Transportation (DOT) classification system, into which of the following classes do missile warheads fall?

1. D
2. C
3. B
4. A

7-27. The Rules and Regulations for Military Explosives and Hazardous Munitions, CG-108, is a NAVSEA manual.

7-28. Which of the following publications contains a listing of UNO/DOD, DOT, and CG classes of ammunition and explosives?

1. OP 4, volume 1
2. OP 4, volume 2
3. OP 5, volume 2
4. OP 5, volume 3

7-29. What is the minimum number of minutes you must wait before approaching a misfired rocket motor?

1. 60
2. 45
3. 30
4. 15

7-30. Which of the following high explosives are normally contained in the missile's self-destruct charges?

1. Tetryl and PETN
2. Composition B and tetryl
3. TNT and fulminate of mercury
4. Composition C and tetryl

---
Learning Objective: Point out the special hazards associated with handling and stowing nuclear weapons.
---

7-31. Conventional high explosives comprise the major hazard associated with accidents involving atomic weapons.

7-32. To what maximum temperature may high explosives in a nuclear weapon be subjected before they ignite or detonate?

1. 300°
2. 350°
3. 400°
4. 450°

7-33. What type of radiation is emitted by plutonium?

1. X-ray
2. Beta
3. Gamma
4. Alpha

---
Learning Objective: Point out the procedures for receipt inspection of guided missiles and the procedures to be used for disposal of unsafe weapons and components.
---

7-34. Which of the following conditions on the warhead of a Terrier BTN missile is cause to reject it?

1. Fungus growth
2. Corrosion
3. Dents
4. Superficial abrasions

7-35. Which of the following officers has custody of the H3114 arming tool?

1. Commanding officer
2. Executive officer
3. Nuclear weapons officer
4. Missile officer

7-36. What is the minimum length of time required to cage the gyro of a misfired BTN missile by applying external power?

1. 4 minutes
2. 3 minutes
3. 2 minutes
4. 1 minute

7-37. A dud Standard ER missile with its battery expended must be handled as a misfire.

7-38. Which of the following conditions is indicated when black smoke exits from the battery vent of a dud or misfired Standard ER missile?

 1. The auto-pilot is leaking
 2. The battery is normal
 3. The battery has failed
 4. The gyro is uncaged

___

Learning Objective: Identify missile components and miscellaneous explosive devices using the standard military color code.

___

7-39. Which of the following colors has no identification color-code significance?

 1. Yellow
 2. White
 3. Brown
 4. Blue

___

In items 7-40 through 7-44, select from column B the color code that identifies the component in column A.

| A. Components | B. Color Codes |
|---|---|
| 7-40. Training | 1. Yellow |
| 7-41. High explosives | 2. Brown |
| 7-42. Low explosives | 3. Green |
| 7-43. Smoke | 4. Blue |
| 7-44. Marker | |

___

7-45. Which of the following publications contains color-coding information on missile and rocket components?

 1. OP 2238
 2. OP 2350
 3. OP 3347
 4. OP 3565

___

Learning Objective: Review the arrangement and construction features of GMLS magazines, including the safety requirements applicable to shipboard stowage areas.

___

7-46. Why are missile magazines constructed so that each missile will be segregated?

 1. For easy handling
 2. For maximum protection of the missile
 3. Both 1 and 2 above
 4. For easy counting

7-47. To eliminate hazards from moving equipment within the magazine areas, remove or position safety switches on the controlling station.

7-48. Before entering a missile magazine for maintenance, what must be done to render the $CO_2$ firefighting system safe?

 1. Close the magazine door
 2. Secure the ventilation
 3. Close the system feed valves
 4. Secure all electrical power

7-49. The two shutoff valves that serve the carbon dioxide ($CO_2$) system are normally secured (but not locked) in the open position.

7-50. How often are missile magazines inspected to verify that the proper temperature exists?

 1. Daily
 2. Weekly
 3. Monthly
 4. Quarterly

7-51. If you have no special breathing equipment available, what is the minimum waiting period after booster or sustainer burnout before you should enter the area?

 1. 10 minutes
 2. 15 minutes
 3. 30 minutes
 4. 45 minutes

---
Learning Objective: Point out the requirements for inspecting and testing $CO_2$ and sprinkling systems, including the hazards involved in these tests. Point out the fusible slug installation procedures and hazards involved.
---

7-52. How often should the sprinkling system control lines be visually inspected for leaks?

1. Quarterly
2. Monthly
3. Weekly
4. Daily

7-53. At what pressure should the pressure tank in a wet-type magazine sprinkler system be maintained?

1. 10 psi
2. 25 psi
3. 50 psi
4. 75 psi

7-54. Which of the following components activates the sprinkling and $CO_2$ systems in the Mk 13 GMLS?

1. Water detector
2. Smoke indicator
3. Heat-sensing device
4. Heat exchanger

7-55. How many types of fusible slugs are used in the Mk 13 GMLS magazine firefighting systems?

1. Two
2. Three
3. Four
4. Five

7-56. At what temperature will the fusible slug in the Mk 13 GMLS magazine sprinkling system melt?

1. 154°F
2. 156°F
3. 158°F
4. 174°F

7-57. What must be done before installing the fusible slug in the heat-sensing device of a sprinkling or $CO_2$ system?

1. Secure the $CO_2$ system
2. Secure the sprinkling system
3. Both 1 and 2 above
4. Secure the ventilation system

7-58. What device, if any, directly releases the compressed spring in the $CO_2$ control head?

1. Diaphragm mechanism
2. Pressure differential
3. Actuating lever
4. None

7-59. Resetting the bellows of a heat-sensing device will automatically reset the tripping mechanism of the $CO_2$ control head.

---
Learning Objective: Identify the functional characteristics of the protective systems and devices that are designed to protect personnel and weapons in GMLS stowage areas.
---

7-60. What is the major design difference between Terrier and Tartar water-injection systems?

1. The Tartar system uses a pressurized saltwater supply tank
2. The Terrier system uses a pressurized saltwater supply tank
3. The Tartar system uses a pressurized freshwater supply tank
4. The Terrier system uses a pressurized freshwater supply tank

7-61. Which of the following GMLSs employs a plenum chamber to vent gases created by accidental missile ignition in the magazine?

1. Mk 10
2. Mk 13
3. Mk 22
4. Both 2 and 3 above

7-62. What is the function, if any, of the anti-icing system on a GMLS?

1. To keep the power drive fluid warm
2. To keep the magazines warm
3. To keep designated exposed surfaces warm
4. None

7-63. The anti-icing fluid in the heat exchanger is composed of what ratio of distilled water to ethylene glycol?

1. 1:1
2. 2:1
3. 3:1
4. 4:1

---

Learning Objective: Identify the records, reports, and inspections required for guided missiles and GMLS magazines.

---

7-64. For which of the following areas are magazine temperature records and magazine logs maintained?

1. All magazines aboard ship
2. Selected magazines ashore
3. All ready-service lockers aboard ship
4. All of the above

7-65. In which OP will you find the orderly and approved manner for storing explosives ashore?

1. OP 4, volumes 1 and 2
2. OP 3347
3. OP 5, volumes 1 and 2
4. OP 3565

7-66. What instruction gives the procedure for reporting lost, stolen, and recovered small arms?

1. OPNAVINST 8110.16 series
2. OPNAVINST 5510.45 series
3. SECNAVINST 5500.4 series
4. NAVSEAINST 8510.5 series

7-67. In which OP will you find the official shipboard rules for care and maintenance, surveillance, stowage, and disposal of various types of pyrotechnics?

1. OP 4, volume 1
2. OP 4, volume 2
3. OP 5, volume 1
4. OP 5, volume 2

7-68. Which of the following instructions details the reporting procedures for accidents, malfunctions, and incidents involving nonnuclear explosive ordnance?

1. NAVSEAINST 8510.3 series
2. OPNAVINST 8110.16 series
3. OPNAVINST 5102.1 series
4. OPNAVINST 5510.45 series

# Assignment 8

Fire Control and Alignment

Textbook, NAVEDTRA 10200-D: Pages 250 through 278

---

Learning Objective: Identify the functions performed by the principal subsystems that constitute a missile weapons system.

8-1. Which of the following weapons system units provide(s) initial detection, location, and identification of the target?

1. Command and control
2. Fire control
3. Delivery
4. Surveillance

8-2. IFF equipment operates in direct association with what equipment?

1. Air search radar
2. Electronic countermeasure gear
3. Fire control director
4. The missile computer

8-3. The transponder in an airborne aircraft responds to an IFF challenge from a friendly ship in which of the following ways?

1. By transmitting a series of predetermined coded pulses
2. By radioing a verbal code word
3. By transmitting a constant amplitude, high power signal
4. All of the above

8-4. Which of the following basic factors of a fire control problem is NOT established by a fire control system?

1. A line of sight
2. A predetermined miss distance
3. The prediction quantities
4. A line of fire

8-5. Which of the following equipment originates a search program if the fire control radar set does NOT acquire the target immediately?

1. The code comparator
2. The air search radar
3. The target designation transmitter
4. The fire control computer

8-6. The computer goes into a ready mode when it receives a "director assigned" signal from the command and control units.

---

In items 8-7 through 8-10, select from column B the function of the search radar in column A.

| A. Search Radars | B. Functions |
|---|---|
| 8-7. AN/SPS-10 | 1. A two-coordinate radar used primarily for surface search |
| 8-8. AN/SPS-37 | |
| 8-9. AN/SPS-43 | 2. A three-coordinate radar used to provide primary air search target data |
| 8-10. AN/SPS-48 | |
| | 3. A two-coordinate radar used to provide secondary air target data |

55

8-11. What information describing the position of an airborne target can be obtained from an air search radar (AN/SPS-48)?

1. Range and azimuth only
2. Azimuth and elevation only
3. Range and elevation only
4. Range, azimuth, and elevation

8-12. IFF interrogation information is synchronized and displayed concurrently with the air search data on the display equipment.

8-13. How many modes of operation are available to the NTDS computer?

1. 12
2. 24
3. 32
4. 64

---

Learning Objective: Point out the functions of the fire control units of the Mk 76 Mod 5 guided missile fire control system.

---

8-14. Which of the following transmitters of the AN/SPS-55B radar set is/are used only with beam-riding missiles?

1. J-band illumination transmitter
2. Guidance transmitter
3. Capture transmitter
4. Both 2 and 3 above

8-15. Which of the following transmitters of the AN/SPS-55B radar set is/are used only with homing missiles?

1. Capture transmitter
2. Guidance transmitter
3. Both 1 and 2 above
4. CW illuminating transmitter

8-16. The radar set receives target designations from the WDS during automatic operation from which of the following units?

1. The target designation transmitter
2. The fire control computer
3. The air search radar
4. The code comparator

8-17. How many beams of RF energy does the CW illumination transmitter provide?

1. One
2. Two
3. Three
4. Four

8-18. What type of computer is the Mk 119?

1. A solid-state digital
2. An electromechanical analog
3. A transistorized analog
4. A mechanical analog

8-19. At what point during operation does the computer switch to the track mode?

1. When a designate alert signal is received
2. When the fire control radar locks on the target
3. When an air ready mode signal is received
4. When the launcher synchronizes to the remote order

8-20. Which of the following missiles is used during the shore bombardment mode?

1. BTN
2. HT
3. HTR
4. SM/ER

8-21. The surface-to-surface mode can be used with either a BTN or a homing missile.

8-22. In the surface homing mode, the director tracks the surface target in ___A___ bearing only, bearing and range, and elevation is ___B___ fixed, programmed by the computer.

1. A - bearing only,
   B - fixed
2. A - bearing and range,
   B - fixed
3. A - bearing only
   B - programmed by the computer
4. A - bearing and range
   B - programmed by the computer

8-23. Which of the following factors is/are combined with the future target position coordinates to obtain aiming position coordinates for the launcher?

1. The missile roll order
2. The seeker head order
3. The ballistic corrections
4. The Doppler predict order

8-24. Which of the following missiles is/are neither guided nor controlled during their boost phase?

1. BTN
2. HTR
3. SM-1
4. All of the above

8-25. It is necessary to compensate for power modulation of the capture beam and beam ratio when supplying ballistic corrections for the SM-1 missile.

In items 8-26 through 8-30, select from column B the functions that match the missile orders listed in column A.

A. Missile Orders

8-26. Variable navigation ratio
8-27. Doppler predict
8-28. Gravity bias
8-29. Sweep select
8-30. Clutter reject band

B. Functions

1. Compensates for the effects of gravity on the missile
2. Regulates the gain of the missile steering system servos
3. Determines position of the start of the speedgate sweep
4. Prevents Doppler search in the clutter frequencies

8-31. Which of the following missiles use(s) the sweep select order to perform basically the same function as the Doppler predict order in the SM-1 missile?

1. Terrier BTN
2. Terrier HTR
3. Tartar ITR
4. Both 2 and 3 above

8-32. What SM-1 or HTR missile component stores vertical reference information prior to launch?

1. The seeker head
2. The gyro system
3. The steering system servo
4. The speedgate

8-33. During what flight phase of an SM-1 missile is the N order used?

1. Boost
2. Separation
3. Midcourse
4. Homing

8-34. What preflight BTN missile order determines when the warhead will detonate?

1. The seeker head
2. The clutter band reject
3. The fuze delay time
4. The sweep select

Learning Objective: Identify the operating procedures and functions for Mk 76 Mods 6, 7, and 8 guided missile fire control systems and the Mk 11 NTDS/WDS from target detection to target destruction.

8-35. What change in the digital conversion to the Mk 76 GMFCS is common to Mods 6, 7, and 8?

1. The installation of a CWAT module in the AN/SPG-55B radar
2. The installation of a track module in the AN/SPG-55B radar
3. The removal of the Mk 119 computer and the installation of the Mk 152 computer and associated equipment
4. The installation of both the CWAT and track modules in the AN/SPG-55B radar

8-36. Which of the following equipment is NOT considered part of the Mk 76 digital guided missile fire control system?

1. The AN/SPG-50 radar set
2. The Mk 77 input/output console
3. The Mk 19 digital data recorder
4. The Mk 75 signal data converter

8-37. Which of the following Mk 76 guided missile fire control systems is/are digital?

1. Mod 5
2. Mod 6
3. Mod 7
4. Both 2 and 3 above

8-38. How many position corrections must the air tracker insert before the computer will establish a firm track?

1. One
2. Two
3. Three
4. Four

8-39. Which of the following data is/are NOT considered valid when establishing the identification of an unknown target?

1. Intelligence reports
2. Current operational orders
3. Orders from the officer in tactical command (OTC)
4. Information obtained from the target designation transmitter

8-40. With whom does the responsibility for overall coordination of the ship's combat system rest?

1. The OTC
2. The SWC
3. The FCSC
4. The RSC

8-41. Special tracking increases GMFCS acquisition time.

8-42. At which of the following positions is the assignment of an air track to a fire control system accomplished?

1. SWC
2. OTC
3. EC
4. FCSC

8-43. At which of the following positions is the final threat and engageability evaluation of targets designated for missile engagement accomplished?

1. EC
2. SWC
3. OTC
4. FCSC

8-44. What position is responsible for reporting missile inventory to the NTDS/WDS operation program?

1. SWC
2. FCSC
3. OTC
4. EC

8-45. At which of the following positions are the launcher clearance lines displayed?

1. FCSC
2. EC
3. SWC
4. SPL TRK

8-46. Target kill information is transmitted from which of the following locations?

1. The EC position
2. The triconsole
3. The FC computer
4. The NTDS computer

---

Learning Objective: Point out the responsibilities for alignment, and identify the sources of information relating to the alignment of missile systems.

---

8-47. Who are responsible for performing the original alignment of all components of a guided missile launching system?

1. GMMs only
2. FTMs only
3. Shipbuilders
4. GMMs and FTMs

8-48. What Navy publication sets forth the procedures for battery alignment for each class of ship?

1. OP 0
2. SW 225/OP 2456
3. OD 3000
4. OD 17425

8-49. What report should be used as the initial document for the alignment smooth log?

1. Shipyard alignment report
2. COMSURFPACINST C8000.2
3. COMSURFLANTINST C8000.1
4. Both 2 and 3 above

Learning Objective: Identify the basic shipyard and shipboard requirements for alignment of a missile weapons system.

8-50. Under which of the following conditions are the final system alignments accomplished?

1. In drydock, 30% loaded
2. Afloat, 30% loaded
3. In drydock, 80% loaded
4. Afloat, 80% loaded

8-51. Which of the following geometric references is/are contained in the reference frame?

1. Reference plane
2. Reference point
3. Reference direction
4. All of the above

8-52. During the construction of a ship, how many base plate(s) is/are installed?

1. Four
2. Three
3. Two
4. One

8-53. What reference is used when machining the foundations for the launchers?

1. The vertical plane
2. The weapons control reference plane
3. The base plane
4. The ship centerline plane

8-54. The weapons control reference plane is parallel to the ship's centerline plane and perpendicular to the ship's base plane.

8-55. After the bench-mark reading is established and recorded, it is used for future battery alignment until

1. the battery guns are fired
2. new waterborne data is obtained
3. one of the elements is disassembled
4. one of the elements is removed for overhaul or replacement

8-56. When must new tram readings be taken and recorded?

1. When new drydock data is obtained
2. Every day as part of your daily checkoff
3. After every firing exercise
4. After making any alignment corrections to mount or launcher and director

8-57. Most of the rotating elements on the DDG-2 class ships are aligned to the Mk 68 director. Which of the following rotating elements on these ships is/are NOT aligned to this director?

1. Missile launcher and the gyrocompass
2. Mk 73 missile fire control system in train
3. Mk 73 missile fire control system in elevation
4. Mk 74 missile fire control system in train

8-58. You should correct an existing synchro and dial error in a mount before performing any battery alignment on a ship afloat.

Learning Objective: Describe the procedures for aligning a launcher in train and elevation.

8-59. What is the best target to use for an accurate battery alignment on a ship afloat?

1. A distant ship or clearly visible landmark
2. The horizon
3. A single star or planet
4. A constellation

8-60. What is the purpose of the T-lug that is installed in the front guide of the Mk 8 launcher during battery alignment?

1. To act as a mount for the peepsight
2. To aid in the alignment of the boresight telescope
3. To prevent the front guide from tripping
4. All of the above

8-61. What simple method may you use to isolate an existing misalignment in the launcher or the fire control system?

1. Compare dial reading of the launcher and the director
2. Take tram readings from the launcher and the director and compare them
3. Hold a bench-mark check on both the launcher and the director
4. Operate the launcher to its fixed mechanical positions and check for correct response

8-62. In addition to having the launcher deenergized for alignment or tests, what must be done with the safety switch?

1. It must be placed in SAFE TO OPERATE position
2. It must be placed in SAFE position only
3. It must be placed in SAFE position and handle removed
4. It must be placed in SAFE TO OPERATE position and handle removed

8-63. In the performance of an elevation horizon check, the angle to which the launcher's line of sight must be depressed below its horizontal position to place its line of sight on the horizon is called the launcher's

1. parallax
2. dip difference
3. dip angle
4. elevation angle

8-64. At which of the following positions in train should readings be taken for the elevation horizon check for a launcher?

1. 090° and 270°
2. 045° and 315°
3. Every 45° of bearing throughout the full arc of launcher train
4. Every 15° of bearing throughout the full arc of launcher train

8-65. Refer to table 8-2 of your textbook and assume that the dip angle of the reference director on your ship is 16 feet above the waterline and the launcher is 5 feet above the waterline. What is the dip difference between the launcher and the director?

1. 11'00"
2. 2'46"
3. 2'41"
4. 1'44"

8-66. Which of the following readings in minutes is equal to zero degrees?

1. 2000
2. 1000
3. 200
4. 0

---

Learning Objective: Explain the requirements for checking and aligning firing cutouts and nonpointing zone systems.

---

8-67. Which of the following components do(es) NOT limit free movement of the launcher?

1. The firing cutout system
2. The train limit-stop system
3. The elevation limit-stop system
4. Both 2 and 3 above

8-68. Which of the following activities is responsible for computing and cutting firing cutout cams?

1. NWC, Seal Beach
2. NAD, Crane
3. NSWC
4. NSWSES

8-69. What is the maximum amount a firing interrupter cam may be out of tolerance with the actual launcher dial setting before the cam must be reset?

1. ±3 seconds
2. ±3 minutes
3. ±1°
4. ±3°

---

Learning Objective: List the final operational checks and tests that are performed after conducting system alignment.

---

8-70. What is the miss distance of a guided missile if the pointing error is 2 mils and the range is 40,000 yards?

1. 20 yards
2. 40 yards
3. 80 yards
4. 120 yards

8-71. If alignment adjustments are made to B-end response in the receiver-regulators, readjustment of which of the following position synchros is necessary?

1. Load
2. Stow
3. Dud jettison
4. All of the above

8-72. In the final alignment operation check, which of the following small errors are common when training right and left?

1. The errors are approximately equal in amplitude and in the same direction
2. The errors are approximately equal in amplitude and in the opposite direction
3. The errors are unequal in amplitude and in the same direction
4. The errors are unequal in amplitude and in the opposite direction

# Assignment 9

Repair and Test

Textbook, NAVEDTRA 10200-D: Pages 279 through 298

---

Learning Objective: Point out the technical, administrative, and maintenance responsibilities of a GMM in the Navy maintenance program.

9-1. Which of the following shipboard maintenance tasks is/are classified as operational maintenance?

1. Replacing unserviceable parts in the equipment
2. Cleaning, lubricating, and preserving the equipment
3. Testing, aligning, and adjusting the equipment
4. All of the above

9-2. Except in an emergency situation, the complete overhaul or rebuilding of a unit or system is accomplished by what type of maintenance?

1. Technical
2. Operational
3. Organizational
4. Intermediate

9-3. Which of the following OPNAV instructions provides the general guidelines for implementing PMS?

1. 5510.19
2. 5100.19
3. 4790.4
4. 3120.32

9-4. Who is responsible for testing, adjusting, overhauling, and repairing the receiver-regulator equipment in the power drive system?

1. The GMM3
2. The GMM2
3. The GMM1
4. The GMMC

9-5. Preventive maintenance involves how many major types of activities?

1. Two
2. Three
3. Four
4. Five

9-6. A visual inspection can be very useful in locating troubles induced by vibrations of the ship.

9-7. Which of the following is a recommended preventive maintenance rule on ordnance equipment?

1. Dismantle the equipment at frequent intervals and inspect for worn parts
2. Make daily adjustments to the equipment
3. Keep the equipment clean and dirt-free at all times
4. Do not operate the equipment unless it is absolutely necessary

9-8. Before you use any solvent to clean a piece of equipment, you must ensure that you have the correct type of solvent by checking the

1. specifications of the solvent
2. equipment OP
3. BUMED instruction on toxic solvents
4. OD 3000

9-9. Which of the following is/are undesirable effects of cleaning solvents?

1. Fumes which can cause ill health or death
2. Fumes which can cause fires or explosions
3. Damage to certain electrical parts
4. All of the above

9-10. Which of the following precautions should you take to prevent the entrance of unnecessary grime into bearings?

1. Keep the lubricating tool clean at all times
2. Clean the unit at the point of application of the lubricant
3. Clean the outside of all units before opening them
4. All of the above

---

Learning Objective: Point out some of the daily, weekly, monthly, quarterly, and periodic tests and checks that are performed on a Mk 13 GMLS.

---

9-11. Which of the following tests is/are conducted daily on the Tartar GMLS?

1. Emergency firing
2. Dud firing
3. Normal firing and misfire
4. All of the above

9-12. Which of the following information is shown on a lubrication chart?

1. The frequency of lubrication
2. The type of lubrication
3. The points requiring lubrication
4. All of the above

9-13. Temperature is one factor that you must always take under consideration in checking the nitrogen pressure of an accumulator unit.

9-14. Before checking the accumulator pressure of a hydraulic unit that has been in operation, you should wait until it has been shut down at least

1. 1 hour
2. 2 hours
3. 4 hours
4. 6 hours

9-15. How often are missile magazine sprinkling systems tested?

1. Daily
2. Weekly
3. Monthly
4. Semiannually

9-16. Which of the following precautions is UNNECESSARY prior to testing the $CO_2$ system?

1. Disconnect the control heads
2. Disconnect the discharge heads
3. Disconnect the main magazine sprinkling valve
4. Cap the connections of the control and discharge heads

9-17. Which of the following equipment is/are required to check the $CO_2$ system control heads?

1. A spanner wrench
2. A hand pump
3. An air gage
4. Both 2 and 3 above

9-18. How often should bladder pressure of the Mk 13 GMLS's anti-icing system be checked?

1. Quarterly
2. Monthly
3. Weekly
4. Daily

9-19. Assume that you are performing time checks on the operation of a component of the Mk 13 GMLS. You have repeated the timing operation three times. How do you determine if the component is operating within the prescribed time limits?

1. By comparing the first and last readings with the prescribed time
2. By comparing the second and third readings with the prescribed time
3. By comparing the second reading only with the prescribed time
4. By comparing the average of all three readings with the prescribed time

9-20. Which of the following tests is NOT valid when performing timing tests on the Mk 13 GMLS?

1. Indexing time between stations on the outer ring
2. Indexing time between stations on the inner ring
3. Initial light-off time
4. Extending time of the retractable rail

9-21. Which of the following timing tests is NOT accomplished by using the electric timer on the Mk 13 GMLS?

1. Extend and retract time of the fin opener cranks
2. Extend and retract time of the electrical contactor
3. Retract time of the jettison piston
4. Arm and disarm time of the arming tool

---

Learning Objective: Describe the operational characteristics of the test equipment used in launcher performance tests.

---

9-22. The test panel on most shipboard launcher systems is the

1. EP1
2. EP2
3. EP3
4. EP4

9-23. What type of test are you performing on the launcher when you use a dummy director with an error recorder?

1. Dynamic
2. Static
3. Closed-loop
4. Stability

9-24. After completion of the launcher's performance tests at installation, where are the records maintained?

1. At NSWSES, Port Hueneme, California
2. On-board ship
3. At NAD, Crane, Indiana
4. At MMFO, Atlantic or Pacific

9-25. What device is used in the Mk 6 Mod 0 dummy director to generate a velocity test signal?

1. A 23TX6A synchro transmitter
2. A 18CT6A synchro control transformer
3. A d.c. tachometer
4. A transistor and potentiometer control circuit

9-26. Which of the following modes of signal generation can be performed with the Mk 6 Mod 0 dummy director?

1. Frequency generation
2. Constant velocity
3. Simple harmonic motion
4. All of the above

9-27. Which of the following GMLS error traces may be taken with the dual channel oscillograph?

1. B-end
2. Velocity
3. Position
4. All of the above

9-28. The signal for the B-end error trace is taken from the rotor of the 36-speed CT in the receiver-regulator. What gearing drives the rotor of this CT?

1. The A-end response
2. The B-end response
3. The position-plus-lead
4. The limit stop

9-29. Which of the following receiver-regulator components provides the velocity trace voltages?

1. The tachometer generator
2. The 36-speed CX
3. The 36-speed CT
4. The 1-speed CX

9-30. Which of the following factors should be considered when reading test traces?

1. The type of test
2. The test conditions
3. The type of trace
4. All of the above

9-31. The position trace voltages are provided by which of the following components in the receiver-regulator?

1. The 1-speed CT
2. The integration potentiometer
3. The 36-speed CT
4. The tachometer generator

9-32. How many degrees will a 36-speed synchro rotate when the launcher moves 1°?

1. 36
2. 24
3. 12
4. 1

9-33. How many degrees of actual launcher movement is required to make a 36-speed synchro rotate one revolution?

1. 100
2. 36
3. 10
4. 1

9-34. If you are reading a position trace that consists of 7-1/2 cycles, how far has the launcher actually moved?

1. 75°
2. 15°
3. 7.5°
4. .75°

9-35. Prior to performing launcher tests, which of the following checkoffs should be performed?

1. Check oil level of the main supply tank
2. Lubricate the launcher components
3. Charge the launcher accumulators
4. All of the above

9-36. How are the maximum limits of the launcher guide arms checked for free and unrestricted movements?

1. By lighting off the guide arm power drive and exercising in local
2. By lighting off the train and elevation power drives and going to the dud-jettison position
3. By using the air drive motors
4. By checking the results of the elevation constant velocity trace

9-37. How many minutes should the train and elevation power drives be warmed up prior to running test traces?

1. 60
2. 30
3. 15
4. 5

9-38. Which of the following traces are calibrated in the same manner?

1. Error and position
2. Error and velocity
3. Position and velocity
4. All of the above

9-39. Unless specified otherwise, how many minutes is the minimum warmup time for test equipment?

1. 30
2. 15
3. 10
4. 5

9-40. Which of the following tests is used only on the train power drive?

1. 25 degrees-per-second constant velocity
2. Simple harmonic motor
3. Static
4. 5-degrees-per-second constant velocity

9-41. How often should you conduct the launcher accuracy test?

1. Weekly
2. Monthly
3. Quarterly
4. Semiannually

9-42. What type of recording is made by the Brush Instrument Company recorder, Mark III, of the Mk 12 Mods 0 and 1 error recorder?

1. A permanent and immediately visible paper chart record
2. A photographic film, negative chart which requires developing and positive copy
3. A photographic paper chart which requires developing
4. An ultraviolet sensitive photographic paper that is self-developing when exposed to artificial light

9-43. When a motion signal is applied to the launcher and the pen of the Mk 12 Mods 0 and 1 error recorder is deflected 35 millimeters, what is the input voltage to the error recorder?

1. 10 millivolts
2. 35 millivolts
3. 350 millivolts
4. 3500 millivolts

9-44. When calibrating the error recorder to zero position, at what position should the launcher be in elevation?

1. 45-degree
2. Zero-degree
3. 10-degree
4. 90-degree

9-45. Which of the following functions can the limiter-demodulator perform?

1. A.c. modulation with 60-hertz ripple frequency
2. A.c. amplification with demodulation and filtering
3. D.c. amplification
4. Both 2 and 3 above

9-46. What do superimposed pips on the position test trace indicate?

1. Bad gearing
2. Misaligned synchro receiver
3. Switch actuation
4. Erratic tachometer generator

9-47. Which of the following components is NOT contained in a limiter and demodulator unit?

1. A modulator
2. An a.c. amplifier
3. An attenuator
4. A d.c. amplifier

9-48. The model G limiter and demodulator is capable of how many basic modes of operation?

1. One
2. Two
3. Three
4. Four

---

Learning Objective: Identify the function, operation, and maintenance of the missile simulators used in GMLSs.

---

9-49. The SM-161/DSM missile simulator that is used to check the Tartar launching system is physically located in the

1. upper section of the EP3 control panel
2. upper section of the EP2 control panel
3. training missile (TSAM)
4. lower section of the EP1 power panel

9-50. What maintenance is required for a simulator which is operating satisfactorily?

1. Open it daily for cleaning and inspection
2. Lubricate the internal working parts daily
3. Periodically clean the air filter
4. All of the above

Learning Objective: Recognize the functional characteristics of the training missile used in personnel training and checking out shipboard launching systems.

9-51. How many Terrier and ASROC training rounds are carried in the Mk 10 Mod 7 GMLS?

1. One and two, respectively
2. One and one, respectively
3. Three and one, respectively
4. Two and one, respectively

9-52. What publication describes the SM-159D/DSM simulator for Terrier missiles used in the Mk 10 GMLS?

1. OP 4093
2. OP 2665
3. OP 2905
4. OP 2831

9-53. Which of the following voltages used by the simulator in the checkout of a Terrier Mk 10 GMLS is/are NOT fed into the simulator through the launcher electrical contactor?

1. Warmup
2. Booster squib
3. Missile identification
4. Changeover

Learning Objective: Identify the functions of the slow run-through (SRT) test.

9-54. What activity determines the detailed requirements for the operational test that equipment and systems must meet during a slow run-through (SRT)?

1. NAVMAT
2. NAVSEASYSCOM
3. The shipbuilder or naval shipyard
4. The equipment or systems builder

9-55. Deficiencies in the equipment or systems that are revealed during the SRT must be corrected by the

1. ship's force
2. installing activity
3. equipment builder
4. NAVSEASYSCOM

9-56. What is the starting phase of the SRT for a new missile launching system installation?

1. Ramming a missile on the launcher
2. Missile system test
3. Missile replenishment-at-sea
4. Strikedown of a missile

9-57. Which of the following auxiliary subsystems aboard ship is/are tested during the SRT?

1. Air-conditioning system
2. Sprinkling system
3. Security alarms
4. All of the above

# Assignment 10

Administration and Supply

Textbook, NAVEDTRA 10200-D: Pages 299 through 322

---

Learning Objective: Point out the basic principles, purposes, functions, and practices in administrating the Navy Maintenance and Material Management (3-M) Systems.

10-1. The primary objective of the Ships' 3-M Systems is to provide

1. the minimum maintenance requirement
2. established methods, procedures, and schedules for corrective maintenance and major repairs
3. for managing maintenance and maintenance support in a manner which will ensure maximum equipment operational readiness
4. standardized procedures for corrective maintenance

10-2. What system(s) aboard Navy ships is/are the nucleus for managing maintenance?

1. PMS
2. 3-M
3. MDS
4. IMMS

10-3. PMS procedures are designed to prevent future equipment failure.

10-4. Some MRCs have equipment guide lists (EGLs) which serve as location guides for

1. parts requiring replacement
2. identical equipment
3. technical standards
4. corrective maintenance action

10-5. PMS documentation is developed in accordance with which of the following references?

1. OPNAVINST 4790.4
2. MIL-P-24534 (Navy)
3. OPNAVINST 3120.32
4. NAVSEAINST 4570.1

10-6. Who uses the master departmental PMS record as a tool to schedule maintenance on the PMS schedule forms?

1. The division officer
2. The work center supervisor
3. The department head
4. Maintenance personnel

10-7. Which of the following schedules is prepared from the cycle schedule?

1. Monthly/semiannual
2. Semiannual/annual
3. Weekly/monthly
4. Quarterly/weekly

10-8. By which of the following means do fleet maintenance personnel request PMS software?

1. OPNAV Form 4790/7B
2. OPNAV Form 4790/2K
3. OPNAV Form 4790/2L
4. OPNAV Form 4790/2Q

10-9. Which of the following systems is used to report corrective maintenance actions on specific categories of equipment by maintenance personnel?

1. MDS
2. PMS
3. IMMS
4. SFOMS

10-10. What is designed for receipt and distribution of maintenance and material data?

1. IMMS data bank
2. CSMP data bank
3. PMS data bank
4. MSOD data bank

10-11. Prior to overhaul and availability periods, automated work requests are generated by

1. MSOD
2. NAVMAT
3. ADP facility
4. TYCOM

10-12. Prior to an inspection by the Board of Inspection and Survey, all deferrals listed in the CSMP file are made up of automated INSERV items. Who assigns the priority numbers to these items?

1. TYCOM
2. INSERV board
3. COMNAVSEASYSCOM
4. COMNAVMATCOM

10-13. What system is used to manage planning, scheduling, production, and monitoring of maintenance workloads of tended ships?

1. MDS
2. IMMS
3. SFOMS
4. PMS

---

Learning Objective: Point out the purpose and fundamentals of inventories, including SECAS and SAIL.

---

10-14. NAVSEA's equipment accounting system has been replaced by

1. CSMP
2. SAIL
3. SECAS
4. ORDLIS

10-15. All commands concerned with armament configuration are furnished a listing of installed shipboard ordnance equipment by what reporting system?

1. MDS
2. ORDLIS
3. SECAS
4. SAIL

10-16. Your ship is scheduled for a yard overhaul in December. In what month should the corrected copy of your SAIL, marked "Prior to Overhaul Report" arrive at the appropriate NAVSEACEN?

1. November
2. August
3. May
4. January

10-17. You have accomplished an ORDALT on your Mk 13 GMLS, and it is a year before your ship is scheduled for the next yard overhaul period. On what form should the report of this ORDALT accomplishment be made?

1. NAVSEA Form 8000/1
2. Marked up copy of SAIL
3. NAVORD Form 8000/2
4. NAVSEA Form 4720/3

Learning Objective: Point out the principles, practices, and procedures for requesting technical assistance.

10-18. When requesting technical assistance by message or telephone from a MOTU, what work request form must follow?

1. 1348/1
2. 4790/1L
3. 4790/2K
4. 1348/2

10-19. Contractor service engineers are specially trained personnel who handle specific equipment.

10-20. During a yard overhaul you request technical assistance. Your request will be directed to

1. TYCOM
2. a NAVSEACEN
3. a MOTU
4. NAVSEASYSCOM

Learning Objective: Describe the purpose of a ship qualification trial (SQT) and the training afforded ship's SMS personnel by the professionals of the ship qualification assistance team (SQAT).

10-21. What is the primary purpose of the ship qualification assistance team (SQAT)?

1. To check out and test the surface missile underway replenishment system
2. To assist ship's company in getting underway
3. To demonstrate weapons system performance as operated and maintained by ship's personnel
4. To train ship's force in the collection of technical information

10-22. Who is responsible for scheduling ship qualification trials (SQTs)?

1. Ship's type commander
2. Ship's weapons officer
3. Ship's gunner
4. Ship's surface missile system chief petty officer

Learning Objective: Outline the responsibility of the Defense Logistics Agency and the sources of supply in the Navy.

10-23. What is the function of the Defense Logistics Agency (DLA)?

1. To consolidate accounting procedures in all the services
2. To consolidate logistic operations in all the services
3. To consolidate information from communication activities in all the services
4. To consolidate budget estimates in all the services

10-24. The DLA publications and directives reflect the policy of the

1. Chief of Naval Operations
2. Secretary of Defense
3. Supply department officer
4. Secretary of the Navy

10-25. Which of the following activities is responsible for regulating the flow of ordnance material?

1. Guided missile support facilities
2. Defense Logistics Agency
3. Navy Ships Parts Control Center
4. Navy distribution centers

10-26. What command is responsible for establishing guided missile support facilities?

1. NAVSEA
2. NAVMAT
3. NAVSUP
4. OPNAV

10-27. What is the primary distribution point for guided missile material on the West Coast?

1. NAD, Seal Beach, Calif.
2. NAD, Mare Island, Calif.
3. NSC, Oakland, Calif.
4. NAD, Bangor, Wash.

Learning Objective: Indicate the purpose, use, and proper maintenance of allowance lists and the breakdown of the COSAL with respect to its contents.

10-28. The coordinated shipboard allowance list (COSAL) is prepared for each individual ship according to the requirements of that particular ship.

10-29. The part of the COSAL that covers electronic test equipment is prepared by the

1. Ships Parts Control Center (SPCC)
2. Naval Ship Engineering Center (NAVSEC)
3. Naval Ship Weapon Systems Engineering Station (NSWSES)
4. Aviation Supply Office (ASO)

10-30. In what part of the COSAL will you find a numerical listing of all APL and AEL identification numbers?

1. I
2. II
3. III
4. IV

10-31. What part of the COSAL contains the alternate number cross-reference to stock numbers?

1. I
2. II
3. III
4. IV

10-32. The ordnance section(s) of the COSAL is/are made up of an introduction and how many parts?

1. Four
2. Three
3. Two
4. One

10-33. Which of the following information is provided in your ship's COSAL?

1. The NSNs for most of the items allotted your ship
2. The OP numbers for your GMLS
3. Stock numbers for your tactical missiles
4. All the above

10-34. Minor revisions to a ship's COSAL are made by pen-and-ink changes.

10-35. You should use NAVSUP Form 1250 or DD Form 1348 to requisition supplies and repair parts from

1. the Ships Parts Control Center
2. a naval shipyard
3. own ship's storeroom
4. a shore-based supply depot

10-36. Why should a file that includes requisition numbers of all requests made to the supply department be maintained in the weapons office?

1. To permit ready access to NSNs of repair parts on board ship
2. To permit determining the exact price paid for an ordered part
3. To permit tracing material delayed in delivery
4. To permit determining the cognizance symbol of the requested part

10-37. Which of the following forms, in addition to a DD Form 1348, should you prepare to requisition a repair part whose NSN is not available?

1. DD 1149
2. DD 1348-6
3. NAVSUP 1250
4. NAVSUP 1250-1

10-38. Assume that a piece of equipment on your ship has been subjected to unusual activity, with the result that the supply of a particular component that is used in the equipment has been exhausted. To order additional quantities of that component, the supply department must prepare the in-excess requisition while your department must prepare a

1. security clearance for the person designated to draw material
2. proposed change to the ship's COSAL
3. letter to CNO
4. written justification of its need

10-39. A major difference between expendable and nonexpendable items is that nonexpendable items are

1. disposed of according to applicable instructions
2. replaced more often than expendable items
3. returned to the supply system when a replacement is requisitioned
4. classified as nonrepairable

---

Learning Objective: List the sources of item identification for ordnance equipment, the sources of stock numbers for publications, forms, and instructions, and the availability of other supply references.

---

10-40. When requisitioning a major component for a Tartar missile launcher, what cognizance symbol is used?

1. 1K
2. 2J
3. 4T
4. 6I

10-41. Which of the following publications shows the relationship of assemblies to one another in a piece of ordnance equipment?

1. Illustrated parts breakdown (IPB)
2. Coordinated shipboard allowance list (COSAL)
3. Federal Supply Catalog (FSC)
4. Afloat Shopping Guide (ASG)

10-42. What reference provides updated stock and reference numbers?

1. Allowance parts list (APL)
2. Federal Supply Catalog (FSC)
3. Coordinated shipboard allowance list (COSAL)
4. Navy Management Data List (NMDL)

10-43. The Federal Supply Code for Manufacturers (FSCM) lists the names of commercial firms manufacturing material for the Department of Defense (DOD) and also identifies each manufacturer by a five-digit code number.

10-44. Training publications, standardized forms, and other printed material of general usage are included under the cognizance symbol

1. Z
2. J
3. I
4. A

10-45. Unless otherwise indicated, changes to OPs should normally be ordered by MILSTRIP form from

1. U.S. Naval Supply Center, Norfolk, Va.
2. Naval Publications and Forms Center, Philadelphia, Pa.
3. Navy Ships Parts Control Center, Mechanicsburg, Pa.
4. U.S. Naval Supply Center, Oakland, Calif.

Learning Objective: Describe the use of weapons department reports, records, and inspections to improve, correct, and evaluate SMS equipment.

10-46. To determine what reports are required by OPNAV and NAVSEASYSCOM, you should consult which of the following instructions?

1. OPNAVINST 5214.1
2. AOINST 5213.2
3. NAVSEAINST 8300.1
4. NAVSEAINST 5214.1

10-47. Detailed information for preparing and submitting a special report normally is obtained from which of the following sources?

1. OPNAVINST 5214.1C
2. NAVORDINST 8000.1
3. AOINST 5213.2
4. The directive issued to direct you to make the report

10-48. Instructions for surface-launched missile firing reports are contained in what instruction?

1. NAVSEAINST 8810.3
2. NAVSEAINST 8810.2
3. NAVSEAINST 8801.3
4. NAVSEAINST 8800.1

10-49. Which of the following NAVSEA firing report forms will be filled out on board a DDG equipped to fire medium range missiles?

1. 8810/1C
2. 8810/1D
3. 8810/1F
4. 8810/1G

10-50. The FXP-3, Ship Exercises Publication, is a valuable aid when you are preparing for

1. weapons exercises
2. physical fitness tests
3. DSOTs
4. system documentation

10-51. Procedures for preparing and completing NAVORD Form 8810/2 are contained in what NAVORD instruction?

1. 8810.2A
2. 8810.3
3. 8810.6B
4. 8830.1

10-52. The commanding officer's narrative report (CONAR), which is the CO's overall appraisal of the weapon system, includes which of the following comments?

1. The quality of the training of personnel
2. Technical assistance requested and received
3. Recommendations for improvement
4. All the above

10-53. The guide for preparing the CONAR is contained in NAVSEAINST 8000.1.

10-54. How does the CONAR help GMMs solve maintenance problems?

1. Through exchange of information among missile launching system personnel
2. Through information received from shore activities
3. Through PMS feedback test designs
4. All of the above

10-55. What does the deficiency corrective action program (DCAP) system provide?

1. Channels for the expeditious flow of problem information
2. Continually updated information on the status of in-service engineering problems and followup action
3. It keeps all levels of management informed through a series of reports
4. All the above

> Learning Objective: Point out the importance of the Guided Missile Service Report and ASROC reports as administrative aids to the operating fleets equipped with surface missile systems.

An up-to-date Guided Missile Service Report, form NAVSEA 4855/1 contains the complete history of a missile aboard ship. All testing, handling, servicing, and other reportable operations are recorded on it in duplicate. The original report is forwarded to Officer in Charge, Fleet Analysis Center (Code 845), Naval Weapons Station, Seal Beach, Corona Annex, Corona, Calif. 91720, while one copy is retained aboard ship.

10-56. Which of the following reports is NOT inserted in the missile log?

1. Handling
2. Testing
3. Servicing
4. Batch

10-57. The Guided Missile Service Report, form NAVSEA 4855/1, is prepared in accordance with NAVSEAINST 5511.28.

10-58. In which of the following ways is an ASROC identified on NAVSEA form 8830/1?

1. By mark and mod
2. By shape and size
3. By assembly identification number (AIN)
4. By color code

> Learning Objective: Outline the rigid requirements of security for nuclear reports and inspections under the control of the Defense Nuclear Agency (DNA).

10-59. Which of the following publications takes precedence over all other technical publications where conflicting information is present regarding nuclear weapons?

1. Navy Security Manual
2. Defense Nuclear Agency (DNA) Personnel Training Format
3. Navy special weapons ordnance publications (SWOPs)
4. Nuclear Weapons Management Manual

10-60. Which of the following SWOPs is an index to publications used for multiple service purpose?

1. 0-1
2. 0.1B
3. 30.19
4. 45.21

10-61. Which of the following technical inspections is performed once a year by a Navy inspection team from a nuclear weapons training center?

1. TSI
2. TPI
3. NTPI
4. NWAI

10-62. What information is attached to the chief inspector's report?

1. All inspector's comments, recommendations, and discrepancies noted
2. All nuclear units operating in the fleet today
3. All action required for comprehensive inspection
4. All matters directly related to weapons

10-63. In accordance with NTPI guides, all matters directly related to the processing, handling, testing, inspecting, maintaining, and storing of nuclear weapons, and all matters and procedures involved in the administration of a nuclear unit are NOT included in every NTPI.

10-64. In accordance with the Nuclear Weapons Management Manual, COMNAVSURFLANT and PACINST C8120.1 series, it is NOT appropriate to have weight test inspection tags for each piece of handling equipment.

10-65. Upon the completion of the NTPI, who makes a report directly to the Chief of Naval Operations?

1. COMNAVSEASYSCOM
2. COMNAVMAT
3. NTPI chief inspector
4. TYCOM

10-66. A prospective nuclear weapons activity undergoes a nuclear weapons acceptance inspection (NWAI). What does this inspection determine?

1. The activity's readiness to perform technical, administrative, and logistical procedures relating to its nuclear weapons capabilities
2. The activity's readiness for a TSI
3. The status of the activity's training
4. The efficiency of the activity's training program and team organizations

10-67. Before a Navy activity can be accepted as a fully operational nuclear weapons activity, it must pass its NWAI with a grade of at least

1. outstanding
2. excellent
3. good
4. satisfactory

10-68. In conducting a technical standardization inspection (TSI), on what criteria will personnel from Field Command, DNA, judge the activity being inspected?

1. On the ability of its personnel to assemble a nuclear warhead into a missile
2. On its capability of handling and maintaining missiles with nuclear warheads
3. On the status of its records, reports, publications, training programs, and team organizations
4. All the above

Learning Objective: Point out the importance of personnel injury reports.

10-69. Why must the weapons department submit written descriptions, photographs, and diagrams to supplement a report of a personnel injury caused by an explosive accident or incident?

1. To determine whether the cause was an accident or an incident
2. To determine a way to prevent the accident or incident from recurring
3. To determine the number of man-hours lost
4. To determine the type of disciplinary action to be taken

10-70. Supervisor's Report of Injury, NAVEXOS Form 180, is an administrative report about any type of injury sustained by civilian personnel.

Learning Objective: Describe the necessity for, and implementation of, security in the Navy.

10-71. Which of the following personnel are included in the security orientation, education, and training program of the GMLS?

1. Personnel in charge of system testing
2. Personnel in charge of maintenance and upkeep
3. All personnel entrusted with classified information regardless of rate or rank
4. All the above

10-72. Who is responsible for safeguarding classified material in a GMLS?

1. Security officer
2. Weapons officer
3. Operations officer
4. Personnel in the GMLS who handle the material

## COURSE DISENROLLMENT

All study materials must be returned. On disenrolling, fill out only the upper part of this page and attach it to the inside front cover of the textbook for this course. Mail your study materials to the Naval Education and Training Program Development Center.

PRINT CLEARLY

| NAVEDTRA NUMBER | COURSE TITLE |
|---|---|
| 10200-D | Gunner's Mate M 1&C |

| Name | Last | First | Middle |
|---|---|---|---|

| Rank/Rate | Designator | Social Security Number |
|---|---|---|

## COURSE COMPLETION

Letters of satisfactory completion are issued only to personnel whose courses are administered by the Naval Education and Training Program Development Center. On completing the course, fill out the lower part of this page and enclose it with your last set of answer sheets. Be sure mailing addresses are complete. Mail to the Naval Education and Training Program Development Center.

PRINT CLEARLY

| NAVEDTRA NUMBER | COURSE TITLE |
|---|---|
| 10200-D | Gunner's Mate M 1&C |

Name

ZIP CODE

MY SERVICE RECORD IS HELD BY:

Activity

Address                                                ZIP CODE

Signature of enrollee

A FINAL QUESTION: What did you think of this course? Of the text material used with the course? Comments and recommendations received from enrollees have been a major source of course improvement. You and your command are urged to submit your constructive criticisms and your recommendations. This tear-out form letter is provided for your convenience. Typewrite if possible, but legible handwriting is acceptable.

Date _____

From: _____
       (RANK, RATE, CIVILIAN)

_____

_____ ZIP CODE _____

To:   Naval Education and Training Program Development Center (PD3)
      Pensacola, Florida 32509

Subj: Gunner's Mate M 1&C, NAVEDTRA 10200-D

1.   The following comments are hereby submitted:

.......................(Fold along dotted line and staple or tape).......................

.......................(Fold along dotted line and staple or tape).......................

**DEPARTMENT OF THE NAVY**
NAVAL EDUCATION AND TRAINING PROGRAM
DEVELOPMENT CENTER (PD 3)
PENSACOLA, FLORIDA 32509

OFFICIAL BUSINESS
PENALTY FOR PRIVATE USE, $300

POSTAGE AND FEES PAID
NAVY DEPARTMENT
DoD-316

**NAVAL EDUCATION AND TRAINING PROGRAM DEVELOPMENT CENTER**
**BUILDING 2435** (PD 3)
**PENSACOLA, FLORIDA 32509**

**PRINT OR TYPE**

NRCC GUNNER'S MATE M 1&C
NAVEDTRA 10200-D

NAME _____ ADDRESS _____
     Last     First    Middle         Street/Ship/Unit/Division, etc.

                                       City or FPO    State              Zip
RANK/RATE _____ SOC. SEC. NO. _____ DESIGNATOR _____ ASSIGNMENT NO. _____

☐ USN   ☐ USNR   ☐ ACTIVE   ☐ INACTIVE   OTHER (Specify) _____ DATE MAILED _____

SCORE

**PRINT OR TYPE**

NRCC GUNNER'S MATE M 1&C
NAVEDTRA 10200-D

NAME _____  ADDRESS _____
       Last    First    Middle    Street/Ship/Unit/Division, etc.

                                  City or FPO    State         Zip
RANK/RATE _____ SOC. SEC. NO. _____ DESIGNATOR _____ ASSIGNMENT NO. _____

☐ USN  ☐ USNR  ☐ ACTIVE  ☐ INACTIVE  OTHER (Specify) _____  DATE MAILED _____

SCORE

**PRINT OR TYPE**

NRCC GUNNER'S MATE M 1&C
NAVEDTRA 10200-D

NAME _____ ADDRESS _____
         Last       First      Middle         Street/Ship/Unit/Division, etc.

                                               City or FPO      State                Zip

RANK/RATE _____ SOC. SEC. NO. _____ DESIGNATOR _____ ASSIGNMENT NO. _____

☐ USN   ☐ USNR   ☐ ACTIVE   ☐ INACTIVE   OTHER (Specify) _____ DATE MAILED _____

| | SCORE |
|---|---|

       1 2 3 4                      1 2 3 4                       1 2 3 4
       T F                          T F                           T F
1  ☐☐☐☐ _____          26 ☐☐☐☐ _____           51 ☐☐☐☐ _____
2  ☐☐☐☐ _____          27 ☐☐☐☐ _____           52 ☐☐☐☐ _____
3  ☐☐☐☐ _____          28 ☐☐☐☐ _____           53 ☐☐☐☐ _____
4  ☐☐☐☐ _____          29 ☐☐☐☐ _____           54 ☐☐☐☐ _____
5  ☐☐☐☐ _____          30 ☐☐☐☐ _____           55 ☐☐☐☐ _____
6  ☐☐☐☐ _____          31 ☐☐☐☐ _____           56 ☐☐☐☐ _____
7  ☐☐☐☐ _____          32 ☐☐☐☐ _____           57 ☐☐☐☐ _____
8  ☐☐☐☐ _____          33 ☐☐☐☐ _____           58 ☐☐☐☐ _____
9  ☐☐☐☐ _____          34 ☐☐☐☐ _____           59 ☐☐☐☐ _____
10 ☐☐☐☐ _____          35 ☐☐☐☐ _____           60 ☐☐☐☐ _____
11 ☐☐☐☐ _____          36 ☐☐☐☐ _____           61 ☐☐☐☐ _____
12 ☐☐☐☐ _____          37 ☐☐☐☐ _____           62 ☐☐☐☐ _____
13 ☐☐☐☐ _____          38 ☐☐☐☐ _____           63 ☐☐☐☐ _____
14 ☐☐☐☐ _____          39 ☐☐☐☐ _____           64 ☐☐☐☐ _____
15 ☐☐☐☐ _____          40 ☐☐☐☐ _____           65 ☐☐☐☐ _____
16 ☐☐☐☐ _____          41 ☐☐☐☐ _____           66 ☐☐☐☐ _____
17 ☐☐☐☐ _____          42 ☐☐☐☐ _____           67 ☐☐☐☐ _____
18 ☐☐☐☐ _____          43 ☐☐☐☐ _____           68 ☐☐☐☐ _____
19 ☐☐☐☐ _____          44 ☐☐☐☐ _____           69 ☐☐☐☐ _____
20 ☐☐☐☐ _____          45 ☐☐☐☐ _____           70 ☐☐☐☐ _____
21 ☐☐☐☐ _____          46 ☐☐☐☐ _____           71 ☐☐☐☐ _____
22 ☐☐☐☐ _____          47 ☐☐☐☐ _____           72 ☐☐☐☐ _____
23 ☐☐☐☐ _____          48 ☐☐☐☐ _____           73 ☐☐☐☐ _____
24 ☐☐☐☐ _____          49 ☐☐☐☐ _____           74 ☐☐☐☐ _____
25 ☐☐☐☐ _____          50 ☐☐☐☐ _____           75 ☐☐☐☐ _____

**PRINT OR TYPE**

NRCC GUNNER'S MATE M 1&C
NAVEDTRA 10200-D

NAME ______________________________ ADDRESS ______________________________
Last First Middle — Street/Ship/Unit/Division, etc.

______________________________
City or FPO State Zip

RANK/RATE ________ SOC. SEC. NO. ________________ DESIGNATOR ________ ASSIGNMENT NO. ________

☐ USN ☐ USNR ☐ ACTIVE ☐ INACTIVE OTHER (Specify) ________________ DATE MAILED ________

SCORE

| No. | 1 T | 2 F | 3 | 4 | | No. | 1 T | 2 F | 3 | 4 | | No. | 1 T | 2 F | 3 | 4 | |
|---|---|---|---|---|---|---|---|---|---|---|---|---|---|---|---|---|---|
| 1 | ☐ | ☐ | ☐ | ☐ | _____ | 26 | ☐ | ☐ | ☐ | ☐ | _____ | 51 | ☐ | ☐ | ☐ | ☐ | _____ |
| 2 | ☐ | ☐ | ☐ | ☐ | _____ | 27 | ☐ | ☐ | ☐ | ☐ | _____ | 52 | ☐ | ☐ | ☐ | ☐ | _____ |
| 3 | ☐ | ☐ | ☐ | ☐ | _____ | 28 | ☐ | ☐ | ☐ | ☐ | _____ | 53 | ☐ | ☐ | ☐ | ☐ | _____ |
| 4 | ☐ | ☐ | ☐ | ☐ | _____ | 29 | ☐ | ☐ | ☐ | ☐ | _____ | 54 | ☐ | ☐ | ☐ | ☐ | _____ |
| 5 | ☐ | ☐ | ☐ | ☐ | _____ | 30 | ☐ | ☐ | ☐ | ☐ | _____ | 55 | ☐ | ☐ | ☐ | ☐ | _____ |
| 6 | ☐ | ☐ | ☐ | ☐ | _____ | 31 | ☐ | ☐ | ☐ | ☐ | _____ | 56 | ☐ | ☐ | ☐ | ☐ | _____ |
| 7 | ☐ | ☐ | ☐ | ☐ | _____ | 32 | ☐ | ☐ | ☐ | ☐ | _____ | 57 | ☐ | ☐ | ☐ | ☐ | _____ |
| 8 | ☐ | ☐ | ☐ | ☐ | _____ | 33 | ☐ | ☐ | ☐ | ☐ | _____ | 58 | ☐ | ☐ | ☐ | ☐ | _____ |
| 9 | ☐ | ☐ | ☐ | ☐ | _____ | 34 | ☐ | ☐ | ☐ | ☐ | _____ | 59 | ☐ | ☐ | ☐ | ☐ | _____ |
| 10 | ☐ | ☐ | ☐ | ☐ | _____ | 35 | ☐ | ☐ | ☐ | ☐ | _____ | 60 | ☐ | ☐ | ☐ | ☐ | _____ |
| 11 | ☐ | ☐ | ☐ | ☐ | _____ | 36 | ☐ | ☐ | ☐ | ☐ | _____ | 61 | ☐ | ☐ | ☐ | ☐ | _____ |
| 12 | ☐ | ☐ | ☐ | ☐ | _____ | 37 | ☐ | ☐ | ☐ | ☐ | _____ | 62 | ☐ | ☐ | ☐ | ☐ | _____ |
| 13 | ☐ | ☐ | ☐ | ☐ | _____ | 38 | ☐ | ☐ | ☐ | ☐ | _____ | 63 | ☐ | ☐ | ☐ | ☐ | _____ |
| 14 | ☐ | ☐ | ☐ | ☐ | _____ | 39 | ☐ | ☐ | ☐ | ☐ | _____ | 64 | ☐ | ☐ | ☐ | ☐ | _____ |
| 15 | ☐ | ☐ | ☐ | ☐ | _____ | 40 | ☐ | ☐ | ☐ | ☐ | _____ | 65 | ☐ | ☐ | ☐ | ☐ | _____ |
| 16 | ☐ | ☐ | ☐ | ☐ | _____ | 41 | ☐ | ☐ | ☐ | ☐ | _____ | 66 | ☐ | ☐ | ☐ | ☐ | _____ |
| 17 | ☐ | ☐ | ☐ | ☐ | _____ | 42 | ☐ | ☐ | ☐ | ☐ | _____ | 67 | ☐ | ☐ | ☐ | ☐ | _____ |
| 18 | ☐ | ☐ | ☐ | ☐ | _____ | 43 | ☐ | ☐ | ☐ | ☐ | _____ | 68 | ☐ | ☐ | ☐ | ☐ | _____ |
| 19 | ☐ | ☐ | ☐ | ☐ | _____ | 44 | ☐ | ☐ | ☐ | ☐ | _____ | 69 | ☐ | ☐ | ☐ | ☐ | _____ |
| 20 | ☐ | ☐ | ☐ | ☐ | _____ | 45 | ☐ | ☐ | ☐ | ☐ | _____ | 70 | ☐ | ☐ | ☐ | ☐ | _____ |
| 21 | ☐ | ☐ | ☐ | ☐ | _____ | 46 | ☐ | ☐ | ☐ | ☐ | _____ | 71 | ☐ | ☐ | ☐ | ☐ | _____ |
| 22 | ☐ | ☐ | ☐ | ☐ | _____ | 47 | ☐ | ☐ | ☐ | ☐ | _____ | 72 | ☐ | ☐ | ☐ | ☐ | _____ |
| 23 | ☐ | ☐ | ☐ | ☐ | _____ | 48 | ☐ | ☐ | ☐ | ☐ | _____ | 73 | ☐ | ☐ | ☐ | ☐ | _____ |
| 24 | ☐ | ☐ | ☐ | ☐ | _____ | 49 | ☐ | ☐ | ☐ | ☐ | _____ | 74 | ☐ | ☐ | ☐ | ☐ | _____ |
| 25 | ☐ | ☐ | ☐ | ☐ | _____ | 50 | ☐ | ☐ | ☐ | ☐ | _____ | 75 | ☐ | ☐ | ☐ | ☐ | _____ |

**PRINT OR TYPE**

NRCC GUNNER'S MATE M 1&C
NAVEDTRA 10200-D

NAME _____ ADDRESS _____
     Last    First    Middle    Street/Ship/Unit/Division, etc.

                                                 City or FPO    State                      Zip

RANK/RATE _____ SOC. SEC. NO. _____ DESIGNATOR _____ ASSIGNMENT NO. _____

☐ USN  ☐ USNR  ☐ ACTIVE  ☐ INACTIVE  OTHER (Specify) _____ DATE MAILED _____

SCORE

**PRINT OR TYPE**

NRCC GUNNER'S MATE M 1&C
NAVEDTRA 10200-D

NAME _____ ADDRESS _____
      Last    First    Middle        Street/Ship/Unit/Division, etc.

                                     City or FPO      State           Zip
RANK/RATE _____ SOC. SEC. NO. _____ DESIGNATOR _____ ASSIGNMENT NO. _____

☐ USN  ☐ USNR  ☐ ACTIVE  ☐ INACTIVE  OTHER (Specify) _____ DATE MAILED _____

SCORE

|     | 1 2 3 4 |     | 1 2 3 4 |     | 1 2 3 4 |
|-----|---------|-----|---------|-----|---------|
|     | T F     |     | T F     |     | T F     |
| 1   | ☐☐☐☐ ____ | 26 | ☐☐☐☐ ____ | 51 | ☐☐☐☐ ____ |
| 2   | ☐☐☐☐ ____ | 27 | ☐☐☐☐ ____ | 52 | ☐☐☐☐ ____ |
| 3   | ☐☐☐☐ ____ | 28 | ☐☐☐☐ ____ | 53 | ☐☐☐☐ ____ |
| 4   | ☐☐☐☐ ____ | 29 | ☐☐☐☐ ____ | 54 | ☐☐☐☐ ____ |
| 5   | ☐☐☐☐ ____ | 30 | ☐☐☐☐ ____ | 55 | ☐☐☐☐ ____ |
| 6   | ☐☐☐☐ ____ | 31 | ☐☐☐☐ ____ | 56 | ☐☐☐☐ ____ |
| 7   | ☐☐☐☐ ____ | 32 | ☐☐☐☐ ____ | 57 | ☐☐☐☐ ____ |
| 8   | ☐☐☐☐ ____ | 33 | ☐☐☐☐ ____ | 58 | ☐☐☐☐ ____ |
| 9   | ☐☐☐☐ ____ | 34 | ☐☐☐☐ ____ | 59 | ☐☐☐☐ ____ |
| 10  | ☐☐☐☐ ____ | 35 | ☐☐☐☐ ____ | 60 | ☐☐☐☐ ____ |
| 11  | ☐☐☐☐ ____ | 36 | ☐☐☐☐ ____ | 61 | ☐☐☐☐ ____ |
| 12  | ☐☐☐☐ ____ | 37 | ☐☐☐☐ ____ | 62 | ☐☐☐☐ ____ |
| 13  | ☐☐☐☐ ____ | 38 | ☐☐☐☐ ____ | 63 | ☐☐☐☐ ____ |
| 14  | ☐☐☐☐ ____ | 39 | ☐☐☐☐ ____ | 64 | ☐☐☐☐ ____ |
| 15  | ☐☐☐☐ ____ | 40 | ☐☐☐☐ ____ | 65 | ☐☐☐☐ ____ |
| 16  | ☐☐☐☐ ____ | 41 | ☐☐☐☐ ____ | 66 | ☐☐☐☐ ____ |
| 17  | ☐☐☐☐ ____ | 42 | ☐☐☐☐ ____ | 67 | ☐☐☐☐ ____ |
| 18  | ☐☐☐☐ ____ | 43 | ☐☐☐☐ ____ | 68 | ☐☐☐☐ ____ |
| 19  | ☐☐☐☐ ____ | 44 | ☐☐☐☐ ____ | 69 | ☐☐☐☐ ____ |
| 20  | ☐☐☐☐ ____ | 45 | ☐☐☐☐ ____ | 70 | ☐☐☐☐ ____ |
| 21  | ☐☐☐☐ ____ | 46 | ☐☐☐☐ ____ | 71 | ☐☐☐☐ ____ |
| 22  | ☐☐☐☐ ____ | 47 | ☐☐☐☐ ____ | 72 | ☐☐☐☐ ____ |
| 23  | ☐☐☐☐ ____ | 48 | ☐☐☐☐ ____ | 73 | ☐☐☐☐ ____ |
| 24  | ☐☐☐☐ ____ | 49 | ☐☐☐☐ ____ | 74 | ☐☐☐☐ ____ |
| 25  | ☐☐☐☐ ____ | 50 | ☐☐☐☐ ____ | 75 | ☐☐☐☐ ____ |

**PRINT OR TYPE**

NRCC GUNNER'S MATE M 1&C
NAVEDTRA 10200-D

NAME _____ ADDRESS _____
        Last       First       Middle              Street/Ship/Unit/Division, etc.

                                                   City or FPO        State                       Zip
RANK/RATE _____ SOC. SEC. NO. _____ DESIGNATOR _____ ASSIGNMENT NO. _____

☐ USN   ☐ USNR   ☐ ACTIVE   ☐ INACTIVE   OTHER (Specify) _____ DATE MAILED _____

SCORE

|      | 1 2 3 4 |      | 1 2 3 4 |      | 1 2 3 4 |
|------|---------|------|---------|------|---------|
|      | T F     |      | T F     |      | T F     |
| 1    | ☐☐☐☐    | 26   | ☐☐☐☐    | 51   | ☐☐☐☐    |
| 2    | ☐☐☐☐    | 27   | ☐☐☐☐    | 52   | ☐☐☐☐    |
| 3    | ☐☐☐☐    | 28   | ☐☐☐☐    | 53   | ☐☐☐☐    |
| 4    | ☐☐☐☐    | 29   | ☐☐☐☐    | 54   | ☐☐☐☐    |
| 5    | ☐☐☐☐    | 30   | ☐☐☐☐    | 55   | ☐☐☐☐    |
| 6    | ☐☐☐☐    | 31   | ☐☐☐☐    | 56   | ☐☐☐☐    |
| 7    | ☐☐☐☐    | 32   | ☐☐☐☐    | 57   | ☐☐☐☐    |
| 8    | ☐☐☐☐    | 33   | ☐☐☐☐    | 58   | ☐☐☐☐    |
| 9    | ☐☐☐☐    | 34   | ☐☐☐☐    | 59   | ☐☐☐☐    |
| 10   | ☐☐☐☐    | 35   | ☐☐☐☐    | 60   | ☐☐☐☐    |
| 11   | ☐☐☐☐    | 36   | ☐☐☐☐    | 61   | ☐☐☐☐    |
| 12   | ☐☐☐☐    | 37   | ☐☐☐☐    | 62   | ☐☐☐☐    |
| 13   | ☐☐☐☐    | 38   | ☐☐☐☐    | 63   | ☐☐☐☐    |
| 14   | ☐☐☐☐    | 39   | ☐☐☐☐    | 64   | ☐☐☐☐    |
| 15   | ☐☐☐☐    | 40   | ☐☐☐☐    | 65   | ☐☐☐☐    |
| 16   | ☐☐☐☐    | 41   | ☐☐☐☐    | 66   | ☐☐☐☐    |
| 17   | ☐☐☐☐    | 42   | ☐☐☐☐    | 67   | ☐☐☐☐    |
| 18   | ☐☐☐☐    | 43   | ☐☐☐☐    | 68   | ☐☐☐☐    |
| 19   | ☐☐☐☐    | 44   | ☐☐☐☐    | 69   | ☐☐☐☐    |
| 20   | ☐☐☐☐    | 45   | ☐☐☐☐    | 70   | ☐☐☐☐    |
| 21   | ☐☐☐☐    | 46   | ☐☐☐☐    | 71   | ☐☐☐☐    |
| 22   | ☐☐☐☐    | 47   | ☐☐☐☐    | 72   | ☐☐☐☐    |
| 23   | ☐☐☐☐    | 48   | ☐☐☐☐    | 73   | ☐☐☐☐    |
| 24   | ☐☐☐☐    | 49   | ☐☐☐☐    | 74   | ☐☐☐☐    |
| 25   | ☐☐☐☐    | 50   | ☐☐☐☐    | 75   | ☐☐☐☐    |

PRINT OR TYPE

NRCC GUNNER'S MATE M 1&C
NAVEDTRA 10200-D

NAME _____ ADDRESS _____
      Last      First     Middle      Street/Ship/Unit/Division, etc.

                                      City or FPO      State              Zip
RANK/RATE _____ SOC. SEC. NO. _____ DESIGNATOR _____ ASSIGNMENT NO. _____
☐ USN  ☐ USNR  ☐ ACTIVE  ☐ INACTIVE  OTHER (Specify) _____ DATE MAILED _____

SCORE

**PRINT OR TYPE**

NRCC GUNNER'S MATE M 1&C
NAVEDTRA 10200-D

NAME _____  ADDRESS _____
      Last    First   Middle       Street/Ship/Unit/Division, etc.

                                   City or FPO    State          Zip
RANK/RATE _____ SOC. SEC. NO. _____ DESIGNATOR _____ ASSIGNMENT NO. _____

☐ USN   ☐ USNR   ☐ ACTIVE   ☐ INACTIVE   OTHER (Specify) _____ DATE MAILED _____

SCORE

**PRINT OR TYPE**

NRCC GUNNER'S MATE M 1&C
NAVEDTRA 10200-D

NAME _____ ADDRESS _____
        Last    First    Middle         Street/Ship/Unit/Division, etc.

                                        City or FPO    State            Zip
RANK/RATE _____ SOC. SEC. NO. _____ DESIGNATOR _____ ASSIGNMENT NO. _____
☐ USN  ☐ USNR  ☐ ACTIVE  ☐ INACTIVE  OTHER (Specify) _____ DATE MAILED _____

SCORE

|     | 1 2 3 4<br>T F |     |     | 1 2 3 4<br>T F |     |     | 1 2 3 4<br>T F |     |
|-----|----------------|-----|-----|----------------|-----|-----|----------------|-----|
| 1   | ☐ ☐ ☐ ☐ _____ |     | 26  | ☐ ☐ ☐ ☐ _____ |     | 51  | ☐ ☐ ☐ ☐ _____ |     |
| 2   | ☐ ☐ ☐ ☐ _____ |     | 27  | ☐ ☐ ☐ ☐ _____ |     | 52  | ☐ ☐ ☐ ☐ _____ |     |
| 3   | ☐ ☐ ☐ ☐ _____ |     | 28  | ☐ ☐ ☐ ☐ _____ |     | 53  | ☐ ☐ ☐ ☐ _____ |     |
| 4   | ☐ ☐ ☐ ☐ _____ |     | 29  | ☐ ☐ ☐ ☐ _____ |     | 54  | ☐ ☐ ☐ ☐ _____ |     |
| 5   | ☐ ☐ ☐ ☐ _____ |     | 30  | ☐ ☐ ☐ ☐ _____ |     | 55  | ☐ ☐ ☐ ☐ _____ |     |
| 6   | ☐ ☐ ☐ ☐ _____ |     | 31  | ☐ ☐ ☐ ☐ _____ |     | 56  | ☐ ☐ ☐ ☐ _____ |     |
| 7   | ☐ ☐ ☐ ☐ _____ |     | 32  | ☐ ☐ ☐ ☐ _____ |     | 57  | ☐ ☐ ☐ ☐ _____ |     |
| 8   | ☐ ☐ ☐ ☐ _____ |     | 33  | ☐ ☐ ☐ ☐ _____ |     | 58  | ☐ ☐ ☐ ☐ _____ |     |
| 9   | ☐ ☐ ☐ ☐ _____ |     | 34  | ☐ ☐ ☐ ☐ _____ |     | 59  | ☐ ☐ ☐ ☐ _____ |     |
| 10  | ☐ ☐ ☐ ☐ _____ |     | 35  | ☐ ☐ ☐ ☐ _____ |     | 60  | ☐ ☐ ☐ ☐ _____ |     |
| 11  | ☐ ☐ ☐ ☐ _____ |     | 36  | ☐ ☐ ☐ ☐ _____ |     | 61  | ☐ ☐ ☐ ☐ _____ |     |
| 12  | ☐ ☐ ☐ ☐ _____ |     | 37  | ☐ ☐ ☐ ☐ _____ |     | 62  | ☐ ☐ ☐ ☐ _____ |     |
| 13  | ☐ ☐ ☐ ☐ _____ |     | 38  | ☐ ☐ ☐ ☐ _____ |     | 63  | ☐ ☐ ☐ ☐ _____ |     |
| 14  | ☐ ☐ ☐ ☐ _____ |     | 39  | ☐ ☐ ☐ ☐ _____ |     | 64  | ☐ ☐ ☐ ☐ _____ |     |
| 15  | ☐ ☐ ☐ ☐ _____ |     | 40  | ☐ ☐ ☐ ☐ _____ |     | 65  | ☐ ☐ ☐ ☐ _____ |     |
| 16  | ☐ ☐ ☐ ☐ _____ |     | 41  | ☐ ☐ ☐ ☐ _____ |     | 66  | ☐ ☐ ☐ ☐ _____ |     |
| 17  | ☐ ☐ ☐ ☐ _____ |     | 42  | ☐ ☐ ☐ ☐ _____ |     | 67  | ☐ ☐ ☐ ☐ _____ |     |
| 18  | ☐ ☐ ☐ ☐ _____ |     | 43  | ☐ ☐ ☐ ☐ _____ |     | 68  | ☐ ☐ ☐ ☐ _____ |     |
| 19  | ☐ ☐ ☐ ☐ _____ |     | 44  | ☐ ☐ ☐ ☐ _____ |     | 69  | ☐ ☐ ☐ ☐ _____ |     |
| 20  | ☐ ☐ ☐ ☐ _____ |     | 45  | ☐ ☐ ☐ ☐ _____ |     | 70  | ☐ ☐ ☐ ☐ _____ |     |
| 21  | ☐ ☐ ☐ ☐ _____ |     | 46  | ☐ ☐ ☐ ☐ _____ |     | 71  | ☐ ☐ ☐ ☐ _____ |     |
| 22  | ☐ ☐ ☐ ☐ _____ |     | 47  | ☐ ☐ ☐ ☐ _____ |     | 72  | ☐ ☐ ☐ ☐ _____ |     |
| 23  | ☐ ☐ ☐ ☐ _____ |     | 48  | ☐ ☐ ☐ ☐ _____ |     | 73  | ☐ ☐ ☐ ☐ _____ |     |
| 24  | ☐ ☐ ☐ ☐ _____ |     | 49  | ☐ ☐ ☐ ☐ _____ |     | 74  | ☐ ☐ ☐ ☐ _____ |     |
| 25  | ☐ ☐ ☐ ☐ _____ |     | 50  | ☐ ☐ ☐ ☐ _____ |     | 75  | ☐ ☐ ☐ ☐ _____ |     |

©2013 Periscope Film LLC
All Rights Reserved
ISBN#978-1-937684-30-3
www.PeriscopeFilm.com

www.ingramcontent.com/pod-product-compliance
Lightning Source LLC
Chambersburg PA
CBHW081343230426
43667CB00017B/2709